Streptomyces: Biological Candidates for Sustainable Agriculture

Talwinder Kaur • Harsh Kumar
Rajesh Kumari Manhas
Editors

Streptomyces: Biological Candidates for Sustainable Agriculture

Springer

Editors
Talwinder Kaur
Department of Microbiology
Guru Nanak Dev University
Amritsar, Punjab, India

Harsh Kumar
Faculty of Materials Science and Technology
VSB-Technical University of Ostrava
Ostrava, Czech Republic

Rajesh Kumari Manhas
Department of Microbiology
Guru Nanak Dev University
Amritsar, Punjab, India

ISBN 978-981-95-5802-5 ISBN 978-981-95-5803-2 (eBook)
https://doi.org/10.1007/978-981-95-5803-2

This Springer imprint is published by the registered company Springer Nature Singapore Pte Ltd.
The registered company address is: 152 Beach Road, #21-01/04 Gateway East, Singapore 189721, Singapore

Preface

Agriculture sector plays an important role in global economy. However, worldwide agricultural production has been affected largely by various biotic (pests, pathogens, weeds) and abiotic stresses such as nutrient deficiency, salinity, drought, and temperature. In addition to these problems, the current 8 billion population of the world is predicted to rise to ~9.7 billion mark around the year 2050. Thus, in the next one or two decades, it will be a herculean task to optimally cater to the nutritional needs of this much number of the world population. Besides increasing human population pressure, the increased concern of public for the environmental problems and consumer demand for pesticide-free food has provoked the researchers to find sustainable, eco-friendly alternatives to traditional agricultural practices that can help us improve crop yields without compromising the health of our planet. This is where the world of beneficial microorganisms, and specifically the genus *Streptomyces*, enters the conversation. These fascinating microbes have long been hailed for their prolific production of secondary metabolites—most famously, antibiotics. However, their potential extends far beyond medicine and can also be harnessed as potent plant growth promoters and robust biocontrol agents, offering a roadmap for their use as safe, renewable bioinoculants in modern sustainable agriculture.

This book, *Streptomyces: Biological Candidates for Sustainable Agriculture*, aims to serve as a valuable resource for researchers, industry professionals and agricultural practitioners, and students engaged in microbiology, agriculture, biotechnology, and plant protection. It examines the diverse mechanisms by which *Streptomyces* species enhance plant health, ranging from nutrient solubilization and plant hormones production to the suppression of wide range of plant pests and pathogens. This book also highlights the isolation of streptomycetes from untapped novel sources, and their successes and the challenges in translating lab-based findings into real-world field applications. We extend our heartfelt gratitude to the contributors whose expertise and dedication have enriched this book, ensuring comprehensive coverage of the subject matter. We sincerely hope that this book will inspire future researchers and practitioners to use biological agents in agriculture, contributing to a more resilient and environmentally sound food system.

Amritsar, Punjab, India — Talwinder Kaur
Ostrava, Czech Republic — Harsh Kumar
Amritsar, Punjab, India — Rajesh Kumari Manhas

Contents

Editors and Contributors

About the Editors

Talwinder Kaur, PhD works as an Assistant Professor in the Department of Microbiology at Guru Nanak Dev University, Amritsar, Punjab, India, and has teaching experience of more than 8 years at the university and college level. She earned her PhD in Microbiology from the Department of Microbiology at Guru Nanak Dev University, Amritsar, Punjab, India. Her area of research is exploration of microbes, especially *Streptomyces* for their antimicrobial, biocontrol, and plant growth-promoting potential. She has participated in many national and international conferences. She has guided 24 MSc students in their project work. She is also a holder of one patent. She has an MSc in Fermentation and Microbial Technology and a BSc in Industrial Microbiology.

Harsh Kumar is currently a postdoctoral researcher in the Faculty of Materials Science and Technology at VSB-Technical University of Ostrava, Czechia. He was former postdoc fellow in the Centre of Advanced Technologies, Faculty of Science, University of Hradec Králové, Czechia. He earned his PhD in Food Technology from the School of Bioengineering and Food Technology at Shoolini University of Biotechnology and Management Sciences, Himachal Pradesh, India. His work involves developing an electrochemical immunosensor for the detection of antibiotics/toxins in food items. In addition, he also has expertise in the isolation and identification of multidrug-resistant bacteria such as *Escherichia coli* and *Staphylococcus aureus* from diverse sources. He has an MSc in Microbiology and a BSc in Industrial Microbiology.

Rajesh Kumari Manhas (MSc, MPhil, PhD) is a retired Associate Professor, Microbiology, from Guru Nanak Dev University, India. She qualified National Level Test (conducted by CSIR) for the award of Junior and Senior Research Fellowships. She has received ten major research project grants from various funding agencies of Govt. of India like DST, UGC, ICAR, DBT, CSIR, and RUSA and a minor research project funded by GNDU. She has guided seven students for the award of PhD. As a dedicated researcher, she has worked in the area of actinobacterial diversity and biotechnology, especially purification and characterization of novel antimicrobial compounds from *Streptomyces* active against drug-resistant human pathogens and fungal phytopathogens, development of *Streptomyces* bioformulations for the protection of plants from phytopathogens, and improvement of plant growth. In addition to antimicrobial activities, she has also reported antioxidant, anticancerous, and probiotic potential of bioactive compounds from *Streptomyces* spp. She has reported a new *Streptomyces* sp. (*Streptomyces amritsarensis* 2A) from Indian soil, and isolated two novel compounds from this novel sp., showing activity against drug-resistant microbes. She has nine MTCC microbial-type culture collection contributions to her credit.

Contributors

Anu Padinhapurath Abhimannue Department of Biotechnology, St Mary's College (Autonomous), Thrissur, Kerala, India

Neethu Asokan Department of Life Sciences, Sri Satya Sai University for Human Excellence, Kalaburagi, Karnataka, India

Thirumurugan Ayyadurai Department of Botany, Bhararathidasan University, Tiruchirappalli, Tamil Nadu, India

Punnen Abraham Bechu Plant Taxonomy and Conservation Lab, Department of Biology, The Gandhigram Rural Institute (Deemed to be University), Dindigul, Tamil Nadu, India

Ashun Chaudhary Department of Plant Sciences, School of Life Sciences, Central University of Himachal Pradesh, Kangra, Himachal Pradesh, India

Sushma Chauhan Department of Chemical and Biochemical Engineering, Dongguk University, Seoul Campus, Seoul, Republic of Korea

Anand Dave Department of Microbiology and Biotechnology Centre, The Maharaja Sayajirao University of Baroda, Vadoadara, Gujarat, India

Kamini Devi Department of Botanical and Environmental Sciences, Guru Nanak Dev University, Amritsar, Punjab, India

Sapna Devi Department of Microbiology, School of Bioengineering and Biosciences, Lovely Professional University, Phagwara, Punjab, India

Nainci Dhiman Department of Biotechnology, Chandigarh Group of Colleges, Landran, Mohali, India

Mohamed Khaled Diab Pest Physiology Department, Plant Protection Research Institute, Agricultural Research Center, Giza, Egypt

Diksha Department of Zoology, Guru Nanak Dev University, Amritsar, Punjab, India

Suhani Kiran Dutta Department of Zoology, Guru Nanak Dev University, Amritsar, Punjab, India

Sahar Ahmed El-Shatoury Botany & Microbiology Department, Faculty of Science, Suez Canal University, Ismailia, Egypt

Garima Department of Medical Laboratory Sciences, Lovely Professional University, Phagwara, Punjab, India

Ashish Gupta Department of Botany, Banaras Hindu University, Varanasi, Uttar Pradesh, India

Manikandan Gurusamy AgriSETA Research Group, College of Human Sciences, School of Research and Graduate Studies, University of South Africa, Pretoria, South Africa

AgriSETA Research Group, College of Human Sciences, Office of Graduate Studies and Research, University of South Africa, Pretoria, South Africa

Pharmaceutical and Biotech Advancement in Africa (PBA2), Department of Pharmaceutical Sciences, Tshwane University of Technology, Arcadia Campus, Pretoria, South Africa

Mekala Jeyaram PG Department of Zoology, Arulmigu Palaniandavar Arts College for Women, Palani, Tamil Nadu, India

Yasotha Jeyaram Department of Botany, Sri Kaliswari College (Autonomous), Sivakasi, Virudhunagar Dist, Tamil Nadu, India

Diksha Jindal Adesh Institute of Pharmacy and Biomedical Sciences, Adesh University, Bathinda, Punjab, India

Nitika Kapoor Department of Botanical and Environmental Sciences, Guru Nanak Dev University, Amritsar, Punjab, India

Jasweer Kaur Department of Chemistry and Biochemistry, Government College for Girls, Ludhiana, Punjab, India

Sumanpreet Kaur Department of Medical Laboratory Sciences, Lovely Professional University, Phagwara, Punjab, India

Shruti Kaushik ICAR-Central Soil Salinity Research Institute, Karnal, Haryana, India

Mareeswari Kesavan PG Department of Microbiology, Syed Ammal Arts and Science College, Kootampuli, Ramanathapuram, Tamil Nadu, India

Satish Koundal Department of Microbiology, DAV University, Jalandhar, Punjab, India

Deepak Kumar Department of Zoology, Guru Nanak Dev University, Amritsar, Punjab, India

PG Department of Zoology, Khalsa College, Amritsar, Punjab, India

Sandeep Kumar Department of Physics, DAV University Sarmastpur, Jalandhar, Punjab, India

Yogesh Kumar Department of Botany, Central University of Jammu, Bagla (Raya-Suchani), Samba, Jammu and Kashmir, India

Ruchika Kumari Department of Plant Sciences, School of Life Sciences, Central University of Himachal Pradesh, Kangra, Himachal Pradesh, India

Tanu Kumari Department of Plant Sciences, School of Life Sciences, Central University of Himachal Pradesh, Kangra, Himachal Pradesh, India

Tendai Makwikwi Department of Laboratory, Diagnostic and Investigative Sciences, Medical Microbiology Unit, Faculty of Medicine and Health Sciences, University of Zimbabwe, Avondale, Harare, Zimbabwe

Mitochondria Research Laboratory, Biomedical and Molecular Metabolism Research (BioMMet), North-West University, Potchefstroom, South Africa

Rajesh Kumari Manhas Department of Microbiology, Guru Nanak Dev University, Amritsar, Punjab, India

Salmiya Marimuthu PG Department of Microbiology, Syed Ammal Arts and Science College, Kootampuli, Ramanathapuram, Tamil Nadu, India

Nsovo Nyeleti Mayimele Department of Pharmaceutical Sciences, Tshwane University of Technology, Arcadia Campus, Pretoria, South Africa

Pharmaceutical and Biotech Advancement in Africa (PBA2), Department of Pharmaceutical Sciences, Tshwane University of Technology, Arcadia Campus, Pretoria, South Africa

Nahida Rehman Mir Department of Botanical and Environmental Sciences, Guru Nanak Dev University, Amritsar, Punjab, India

Keshavamurthy Mukunda Department of Biotechnology, School of Applied Sciences, REVA University, Bengaluru, Karnataka, India

Center for Biotechnology and Molecular Biology, REVA Research Centre, REVA University, Bengaluru, Karnataka, India

Ranjithkumar Murugesan PG Department of Microbiology, Syed Ammal Arts and Science College, Kootampuli, Ramanathapuram, Tamil Nadu, India

Puja Ohri Department of Zoology, Guru Nanak Dev University, Amritsar, Punjab, India

Preeti K. Pathania Department of Microbiology, Amity University, Mohali, Punjab, India

Geenat Paul Department of Biotechnology, St Mary's College (Autonomous), Thrissur, Kerala, India

Maya Mundanattu Rajan Department of Biotechnology, Administrative Management College, Bengaluru, Karnataka, India

B. P. Rakshitha Department of Life Sciences, Sri Satya Sai University for Human Excellence, Kalaburagi, Karnataka, India

Raju Ramasubbu Plant Taxonomy and Conservation Lab, Department of Biology, The Gandhigram Rural Institute (Deemed to be University), Dindigul, Tamil Nadu, India

Ramandeep Saini Department of Biotechnology, Chandigarh Group of Colleges, Landran, Mohali, India

Alisha Sharma Department of Zoology, Guru Nanak Dev University, Amritsar, Punjab, India

Ashutosh Sharma Faculty of Agricultural Sciences, DAV University, Jalandhar, Punjab, India

Avinash Sharma Department of Microbiology, DAV University, Jalandhar, Punjab, India

Deepika Sharma Amity Institute of Organic Agriculture, Amity University, Noida, Uttar Pradesh, India
Department of Biotechnology, Chandigarh Group of Colleges, Landran, Mohali, India

Indu Sharma Department of Life Sciences, University Institute of Sciences, Sant Baba Bhag Singh University, Jalandhar, Punjab, India

Rohit Sharma Department of Forest Products, College of Forestry, Dr. Yashwant Singh Parmar University of Horticulture and Forestry, Solan, Himachal Pradesh, India

Roohi Sharma Department of Zoology, Guru Nanak Dev University, Amritsar, Punjab, India

Sonika Sharma Faculty of Agricultural Sciences, DAV University, Jalandhar, Punjab, India

Vipasha Sharma Department of Plant Sciences, School of Life Sciences, Central University of Himachal Pradesh, Kangra, Himachal Pradesh, India

Amandeep Singh Department of Botanical and Environmental Sciences, Guru Nanak Dev University, Amritsar, Punjab, India

Anchita Singh Department of Zoology, Guru Nanak Dev University, Amritsar, Punjab, India

Arun Dev Singh Department of Botanical and Environmental Sciences, Guru Nanak Dev University, Amritsar, Punjab, India

Randeep Singh PG Department of Zoology, Khalsa College, Amritsar, Punjab, India

Palak Thakur Department of Plant Sciences, School of Life Sciences, Central University of Himachal Pradesh, Kangra, Himachal Pradesh, India

Raman Thakur Department of Medical Laboratory Sciences, Lovely Professional University, Phagwara, Punjab, India

Rajeshwari Ullagaddi Department of Life Sciences, Sri Satya Sai University for Human Excellence, Kalaburagi, Karnataka, India

Greeshma Pana Valappil Department of Microbiology, St Mary's College (Autonomous), Thrissur, Kerala, India

Sneha Varghese Plant Taxonomy and Conservation Lab, Department of Biology, The Gandhigram Rural Institute (Deemed to be University), Dindigul, Tamil Nadu, India

Balasubramanian Velramar Amity Institute of Biotechnology, Amity University Chhattisgarh, Raipur, Chhattisgarh, India

Jaya Verma Department of Microbiology, Guru Nanak Dev University, Amritsar, Punjab, India

Sakshi Verma Department of Zoology, Hans Raj Mahila MahaVidyalaya, Jalandhar, Punjab, India

Vikram Department of Zoology, Guru Nanak Dev University, Amritsar, Punjab, India

Monika Yadav Department of Applied Sciences and Humanities, Faculty of Engineering and Technology, Rama University, Kanpur, Uttar Pradesh, India

Diversity, Physiology, and Classification of Streptomycetes 1

Sneha Varghese, Punnen Abraham Bechu, and Raju Ramasubbu

Abstract

The genus *Streptomyces* is characterized by a unique assemblage of Gram-positive bacteria, which exhibit a multifaceted spectrum of pigment secretion, colony color, and other characteristic traits. *Streptomyces* plays a significant role in producing physiologically active compounds, including antivirals, immunosuppressants, antibiotics, antihypertensives, fungicides, and anticancer agents. *Streptomyces* synthesize compounds such as auxins (indole acetic acid (IAA)), cytokinins, and gibberellins. Ecologically, these bacteria also play a crucial role in decomposing organic matter, biogeochemical cycling, and enhancing soil fertility. The genus *Streptomyces* is widely distributed in terrestrial and marine ecosystems, as well as in healthy plant tissues (as endophytes), and exhibits a high degree of diversity. The classification of *Streptomyces* is based on several characteristics, including the color of aerial and substrate mycelium, pigment production, other morphological traits, and structural features such as the arrangement of spore chains. Their classification also involves the use of a variety of modern methods, including molecular tools, chemical profiling, genomic sequencing (phylogenetic approach), and numerical taxonomy. Physiologically, Streptomycetes undergo programmed cell death (PCD), during which the vegetative mycelium lyses to provide nutrition to the secondary mycelium. These released nutrients may also attract other microbes.

Keywords

Streptomyces · Taxonomy · Physiology · Phylogeny · Mutualism · Chemotaxonomy

S. Varghese · P. A. Bechu · R. Ramasubbu (✉)
Plant Taxonomy and Conservation Lab, Department of Biology, The Gandhigram Rural Institute (Deemed to be University), Dindigul, Tamil Nadu, India

T. Kaur et al. (eds.), *Streptomyces: Biological Candidates for Sustainable Agriculture*,
https://doi.org/10.1007/978-981-95-5803-2_1

1.1 Introduction

The genus *Streptomyces* was established by Waksman and Henrici (1943) from spore-producing, aerobic actinobacteria (Krainsky 1914; Waksman and Curtis 1916; Buchanan 1918; Waksman 1919, 1940; Lieske 1921; Orskov 1923; Jensen 1930, 1931; Krasilnikov 1938; Baldacci 1939; Williams et al. 1983a, b, c, d). *Streptomycetes* are high G + C (70%) Gram-positive bacteria that belong to the Streptomycetaceae, which is the largest genus among Actinobacteria (Watve et al. 2001; Berdy 2005; Botas Muñoz 2013; Law et al. 2017; Hoskisson and van Wezel 2019). These non-acid-fast bacteria produce branched aerial and substrate mycelia that are rarely septate and typically have elongated tips. These bacteria thrive in temperatures ranging from 25 to 35 °C and form discrete colonies, which are butyrous, leathery, or lichenoid in appearance. The life cycle of *Streptomycetes* is complex, starting with spore germination. The hyphal tips lengthen as a result of the germ tubes, and they subsequently start to branch. The development of the vegetative mycelium is the outcome of this branching, which eventually creates a solid network of vegetative cells (Jones and Elliot 2017).

Streptomyces is the most diverse genus in both archaea and bacteria (Euzeby 2012). The rich diversity of *Streptomyces* is reflected by the presence of 793 sub-taxa with valid published and correct names (accessed on October 27, 2025, http://www.bacterio.net/streptomyces.html). *Streptomycetes* genomes are unique among bacteria due to their long linear chromosomes with terminal inverted repeats. The genome sizes range from 6 to 12 Mb and encode 5300–11,000 proteins. On a circular chromosome, a typical bacterial genome has 5 Mb size and encodes around 5000 proteins. Numerous genes encoding a diverse array of secondary metabolites of agricultural and pharmaceutical potential have been identified in *Streptomyces* due to their diverse physiological activities. Due to the plasticity of their genetic material, streptomycetes can quickly adapt to their surroundings. As soil dwellers, their primary responsibility is to break down organic particles, thereby improving soil fertility (Komaki 2023). *Streptomycetes* are thought to have first appeared around 400 million years ago, most likely when the flora began to dominate the Earth. Breaking down the structural components of plants, fungi, and insects suggests their involvement in the decomposition of early terrestrial vegetation, which contributes considerably to the development of ancient soils (Chater et al. 2010; Chater 2016). Most agricultural fields contain almost 10^6–10^7 CFU per gram of soil (Janssen 2006).

Streptomyces are the primary natural source of a wide range of bioactive and secondary metabolites, including antibiotics, antifungals, antivirals, antitumor agents, antihypertensives, and immunosuppressants (McKenzie et al. 2010, de Lima Procópio et al. 2012). Two-thirds of the total discovered secondary metabolites are derived from Actinomyces, with 75% comprising *Streptomyces*, including over 7600 compounds (Franco-Correa et al. 2010). More than 100,000 antibiotics have been produced from existing studies of *Streptomyces*, and many species are still unidentified on Earth (Milind et al. 2001). Significant contributions of *Streptomycetes* in the discovery of antibiotics such as streptomycin, actinomycin D, and streptothricin were made possible by Waksman (1940) (Kämpfer 2012). Their secondary metabolites, likely evolved through interactions with other organisms, include geosmin, which gives soil its characteristic scent but lacks antibiotic activity, highlighting the widespread presence of these bacteria (Seipke et al. 2012). These bacteria have also been reported to promote plant growth, particularly by localizing in the rhizosphere by producing diverse metabolites. They promote plant growth at root and shoot apicals, and also regulate it by inhibiting processes such as ethylene production (Worsley et al. 2020).

Reports of symbiotic interactions between *Streptomyces* species with insects, marine invertebrates, marine sponges, and cone snails (Piel 2004, 2009; Gulder and Moore 2009) are also reported in symbiotic studies. *Streptomycetes* have been found to impact mycorrhiza and root nodulation, two processes that sustain plant root symbioses. By promoting fungal growth and inhibiting plant defense mechanisms, these bacteria enhance mycorrhizal growth (Tarkka et al. 2008). In contrast, common scab in roots and tubers is caused by *Streptomyces* species, which are pathogenic and result in agricultural loss (Hiltunen et al. 2009; Labeda 2011; Seipke et al. 2012). To prevent self-destruction, *Streptomyces* bacteria resist their own antibiotics.

Clarities in taxonomic classification characters, as well as in their complex life cycle, are not sufficiently met with quality research in these bacteria. Moreover, the morphology, taxonomy, phylogeny, classification, and identification of *Streptomycetes* have not been well understood to realize their life cycle and physiological approaches to various stressful conditions. Therefore, the present study documented the morphology, taxonomy, phylogeny, and classification (ecological and economic importance) of *Streptomycetes*. This chapter provides a detailed idea of the growth and survival of *Streptomycetes* in natural habitats and environmental conditions, and the diversity of *Streptomycetes* across different ecological niches (soil, marine, plant-associated, etc.). Additionally, this manuscript discusses traditional taxonomic approaches in conjunction with modern molecular-based classification systems and physiological processes that support growth and secondary metabolite production of *Streptomycetes*.

1.2 Ecological Distribution of *Streptomycetes*

Streptomyces may be found in various environments, but they are mostly distributed in soil, accounting for 1–20% of the total viable count (Ravi and Jadeja 2016). *Streptomyces* usually prefer to grow at neutral-alkaline pH and their populations from acidic soils differed substantially from natural soils (Williams et al. 1971). Furthermore, streptomycetes exhibit a greater diversity, particularly in rhizospheric soils of leguminous and grass species, and also show plant specificity, highlighting species variation (Bakker et al. 2013). These soil dwelling streptomycetes perform a considerable role in decomposition, nutrient cycling, and plant growth. As saprophytes they break down complex organic matter originating from dead vegetation and animals. This practice supports the characteristic smell of soil, caused by "geosmin" produced by these bacteria.

Streptomyces were isolated from freshwater and aquatic habitats (Simpson et al. 1987), from drinking water reservoirs, which might degrade the water's quality by producing an unpleasant odor due to the production of geosmin (Bentley and Meganathan 1981; Asquith et al. 2018).). After heavy rainfalls, these bacteria transport to waterbodies, even reach oceans due to soil runoff (Kämpfer 2012). *Streptomycetes* are also occasionally closely associated with sediments and debris of aquatic bodies. In sediments at depth of 70–520 m of shallow oceans, the count of these bacteria ranges from 300 to 1270 colonies per cm^3 (Okami and Okazaki 1972). These bacteria produce crucial secondary metabolites, even from seaweeds (Nissen 1963). Certain *Streptomyces* strains were also isolated from manures and certain food crops, particularly from legumes that had undergone spontaneous combustion (Goodfellow and Simpson 1987). Several *Streptomyces* strains were collected from different locations around Greece and these locations included preserved woods and spoiled agricultural areas (Derry et al. 1999).

The isolation of novel strains from *Streptomyces* was conducted in diverse environments, including mangroves, plants, caves, marine sediments, peat swamps, deserts, and lakes (Law et al. 2018). Deep-sea sediments, marine organisms such as sponges and molluscs, and mangrove ecosystems have also been reported as habitats of *Streptomyces* species. *Streptomyces* strains have the ability to survive under adverse conditions, such as salinity and low oxygen and help to decompose mangrove leaf litter. Interestingly, desert soils have also been reported as habitats for streptomycetes. A few species of *Streptomyces* can survive the low freeze conditions of the Arctic and Antarctic and produce specific secondary metabolites to survive in cold conditions. Volcanic soils and springs are extreme environments for streptomycetes, in which they adapt to higher temperatures and mineral compositions (Sivalingam et al. 2024). Some species live in symbiotic relationships and are found to associate with ants and beetles and secrete unique compounds to protect them from predators and pathogens. Strains of streptomycetes were also found in gamma-irradiated locations and highly thermal springs (Seipke et al. 2012).

As endophytes, *Streptomyces* reside in internal healthy tissues of plants, and protect them from pathogens, and thus help them grow (Shepherdson et al. 2023). Both the vegetative and sexual sections of plants (land and water plants) are dominated by these bacteria (Hamedi and Mohammadipana 2015; Jog et al. 2014). They mostly occur in the lumen (xylem and sclerenchyma) and intercellular gaps of the host plant parts (leaves, stems, roots, fruits, flowers, and seeds); intracellular occurrences appear to be less common (Qin et al. 2011). The other parts include root hair fissures, lenticels, stomatal apertures, hydathodes, fractured trichomes, and wounds created by lateral root emergence zones, which allow *Streptomyces* to infiltrate plant tissues (Shimizu 2011). Rarely, they cause infections in humans and animals. *Streptomyces somaliensis* cause chronic subcutaneous infection actinomycetoma (Keijser et al. 2000; Bibb 2013).

A significant evaluation of this group revealed variability in phenotypic characteristics of *Streptomyces* across various habitats, which might triggered by interactions between distinct species (Schlatter and Kinkel 2014). Even though *Streptomyces* play a vital role in both the environment and human health, there is very little structured evidence about their large-scale distribution and variety outside of local and regional studies. This has to be looked into further, utilizing a standardized process on a global scale (Pombubpa et al. 2023).

1.3 Morphology of *Streptomycetes*

Streptomycetes exhibit differential morphological expression throughout their life cycle. Their morphological studies include features such as spore-pigmentation, substrate mycelium, and diffusible-exopigments, along with colony morphology and aerial mycelia texture. The interpretation via color might feel difficult as the spore might change its color due to age, cultured medium, and growth (Kutzner 1981). The vegetative mycelium of *Streptomyces* is characterized by dense filaments due to its repetitive multiplication in branching and polar growth, in which the mycelium is anchored to the substrate (Flärdh 2003; Hempel et al. 2008; Holmes et al. 2013). Spores of the *Streptomycetes* are of taxonomic importance when observing their surface. They are classified into smooth, spiny, hairy, rugose, and warty, but are only distinguishable with keen observation and sophisticated microscopes (Dietz and Mathews 1971; Kämpfer 2012).

1.3.1 Life Cycle of *Streptomycetes*

Actinobacteria exhibit a complex life cycle with three distinguished phases: vegetative mycelium, aerial mycelium, and arthrospores (Kutzner 1981). Similarly, the *Streptomyces* develops primary mycelium, initial hyphae, secondary mycelium, and spores during its life cycle (Fig. 1.1). The primary mycelium is formed only after the nuclear division, followed by hyphal enlargement, giving rise to secondary mycelium and producing aerial hyphae with a chain of single-nucleated spores (McGregor 1954). The arthrospores are developed from the existing mycelium through extended growth and septation. The parental hyphae help form spore walls or holothallic development (Locci and Sharples 1984).

Streptomyces becomes much visible at the stages of spore germination and filament elongation (van Dissel et al. 2014). *Streptomyces* rely on favorable conditions and nutritional availability to initiate sporulation (Flardh and Buttner 2009). The substrate mycelia produce feeding hyphae for their nutritional intake. Hyphae containing mycelium enter the medium and use extracellular hydrolytic enzymes to break down the organic compounds (Miguélez et al. 2010). At this glance, stress is generated due to a decline in the nutritional medium, leading to cell apoptosis, and the emergence of a sporulating structure known as aerial hyphae, driven by the cell core (Dyson 2009; Hasani et al. 2014). Crucially, at the stage of apoptosis, the secondary metabolites are released to hinder the uptake of remaining limited nutrient substrate by the other microbes (Claessen et al. 2014). When the

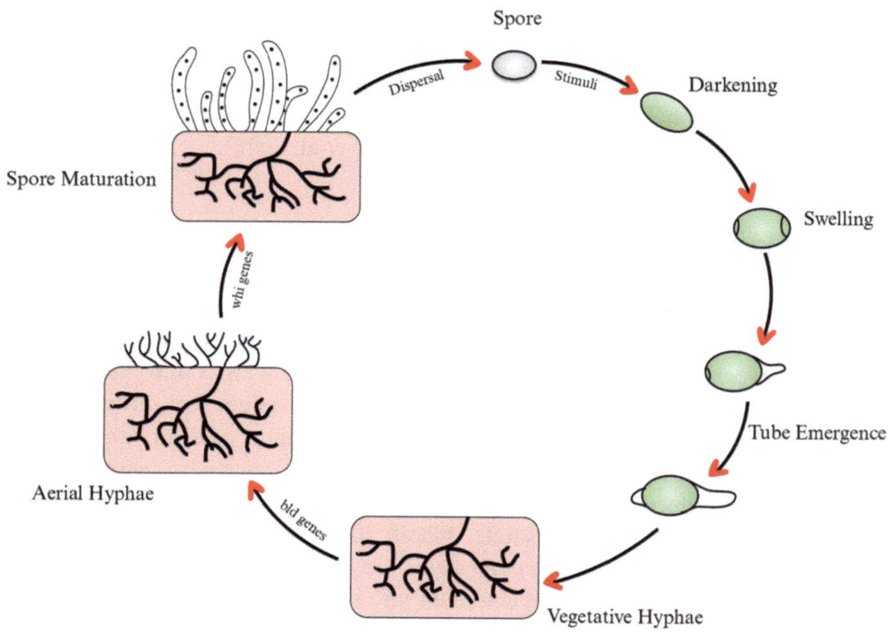

Fig. 1.1 Life cycle of *Streptomyces*

colony matures or the nutrients are exhausted, the bacterial colonial surface develops aerial hyphae, which are then released into the atmosphere (Hu et al. 2012). These hyphae are substantially different, containing a hydrophobic cell wall compared to hydrophilic one in feeding hyphae (Petrus and Claessen 2014). Rodlet layer serves as the outer layer on the aerial mycelium, with rodlin and chaplin proteins, which feature a hydrophobic nature on reducing surface tension and apical growth (Wildermuth et al. 1971; Claessen et al. 2004; Elliot and Talbot 2004; Claessen et al. 2006). These aerial hyphae undergoes elongation and subaerial branching (Mistry 2008). These evolved apical cells of the aerial hyphae then differentiate into a sporular chain (Flardh and Buttner 2009), after controlled multiple cell divisions occurred on the apical cell with an extension at the tip (Fig. 1.2) (Willemse et al. 2011). The hydrophobic aerial hyphal growth and their sporulation septation to form pigmented spores is regulated by *bld* and *whi* genes, respectively (Jones and Elliot 2018). The size of these spores has been recorded between 0.7 and 1.2 μm (Pimentel-Elardo et al. 2009).

1.3.2 Spore Germination and Mycelial Development

Streptomycetes are extremely competitive in soil, accept physiological cues, and adapt well to various environmental circumstances (Paradkar et al. 2003). However, variabilities in *Streptomyces* are due to the direct influence and effect on the

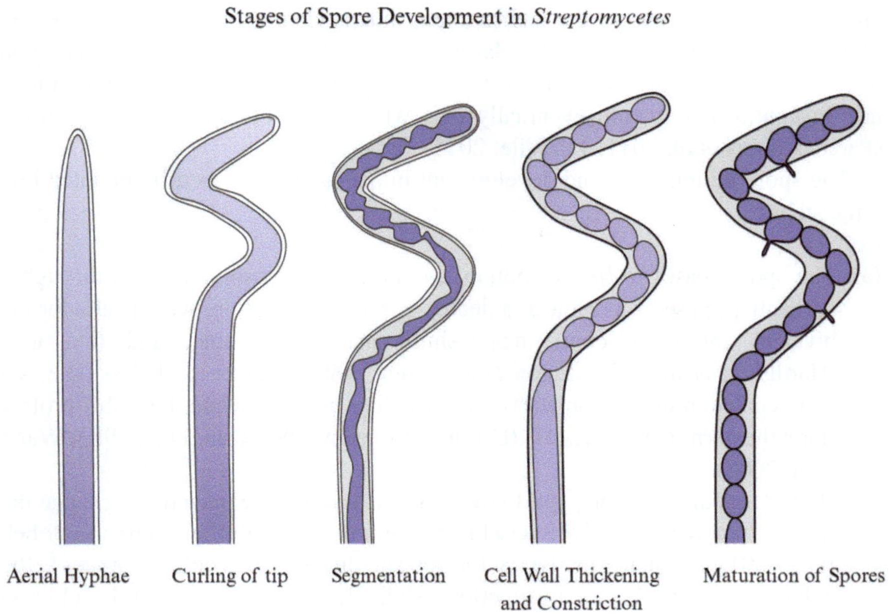

Fig. 1.2 Spore developmental stages of *Streptomyces*

environment, ageing, unknown causes, and differences in substratum. It can be predicted by observing the pigmentation, odor, location, and temperature thrived by the spore (Jones 1954). They are able to adapt to the environment and rapidly changing physical conditions such as pH, aeration, hydrostatic pressure, drought, anaerobic conditions, frost, and moisture tension (Ensign 1978; Mayfield et al. 1972). By forming partially dormant spores in their life cycle, *Streptomycetes* can survive in soil for extended periods of time. *Streptomycetes* spores undergo dormancy, keeping haploid condition, not utilizing the nutrients, and conserving genetic materials (Bobek et al. 2017). The spores' dehydrated stage preserves all macromolecules, including resistance, genetic material condensation, mRNA, ribosomes, translational machinery, and dipicolinic acid (Setlow et al. 2006; Kalakoutskii and Agre 1976; Mikulik et al. 2002). NepA protein balances the spore germination and dormancy metabolism, and it is crucial for dormancy (de Jong et al. 2009). *Streptomyces* spores germinate faster while partially maturing, and they do not show much viability, and an extended lag phase (Hirsch and Ensign 1976). The thickness of the spore wall is determined by the *Streptomyces* spore wall synthesizing complex (Kleinschnitz et al. 2011). Condensation of DNA along with pigmentation of spores is attained during its maturity (Kelemen and Buttner 1998). Spore wall on maturation bears a basket-like fibrous covering (Claessen et al. 2002, 2003; Elliot et al. 2003), and this cell wall undergoes hydrolysis by SsgE and SsgF and releases the spores to the environment independently (Traag and van Wezel 2008; Flardh and Buttner 2009; Haiser et al. 2009). The spore germination in *Streptomycetes* is faster by the availability of nutrients and a favorable environment, and in contrast, the spore eventually dies when the conditions are not in favor (Paredes-Sabja et al. 2011; van Vliet 2015).

The spore germination and development in *Streptomycetes* is differentiated into three steps.

(a) The spore ceases to be hydrophobic, undergoes osmosis, and goes through a darkening phase as a result of a decrease in light refraction, which calls for the bivalent cations calcium, magnesium, manganese, zinc, and iron ions (Hardisson et al. 1978; Eaton and Ensign 1980; Salas et al. 1983). The cell wall's calcium deposition starts the uncoating process, while the CabC protein starts the germination process (Eaton and Ensign 1980; Salas et al. 1983; Wang et al. 2008).
(b) Hydrolases dissolve the peptidoglycan layer's covalent connections, change the spore structure and produce aerial hyphae for the extraction of nutrients (Bobek et al. 2017). When glucose is introduced during the swelling process, the trehalose enzyme begins to function (McBride and Ensign 1987). In addition to reviving the dormant proteins and ribosomes, the water in spores triggers the activation of the chaperones DnaK, Trigger factor, and GroEL (Cowan et al. 2003; Bobek et al. 2004). The germination occurs within a minute (Strakova et al. 2013), whereas DNA replication and the germ tube elongation and protein synthesis begin in 60 min (Mikulik et al. 1977).
(c) The swelling induced by the SsgA emerges from the rodlet layer and forms germ tubes (Noens et al. 2007). It takes one and a half hours for the germination tube to emerge (Ruban-Osmialowska et al. 2006). DivIVA controls the vegetative mycelial development in *Streptomycetes* (Strakova et al. 2013).

BldD-c-di-GMP complex initiates the sporulation genetic expression (Tschowri et al. 2014; Bush et al. 2015). *Streptomyces* sporulation is triggered by a family of SsgA-like proteins (SALP), which control oversporulation and the metabolic transfer to emerging germ tube. Also, they mark the germination sites in the future spores. Septation sites are regulated by SsgB, and the spore production is regulated by SsgC. Deletion of SsgC leads to hypersporulation, and suggested to function as an antagonist of SsgA. The thick spore wall development is regulated by SsgD, another SAPL protein, and the localization of divisional sites is controlled by SsgG (Noens et al. 2005, 2007; Traag and van Wezel 2008; Willemse et al. 2011). The sporulating behavior in *Streptomycetes* has triggered their vast distribution and stability in harsh environments (Seipke et al. 2013). Spores of *Streptomyces* resist the moist content to form arthrospores, but they will become susceptible if water clogs (Williams et al. 1972).

1.4 Taxonomy of *Streptomyces*

Phenotypic characteristics, shape, and prearrangement of aerial-filaments and spore-bearing structure were used as the basis of earlier taxonomy of *Streptomyces*, and they were classified based on the noted similarities and distinctions. The conidial shape and pigmented sporophore structure contribute a characteristic color to the mature colony. The dirty look of the developed colony, its compact nature, and the color help in the detection of *Streptomyces* colonies on agar plates easily (Madigan and Martinko 2005). The systematic treatment by comparing the morphology of the spore chains, the ornamentation on the spore surface, the color of the spores, the mycelial substrate, and the soluble pigments, as well as the formation of melanin, was used. *Streptomycetes* were further separated based on patterns of carbon source consumption (Pridham 1974). To enable the simultaneous assessment of several phenotypic characteristics, numerical taxonomy was developed (Sneath and Johnson 1972). Using phenetic traits, a numerical taxonomic analysis of *Streptomycetes* and allied taxa with cell wall chemotype I (Lechevalier and Lechevalier 1970; Williams et al. 1983a) was carried out. Based on this assessment, 475 strains were described using 139 unit characteristics. The average linkage procedure and a standard similarity coefficient were used to analyze the data. Additionally, *Streptomyces* was separated into species groups.

1.4.1 Classification

Numerous methods, including genomic sequence (phylogenetic approach), molecular techniques, chemical properties, and numerical taxonomy, have been used to classify *Streptomyces* (Fig. 1.3) (Hungund et al. 2022). Historical reviews of *Streptomyces*' taxonomic system from Waksman et al.'s system to Ettlinger et al.'s system show that morphological features have progressively gained weight as taxonomic criteria; however, the author believes that even in Ettlinger's system, taxonomic thought of the morpho-traits is not always adequate (Ryosaku Nomi 1960).

1.4.2 Techniques Used for Classifying *Streptomyces*

1.4.2.1 International *Streptomyces* Project (ISP)

The International *Streptomyces* Project (ISP) was established in 1966 with the cooperation of two committees such as the American Academy of Microbiology's taxonomical and bacteriological nomenclature subcommittees (Hungund et al. 2022). ISP was sponsored by the International Subcommittee on Taxonomy of the Actinobacteria. Using established techniques, researchers analyzed 300 strains of identified *Streptomyces* species, many of which were type strains. Three articles (Shirling and Gottlieb 1968a, b, 1969) provide the descriptions produced due to this collaborative effort. The following characteristics were picked and employed in that study (Shirling and Gottlieb 1966): the morphology of sporophores, the morphology

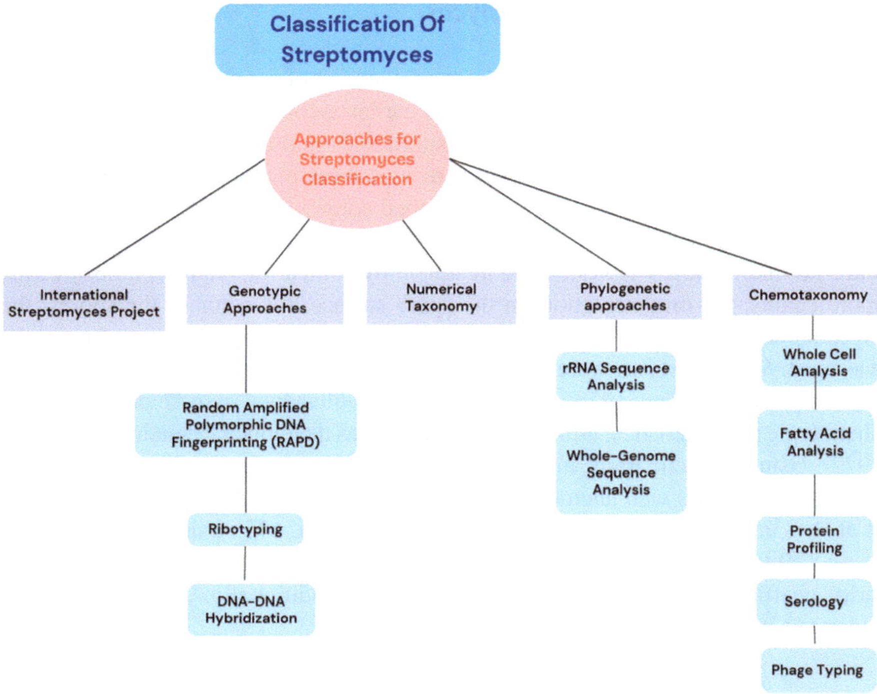

Fig. 1.3 Classification of *Streptomyces* by morphological, chemical, and molecular approaches

of spores, the color of the aerial mycelium, the color of the vegetative mycelium's reverse side, the melanin reaction, the creation of soluble pigment, and the employment of nine distinct sugars. Based on an examination of an earlier worldwide joint study, these traits were considered consistent and dependable (Eberhard Kuster 1972). Eberhard Kuster (1972) developed a diagnostic key to classify and identify 274 named taxa of the International *Streptomyces* Project (ISP).

1.4.2.2 Numerical Classification

Williams (1982) investigated 475 strains from 394 type-cultures of *Streptomyces* and 14 other actiomycete genera. Jaccard (SJ) and Simple Matching (SSM) coefficients and clustering by the UPGMA algorithm were used to determine the overall similarities of these strains. Based on the phonetic data obtained and data from the previous study, *Streptomyces*, *Intrasporangium*, *Nocardioides*, and *Streptoverticillium* were treated as distinct genera, and *Actinosporangium*, *Kitasatoa*, *Actinopycnidium*, *Chainia*, *Elytrosporangium*, and *Microellobosporia* were synonymized.

To improve the resolution between the subgroups, the modification was carried out by combining the main or secondary groups with the minor or primary groupings. Williams et al. (1983b) developed a matrix in which the principal or minor categories were statistically less plausible because they were described using not more than two strains. Kämpfer et al. (1991) reevaluated 821 strains of *Streptoverticillium* and *Streptomyces* using about 329 physiological tests. The strains utilized by Williams et al. (1983a) were also reevaluated by Kämpfer et al. (1991). A total of 40 individual groupings, 34 minor groups, and 15 primary or large groups were described using an 81.5% similarity index.

1.4.2.3 Chemotaxonomy of *Streptomyces*

A comprehensive description based on the similarities and contrasts of the biochemical characteristics of bacteria is known as chemotaxonomy. Chemotaxonomic techniques have also been extensively employed to differentiate *Streptomyces* from other Actinobacteria (Lechevalier and Lechevalier 1970). Alternative chemotaxonomic methodologies for *Streptomycete* categorization have emerged as a result of spectroscopy, chromatography, and electrophoresis (Goodfellow and O'Donnell 1993). Based on the assessment, the straight-chain, iso-branched chain, and ante isobranched chain fatty acids with 14–18 carbon chain length were identified in *Streptomyces* members (Hofheinz and Grisebach 1965; Lechevalier 1977; Popisil et al.1985; Saddler et al. 1986, 1987; Grisebach 1975). Gas chromatography was used to characterize fatty acid methyl esters (FAMEs), as well as to analyze the results. SIMCA, the statistical tool, was also used. *Streptomyces cyaneus* strains were mostly found in the first cluster, followed by the second cluster, consisting of *Streptomyces cyaneus* and other *Streptomyces* strains, and the blue-spored soil *Streptomycetes* were seen in the third cluster. FAME analysis failed to distinguish 11 *Streptomyces violaceoniger* strains from 74 type strains and wild isolates (Phillips 1992). Lee et al. (2015) chemotaxonomically classified utilizing multivariate statistical analysis, quadrupole time-of-flight mass spectrometry, and ultraperformance liquid chromatography.

Protein profiling is crucial for Actinobacteria differentiation. PAGE was used to assess this profile. To distinguish species and subspecies, the PAGE protein banding pattern is required. Goodfellow and O'Donnell (1993) reported that one-dimensional (1D) protein electrophoresis can quickly compare and replicate a large number of bacterial strains (Manchester et al. 1990). It takes a long time to prepare samples for bidirectional PAGE of ribosomal-proteins, and standardizing the extremely complicated patterns of protein for interpretation is challenging (Wang and Jiang 2016).

The taxonomical study of *Streptomyces* from potato was carried out by Paradis et al. (1994) using the PAGE and DNA hybridization techniques. Fatty acid analysis, hybridization, and PAGE were used to examine the isolates. The procedure for scanning and normalizing the electrophoretic analysis of whole-cell protein using SDS-PAGE was formulated (Pot et al. 1994). Williams et al. (1983a) used SDS-PAGE for whole cell protein under denaturing conditions to investigate the taxonomical consistency of the *Streptomyces cyaneus*, *Streptomyces albidoflavus*, and *Streptomyces anulatus*. Ochi and Hiranuma (1994) reclassified *Kitasatospora* and *Streptoverticillium* in *Streptomyces* by determining that their terminal sequences are comparable to *Streptomyces exfoliatus*.

1.4.2.4 Biochemical Assays for the Classification of Streptomycetes

Goodfellow et al. (1987) used a fluorogenic probe coupled to fluoresce upon hydrolysis by the target enzyme to study the utility of quick biochemical assays for *Streptomycete* taxonomy. For the *Streptomycete* strains belonging to 17 taxonomic groupings, 14 distinct enzyme activities were examined (Williams et al. 1983a, b, c, d). All test strains showed a discontinuous distribution of enzymes, which raises questions about the method employed as a classification implement to *Streptomyces*.

1.4.2.5 Phylogenetic Approach

The phylogenetic approach mainly includes rRNA Sequence Analysis and Whole-Genome Sequence Analysis of *Streptomyces*. Labeda et al. (2012) examined the phylogenic relationship of 615 taxa within the family Streptomycetaceae by 16S rRNA gene sequences, which resulted in almost 130 clades with greater than 60% bootstrap values. For *Streptomyces* taxonomy, rRNA sequences are primarily an efficient method. It also assisted in providing answers to some of the concerns pertaining to lateral gene transfer between genera (Huddleston et al. 1997). Whole-genome sequencing (WGS) verifies an organism's whole genome sequence. The complete genomes' sequences may be determined using this method, and the outcomes can then be compared to those of other genomes. The analysis reflects better evolutionary changes of the characteristics, such as gene content, genome structure, and average identity, which are reflected in whole-genome sequencing analysis. Whole-genome sequencing analysis gives an insight into pathogenicity, metabolism, pathway regulation, and structure (Lanoot 2004; Procópio et al. 2012).

1.4.2.6 Genotypic Approach

Gene-specific sequence techniques aided in the comprehension of the investigation of evolutionary links between prokaryotes at different taxonomic ranks (Hungund et al. 2022). The major genotypic approaches used for classification of *Streptomyces* include RAPD, DNA–DNA hybridization, and ribotyping. The most practical technique for rapidly screening different *Streptomyces* strains based on similarity is RAPD-PCR. However, this approach necessitates strong regularization of response limits. Parameters such as buffer composition, primer sequence, DNA template quality and concentration, and annealing temperature are the boundaries. A number of *Streptomyces* species, including *Streptomyces glaucescens*, *Streptomyces ambofaciens*, *Streptomyces coelicolor*, *Streptomyces rimosus*, *Streptomyces lividans*, and *Streptomyces pristinaespiralis*, are classified using this technique (Martin et al. 2000).

The DNA–DNA hybridization techniques are used to enhance the classification of bacteria and to determine the scale of similarity of genomes. The same species within the genus of *Streptomyces* were identified using the DNA–DNA hybridization technique for total chromosomal DNA (Stackbrandt and Goebel 1994; Wayne et al. 1987). The process of DNA–DNA hybridization (DDH) involves analyzing how individual DNA strands from different species reassemble. When ssDNA from two unique bacterial strains is used in the DNA–DNA hybridization process, the two single strands of DNA hybridize to generate dsDNA under various temperature and ionic concentration conditions (Hungund et al.2022). The DDH value greater than 70% is generally accepted as the threshold for considering two strains belong to the same species.

A molecular method known as ribotyping can be used to identify and classify bacteria. Phylogenetic analyses that incorporate rRNA are known as ribotyping (Madigan 2014). Ribotyping technique involves four steps: the first one involves restriction digestion method to digest the bacterial chromosome using an endonuclease, such as *Eco*RI, *Sal*I, or *Cla*I; the second step involves separating the digests or their fragments using gel electrophoresis; the third involves transferring the restriction digests to a nylon membrane using the Southern blotting technique; and finally hybridizing the gel (Grimont and Grimont 1986), with a marked probe that is complementary to 16S and 23S rDNA.

1.5 Physiology of *Streptomycetes*

Streptomycetes recycle organic matter and produce around 800 proteins (Bentley et al. 2002; Ventura et al. 2007) and even grow well in association with other microorganisms like yeast cells. They rely on their spores for dispersion as they are nonmotile. In the presence of competitive organisms like fungi (yeast), streptomycetes undergo exploratory mode of growth, where cells can rapidly traverse both biotic and abiotic surfaces by adopting a non-branching vegetative hyphal conformation. In this mode, the explorer cells of streptomycetes communicate with nearby *Streptomyces* using an airborne signal, a volatile organic compound (trimethylamine), for mutualistic growth and survivability, enabling contact with nutrient-rich surroundings (Jones et al. 2017; Jones and Elliot 2017). Understanding the morphology and dormancy of spores, as well as their life cycle, germination, and secondary metabolites production, is fundamental in elucidating the importance of *Streptomycetes*. Relatively, actions on generating stress, habitat dispersal, co-existence, and application of *Streptomycetes* are very different from other bacteria.

1.5.1 Secondary Metabolites as a Symptom of Stress

During stressed conditions, the *Streptomycetes* undergo programmed cell death (PCD), which lyses the vegetative mycelium to provide nutrition to the secondary mycelium. This accumulates nutrients that attract the other microbes. In this scenario, secondary metabolites are forced to secrete to inhibit the foreign microbes (Méndez et al. 1985; Miguélez et al. 1999; Wildermuth 1970; Barka et al. 2016). The secondary metabolites secreted by the *Streptomyces* in harsh environments are also called stress metabolites. The metabolites produced by *Streptomycetes* are used to inhibit the microbes, and degrade lignin, which occurs naturally with cellulose and xylan (hemicellulose) in a lignocellulose complex (Crawford 1978). Mesophilic and thermophilic *Streptomycetes* can decompose natural and synthetic matter (Kämpfer 2012). Teichoic acids affect physiological metabolism and cell co-aggregation via influencing the negative charge (Hamid 2013; Lechevalier and Lechevalier 1970).

Fungicides, antibiotics, cytostatics, immune response modulators, and plant growth promoting substances are the main secondary metabolites produced by streptomycetes (Kämpfer 2012). Geosmin, siderophores, and pentaleno–lactone are often secreted and usually occur in syntenous positions inside the genome (Bentley et al. 2002; Ikeda et al. 2003). Antibiotics, cytokinins, auxins, volatile chemicals, gibberellins, ethylene, and enzymes are all produced for their survival (Olanrewaju et al. 2017), which are also used in plant growth promotion. Glucose deficiency activates plant growth promoters through yeast stimulation (Jones et al. 2017); and these promoters pave the way to secrete volatile organic compounds, which can handle the entire environment, suppress other microbes, acting as elicitors, taking complete control over the gene expression of surrounding bacteria. This volatile, trimethylamine, can be treated as an antibacterial and antifungal agent (Jones et al. 2017).

1.5.2 Physiology of Mutualism

Streptomycetes have a specialized mechanism to promote mycorrhizal symbiosis and root nodulation by limiting plant defense mechanisms and promoting fungal growth in root portions (Tarkka et al. 2008). Nutritional resources include rhizosphere-derived ionic secretions, enzymes, mucilage, free oxygen and water, proteins, organic acids, amino acids, phenolics, and carbohydrates (Bais et al. 2006). The endophytic *Streptomyces* produce potent antibiotics and other secondary metabolites that help control pathogens, enhance plant's defense response, and protect the plant (Lehr et al. 2008). The mycorrhizae exudate attracts *Streptomyces*, which affects its chemotaxis migration to the rhizosphere. The rhizosphere support only certain bacteria to survive, whereas certain *Streptomyces* are adapted to colonize true roots (Olanrewaju and Babalola 2019).

Streptomyces are nonmotile, but they disperse spores through a vector or medium. The transport of the *Streptomyces globisporus* via honeybees to tackle down gray mold disease caused by *Botrytis cinerea* was facilitated in strawberry plants. The yield of strawberry was conserved, and the mortality of bees was declined due to the reduced fungal attack on pollens (Kim et al. 2019). By secreting particular compounds, *Streptomyces* provides honeybees and even the pollen stock a healthy environment (Menegatti et al. 2020; Grubbs et al. 2021). The mobility of the spores is also ensured by associating with other microbes. *Bacillus subtilis* was found carrying *Streptomyces* species using the flagellar structures and reaching the plant rhizome region and translocating the spore (Muok et al. 2021). Cycloheximide, an antifungal compound produced by a *Streptomyces* sp., inhibited the fungal attack and prevented the fungal colony from growing by the mutualistic beetle (Grubbs et al. 2020). *Streptomyces* association is also found inside the gut of Passalid beetles, which are reached through feces. By producing antifungal chemicals, the bacteria counteract fungal diseases that are particular to insects (Pessotti et al. 2021). They are mutualistic and manipulate organisms like nematode *Caenorhabditis elegans*, which facilitated through the production of volatile compounds by the *Streptomyces* (Zaroubi et al. 2022). Insects that feed on plants likely choose symbiotic *Streptomyces* for their improved cellulose-degrading process (Book et al. 2016).

1.6 Ecological, Industrial, Agricultural, Medical, and Commercial Importance

1.6.1 Ecological Importance

About 600 species of *Streptomyces* produce several hydrolytic exoenzymes to digest macromolecules. Recalcitrant biopolymers such as cellulose, hemicelluloses, or lignocellulose in wood are broken by some *Streptomyces* isolates (Gottschalk et al. 1999; Bruce et al. 2010; Adams et al. 2011; Amore et al. 2012; Li et al. 2012). Most *Streptomyces* lineages can break down wood and demonstrate hemicellulolytic and cellulolytic properties, and molecular screening for a particular hydrogenase revealed a broad capability for atmospheric hydrogen absorption (Bontemps et al. 2013). *Streptomyces* are nonpathogenic bacteria that have also been involved in the bioabsorption and bioaccumulation of heavy metals (Kumar et al. 2020). In addition, antifungal and antibiotic compounds derived from *Streptomyces* are effective in controlling both pathogenic and nonpathogenic bacteria and fungi in wastewater. A variety of insect-specific niches have been occupied by *Streptomyces* species. Initially, they were found in association with ants, and these symbiotic connections have subsequently spread to several other insects. They produce compounds that benefit their host insect, conferring protective benefits (Currie et al. 1999; Seipke et al. 2012; Chevrette et al. 2019).

1.6.2 Industrial Importance

Some of the *Streptomyces* species produce pigments, including metacycloprodigiosin, undecylprodigiosin, rhodomycin, granaticin, and so on (Hungund et al. 2022). Moreover, 60% of the commercial antibiotics used in industry and medicines are produced by *Streptomyces* (Wanner and Kirk 2015). *Streptomyces* produces extracellular enzymes that may replace hazardous chemicals used in various industries (Kumar et al. 2020). The extracellular enzymes of *Streptomyces* are more diverse than those of other sources, such as fungi and other bacteria. The extracellular enzymes of *Streptomyces* used in cotton pretreatment procedures such bleaching to increase the whiteness, scouring to increase its hydrophilicity, and desizing to reduce its size (Kumar et al. 2020). The fermentation technique employing the mangrove-associated *Streptomyces olivaceus* was researched to produce bioethanol from agricultural leftovers (Muthusamy et al. 2017).

1.6.3 Agricultural Importance

Rhizobacteria have replaced chemical fertilizers in promoting plant growth, without health issues to plants and humans (Olanrewaju and Babalola 2019). *Streptomycetes* also contribute to soil disease resistance and protect plants from potential root infections by producing a wide array of secondary metabolites and other bioactive compounds (Weller et al. 2002). Antibiotics produced by *Streptomyces* function as chemical weapons that kill or inhibit pathogens in soil, and *Streptomyces* also frequently coexist as saprotrophs with decomposing plant matter (Pombubpa et al. 2023; Norte et al. 2025). Introducing *Streptomycetes* to plants can improve their salt content and heat tolerance by increasing their enzymatic activity (Fig. 1.4) (Manullang and Chuang 2020). Stable chitinases and bioactive chemicals produced by *Streptomyces* species are utilized in agricultural applications as efficient pesticides (Vurukonda et al. 2018). *Streptomyces* species prevent a variety of fungal pathogenic assaults. For instance, *Streptomyces griseorubiginosus* produced siderophores that effectively combat *Fusarium oxysporum cubense*-induced banana wilt (Barka et al. 2016). *Streptomyces kasugaensis* promotes the growth of seedlings of *Pinus taeda* and prevents the formation of *Fusarium* and *Armillaria* pine rot (de Vasconcellos and Cardoso 2009). Some *Streptomyces* strains were isolated from the roots of *Arabidopsis thaliana*, and these isolates showed a significant role in increased biomass production in the plant (Shepherdson et al. 2023). The use of antagonistic *Streptomyces* spp. in tomato seed coating considerably increased the production. Although this has not yet been properly validated, it was suggested that the improvement in tomato growth was due to an enhanced availability of growth regulators generated by the inoculum (Barka et al. 2016).

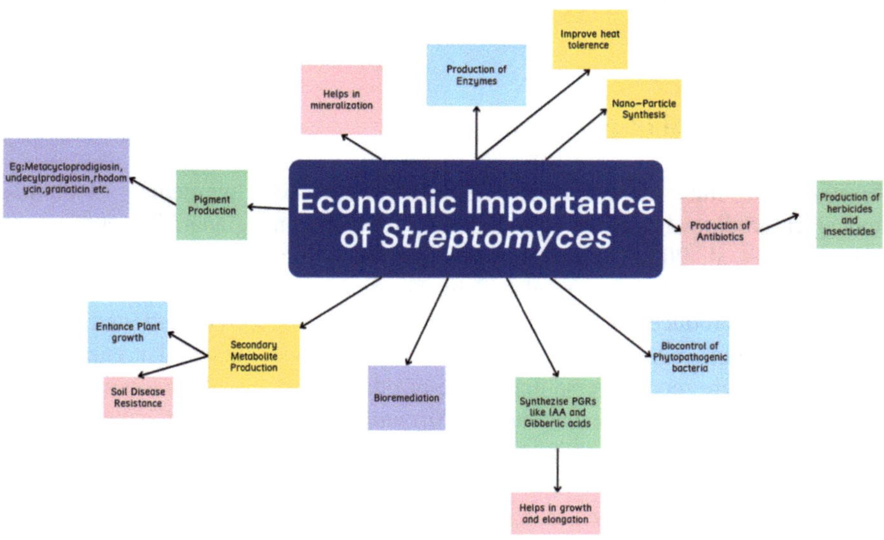

Fig. 1.4 Economic importance of *Streptomyces*

Phosphate-solubilizing *Streptomyces* isolated from a wheat field was also used to increase the N, Fe, P, and Mn content in soil (Sadeghi et al. 2012). *Streptomyces* isolated from marine sources synthesize compounds such as IAA, auxins, and gibberellic acid, which directly impart growth and elongation in agricultural crops (Igarashi 2004; Rashad et al. 2015). *Streptomyces* is also a viable option for controlling phytopathogenic bacteria and offers numerous biotechnological possibilities. According to phylogenetic research, *Streptomyces*' antimicrobial efficacy is correlated with soil organic matter, exhibiting both site-specific and lineage-dependent differences in antibiotic synthesis (Norte et al. 2025). *Streptomyces* produces 60% of the herbicides and insecticides because of its strong antibiotic properties and greater capacity to dominate the creation of biofertilizer and biocontrol in agricultural fields (Pacios-Michelena et al. 2021). Furthermore, *Streptomyces* enhance plant growth by producing phytohormones, siderophores (that chelate ferric iron from the environment), fixing atmospheric nitrogen, and mitigating plant stress via 1-aminocyclopropane-1-carboxylate (ACC) deaminase activity (Verma et al. 2011; Sadeghi et al. 2012). They also contribute to improved plant growth through emission of antimicrobial volatile organic compounds (Jones and Elliot 2017), as well as competition for nutrients with potential pathogens.

1.6.4 Medical Importance

Streptomyces produces antibiotics that can be used as a weapon for microbial competitiveness, spore survival, cohost protection, and interspecies communication between microorganisms (Ueda et al. 2000; Currie 2001; Linares et al. 2006). *Streptomyces* species are renowned as a key generator of antibiotics, in which they synthesize bioactive small molecules used in the pharmaceutical sector (Table 1.1) (Pombubpa et al. 2023). *Streptomyces* are the source of most antibiotics used in medical, veterinary care, and agriculture sectors (Chater 2016). *Streptomyces hyalurolyticus* produces hyaluronidase enzyme, which is used in ophthalmic treatments (Schuler and Schuler 2005). *Streptomyces* may offer a viable substitute platform for synthesizing recombinant proteins. Secondary metabolites of *Streptomyces*, such as enzymes or antibiotics, are produced to catabolize organic molecules (Adegboye and Babalola 2012; Tyc et al. 2017). The utilization of *Streptomyces pulcher*, *S. canescens*, and *Streptomyces citreofluorescens* has effectively controlled the other pathogens such as *Alternaria solani*, *Pseudomonas solanacearum*, *Fusarium oxysporum*, *Clavibactermichiganensis*, and *Verticillium albo-atrum* (El-Abyad et al. 1993). Blasticidin S and Hydrochloride are also commercially available antibiotics derived from *Streptomyces griseochromogenes*, which act against viral infections and are widely used in research labs for selecting cells with blasticidin resistance (Atala and Lanza 2012).

Table 1.1 Economic importance of *Streptomyces* in the production of useful compounds (Madigan 2005; Willey et al. 2014)

Streptomyces sp.	Natural product isolated	Application
Streptomyces verticillus	Bleomycin	Anticancer
Streptomyces clavuligerus	Clavulanic acid	Beta-lactamase inhibitor
Streptomyces peucetius	Daunorubicin Epirubicin Doxorubicin	Anticancer
Streptomyces noursei	Nysatin	Antifungal
Streptomyces mediterranei	Rifamycin	Antibiotic; transcription inhibitor
Streptomyces rimosus	Tetracycline	Antibiotic: protein synthesis inhibitor
Streptomyces nodosis	Amphotericin B	Antifungal
S. venezuelae	Chloramphenicol	Antibiotic: protein synthesis inhibitor drug used for typhoid fever
S. griseus	Streptomycin	Antibacterial
S. fradiae	Neomycin	Used in topical applications because of toxicity
S. aureofaciens	Tetracycline Chlortetracycline	Antibacterial against most gram-positive and gram-negative bacteria.
S. linolnensis	Clindamycin	Effective against obligate anaerobes, e.g., *Bacteroides fragilis*

1.6.5 Commercial Applications/Commercial Products

Mycostop and Actinovate are biocontrol agents derived from *Streptomyces* species (Minuto et al. 2006; Zeng et al. 2012). Mykocide (*S. colombiensis*), Safegrow (*S. kasugaensis*), Actofit, Astur (*S. avermitilis*), Bactophil (*S. albus*), BialaIncide SP (*S. hygroscopicus*), Toku-E (*S. viridochromogenes*), Actin, Incide SP (*S. atrovirens*) are some of the other registered commercial products developed with due incorporation of *Streptomyces* that are available around the world to control various plant diseases (Vurukonda et al. 2018). *Streptomyces phaeochromogenes* and *Streptomyces setonii* are found to secrete phosphotriesterases, which are commercially implemented for treating soil and waterbodies, food contaminations, and to break down pesticides and other petroleum products (Santillan et al. 2020). Milbemycins produced by *Streptomyces* are eco-friendly biopesticides with acaricidal and insecticidal properties (Yang et al. 2024).

1.7 Conclusion and Future Prospects

Streptomycetes are filamentous, gram-positive bacteria that have a mutualistic interaction with other organisms and are crucial sources of important secondary metabolites, which have wide applications in commercial, agricultural, ecological, and medicinal fields. The future of *Streptomyces* production for commercial use is promising, driven by advancements in synthetic biology and a renewed focus on natural products to combat antibiotic resistance. The field is moving beyond traditional antibiotic manufacturing to encompass a wider range of high-value compounds for applications in medicine, agriculture, and other industries. *Streptomyces* production for commercial purposes is a promising line driven by modernization in biology and a transformed focus on natural drugs to combat antibiotic resistance. Genetic engineering and metabolic regulation strategies can be adopted to optimize the appropriate strains of *Streptomyces* to improve bioactive compound yield. *Streptomyces* can be used as cell factories for the mass production of existing drugs and the discovery of novel drugs. Progressive genome mining methodologies may enable researchers to identify and activate "silent" or cryptic biosynthetic gene clusters in *Streptomyces*, which may synthesize undiscovered antibiotics and drugs. *Streptomyces* enzymes may also be used as powerful biocatalysts to perform regio- and stereoselective biotransformation processes. To improve the strain, CRISPR/Cas systems and other targeted genome editing tools can be used by allowing for the rapid activation of silent gene clusters and the optimization of metabolic pathways. Genetic engineering approaches cum the fermentation process have to be accelerated to improve yields. Interestingly, genetic modification in *Streptomyces* can be used to eliminate the geosmin and 2-methylisoborneol (MIB) responsible for the basic flavor in fish flesh.

References

Adams AS, Jordan MS, Adams SM, Suen G et al (2011) Cellulose-degrading bacteria associated with the invasive woodwasp *Sirex noctilio*. ISME J 5:1323–1331

Adegboye MF, Babalola OO (2012) Taxonomy and ecology of antibiotic producing Actinobacteria. Afr J Agric Res 7:2255–2261

Amore A, Pepe O, Ventorino V, Birolo L et al (2012) Cloning and recombinant expression of a cellulase from the cellulolytic strain *Streptomyces* sp. G12 isolated from compost. Microb Cell Fact 11:164

Asquith E, Evans C, Dunstan RH, Geary P, Cole B et al (2018) Distribution, abundance and activity of geosmin and 2-methylisoborneol-producing *Streptomyces* in drinking water reservoirs. Water Res 145: 30–38

Atala A, Lanza R (2012) Handbook of stem cells, vol 2. Academic Press

Bais HP, Weir TL, Perry LG, Gilroy S, Vivanco JM et al (2006) The role of root exudates in rhizosphere interactions with plants and other organisms. Annu Rev Plant Biol 57: 233–266

Bakker MG, Otto-Hanson L, Lange AJ et al (2013) Plant monocultures produce more antagonistic soil *Streptomyces* communities than high-diversity plant communities. Soil Biol Biochem 65: 304–312

Baldacci E (1939) Introduzioneallo studio degliAttinomiceti. Mycopathologia 2:84–106

Barka EA, Vatsa P, Sanchez L et al (2016) Taxonomy, physiology, and natural products of Actinobacteria. Microbiol Mol Biol Rev 80:1–43
Bentley R, Meganathan R (1981) Geosmin and methylisoborneol biosynthesis in streptomycetes: evidence for an isoprenoid pathway and its absence in non-differentiating isolates. FEBS Lett 125:220–222
Bentley SD, Chater KF, Cerdeño-Tárraga AM et al (2002) Complete genome sequence of the model Actinobacteria *Streptomyces* coelicolor A3(2). Nature 417:141–147
Berdy J (2005) Bioactive microbial metabolites. J Antibiot 58:1–26
Bibb MJ (2013) Understanding and manipulating antibiotic production in actinomycetes. Biochem Soci Transa 41:1355–1364
Bobek J, Halada P, Angelis J et al (2004) Activation and expression of proteins during synchronous germination of aerial spores of *Streptomyces granaticolor*. Proteomics 4:3864–3880
Bobek J, Šmídová K, Čihák M et al (2017) A waking review: old and novel insights into the spore germination in *Streptomyces*. Front Microbiol 8: 2205
Bontemps C, Toussaint M, Revol PV et al (2013) Taxonomic and functional diversity of *Streptomyces* in a forest soil. FEMS Microbiol Lett 342:157–167
Book AJ, Lewin GR, McDonald BR et al (2016) Evolution of high cellulolytic activity in symbiotic *Streptomyces* through selection of expanded gene content and coordinated gene expression. PLoS Biol 14:e1002475
Botas Muñoz AM (2013) Regulación del metabolismoen *Streptomyces*: control porArgR. Doctoral dissertation, Universidad de León
Bruce T, Martinez IB, Neto OM et al (2010) Bacterial community diversity in the Brazilian Atlantic forest soils. MicrobEcol 60:840–849
Buchanan RE (1918) Studies in the nomenclature and classification of the bacteria: VIII. The subgroups and genera of the Actinomycetales. J Bacteriol 3:403–406
Bush MJ, Tschowri N, Schlimpert S et al (2015) c-di-GMP signalling and the regulation of developmental transitions in streptomycetes. Nat Rev Microbiol 13:749–760
Chater KF (2016) Recent advances in understanding *Streptomyces*. F1000Res 5:2795
Chater KF, Biró S, Lee KJ et al (2010) The complex extracellular biology of *Streptomyces*. FEMS Microbiol Rev 34:171–198
Chevrette MG, Carlson CM, Ortega HE et al (2019) The antimicrobial potential of *Streptomyces* from insect microbiomes. Nat Commun 10(1):516
Claessen D, Wösten HA, Van Keulen G et al (2002) Two novel homologous proteins of *Streptomyces coelicolor* and *Streptomyces lividans* are involved in the formation of the rodlet layer and mediate attachment to a hydrophobic surface. Mol Microbiol 44:1483–1492
Claessen D, Rink R, De Jong W et al (2003) A novel class of secreted hydrophobic proteins is involved in aerial hyphae formation in *Streptomyces coelicolor* by forming amyloid-like fibrils. Genes Dev 17:1714–1726
Claessen D, Stokroos I, Deelstra HJ, Penninga NA, Bormann C, Salas JA, Wösten HA et al (2004) The formation of the rodlet layer of streptomycetes is the result of the interplay between rodlins and chaplins. Mol Microbiol 53:433–443
Claessen D, De Jong W, Dijkhuizen L et al (2006) Regulation of *Streptomyces* development: reach for the sky! Trends Microbiol 14:313–319
Claessen D, Rozen DE, Kuipers OP et al (2014) Bacterial solutions to multicellularity: a tale of biofilms, filaments and fruiting bodies. Nat Rev Microbiol 12:115–124
Cowan AE, Koppel DE, Setlow B, Setlow P et al (2003) A soluble protein is immobile in dormant spores of *Bacillus subtilis* but is mobile in germinated spores: implications for spore dormancy. Proc Natl Acad Sci USA 100:4209–4214
Crawford DL (1978) Lignocellulose decomposition by selected *Streptomyces* strains. Appl Environ Microbiol 35:1041–1045
Currie CR (2001) A community of ants, fungi, and bacteria: a multilateral approach to studying symbiosis. Ann Rev Microbiol 55:357–380

Currie CR, Scott JA, Summerbell RC, Malloch D et al (1999) Fungus-growing ants use antibiotic-producing bacteria to control garden parasites. Nature 398: 701–704

De Jong W, Wösten HA, Dijkhuizen L, Claessen D et al (2009) Attachment of *Streptomyces coelicolor* is mediated by amyloidal fimbriae that are anchored to the cell surface via cellulose. Mol Microbiol 73: 1128–1140

Derry AM, Staddon WJ, Kevan PG, Trevors JT et al (1999) Functional diversity and community structure of micro-organisms in three arctic soils as determined by sole-carbon-source-utilization. Biodivers Conserv 8(2):205–221

De Vasconcellos RLF, Cardoso EJBN (2009) Rhizospheric streptomycetes as potential biocontrol agents of Fusarium and Armillaria pine rot and as PGPR for *Pinus taeda*. BioControl 54:807–816

Dietz A, Mathews J (1971) Classification of *Streptomyces* spore surfaces into five groups. Appl Microbiol 21:527–533

Dyson P (2009) Streptomyces. In: Schaechter M (ed) Encyclopedia of microbiology, 3rd edn. Academic, Oxford, pp 318–332

Eaton D, Ensign JC (1980) Streptomyces viridochromogenes spore germination initiated by calcium ions. J Bacteriol 143:377–382

El-Abyad MS, El-Sayed MA, El-Shanshoury AR, El-Sabbagh SM et al (1993) Towards the biological control of fungal and bacterial diseases of tomato using antagonistic *Streptomyces* spp. Plant Soil 149:185–195

Elliot MA, Talbot NJ (2004) Building filaments in the air: aerial morphogenesis in bacteria and fungi. Curr Opin Microbiol 7:594–601

Elliot MA, Karoonuthaisiri N, Huang J, Bibb MJ, Cohen SN, Kao CM, Buttner MJ et al (2003) The chaplins: a family of hydrophobic cell-surface proteins involved in aerial mycelium formation in *Streptomyces coelicolor*. Genes Dev 17:1727–1740

Ensign JC (1978) Formation, properties, and germination of bacterial spores. Ann Rev Microbiol 32:185–219

Euzeby JP (2012) List of prokaryotic names with standing in nomenclature—genus *Streptomyces*. http://www.bacterio.cict.fr/s/*Streptomyces*b.html. Accessed 20 Feb 2012

Flärdh K (2003) Growth polarity and cell division in *Streptomyces*. Curr OpinMicrobiol 6(6): 564–571

Flärdh K, Buttner MJ (2009) *Streptomyces* morphogenetics: dissecting differentiation in a filamentous bacterium. Nat Rev Microbiol 7(1):36–49

Franco-Correa M, Quintana A, Duque C, Suarez C, Rodríguez MX, Barea JM et al (2010) Evaluation of Actinobacteria strains for key traits related with plant growth promotion and mycorrhiza helping activities. Appl Soil Ecol 45(3):209–217

Goodfellow M, O'Donnell AG (1993) Roots of bacterial systematic. In: Handbook of new bacterial systematics. Academic, London, pp 3–56

Goodfellow M, Lacey J et al (1987) Numerical classification of thermophilic streptomycetes. J Gen Microbiol 133:3135–3149

Gottschalk LM, Macedo JM et al (1999) Lignin peroxidase production by *Streptomyces viridosporus* T7A: use of corn oil as a carbon source. Appl Biochem Biotechnol 77–79:771–778

Grimont F, Grimont P (1986) Ribosomal ribonucleic acid gene restriction patterns as potential taxonomic tools. Annales de l'Institut Pasteur/Microbiologie 137:165–175

Grisebach H (1975) Neuereuntersuchungen Zur Biosynthese Einiger Antibiotika. Planta Med 28: 232–250

Grubbs KJ, Surup F, Biedermann PH, McDonald BR, Klassen JL, Carlson CM, Clardy J, Currie CR et al (2020) Cycloheximide-producing *Streptomyces* associated with *Xyleborinus saxesenii* and *Xyleborus affinis* fungus-farming ambrosia beetles. Front Microbiol 11:562140

Grubbs KJ, May DS, Sardina JA, Dermenjian RK, Wyche TP, Pinto-Tomás AA, Clardy J, Currie CR et al (2021) Pollen *Streptomyces* produce antibiotic that inhibits the honey bee pathogen Paenibacillus larvae. Front Microbiol 12:632637

Gulder TA, Moore BS (2009) Chasing the treasures of the sea—bacterial marine natural products. Curr Opin Microbiol 12(3):252–260

Haiser HJ, Yousef MR, Elliot MA et al (2009) Cell wall hydrolases affect germination, vegetative growth, and sporulation in *Streptomyces coelicolor*. J Bacteriol 191(21):6501–6512

Hamedi J, Mohammadipanah F (2015) Biotechnological application and taxonomical distribution of plant growth promoting actinobacteria. J Ind Microbiol Biotechnol 42(2):157–171

Hamid ME (2013) The use of morphological and cell wall chemical markers in the identification of *Streptomyces* species associated with Actinomycetoma. Afr J Clin Exp Microbiol 14(2):45–50

Hardisson C, Manzanal MB, Salas JA, Suarez JE et al (1978) Fine structure, physiology and biochemistry of arthrospore germination in *Streptomyces* antibioticus. J Gen Microbiol 105: 203–214

Hasani A, Kariminik A, Issazadeh K et al (2014) Streptomycetes: characteristics and their antimicrobial activities. Int J Adv Biol Biomed Res 2(1):63–75

Hempel AM, Wang S, Letek M, Gil JA, Flärdh K et al (2008) Assemblies of DivIVA mark sites for hyphal branching and can establish new zones of cell wall growth in *Streptomyces coelicolor*. J Bacteriol 190:7579–7583

Hiltunen LH, Ojanperä T, Kortemaa H, Richter E, Lehtonen MJ, Valkonen JPT et al (2009) Interactions and biocontrol of pathogenic *Streptomyces* strains co-occurring in potato scab lesions. J Appl Microbiol 106(1):199–212

Hirsch CF, Ensign JC (1976) Nutritionally defined conditions for germination of *Streptomyces viridochromogenes* spores. J Bacteriol 126(1):13–23

Hofheinz W, Grisebach H (1965) Die Fettsäuren von Streptomyces erythreus and Streptomyces halstedii. Z Naturforsch 20B:43

Holmes NA, Walshaw J, Leggett RM, Thibessard A, Dalton KA, Gillespie MD, Kelemen GH et al (2013) Coiled-coil protein Scy is a key component of a multiprotein assembly controlling polarized growth in *Streptomyces*. Proc Natl Acad Sci USA 110(5):E397–E406

Hoskisson PA, van Wezel GP (2019) Streptomyces coelicolor. Trends Microbiol 27(5):468–469

Hu H, Lin HP, Xie Q, Li L, Xie XQ, Hong K et al (2012) Streptomyces qinglanensis sp. nov., isolated from mangrove sediment. Int J Syst Evol Microbiol 62(Pt 3) 596–600

Huddleston A, Cresswell N, Neves M et al (1997) Molecular detection of streptomycin-producing Streptomycetes in Brazilian soils. Appl Environ Microbiol 63:1288–1297

Hungund BS, Honnangi S, Desai SS et al (2022) Diversity and classification of *Streptomyces*. In: Actinobacteria: ecology, diversity, classification and extensive applications. Springer Nature Singapore, Singapore, pp 89–116

Igarashi T (2004) Deletion in sortase gene of *Streptococcus mutans* Ingbritt. Oral Microbiol Immunol 19(3):210–213

Ikeda H, Ishikawa J, Hanamoto A, Shinose M, Kikuchi H, Shiba T, Sakaki Y, Hattori M, Ōmura S et al (2003) Complete genome sequence and comparative analysis of the industrial microorganism *Streptomyces avermitilis*. Nat Biotechnol 21(5):526–531

Janssen PH (2006) Identifying the dominant soil bacterial taxa in libraries of 16S rRNA and 16S rRNA genes. Appl Environ Microbiol 72(3):1719–1728

Jensen HL (1930) Actinobacteria in Danish soils. Soil Sci 30:59–77

Jensen HL (1931) Contribution to our knowledge of the Actinomycetales. II. The definition and subdivision of the genus Actinomyces, with a preliminary account of Australian soil Actinobacteria. Proc Linnean Soc NSW 56:345–370

Jog R, Pandya M, Nareshkumar G, Rajkumar S et al (2014) Mechanism of phosphate solubilization and antifungal activity of *Streptomyces* spp. isolated from wheat roots and rhizosphere and their application in improving plant growth. Microbiology 160(4) 778–788

Jones KL (1954) Variation in *Streptomyces*. Ann N Y Acad Sci 60:124

Jones SE, Elliot MA (2017) *Streptomyces* exploration: competition, volatile communication and new bacterial behaviours. Trends Microbiol 25(7):522–531

Jones SE, Elliot MA (2018) 'Exploring' the regulation of *Streptomyces* growth and development. Curr Opin Microbiol 42:25–30

Jones SE, Ho L, Rees CA, Hill JE, Nodwell JR, Elliot MA et al (2017) *Streptomyces* exploration is triggered by fungal interactions and volatile signals. Elife 6 e21738

Kalakoutskii LV, Agre NS (1976) Comparative aspects of development and differentiation in Actinobacteria. Bacteriol Rev 40(2):469–524

Seipke RF, Kaltenpoth M, Hutchings MI et al (2012) *Streptomyces* as symbionts: an emerging and widespread theme. FEMS Microbiol Rev 36:862–876

Kämpfer P (2012) *Streptomyces* Waksman and Henrici 1943, 339 AL. In: Bergey's manual of systematic bacteriology, 2nd edn. Springer, New York, pp 1679–1680

Kämpfer P, Kroppenstedt RM, Dott W et al (1991) A numerical classification of the genera Streptomyces and Streptoverticillium using miniaturized physiological tests. Microbiology 137(8):1831–1891

Keijser BJ, van Wezel GP, Canters GW, Kieser T, Vijgenboom E et al (2000) The ram-dependence of *Streptomyces lividans* differentiation is bypassed by copper. J Mol Microbiol Biotechnol 2: 565–574

Kelemen GH, Buttner MJ (1998) Initiation of aerial mycelium formation in *Streptomyces*. Curr Opin Microbiol 1(6):656–662

Kim DR, Cho G, Jeon CW, Weller DM, Thomashow LS, Paulitz TC, Kwak YS et al (2019) A mutualistic interaction between *Streptomyces* bacteria, strawberry plants and pollinating bees. Nat Commun 10(1):4802

Kleinschnitz EM, Heichlinger A, Schirner K, Winkler J, Latus A, Maldener I, Wohlleben W, Muth G et al (2011) Proteins encoded by the mre gene cluster in *Streptomyces coelicolor* A3 (2) cooperate in spore wall synthesis. Mol Microbiol 79(5):1367–1379

Komaki H (2023) Recent progress of reclassification of the genus *Streptomyces*. Microorganisms 11(4):831

Krainsky A (1914) Die Aktinomyceten und ihreBedeutung in der Natur. ZentralblBakteriolParasitenkdInfektionskrHyg II Abt 41:649–688

Krasilnikov NA (1938) Ray fungi and related organisms. Actinomycetales. Izdatel'stvoAkademiiNauk SSSR, Moscow

Kumar M, Kumar P, Das P et al (2020) Potential applications of extracellular enzymes from *Streptomyces* spp. in various industries. Arch Microbiol 202:1597–1615

Kuster E (1972) Simple working key for the classification and identification of named taxa included in the International *Streptomyces* Project. Int J Syst Evol Microbiol 22:139–148

Kutzner HJ (1981) The family Streptomycetaceae. In: Starr MP, Stolp H, Trüper HG, Balows A, Schlegel HG (eds) The prokaryotes: a handbook on habitats, isolation and identification of bacteria, vol II. Springer, Berlin, pp 2028–2090

Labeda DP (2011) Multilocus sequence analysis of phytopathogenic species of the genus *Streptomyces*. Int J Syst Evol Microbiol 61(10):2525–2531

Labeda DP, Goodfellow M, Brown R, Ward AC, Lanoot B, Vanncanneyt M et al (2012) Phylogenetic study of the species within the family Streptomycetaceae. Antonie Van Leeuwenhoek 101: 73–104

Lanoot B (2004) Improved taxonomy of the genus *Streptomyces*. PhD thesis. University Gent, pp 1–321

Law JWF, Ser HL, Duangjai A, Saokaew S, Bukhari SI, Khan TM, Lee LH et al (2017) *Streptomyces colonosanans* sp. nov., a novel actinobacterium isolated from Malaysia mangrove soil exhibiting antioxidative activity and cytotoxic potential against human colon cancer cell lines. Front Microbiol 8:877

Law JWF, Tan KX, Wong SH et al (2018) Taxonomic and characterization methods of *Streptomyces*: a review. Prog Microbes Mol Biol 1(1):a0000009

Lechevalier MP (1977) Lipids in bacterial taxonomy- a taxonomist's view. CRC Crit Rev Microbiol 5:109–210

Lechevalier M, Lechevalier H (1970) Chemical composition as a criterion in the classification of aerobic Actinobacteria. Int J Syst Bacteriol 20:435–443

Lee MY, Kim HY, Lee S et al (2015) Metabolomics-based chemotaxonomic classification of *Streptomyces* spp. and its correlation with antibacterial activity. J Microbiol Biotechnol 25: 1265–1274

Lehr NA, Schrey SD, Hampp R, Tarkka MT et al (2008) Root inoculation with a forest soil streptomycete leads to locally and systemically increased resistance against phytopathogens in Norway spruce. New Phytol 177(4):965–976

Li X, Li E, Zhu Y, Teng C et al (2012) A typical endo-xylanase from *Streptomyces rameus* L2001 and its unique characteristics in xylooligosaccharide production. Carbohydr Res 359:30–36

Lieske R (1921) Morphologie und Biologie der Strahlenpilze. Bontraeger, Leipzig

Linares JF, Gustafsson I, Baquero F, Martinez JL et al (2006) Antibiotics as intermicrobial signaling agents instead of weapons. Proc Natl Acad Sci USA 103:19484–19489

Locci R, Sharples GP (1984) Morphology. In: Goodfellow M, Mordarski M, Williams ST (eds) The biology of the Actinobacteria. Academic Press, London, pp 165–199

Madigan MT, Martinko J (2005) Brock biology of microorganisms, 11th edn. Pearson Prentice Hall, Upper Saddle River

Madigan MT (2014) Bacteriology. In eLS, John Wiley & Sons, Ltd (ed.) https://doi.org/10.1002/9780470015902.a0000449.pub2

Madigan MT, Martinko JM, Bender KS et al (2014) Brock biology of microorganisms. Pearson, London

Manchester L, Pot B, Kersters K et al (1990) Classification of *Streptomyces* and *Streptoverticillium* species by numerical analysis of electrophoretic protein patterns. Syst Appl Microbiol 13:333–337

Manullang W, Chuang HW (2020) *Streptomyces* sp. mitigates abiotic stress response and promotes plant growth. J Plant Prot Res 60(3):263–274

Martin P, Dary A, André A et al (2000) Identification and typing of *Streptomyces* strains: evaluation of interspecific, intraspecific and intraclonal differences by RAPD fingerprinting. Res Microbiol 151:853–864

Mayfield CI, Williams ST, Ruddick SM, Hatfield HL et al (1972) Studies on the ecology of Actinobacteria in soil IV. Observations on the form and growth of streptomycetes in soil. Soil Biol Biochem 4(1):79–91

McBride MJ, Ensign JC (1987) Effects of intracellular trehalose content on *Streptomyces griseus* spores. J Bacteriol 169(11):4995–5001

McGregor JF (1954) Nuclear division and the life cycle in a *Streptomyces* sp. Microbiology 11(1): 52–56

McKenzie NL, Thaker M, Koteva K, Hughes DW, Wright GD, Nodwell JR et al (2010) Induction of antimicrobial activities in heterologous streptomycetes using alleles of the *Streptomyces coelicolor* gene absA1. J Antibiot (Tokyo) 63(4) 177–182

Méndez C, Brana AF, Manzanal MB, Hardisson C et al (1985) Role of substrate mycelium in colony development in *Streptomyces*. Can J Microbiol 31(5):446–450

Menegatti C, Lourenzon VB, Rodríguez-Hernández D, Da Paixão Melo WG, Ferreira LLG, Andricopulo AD et al (2020) Meliponamycins: antimicrobials from stingless bee-associated *Streptomyces* sp. J Nat Prod 83:610–616

Miguélez EM, Hardisson C, Manzanal MB et al (1999) Hyphal death during colony development in *Streptomyces antibioticus*: morphological evidence for the existence of a process of cell deletion in a multicellular prokaryote. J Cell Biol 145(3):515–525

Miguélez EM, Hardisson C, Manzanal MB et al (2010) Streptomycetes: a new model to study cell death. Int Microbiol 3(3):153–158

Mikulik K, Janda I, Mašková H, Šťastná J, Jiráňová A et al (1977) Macromolecular synthesis accompanying the transition from spores to vegetative forms of *Streptomyces granaticolor*. Folia Microbiol 22:252–261

Mikulik K, Bobek J, Bezouskova S, Benada O, Kofronova O et al (2002) Expression of proteins and protein kinase activity during germination of aerial spores of *Streptomyces granaticolor*. Biochem Biophys Res Commun 299:335–342

Milind G, Watve TR, Jog M, Bhalachandra D et al (2001) How many antibiotics are produced by the genus *Streptomyces*. Arch Microbiol 176(5):386–390

Minuto A, Spadaro D, Garibaldi A, Gullino ML et al (2006) Control of soilborne pathogens of tomato using a commercial formulation of *Streptomyces* griseoviridis and solarization. Crop Prot 25(5):468–475

Mistry BV (2008) FtsW is a dispensable cell division protein required for Z-ring stabilization during sporulation septation in *Streptomyces* coelicolor. J Bacteriol 190(16):5555–5566

Muok AR, Claessen D, Briegel A et al (2021) Microbial hitchhiking: how *Streptomyces* spores are transported by motile soil bacteria. ISME J 15:2591–2600

Muthusamy S, Selvan ST, Grasian I et al (2017) Bioconversion and bioethanol production from agro-residues through fermentation process using mangrove-associated actinobacterium *Streptomyces olivaceus* (MSU3). Biofuels 10(2):167–179. https://doi.org/10.1080/17597269.2017.1309853

Nissen TV (1963) Distribution of antibiotic-producing Actinobacteria in Danish soil. Experientia 19:470–471

Noens EE, Mersinias V, Traag BA, Smith CP, Koerten HK, van Wezel GP et al (2005) SsgA-like proteins determine the fate of peptidoglycan during sporulation of *Streptomyces* coelicolor. Mol Microbiol 58:929–944

Noens EE, Mersinias V, Willemse J, Traag BA, Laing E, Chater KF, Smith CP, Koerten HK, Van Wezel GP et al (2007) Loss of the controlled localization of growth stage-specific cell-wall synthesis pleiotropically affects developmental gene expression in anssgA mutant of *Streptomyces* coelicolor. Mol Microbiol 64:1244–1259

Norte DM, Avitia-Dominguez LA, Rozen DE (2025) Evolution and ecology of *Streptomyces*. Annu Rev Microbiol 79:383–403

Ochi K, Hiranuma H (1994) A taxonomic review of the genera Kitasatosporia and Streptoverticillium by analysis of ribosomal protein AT-L30. Int J Syst Evol Micro 44:285–292

Okami Y, Okazaki T (1972) Studies on marine microorganisms. I: Actinobacteria in Sagami Bay and their antibiotic substances. J Antibiot 25:456–460

Olanrewaju OS, Babalola OO (2019) *Streptomyces*: implications and interactions in plant growth promotion. Appl Microbiol Biotechnol 103:1179–1188

Olanrewaju OS, Glick BR, Babalola OO et al (2017) Mechanisms of action of plant growth promoting bacteria. World J Microbiol Biotechnol 33:1–16

Orskov J (1923) Investigations into the morphology of the ray fungi. Levin & Munksgaard, Copenhagen

Pacios-Michelena S, Aguilar González CN, Alvarez-Perez OB, Rodriguez-Herrera R, Chávez-González M, Arredondo Valdés R, Ascacio Valdés JA, Govea Salas M, Ilyina A et al (2021) Application of *Streptomyces* antimicrobial compounds for the control of phytopathogens. Front Sustain Food Syst 5:696518

Paradis E, Goyer C, Hodge N et al (1994) Fatty acid and protein profiles of *Streptomyces* scabies strains isolated in eastern Canada. Int J Syst Bacteriol 44:561–564

Paradkar A, Trefzer A, Chakraburtty R, Stassi D et al (2003) *Streptomyces* genetics: a genomic perspective. Crit Rev Biotechnol 23:1–27

Paredes-Sabja D, Setlow P, Sarker MR et al (2011) Germination of spores of Bacillales and Clostridiales species: mechanisms and proteins involved. Trends Microbiol 19:85–94

Pessotti RDC, Hansen BL, Reaso JN, Ceja-Navarro JA, El-Hifnawi L, Brodie EL, Traxler MF et al (2021) Multiple lineages of *Streptomyces* produce antimicrobials within passalid beetle galleries across eastern North America. Elife 10:e65091

Petrus MLC, Claessen D (2014) Pivotal roles for *Streptomyces* cell surface polymers in morphological differentiation, attachment and mycelial architecture. Antonie Van Leeuwenhoek 106:127–139

Phillips L (1992) The distribution of phenotypic and genotypic characters within Streptomycetes and their relationship to antibiotic production. Doctoral dissertation, University of Warwick

Piel J (2004) Metabolites from symbiotic bacteria. Nat Prod Rep 21:519–538

Piel J (2009) Metabolites from symbiotic bacteria. Nat Prod Rep 26:338–362

Pimentel-Elardo SM, Scheuermayer M, Kozytska S, Hentschel U et al (2009) *Streptomyces* axinellae sp. nov., isolated from the Mediterranean sponge Axinellapolypoides (Porifera). Int J Syst Evol Microbiol 59:1433–1437

Pombubpa N, Lakmuang C, Tiwong P et al (2023) *Streptomyces* diversity maps reveal distinct high-specificity biogeographical and environmental patterns compared to the overall bacterial diversity. Life 14: 11

Pospíšil S, Řezanka T, Viden I et al (1985) Altered fatty acid composition in regulatory mutants of *Streptomyces* cinnamonensis. FEMS Micro Lett 27:41–43

Pot B, Vandamme P et al (1994) Analysis of electrophoretic whole-organism protein f ingerprints. In: Goodfellow M, O'Donnell AG (eds) Modern microbial methods. Chemical methods in prokaryotic systematics. Wiley, Chichester, pp 493–521

Pridham TG (1974) Genus I. Streptomyces Waksman and Henrici. 1943. Bergey's Manual of Determinative Bacteriology 748–829

Procópio RE, Silva IR, Martins MK et al (2012) Antibiotics produced by *Streptomyces*. Braz J Infect Dis 16:466–471

Qin S, Xing K, Jiang JH, Xu LH, Li WJ et al (2011) Biodiversity, bioactive natural products and biotechnological potential of plant-associated endophytic actinobacteria. Appl MicrobiolBiotechnol 89:457–473

Rashad FM, Fathy HM, El-Zayat AS, Elghonaimy AM et al (2015) Isolation and characterization of multifunctional *Streptomyces* species with antimicrobial, nematicidal and phytohormone activities from marine environments in Egypt. Microbiol Res 175:34–47

Ravi RK, Jadeja VJ (2016) Isolation of Actinomycetes: A complete approach. Int J Curr Microbiol App Sci 5(5):606–618

Ruban-Osmialowska B, Jakimowicz D, Smulczyk-Krawczyszyn A, Chater KF, Zakrzewska-Czerwinska J et al (2006) Replisome localization in vegetative and aerial hyphae of *Streptomyces* coelicolor. J Bacteriol 188:7311–7316

Ryosaku N (1960) On the classification of *Streptomyces*. J Antibiot Ser A 13:236–247

Saddler GS, Goodfellow M, Minnikin DE et al (1986) Influence of the growth cycle on the fatty acid and menaquinone composition of *Streptomyces* cyaneus NCIB 9616. J Appl Bacteriol 60: 51–56

Saddler GS, O'donnell AG, Goodfellow M, & Minnikin DE et al (1987) SIMCA pattern recognition in the analysis of streptomycete fatty acids. Microbiology 133(5):1137–1147

Sadeghi A, Karimi E, Dahaji PA, Javid MG, Dalvand Y, Askari H et al (2012) Plant growth promoting activity of an auxin and siderophore producing isolate of *Streptomyces* under saline soil conditions. World J Microbiol Biotechnol 28:1503–1509

Salas JA, Guijarro JA, Hardisson C et al (1983) High calcium content in *Streptomyces* spores and its release as an early event during spore germination. J Bacteriol 155:1316–1323

Santillan JY, Muzlera A, Molina M, Lewkowicz ES, Iribarren AM et al (2020) Microbial degradation of organophosphorus pesticides using whole cells and enzyme extracts. Biodegradation 31: 423–433

Schlatter DC, Kinkel LL (2014) Global biogeography of *Streptomyces* antibiotic inhibition, resistance, and resource use. FEMS Microbiol Ecol 88:386–397

Schuler E, Schuler C (2005) Use of *Streptomyces hyalurolyticus* enzyme in ophthalmic treatments. U.S. Patent 6: 902–548

Seipke RF, Kaltenpoth M, Hutchings MI et al (2012) Streptomyces as symbionts: an emerging and widespread theme? FEMS Microbiology Reviews 36(4):862–876

Seipke RF, Barke J, Heavens D, Yu DW, Hutchings MI. et al (2013) Analysis of the bacterial communities associated with two ant-plant symbioses. MicrobiologyOpen 2(2):276–283

Setlow B, Atluri S, Kitchel R, Koziol-Dube K, Setlow P et al (2006) Role of dipicolinic acid in resistance and stability of spores of Bacillus subtilis with or without DNA-protective alpha/beta-type small acid-soluble proteins. J Bacteriol 188:3740–3747

Shepherdson EM, Baglio CR, Elliot MA et al (2023) *Streptomyces* behavior and competition in the natural environment. Curr Opin Microbiol 71:102257

Shimizu M (2011) Endophytic Actinobacteria: biocontrol agents and growth promoters. In: Bacteria in agrobiology: plant growth responses, pp 201–220

Shirling EB, Gottlieb D (1966) Methods for characterization of *Streptomyces* species. Int J Syst Evol Microbiol 16:313–340

Shirling EB, Gottlieb D (1968a) Cooperative description of type cultures of *Streptomyces*. II. Species descriptions from first study. Int J Syst Evol Microbiol 18:69–189

Shirling EB, Gottlieb D (1968b) Cooperative description of type cultures of *Streptomyces* III. Additional species descriptions from first and second studies. Int J Syst Evol Microbiol 18:279–392

Shirling EB, Gottlieb D (1969) Cooperative description of type cultures of *Streptomyces*. IV. Species descriptions from the second, third and fourth studies. Int J Syst Evol Microbiol 19:391–512

Simpson NE, Kidd KK, Goodfellow PJ, McDermid H, Myers S, Kidd JR, Jackson CE, Duncan AM, Farrer LA, Brasch K, et al (1987) Assignment of multiple endocrine neoplasia type 2A to chromosome 10 by linkage. Nature 328(6130):528–530

Sivalingam P, Easwaran M, Ganapathy D, Basha SF, Poté J et al (2024) Endophytic *Streptomyces*: an underexplored source with potential for novel natural drug discovery and development. Arch Microbiol 206:442

Sneath PH, Johnson R (1972) The influence on numerical taxonomic similarities of errors in microbiological tests. J Gen Microbiol 72:377–392

Stackbrandt E, Goebel B (1994) Taxonomic note: a place for DNA-DNA reassociation and 16S rRNA sequence analysis in the present species definition in bacteriology. Int J Syst Evol Microbiol 44:846–849

Strakova E, Bobek J, Zikova A, Rehulka P, Benada O, Rehulkova H, Kofronova O, Vohradsky J et al (2013) Systems insight into the spore germination of *Streptomyces* coelicolor. J Proteome Res 12:525–536

Tarkka MT, Lehr NA, Hampp R, Schrey SD et al (2008) Plant behavior upon contact with Streptomycetes. Plant Signal Behav 3:917–919

Traag BA, van Wezel GP (2008) The SsgA-like proteins in Actinobacteria: small proteins up to a big task. Antonie Van Leeuwenhoek 94:85–97

Tschowri N, Schumacher MA, Schlimpert S, Chinnam NB, Findlay KC, Brennan RG, Buttner MJ et al (2014) Tetrameric c-di-GMP mediates effective transcription factor dimerization to control *Streptomyces* development. Cell 158:1136–1147

Tyc O, Song C, Dickschat JS, Vos M, Garbeva P et al (2017) The ecological role of volatile and soluble secondary metabolites produced by soil bacteria. Trends Microbiol 25:280–292

Ueda K, Kawai S, Ogawa HO, Kiyama A, Kubota T, Kawanobe H, Beppu T et al (2000) Wide distribution of interspecific stimulatory events on antibiotic production and sporulation among *Streptomyces* species. J Antibiot 53:979–982

Van Dissel D, Claessen D, van Wezel GP et al (2014) Morphogenesis of *Streptomyces* in submerged cultures. Adv Appl Microbiol 89:1–45

Van Vliet S (2015) Bacterial dormancy: how to decide when to wake up. Curr Biol 25:R753–R755

Ventura M, Canchaya C, Tauch A, Chandra G, Fitzgerald GF, Chater KF, van Sinderen D et al (2007) Genomics of Actinobacteria: tracing the evolutionary history of an ancient phylum. Microbiol Mol Biol Rev 71:495–548

Verma VC, Singh SK, Prakash S et al (2011) Bio-control and plant growth promotion potential of siderophore-producing endophytic *Streptomyces* from Azadirachta indica A. Juss. J Basic Microbiol 51:550–556

Vurukonda SSKP, GiovanardiD, Stefani E et al (2018) Plant growth promoting and biocontrol activity of *Streptomyces* spp. as endophytes. Int J Mol Sci 19: 952. https://doi.org/10.3390/ijms19040952

Waksman SA (1919) Cultural studies of species of actinomyces. Soil Sci 8:71–215

Waksman SA (1940) On the classification of Actinobacteria. J Bacteriol 39:549–558

Waksman SA, Curtis RE (1916) The actinomyces of the soil. Soil Sci 1:99–134

Waksman SA, Henrici AT (1943) The nomenclature and classification of the Actinomycetes. Journal of bacteriology 46(4):337–341

Wang Y, Jiang Y (2016) Chemotaxonomy of actinobacteria. In: Dhanasekaran D (ed) Actinobacteria—basics and biotechnological applications. Intech Open, London, p 61482

Wang SL, Fan KQ, Yang X, Lin ZX, Xu XP, Yang KQ et al (2008) CabC, an EF-hand calcium-binding protein, is involved in Ca2+-mediated regulation of spore germination and aerial hypha formation in *Streptomyces* coelicolor. J Bacteriol 190:4061–4068

Wanner LA, Kirk WW (2015) *Streptomyces*–from basic microbiology to role as a plant pathogen. Am J Potato Res 92:236–242

Watve MG, Tickoo R, Jog MM, Bhole BD et al (2001) How many antibiotics are produced by the genus *Streptomyces*? Arch Microbiol 176:386–390

Wayne L, Moore W, Stackebrandt E et al (1987) Report of the Ad Hoc Committee on reconciliation of approaches to bacterial systematics. Int J Syst Evol Microbiol 37:463–464

Weller DM, Raaijmakers JM, Gardener BBM, Thomashow LS et al (2002) Microbial populations responsible for specific soil suppressiveness to plant pathogens. Annu Rev Phytopathol 40:309–348

Wildermuth H (1970) Development and organization of the aerial mycelium in *Streptomyces coelicolor*. Microbiology 60:43–50

Wildermuth H, Wehrli E, Horne RW et al (1971) The surface structure of spores and aerial mycelium in *Streptomyces* coelicolor. J Ultrastruct Res 35:168–180

Willemse J, Borst JW, de Waal E, Bisseling T, van Wezel GP et al (2011) Positive control of cell division: FtsZ is recruited by SsgB during sporulation of *Streptomyces*. Genes Dev 25:89–99

Willey JM, Sherwood LM et al (2014) Prescott's microbiology, 9th edn. McGraw-Hill

Williams ST (1982) Are antibiotics produced in soil? Pedobiologia 23(4):427–435

Williams ST, Bradshaw RM, Costerton JW, Forge A et al (1972) Fine structure of the spore sheath of some *Streptomyces* species. Microbiology 72:249–258

Williams ST, Davies FL, Mayfield CI, Khan MR et al (1971) Studies on the ecology of actinomycetes in soil II. The pH requirements of streptomycetes from two acid soils. Soil Biology and Biochemistry 3(3):187–195

Williams ST, Goodfellow M, Alderson G, Wellington EMH, Sneath PHA, Sackin MJ et al (1983a) Numerical classification of *Streptomyces* and related genera. J Gen Microbiol 129:1743–1813

Williams ST, Goodfellow M, EMH W, Vickers JC, Alderson G, PHA S, Sackin MJ, Mortimer AM et al (1983b) A probability matrix for identification of some streptomycetes. J Gen Microbiol 129:1815–1830

Williams S, Goodfellow M, Alderson G et al (1983c) Numerical classification of *Streptomyces* and related genera. Microbiology 129:1743–1813

Williams S, Goodfellow M, Wellington E et al (1983d) A probability matrix for identification of some Streptomycetes. Microbiology 129:1815–1830

Worsley SF, Newitt J, Rassbach J, Batey SF, Holmes NA, Murrell JC, Hutchings MI et al (2020) *Streptomyces* endophytes promote host health and enhance growth across plant species. Appl Environ Microbiol 86:e01053–e01020

Yang X, Du G, Li X, Xiang W, Li S, Zhang Y et al (2024) Recent advances of natural product pesticide milbemycins from *Streptomyces*. New Plant Protection 1:e18

Zaroubi L, Ozugergin I, Mastronardi K, Imfeld A, Law C, Gélinas Y, Piekny A, Findlay BL et al (2022) The ubiquitous soil terpene geosmin acts as a warning chemical. Appl Environ Microbiol 88:e00093–e00022

Zeng W, Wang D, Kirk W, Hao J et al (2012) Use of *Coniothyriumminitans* and other microorganisms for reducing *Sclerotinia sclerotiorum*. Biol Control 60:225–232

Streptomyces and Sustainable Agriculture

2

Maya Mundanattu Rajan, Keshavamurthy Mukunda, Geenat Paul, Greeshma Pana Valappil, and Anu Padinhapurath Abhimannue

Abstract

The plant pathogens result in crop damage of 30–40%, and dependence on chemical pesticides to combat these pathogens has led to resistance problem in phytopathogens. The various side effects of use of agrochemicals have sparked the interest of scientists toward sustainable agriculture that favors biological pesticides over chemical pesticides. Among microbes, having potential to use in sustainable agriculture as safe alternatives to chemical pesticides or fertilizers, members of genus *Streptomyces* in phylum Actinobacteria hold significant position. *Streptomyces* species produce a wide range of antimicrobial compounds, including antibiotics, volatile organic compounds, cyclic peptides and alkaloid derivatives, and lytic enzymes, which effectively mitigate bacterial and fungal pathogens in agriculture. Using diverse modes of operation (synthesis of different classes of antibiotics, lytic enzymes, volatiles, etc.), *Streptomycetes* has gained the status of good Microbial Biological Control Agents (MBCA). Metabolomics profiling of several *Streptomyces* species has exemplified the presence of

M. M. Rajan
Department of Biotechnology, Administrative Management College, Bengaluru, Karnataka, India

K. Mukunda
Department of Biotechnology, School of Applied Sciences, REVA University, Bengaluru, Karnataka, India

Center for Biotechnology and Molecular Biology, REVA Research Centre, REVA University, Bengaluru, Karnataka, India

G. Paul · A. P. Abhimannue (✉)
Department of Biotechnology, St Mary's College (Autonomous), Thrissur, Kerala, India
e-mail: anu.pa@smctsr.ac.in

G. P. Valappil
Department of Microbiology, St Mary's College (Autonomous), Thrissur, Kerala, India

T. Kaur et al. (eds.), *Streptomyces: Biological Candidates for Sustainable Agriculture*,
https://doi.org/10.1007/978-981-95-5803-2_2

bioactive metabolites with both growth-promoting and antifungal properties, such as tetrangulol, indole-3-acetamide, nicotinic acid, and N-succinyl-L-glutamate. *Streptomyces* metabolites promote overall plant growth, nutrient uptake, and phytohormone regulation. Although many *Streptomyces* species have been explored and employed in the manufacturing of commercial products, their role in the biocontrol of plant pathogens and mechanisms of functional metabolite synthesis remain poorly understood. This chapter further emphasizes the significance of bioactive compounds produced by *Streptomyces* in combating plant pathogens that threaten crop productivity, while simultaneously enhancing soil health.

Keywords

Streptomyces · Plant growth promoting bacteria · Sustainable agriculture · Plant growth promotion · Bioactive metabolites · Soil health

2.1 Introduction

The Food and Agriculture Organization (FAO) of the United Nations emphasizes the importance of soil health, where soil biodiversity plays a crucial role in nutrient cycling, pathogen suppression, and climate change mitigation. The FAO warns that a decline in soil health poses a risk to global food security and ecosystem services (FAO 2020). The agricultural sector is facing immense challenges to satisfy the food demands of a rapidly growing population while protecting natural resources and ensuring ecological sustainability simultaneously. Although the use of chemical fertilizers and pesticides has led to increased yields and short-term food security, the excessive and indiscriminate use of this practice has also resulted in reduced soil fertility, loss of biodiversity, enhanced pest resistance, and environmental pollution over time. Thus, the increasing pressure on agriculture has intensified the movement toward sustainable agricultural practices that are productive, resilient, and environment friendly. In this context, microorganisms play a vital role in reshaping sustainable agriculture.

2.2 Role of Microbes in Sustainable Agriculture

A plethora of microbial communities have been identified to be associated with the plants, and a symbiotic relationship exists between the microbial population and plants that significantly contribute toward sustainable agriculture. Often the plant–microbial relationship results in the production of beneficial secondary metabolites that contribute indirectly toward sustainable agriculture. Exploiting this plant–microbe connection toward sustainable agriculture is a promising strategy for the future. Increasing the yield of crop, preventing plant diseases and imparting abiotic stress tolerance are important parameters to be considered in sustainable agricultural practices. Employing suitable microbial communities for improving nutrient availability to the plants through nitrogen fixation and phosphate solubilization, enzyme and hormone production is the core direct mechanisms involved in sustainable agricultural practices. The phytoprotection enabled through secondary metabolites, antibiotics, and so on are considered to have indirect effect on sustainable agriculture (Nadarajah and Abdul Rahman 2023).

Among the various microbial communities found in soil, the genus *Streptomyces* stands out (Larissa et al. 2025). These filamentous actinobacteria are prevalent in soils worldwide but they are well-suited for life in the rhizosphere, where they closely interact with plants and other soil organisms. Their ecological function goes beyond mere existence; they actively engage in nutrient cycling, promote soil health, and help suppress harmful plant pathogens.

Genus *Streptomyces* is particularly renowned for its extraordinary metabolic capabilities. They produce a wide range of natural compounds that can enhance both the health and productivity of plants. While this metabolic diversity has been historically exploited for the discovery of antibiotics in medicine, its significance for agriculture is equally important. Their ability to influence plant–microbe interactions positions them as potential allies in creating agricultural practices that depend less on synthetic inputs. The growing interest in *Streptomyces* in the realm of sustainable agriculture is also linked to broader shifts in how we understand crop production (Zhang et al. 2024). Today, agriculture is being recognized not solely as a chemical or mechanical process but as an ecological system where soil organisms, plants, and environmental factors interact with one another. *Streptomyces* exemplify this ecological viewpoint: they can naturally enhance plant growth, strengthen resistance to various stress factors, and support the maintenance of soil diversity (Olanrewaju and Babalola 2018). As a result, studying *Streptomyces* goes beyond simply identifying new microbial compounds; it entails rethinking the core principles of agriculture. By integrating their natural functions into farming practices, it is possible to reduce environmental harm, maintain sustainable productivity, and meet the rising global food demands. Thus, the exploration of *Streptomyces* is vital for understanding of how microbial resources can shape the future of sustainable agriculture.

2.3 Diversity and Taxonomy of *Streptomyces*

The genus *Streptomyces* represents one of the most taxonomically diverse and ecologically significant bacterial lineages within the phylum Actinobacteria. Recent advances in whole-genome sequencing and phylogenomics have greatly refined its classification, with large-scale genome analyses consolidating the genus to over 540 validly published species (Nouioui et al. 2023). Taxonomically, *Streptomyces* is positioned within the class Actinomycetia, order Streptomycetales, and family Streptomycetaceae, which is distinguished by filamentous growth and the production of chains of spores—features that differentiate this genus from closely related actinobacteria such as *Kitasatospora* and *Streptacidiphilus* (Nouioui et al. 2023). Genome-based approaches have revealed the dynamic and fluid taxonomy of the genus. Over the past decade, several species have been synonymized or reassigned to other genera based on genome similarity, while phylogenomic frameworks continue to identify novel species with unique biosynthetic capabilities (Shukla et al. 2024; Voblikova et al. 2024). This reclassification underscores not only the complexity of *Streptomyces* taxonomy but also its immense evolutionary diversification.

Morphologically, *Streptomyces* resemble filamentous fungi in their growth patterns. They develop a branching substrate mycelium and a distinct aerial mycelium, which differentiates into chains of nonmotile spores, a hallmark of the genus. Colonies typically appear chalky or powdery and produce a wide variety of pigments, including actinorhodin and melanin, which contribute to their vivid coloration (Shukla et al. 2024). Their cultural characteristics remain central to identification, they grow well on nutrient-rich media, reflecting their adaptability (Zhang et al. 2024). Additionally, colonies produce geosmin, a compound responsible for the earthy smell of soil, which is often used as a diagnostic trait.

At the genomic level, *Streptomyces* species harbor large genomes (8–10 Mb) rich in biosynthetic gene clusters (BGCs), which encode antibiotics, enzymes, and other secondary metabolites of medical and agricultural importance (Voblikova et al. 2024). Horizontal gene transfer and genomic rearrangements play a central role in their metabolic innovation and ecological versatility. Comparative genomics and Average Nucleotide Identity (ANI) analyses now provide higher taxonomic resolution than traditional 16S rRNA sequencing, which often fails to differentiate closely related species (Nouioui et al. 2023). These phylogenomic insights confirm that *Streptomyces* form a coherent clade within Streptomycetaceae, while also exhibiting deep branching patterns linked to ecological niches such as soil, marine, and endophytic habitats (Shukla et al. 2024).

Ecologically, *Streptomyces* demonstrate remarkable adaptability across diverse soils; they adapt to diverse habitats, from fertile soils to extreme environments (El-Naggar et al. 2023). Such versatility allows them to act as key contributors to nutrient cycling, organic matter decomposition, and plant health promotion, while maintaining dominance in microbial communities across terrestrial ecosystems. *Streptomyces* represent one of the largest and most complex genera of *Actinomycetota*, combining morphological distinctiveness, genetic richness, and ecological versatility. Their taxonomic fluidity, driven by genomic discoveries, reflects not only their evolutionary complexity but also their unparalleled biosynthetic potential that continues to drive interest in biotechnology, agriculture, and medicine (Zhang et al. 2024).

2.4 Bioactive Compounds from *Streptomyces*

Streptomyces has been recognized as the most significant contributor to antibiotics developed so far, accounting for nearly two-thirds of known antibiotics due to their diverse secondary metabolites with potent bioactivity. The secondary metabolites of *Streptomyces* have been reported to exhibit antimicrobial, insecticidal, immunomodulatory, apoptosis-inducing, antioxidant, and cytotoxic properties (Salwan and Sharma 2020).

The *Streptomyces* genus is of paramount significance, with numerous antimicrobials synthesized and commercialized from its secondary metabolites. Many antibiotics such as tetracyclines, chloramphenicol, aminoglycosides, and rifamycin have shown clinical efficacy (Gopalakrishnan et al. 2020). Streptomycin, synthesized by *Streptomyces griseus*, belongs to the aminoglycoside family and has proven its efficiency against spirochetes that can cause plant diseases (El-Naggar 2021). Members of the *Streptomyces* genus has been reported to be good Microbial Biological Control Agents (MBCA), targeting phytopathogens using diverse modes of operation (synthesis of different classes of antibiotics, lytic enzymes, volatiles, etc.). Thus, application of streptomycetes as biocontrol agents in the soil holds significant potential as safe alternatives to harmful agrochemicals in sustainable agriculture practices (Khan et al. 2023). The anti-fungal property of *Streptomyces* spp. is much more pronounced with inhibitory potential against a plethora of species such as *Aspergillus*, *Phytophthora capsici*, *Fusarium oxysporum* f.sp. *lycopersici*, *Collectotrichum musae*, *Fusarium oxysporum*, and so on. The extracellular enzymes released by *Streptomyces* often target vital biomolecules such as lipids, proteins, chitin, and β-1,3-glucanases that are important for the growth and development of phytopathogenic fungi (Rios-Muñiz and Evangelista-Martínez 2022). Streptomycetes also produce volatile organic compounds (VOCs) that are low molecular weight (100–500 Da) lipophilic components having low boiling point but high vapor pressure. Chromatographic studies revealed that VOCs basically belong to seven chemical entities including alkenes, alcohols, aromatic naphthalenes, esters, ketones, furan, and heterocyclic benzoxepine. VOCs can exert significant biological properties such as promoting plant growth, stress resistance, and phytopathogenicity. Phytopathogenicity is induced via different mechanisms, such as facilitating inter- and intraspecific interactions and quorum quenching (He et al. 2022). Anti-fungal activity by VOCs released from streptomycetes is mediated by the inhibition of mycelial growth, sporulation, and conidial germination (Lyu et al. 2020).

The alkaloid metabolites released from *Streptomyces* are classified into nine types: indole, pyrrole, pyridine, amide, oxazole & thiazole, pyrazine & piperazine, phenazine & phenoxazine, indolizidine & pyrrolizidine and finally miscellaneous alkaloids (Liu et al. 2024). An alkaloid, Benhamycin, has been reported to possess antibacterial activity against both Gram-positive and Gram-negative strains and has anti-fungal property (Shaaban and Abdel-Aziz 2007). Alkaloids legonimide and 1H-indole-3-carbaldehyde have been studied to have anti-fungal properties (Fang et al. 2020).

2.5 *Streptomyces* in Plant Disease Management

Microbial based biocontrol agents have gained special emphasis among researchers to protect crops from pathogen infections. A broad range of bacterial genera such as *Agrobacterium*, *Alcaligenes*, *Arthrobacter*, *Bacillus*, *Burkholeria*, *Enterobacter*, *Erwinia*, *Klebsiella*, *Lactobacillus*, *Pantoea*, *Pseudomonas*, *Rhizobium*, *Serratia*,

Stenotrophomonas, *Streptomyces*, and *Xanthomonas* have shown remarkable applications in agriculture against various pathogens (Kohl et al. 2019; Bonaterra et al. 2022). The composite mechanisms of biocontrol activity include antagonistic interactions through invasion and destruction of bacterial and fungal cell structures and spores by hyperparasitism and mycoparasitism. Biocontrol agents also resist pathogens through competitive exclusion by suppressing their proliferation, secreting secondary metabolites with antimicrobial properties, and enhance immunity against various plant pathogens. Additionally, they degrade quorum sensing signal molecules that are used by bacterial pathogens to initiate infection (Helman and Chernin 2015).

Streptomyces are soil inhabitants that offer a promising and sustainable agriculture solution by colonizing plant roots and producing anthelmintic compounds. Their antimicrobial secondary metabolites include macrolide, benzoquinones, aminoglycosides, polyenes, nucleosides, and enzymes such as cellulases, chitinases, and proteases to control fungal phytopathogens. *Streptomycetes* can inhibit plant pathogens, such as *Fusarium*, *Rhizoctani*, *Gaeumannomyces*, and *Magnaporthe* (Table 2.1) (Lee et al. 2023).

Certain strains of *Streptomyces* can induce systemic resistance through activation of the jasmonic acid/ethylene and salicylic acid pathways. This immune response reduces the accumulation of reactive oxygen species (ROS) and other toxic chemicals while enhancing plant protection through cell wall strengthening (Viaene et al. 2016; Bonaterra et al. 2022). Different isolates of *Streptomyces* spp. are adaptable and employed as consortia, showing up to 85% suppression of various plant diseases (Vijayabharathi et al. 2018).

2.6 *Streptomyces* in Plant Growth Promotion

Plant growth-promoting *Streptomyces* (PGPS) in the rhizosphere can stimulate and improve various biosynthetic pathways in plants (Fig. 2.1). These include enhanced solubilization of inorganic phosphate, potassium, and other micronutrients; biosynthesis of metal chelating compounds; synthesis of phytohormones such as IAA, gibberellins, cytokines; and alleviation of various abiotic and biotic stresses (Fig. 2.1). PGPS also contribute to organic matter decomposition, nutrient acquisition through direct and indirect mechanisms, and maintaining the sustainability of soil health (Lou et al. 2011). Phosphate solubilization is a mineralization process in which phosphatases and phosphohydrolases transform the complex, insoluble forms of phosphate into forms that plants can readily absorb.

Streptomyces such as *Streptomyces thermolilacinus*, *Streptomyces rochei*, and *Streptomyces carpinensis* have been reported as effective phosphate solubilizers, releasing various organic acids, phosphatases, and metal chelator-like substances (Ben Farhat et al. 2009; Hamdali et al. 2011; Jog et al. 2012; Rungin et al. 2012; Sadeghi et al. 2012). Similarly, potassium plays an important role in enzyme activation, synthesis of proteins and photosynthesis. *Streptomyces* can degrade

Table 2.1 Mode of action of selected *Streptomyces* spp. as biocontrol agents of plant diseases

Streptomyces strains	Target pathogen or disease	Mechanism of action	Reference
Streptomyces rimosus subsp. *rimosus* LMG19352	*Fusarium graminearum* (Fusarium Head Blight)	FHB is reduced by outcompeting fungal spicosphere microbiome	Troyer et al. (2025)
Streptomyces lydicus WYEC108	*Pythium* or *Aphanomyces* species	Produces chitinase to break down the cell wall of fungi	Mahadevan and Crawford (1997)
Streptomyces strains CBQ-EA-2 and CBQ-B-8	*Macrophomina phaseolina* and *Rhizoctonia solani*	Chitinolytic, cellulolytic, and proteolytic activity	Díaz-Díaz et al. (2022)
S. lydicus M01	*Alternaria alternata* (Foliar disease)	Antioxidases activity reduced the reactive oxygen species (ROS) accumulation	Wang et al. (2020)
Streptomyces spp. AcH505	Oak powdery mildew infection	Activate plant defense responses by increasing phenylalanine ammonia lyase activity	Kurth et al. (2014)
Streptomyces spp. VV/E1, VV/R1, VV/R4	*Diplodia seriata*, *Dactylonectria macrodidyma* (Grapevine trunk diseases)	Establish an endophytic relation with plant roots produce a wide range of secondary metabolites and antifungal compounds	Álvarez-Pérez et al. (2017)
Streptomyces setonii WY228	*Ceratocystis fimbriata* (Black spot disease)	Volatile compound 2-ethyl-5-methylpyrazineinhibited the mycelial growth and spore germination	Gong et al. (2022)
Streptomyces yanglinensis	*A. flavus*, *Aspergillus parasiticus*	VOCs can reduce the disease severity and inhibit the aflatoxins production, suppression of mycelial growth	Lyu et al. (2020)
Streptomyces sp. AN090126	*Ralstonia solanacearum* and *Xanthomonas euvesicatoria* (Leaf spot in red pepper) *Sclerotinia homoeocarpa* (Dollar spot in bent grass)	Antibacterial and antifungal activities	Duy Le et al. (2022)
Streptomyces sp. strain FJAT-31547	*Fusarium oxysporum*cause (*Fusarium* wilt) *Ralstonia solanacearum* (bacterial wilt)	*n*-Hexadecanoic acid have broad-spectrum antifungal and antibacterial activities	Zheng et al. (2019)
Streptomyces sp. isolates (ATIRS43, ATIRS65, and ARRS10)	*Botrytis cinerea*, causes (*Botrytis* gray mold (BGM) disease)	Induce host–plant resistance, secret various antioxidant enzymes and phenolics	Vijayabharathi et al. (2018)

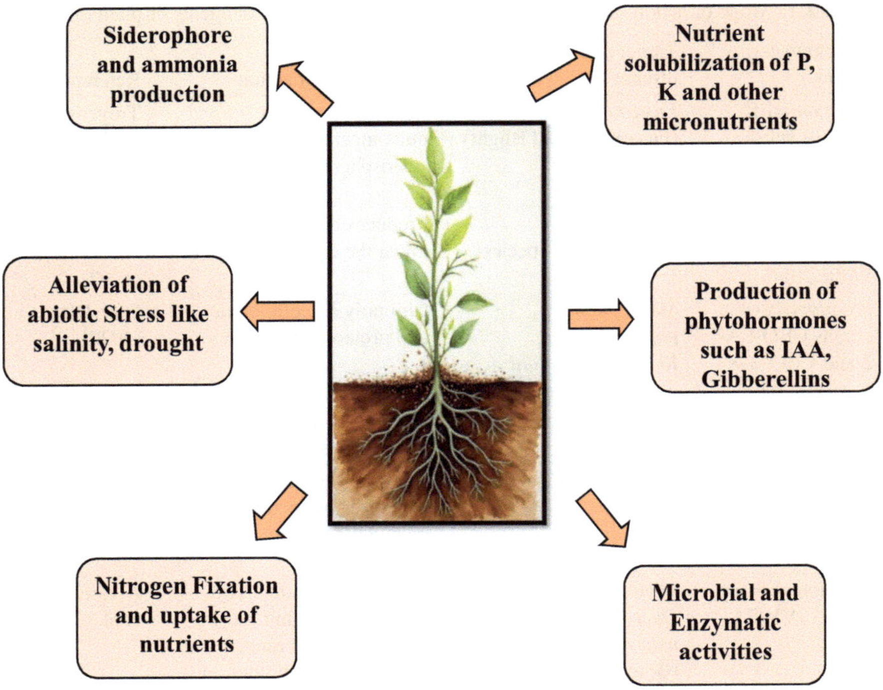

Fig. 2.1 Different traits of *Streptomyces* responsible for plant growth promotion

potassium containing rocks and ultimately release potassium from silicate minerals (Armengaud et al. 2009; Basak and Biswas 2009; Liu et al. 2016).

Production of siderophore compounds is essential to survive under iron stressed conditions, as they enhance plant growth and suppress disease. Various strains such as *S. rochei* and *S. thermolilacinus* have been reported to produce hydroxymate-type siderophores and catechols, which prevent growth of plant pathogens in the rhizosphere (Khamna et al. 2009). Phytohormones such as auxins, gibberellins, cytokinins, ethylene, and abscisic acid play vital roles in regulating plant's physiological processes, including seed germination, growth of leaves and stems, floral induction, and fruits and flowers development. Several species of *Streptomyces*, including *Streptomyces caviscabies*, *Streptomyces nobilis*, *Streptomyces rimosus*, and *Streptomyces griseoviridis* have been reported for their ability to produce phytohormones (Mahadevan and Crawford 1997; Anwar et al. 2016). Among these, auxin production stimulates plant growth by ensuring cell division, releasing saccharides during cell elongation, and stimulating the formation of lateral roots and root hairs, thereby improving water uptake by plants (Duca and Glick 2020).

Plant-associated streptomycetes (endophytes) can also benefit the host–plant by mitigating abiotic stresses such as salinity, heat, cold, drought, waterlogging, metal toxicity, nutrient depletion, and osmotic stress, thus reducing their negative impacts and consequently increasing plant growth (Atkinson and Urwin 2012). *Streptomyces*

Table 2.2 Different plant growth promoting traits of *Streptomyces* species

Species	Source	PGP traits	References
Sreptomyces fradiae	Tomato	IAA synthesis	Rafik et al. (2014)
Streptomyces rochei	Wheat	Gibberellin biosynthesis	Goudjal et al. (2015)
Streptomyces tendae	Soil	Siderophore production	Lee et al. (2012)
Streptomyces thermolilacinus	Rice	Phosphate solubilization	Jog et al. (2012)
Streptomyces acidiscabies	Potato	Auxin production	Akram et al. (2012)
Streptomyces psammoticus	Water and soil from mangrove areas	ACC deaminase activity	Suksaard et al. (2017)
Streptomyces mutabilis	Rhizosphere	Chitinase production	Dede et al. (2020)
Streptomyces griseorubiginosus	Sugarcane Rhizosphere	Tolerance to biotic stresses such as pests and pathogens	Wang et al. (2021)
Streptomyces pactum	Wheat	Tolerance to abiotic stresses such as salt and drought	Li et al. (2020)

pada, *Streptomyces olivaceus*, *Streptomyces niveus*, and *Streptomyces geysiriensis* are especially important for reducing salinity and drought stress in various crops by altering the plants' physiological responses (Yandigeri et al. 2012; Shrivastava and Kumar 2015). Table 2.2 represents the plant growth promoting traits exhibited by different *Streptomyces* species.

2.7 Soil Health Improvement Through *Streptomyces*

Streptomyces is a dominant genus of actinobacteria with branched hyphae that play notable role in mineralization of dead plant mass and decomposition of complex polysaccharides such as lignin, xylan, ligo cellulose, and cellulose, which are dominant in soil organic matter. They significantly take part in recycling of organic matter by the production of hydrolytic enzymes such as amylases, cellulases, chitinases, lipases, and proteases, which breakdown heterogeneous biomaterials and hence help in maintaining soil vitality and fertility (Javed et al. 2021).

Streptomyces viridobrunneus strain displayed magnificent potency to decompose cellulose (Da Vinha et al. 2010). *Streptomyces argenteolus* AE58P acts as an organic catalyst in composting and transformation of lingocellulosic material (Ventorino et al. 2016). Thermo tolerant actinobacteria are particularly effective at high temperature for biodegradation of hydrocarbons. *Streptomyces thermoalkaliphilus* secretes cellulase for the degradation of carboxy methyl cellulose (Wu et al. 2018). *Streptomyces avermitilis* is notable for biodegradation of lignocellulose (Utarti et al. 2021), while *S. avermitilis* and *Streptomyces labedae* exhibit L-glutaminase activity (Abdallah et al. 2012). *Streptomyces amycolatopsis* produces CMCase, xylanase, amylase, pectinase, chitinase, and protease enzymes, contributing to soil structure improvement, decomposition of organic waste, and nutrient cycling (Sonia et al. 2011).

Rhizosphere microbiome is considered as the main indicator of plant health and defense against plant pathogens (Bakker et al. 2020). A diverse microbiome in the rhizosphere negatively affects the pathogen levels. Disease suppression by *Streptomyces* is mediated by interactions among rhizospheric microbiome, plants, pathogens, and other soil conditions (Gouda et al. 2018). *Streptomyces* spp. are metabolically versatile capable of producing many inhibitory substances such as antibiotics, toxins, and lytic enzymes against various fungal and bacterial pathogens. Soil inoculation with *Streptomyces* has shown to suppress diseases such as *Fusarium* wilt, tomato yellow leaf curl, and potato late blight disease, showing their great potential as biocontrol agents (Abbasi et al. 2019).

Streptomycetes also secrete useful secondary metabolites such as 2,4-diacetylphloroglucinol (DAPG), an inhibitory compound, playing a key role in suppressing soil-borne pathogens, along with volatile organic compounds (VOC_S) and siderophores that promote plant defense and reduce disease severity by inducing morphological and biological defects in pathogens (Mousa and Raizada 2016; Salla et al. 2014). Application of *Streptomyces* enriches beneficial microbial communities, boosts soil health, and enhances plant health, contributing to sustainable agriculture (Fig. 2.2).

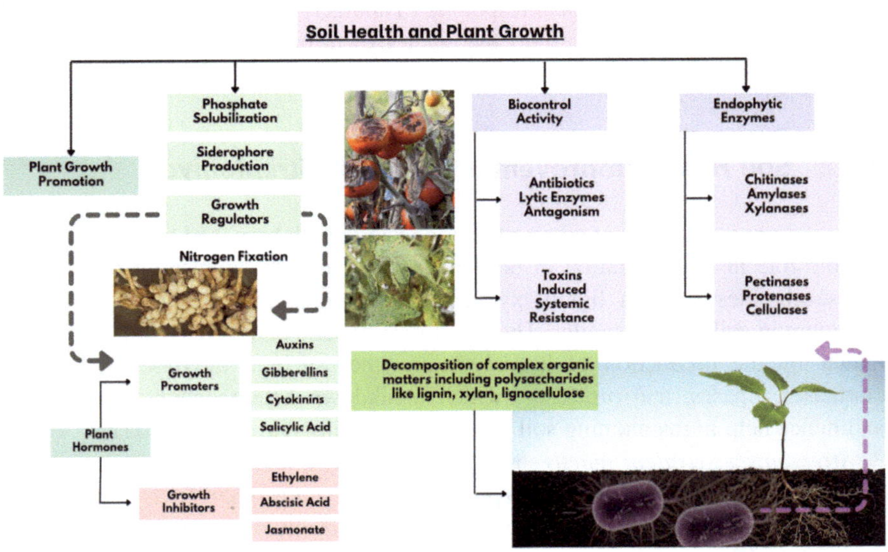

Fig. 2.2 *Streptomyces* in soil ecosystem

2.8 Metabolomics Approaches in *Streptomyces* Research

A sophisticated high-throughput screening that employs combination of chromatographic approaches with mass spectrometry, called as metabolomics, is crucial for identifying microbial metabolites such as amino acids, vitamins, antibiotics, carbohydrates, and so on. Metabolomic profiling of *Streptomyces* employs techniques such as nuclear magnetic resonance (NMR) spectroscopy and liquid chromatography-high resolution mass spectrometry (LC-HRMS). Combining multiple analytical techniques enhances accuracy in identifying compounds, even in trace amounts. However, the combination of analytical techniques needs to be chosen wisely for structural elucidation of all classes of compounds (Ahmad et al. 2021). The components identified through metabolomic profiling can be further analyzed using multivariate methods and referenced against natural product databases. Many studies have revealed the presence of targeted and untargeted metabolites upon profiling using techniques such as GC-MS, LC-MS, and ESI-LC–MS/MS, supported by NIST-MS library data (Managamuri et al. 2017).

Metabolomic profiling of *Streptomyces* has revealed a plethora of bioactive compounds. Some components such as tetrangulol and cyclohexane exhibit antifungal activity, while others, such as N-succinyl-L-glutamate, nicotinic acid, indole-3-acetamide, and so on promote plant growth (Mohamad et al. 2024). The biosynthesis of tetrangulol begins with the condensation of acetyl-CoA and malonyl-CoA, leading to a polyketide intermediate (tetrangomycin), which finally dehydrogenated to form tetrangulol (Trejo-Alarcon et al. 2025). Indole-3-acetic acid (IAA), a common metabolite in various *Streptomyces* species such as *Streptomyces scabies*, *S. griseus*, *Streptomyces violaceus*, and so on is derived from L-tryptophan via indole-3-acetamide (IAM) as an intermediate, and functions as a phytohormone that regulates growth and suppresses weeds and pathogens (Manulis et al. 1994). NMR-based metabolomic profiling of marine *Streptomyces* has also identified pyridinopyrones, which is synthesized from nicotinic acid (Fukuda et al. 2011).

Metabolo-genomics efficiently integrates chromatographic and spectroscopic techniques with genome mining tools to link metabolites to their biosynthetic genes. Recent studies have identified and characterized gene clusters responsible for biosynthesis of the tetrangulol, a notable secondary metabolite of *Streptomyces* (Trejo-Alarcon et al. 2025). Many secondary metabolites were discovered through analyzing biosynthetic gene clusters (BGCs) primarily with the help of metabolo-genomics tools. The common bioinformatics tools in metabolo-genomics for assessing BGCs are BAGEL, ClustScan, CLUSEAN, PRISM, and so on and has revealed about 30 BGCs in streptomycetes (Table 2.3) (Lee et al. 2020).

2.9 Commercial Applications and Future Prospects

Commercial bioinoculants and biopesticides have expanded quickly, with the most popular products derived from *Streptomyces* species, fungi, and plant growth-promoting rhizobacteria. Consumer demand for residue-free produce, regulatory

Table 2.3 Summary of key metabolites and their biosynthetic gene clusters

Streptomyces species	Gene cluster	Secondary metabolite	Role of metabolite in agriculture	References
Streptomyces cinnamonensi	Monensin BGC	Monensin	Growth-promoter and herbicide	Olanrewaju and Babalola (2018)
Streptomyces kasugaensis M338-M1	Kasugamycin BGC	Kasugamycin	Antibiotic against rice blast disease	Ikeno et al. (2006)
Streptomyces hygroscopicus subsp. *jinggangensis*	Acarbose BGC	Validamycin	Antibiotic against rice sheath blight disease	Yu et al. (2005)
Streptomyces cacaoi	Polyoxin BGC	Polyoxin	Antimycotic agent	Chen et al. (2009)
Streptomyces avermitilis	Avermectin BGC	Avermectin	Anthelmintic agent	Ikeda et al. (1999)
Streptomyces spectabilis	Spectinabilin BGC	Spectinabilin	Nematicidal component	Kang et al. (2023) and Choi et al. (2009)

support, and sustainable crop protection strategies are the main drivers of market expansion. However, variable field efficacy and product consistency are the major constraints. To improve stability and shelf life, commercial formulations often use microbial consortia, advanced carriers, and encapsulation technologies (Díaz-Rodríguez et al. 2025). Botanical-based products are reemerging, particularly in regions with large-scale extraction and formulation capacity, though challenges such as cold-chain requirements and farmers willingness persist. Standardized efficacy testing, regulatory harmonization, and public–private partnerships are necessary to fill these gaps. In some regions, full chemical-level dossiers are still required for biologicals, creating hurdles for global expansion. Numerous *Streptomyces*-based bioinoculants are currently on the market and growing all over the world. Actinovate® (*Streptomyces lydicus* WYEC 108, Novozymes BioAg) and Mycostop® (*Streptomyces griseoviridis* K61, Verdera Oy/Certis USA) are well-known biopesticides that work well against *Rhizoctonia*, *Pythium*, and *Fusarium* in ornamentals and vegetables. While BioRoot® and StreptoLife® products are marketed as biofertilizers that enhance soil health and root vigor, YANTAS® (*Streptomyces anulatus*) and Rhizocell® formulations show phosphate solubilization and plant growth promotion under field conditions (Barka et al. 2024). These business achievements highlight how feasible it is to turn *Streptomyces* research into environmentally friendly, farmer-ready bioformulations.

Prospects for the future focus on creating multi-strain formulations with improved spore stability, extended shelf life, and complementary plant growth-promoting and biocontrol characteristics. By combining *Streptomyces*-based bioinoculants with agroecological techniques such as crop rotation, organic soil amendments, and integrated pest management (IPM), chemical inputs could be significantly reduced while soil resilience to climate stress and biodiversity are restored. To fully utilize *Streptomyces* as a sustainable substitute in next-generation agriculture, it will be essential to make use of synthetic biology, metagenomics, and field-based experiments.

Fermentation on an industrial scale and quality control measures have cut down the costs, but stability in tropical conditions and farmers awareness remain limiting factors (Vurukonda et al. 2018). Integrating bioinoculants and biopesticides into Integrated Pest Management (IPM) systems and co-formulating with fertilizers or biofertilizers are becoming more popular business strategies for wider farmers adoption. Next-generation microbial consortia and bio-based pesticides can serve as key players in climate-smart farming systems. Smart delivery systems, precision agriculture tools, sensor-based monitoring, IoT devices, and GIS mapping allow real-time tracking of soil health, pest pressure, and plant responses, helping optimize dosage and timing for maximum efficiency.

2.10 Conclusion

Streptomyces, a diverse group of soil actinobacteria, plays a crucial role in sustainable agriculture by improving soil health, promoting plant growth, and managing pathogens. These microorganisms produce a wide range of bioactive compounds, including antibiotics, enzymes, plant hormones, and volatile organic compounds, that enhance nutrient availability, stimulate plant development, and mitigate both biotic and abiotic stresses. Their capacity to suppress harmful pathogens strengthens agricultural practices, while advancements in metabolomics and genomics provide new opportunities for discovering new metabolites useful in farming. Bioinoculants and biopesticides obtained from *Streptomyces* offer eco-friendly alternatives to synthetic chemicals. When integrated with modern farming methods, they can promote resilient crops, sustainable production, and healthier soil ecosystems, thereby supporting global food security and ecological sustainability.

References

Abbasi S et al (2019) *Streptomyces* strains induce resistance to *Fusarium oxysporum* f. sp. lycopersici race 3 in tomato through different molecular mechanisms. Front Microbiol 10:1505

Abdallah NA, Amer SK, Habeeb MK (2012) Screening of L-Glutaminase produced by actinobacteria isolated from different soils in Egypt. Int J ChemTech Res 4:1451–1460

Ahmad SJ, Zin NM, Mazlan NW et al (2021) Metabolite profiling of endophytic *Streptomyces* spp. and its antiplasmodial potential. PeerJ 9:e10816

Akram S, Ebrahim K et al (2012) Plant growth promoting activity of an auxin and siderophore producing isolate of Streptomyces under saline soil conditions. World J Microbiol Biotechnol 28:1503–1509

Anwar S, Ali B, Sajid I (2016) Screening of rhizospheric actinobacteria for various in-vitro and in-vivo plant growth promoting (PGP) traits and for agroactive compounds. Front Microbiol 7: 1334

Álvarez-Pérez JM, González-García S, Cobos R et al (2017) Use of endophytic and rhizosphere actinobacteria from grapevine plants to reduce nursery fungal graft infections that lead to young grapevine decline. Appl Environ Microbiol 83(24)

Armengaud P, Sulpice R, Miller AJ et al (2009) Multilevel analysis of primary metabolism provides new insights into the role of potassium nutrition for glycolysis and nitrogen assimilation in arabidopsis roots. Plant Physiol 150:772–785

Atkinson NJ, Urwin PE (2012) The interaction of plant biotic and abiotic stresses: from genes to the field. J Exp Bot 63:3523–3543

Barka EA, Vatsa P, Sanchez L, Gavaut-Vaillant N, Cormier A (2024) Commercial applications of *Streptomyces*-based biocontrol agents and biofertilizers in sustainable agriculture. Appl Microbiol Biotechnol 108(7):3121–3138. https://doi.org/10.1007/s00253-024-12942-7

Basak BB, Biswas DR (2009) Influence of potassium solubilizing microorganism (*Bacillus mucilaginosus*) and waste mica on potassium uptake dynamics by Sudan grass (*Sorghum vulgare* Pers.) grown under two Alfisols. Plant Soil 317:235–255

Ben Farhat M, Farhat A et al (2009) Characterization of the mineral phosphate solubilizing activity of *Serratia marcescens* CTM 50650 isolated from the phosphate mine of Gafsa. Arch Microbiol 191:815–824

Bonaterra A, Badosa E, Daranas N et al (2022) Bacteria as biological control agents of plant diseases. Microorganisms 10(9):1759

Chen W, Huang T, He X et al (2009) Characterization of the Polyoxin biosynthetic gene cluster from *Streptomyces cacaoi* and engineered production of Polyoxin H. J Biol Chem 284:10627–10638

Choi YS, Johannes TW, Simurdiak M et al (2009) Cloning and heterologous expression of the spectinabilin biosynthetic gene cluster from *Streptomyces spectabilis*. Mol BioSyst 6:336–338

Da Vinha FNM, Gravina-Oliveira MP, Franco MN et al (2010) Cellulase production by *Streptomyces viridobrunneus* SCPE-09 using lignocellulosic biomass as inducer substrate. Appl Biochem Biotechnol 164:256–267

Dede A, Güven K, Şahin N (2020) Isolation, plant growth-promoting traits, antagonistic effects on clinical and plant pathogenic organisms and identification of actinomycetes from olive rhizosphere. Microb Pathog 143:104134

Duca DR, Glick BR (2020) Indole-3-acetic acid biosynthesis and its regulation in plant-associated bacteria. Appl Microbiol Biotechnol 10420:8607–8619

Duy Le K, Hee YN, Ran Park A et al (2022) *Streptomyces* sp. AN090126 as a biocontrol agent against bacterial and fungal plant diseases. Microorganisms 10(4):791

Díaz-Díaz M, Bernal-Cabrera A, Trapero A et al (2022) Characterization of actinobacterial strains as potential biocontrol agents against *Macrophomina phaseolina* and *Rhizoctonia solani*, the main soil-borne pathogens of *Phaseolus vulgaris* in Cuba. Plants (Basel) 11(5):645

Díaz-Rodríguez AM, Parra Cota FI, Cira Chávez LA et al (2025) Microbial inoculants in sustainable agriculture: advancements, challenges, and future directions. Plants 14(2):191

El-Naggar NE-A (2021) Streptomyces-based cell factories for production of biomolecules and bioactive metabolites. In: Elsevier eBooks. pp 183–234

El-Naggar M et al (2023) Ecological resilience of *Streptomyces* in saline and stress-prone environments. Appl Soil Ecol, Elsevier

Fang Q, Maglangit F, Mugat M et al (2020) Targeted isolation of indole alkaloids from *Streptomyces* sp. CT37. Molecules 25:1108

FAO (2020) The state of knowledge of soil biodiversity. Food and Agriculture Organization of the United Nations, Rome

Fukuda T, Miller ED, Clark BR et al (2011) Structures and biosynthesis of the pyridinopyrones, polyenepyrones from a marine-derived *Streptomyces* species. J Nat Prod 74:1773–1778

Gong Y, Jia-Q Liu J, Xu M et al (2022) Antifungal volatile organic compounds from *Streptomyces setonii* WY228 control black spot disease of sweet potato. Appl Environ Microbiol 88(6)

Gopalakrishnan S, Srinivas V, Prasanna SL (2020) Streptomyces. In: Elsevier eBooks. pp 55–71

Gouda S et al (2018) Revitalization of plant growth promoting rhizobacteria for sustainable development in agriculture. Microbiol Res 206:131–140

Goudjal Y, Zamoum M et al (2015) Plant growth-promoting potential of endosymbiotic actinobacteria isolated from sand truffles (TerfezialeonisTul.) of the Algerian Sahara. Ann Microbiol 66:91–100

Hamdali H, Virolle MJ et al (2011) *Streptomyces youssoufiensis* sp. nov., isolated from a *Moroccan phosphate* mine. Int J Syst Evol Microbiol 61:1104–1108

He Y, Guo W, Peng J et al (2022) Volatile organic compounds of *Streptomyces* sp. TOR3209 stimulated tobacco growth by up-regulating the expression of genes related to plant growth and development. Front Microbiol

Helman Y, Chernin L (2015) Silencing the mob: disrupting quorum sensing as a means to fight plant disease. Mol Plant Pathol 16(3):316–329

Ikeda H, Nonomiya T, Usami M et al (1999) Organization of the biosynthetic gene cluster for the polyketide anthelmintic macrolide avermectin in *Streptomyces avermitilis*. Proc Natl Acad Sci 96:9509–9514

Ikeno S, Aoki D, Hamada M et al (2006) DNA sequencing and transcriptional analysis of the Kasugamycin biosynthetic gene cluster from *Streptomyces kasugaensis* M338-M1. J Antibiot 59:18–28

Javed Z, Tripathi GD, Mishra M et al (2021) Actinobacteria—the microbial machinery for the organic-cycling, plant growth, and sustainable soil health. Biocatal Agric Biotechnol 31

Jog R, Nareshkumar G, Rajkumar S (2012) Plant growth promoting potential and soil enzyme production of the most abundant *Streptomyces* spp. from wheat rhizosphere. J Appl Microbiol 113:1154–1164

Kang M-K, Kim J-H, Moon KH et al (2023) Development of spectinabilin over-producing mutant *Streptomyces* sp. S-N87 having nematicidal activities. Res Square

Khamna S, Yokota A, Lumyong S (2009) Actinobacteria isolated from medicinal plant rhizosphere soil: diversity and screening of antifungal compound, indole-3-acetic acid and siderophore production. World J Microbiol Biotechnol 2:649–655

Khan S, Srivastava S, Karnwal A, Malik T (2023) *Streptomyces* as a promising biological control agents for plant pathogens. Front Microbiol 14

Kohl J, Kolnaar R, Ravensberg WJ et al (2019) Mode of action of microbial biological control agents against plant diseases: relevance beyond efficacy. Front Plant Sci 19(10):845

Kurth F, Mailänder S, Bönn M et al (2014) Streptomyces-induced resistance against oak powdery mildew involves host plant responses in defense, photosynthesis, and secondary metabolism pathways. Mol Plant Microbe Interact 27(9):891–900

Larissa DT, Kris A, Sarah O et al (2025) The biocontrol agent *Streptomyces rimosus* subsp. *rimosus* tempers shifts in the wheat spicosphere microbiome induced by Fusarium Head Blight. Front Plant Sci 16

Lee N, Hwang S, Kim J et al (2020) Mini review: genome mining approaches for the identification of secondary metabolite biosynthetic gene clusters in Streptomyces. Comput Struct Biotechnol J 18:1548–1556

Lee J, Kim S, Jung H et al (2023) Exploiting bacterial genera as biocontrol agents: mechanisms, interactions and applications in sustainable agriculture. J Plant Biol 66:485–498

Lee J, Postmaster A, Soon HP et al (2012) Siderophore production by Actinomycetes isolates from two soil sites in Western Australia. Biometals 25:285–296

Li H, Guo Q, Jing Y et al (2020) Application of Streptomyces pactum Act12 enhances drought resistance in wheat. J Plant Growth Regul 39:122–132

Liu D, Lian B, Wang B (2016) Solubilization of potassium containing minerals by high temperature resistant Streptomyces sp. isolated from earthworm's gut. Acta Geochimica 35(3):262–270

Liu Z, Sun W, Hu Z et al (2024) Marine Streptomyces-Derived novel alkaloids discovered in the past decade. Marine Drugs 22(1):51–62

Lou Y, Liang W, Xu M et al (2011) Straw coverage alleviates seasonal variability of the topsoil microbial biomass and activity. Catena 86:117–120

Lyu A, Yang L, Wu M et al (2020) High efficacy of the volatile organic compounds of *Streptomyces yanglinensis* 3-10 in suppression of aspergillus contamination on peanut kernels. Front Microbiol 11:19

Mahadevan B, Crawford DL (1997) Properties of the chitinase of the antifungal biocontrol agent *Streptomyces lydicus* WYEC108. Enzym Microb Technol 20:489–493

Managamuri U, Vijayalakshmi M, Ganduri VSRK et al (2017) Isolation, identification, optimization, and metabolite profiling of *Streptomyces sparsus* VSM-30. 3 Biotech 7:12–17

Manulis S, Shafrir H, Epstein E et al (1994) Biosynthesis of indole-3-acetic acid via the indole-3-acetamide pathway in *Streptomyces* spp. Microbiology 140:1045–1050

Mohamad OAA, Liu Y-H, Huang Y et al (2024) Metabonomic analysis to identify exometabolome changes underlying antifungal and growth promotion mechanisms of endophytic Actinobacterium *Streptomyces albidoflavus* for sustainable agriculture practice. Front Microbiol 15:21

Mousa WK, Raizada M (2016) Natural disease control in cereal grains. In: Wrigley CW, Corke H, Seetharaman K, Faubion J (eds) Encyclopedia of food grains, vol 2, pp 257–263

Nadarajah K, Abdul Rahman NSN (2023) The microbial connection to sustainable agriculture. Plants 12(12):2307

Nouioui I et al (2023) Advances in genome-based taxonomy of *Streptomyces* and related actinobacteria. Antonie van Leeuwenhoek, Springer

Olanrewaju OS, Babalola OO (2018) Streptomyces: implications and interactions in plant growth promotion. Appl Microbiol Biotechnol 103:1179–1188

Rafik E, Rahal E, Ahmed L (2014) Isolation and screening of actinomycetes strains producing substances plant growth-promoting. Indo-Am J Agric Vet Sci 2:1–12

Rios-Muñiz DE, Evangelista-Martínez Z (2022) Antifungal activity of *Streptomyces* sp. CACIS-2.15CA, as a potential biocontrol agent, against some soil-borne fungi. Egypt J Biol Pest Control 32:38

Rungin S, Indananda C et al (2012) Plant growth enhancing effects by a siderophore producing endophytic Streptomycete isolated from a Thai jasmine rice plant (*Oryza sativa* L. cv. KDML105). Antonie Van Leeuwenhoek 102:463–472

Sadeghi A, Karimi E et al (2012) Plant growth-promoting activity of an auxin and siderophore producing isolate of *Streptomyces* under saline soil conditions. World J Microbiol Biotechnol 28:1503–1509

Salla TD et al (2014) *Streptomyces rhizobacteria* modulate the secondary metabolism of Eucalyptus plants. Plant Physiol Biochem 85:14–20

Salwan R, Sharma V (2020) Bioactive compounds of Streptomyces: biosynthesis to applications. In: Studies in natural products chemistry, pp 467–491

Shaaban M, Abdel-Aziz MS (2007) Benhamycin, novel alkaloid from terrestrial *Streptomyces* sp. Nat Prod Res 21:1205–1211

Shrivastava P, Kumar R (2015) Soil salinity: a serious environmental issue and plant growth-promoting bacteria as one of the tools for its alleviation. Saudi J Biol Sci 22:123–131

Shukla R et al (2024) Morphological, cultural, and ecological characterization of novel *Streptomyces* from Indian soils. Springer, Arch Microbiol

Sonia MT, Hafedh B, Abdennaceur H, Ali G (2011) Studies on the ecology of actinobacteria in an agricultural soil amended with organic residues: II. Assessment of enzymatic activities of Actinomycetales isolates. World J Microbiol Biotechnol 27:2251–2259

Trejo-Alarcon LM, Cano-Prieto C, De Carvalho AC et al (2025) Integrative metabolo-genomics suggests a biosynthetic pathway for tetrangulol in *Streptomyces* sp. KL110A. World J Microbiol Biotechnol 41:44

Troyer LD, Audenaert K, Ommeslag S et al (2025) The biocontrol agent *Streptomyces rimosus* subsp. *rimosus* tempers shifts in the wheat spicosphere microbiome induced by Fusarium Head Blight. Front Plant Sci 16

Utarti E, Suwanto A, Suhartono MT et al (2021) Lytic polysaccharide monooxygenase of soil actinomycete with potential use for lignocellulose biodegradation. Malays J Microbiol 17(1): 60–68

Suksaard P, Pathom-aree W, Duangmal K (2017) Diversity and plant growth promoting activities of actinomycetes from mangroves. Chiang Mai J Sci 44:1210–1223

Ventorino V, Ionata E, Birolo L et al (2016) Lignocellulose-adapted endo-cellulase producing *Streptomyces* strains for bioconversion of cellulose-based materials. Front Microbiol 7

Viaene T, Langendries S, Beirinckx S et al (2016) *Streptomyces* as a plant's best friend? FEMS Microbiol Ecol 92(8)

Vijayabharathi R, Gopalakrishnan S, Sathya A et al (2018) *Streptomyces* sp. as plant growth-promoters and host-plant resistance inducers against *Botrytis cinerea* in chickpea. Biocontrol Sci Tech 28(12):1140–1163

Voblikova V et al (2024) Comparative genomics and phylogenomics of Streptomyces: implications for taxonomy and secondary metabolism. BMC Genomics, Springer Nature

Vurukonda SSKP, Giovanardi D, Stefani E (2018) Plant growth promoting and biocontrol activity of *Streptomyces* spp. as endophytes. Int J Mol Sci 19(4):952

Wang M, Xue J, Ma J et al (2020) *Streptomyces lydicus* M01 regulates soil microbial community and alleviates foliar disease caused by *Alternaria alternata* on cucumbers. Front Microbiol 11: 942

Wang Z, Solanki MK, Yuz K et al (2021) Genome characteristics reveal the biocontrol potential of actinobacteria isolated from sugarcane rhizosphere. Front Microbiol 12:797889

Wu H, Liu B, Ou X et al (2018) Streptomyces thermoalkaliphilus sp. nov., an alkaline cellulase producing thermophilic actinomycete isolated from tropical rainforest soil. Antonie Van Leeuwenhoek 111(3):413–422

Yandigeri MS, Meena KK, Singh D et al (2012) Drought-tolerant endophytic actinobacteria promote growth of wheat (*Triticum aestivum*) under water stress conditions. Plant Growth Regul 6:411–420

Yu Y, Bai L, Minagawa K et al (2005) Gene cluster responsible for validamycin biosynthesis in *Streptomyces hygroscopicus* subsp. jinggangensis 5008. Appl Environ Microbiol 71:5066–5076

Zhang Y et al (2024) Genomic and ecological insights into *Streptomyces* diversity across soil ecosystems. Elsevier, Soil Biology and Biochemistry

Zheng X, Wang J, Chen Z et al (2019) A *Streptomyces* sp. strain: isolation, identification, and potential as a biocontrol agent against soilborne diseases of tomato plants. Biol Control 136

Bioactive Compounds from *Streptomycetes* for Applications in Diverse Fields 3

Garima, Jasweer Kaur, Raman Thakur, and Sumanpreet Kaur

Abstract

Streptomycetes, a remarkable genus of filamentous actinobacteria, are well known for their prolific production of bioactive compounds. These microorganisms have long been central to antibiotic discovery, with more than two-thirds of naturally derived antibiotics originating from them. Beyond antimicrobials, *Streptomyces* species produce a wide array of compounds with antiviral, antifungal, anticancer, antioxidant, and immunomodulatory properties. This chapter explores the diversity and biological potential of *Streptomycete*-derived metabolites, emphasizing their relevance in pharmaceuticals, agriculture, and environmental sustainability. With the rise of multidrug-resistant pathogens and the urgent need for green alternatives in industry, the untapped biosynthetic potential of *Streptomycetes* is gaining renewed attention. Advances in genome mining and metabolic engineering are enabling researchers to unlock previously silent gene clusters, leading to the discovery of novel molecules. By highlighting recent findings and applications, this chapter underlines the growing significance of *Streptomycetes* as a source of innovative solutions across multiple domains of human and environmental health.

Keywords

Streptomycetes · Bioactive compounds · Antiviral · Antibiotics · Antifungal · Metabolites

Garima · R. Thakur · S. Kaur (✉)
Department of Medical Laboratory Sciences, Lovely Professional University, Phagwara, Punjab, India

J. Kaur
Department of Chemistry and Biochemistry, Government College for Girls, Ludhiana, Punjab, India

T. Kaur et al. (eds.), *Streptomyces: Biological Candidates for Sustainable Agriculture*,
https://doi.org/10.1007/978-981-95-5803-2_3

3.1 Introduction

Over the years since mankind has come into existence, different organic molecules have been playing a crucial role in maintaining the life of terrestrial and aquatic organisms. These organic molecules in the form of antibiotics, antiviral, anticancer medications are produced by the microorganisms. Low molecular weight extracellular compounds are produced by microbes during the late development stage and considered as secondary metabolites. In clinical practice, resistance of pathogenic microorganisms to the pharmaceutically synthesized drugs is a considerable phenomenon. This results in an elevation in mortality rates because of the sustained infections caused by the pathogens. Despite the successful discovery and manufacture of antibiotics, infectious illnesses continue to rank as the second leading cause of death worldwide (Laxminarayan et al. 2013). Genetic alterations, such as mutation and the horizontal gene transfer of resistant genes via plasmids, are the primary causes of the acquired resistance. However, certain physiological states, including biofilm formation, also contribute to the development of antimicrobial resistance. Actinobacteria, formerly known as *Actinomyces*, are Gram-positive bacteria having high G + C content and several traits similar to fungi. Although they lack a recognizable cell wall, they develop thin, nonseptate structures that resemble mycelium, which is a feature similar to of fungi. Actinobacteria colonies produce reproductive structures that resemble conidia or sporangia when grown in culture media. According to Gitari et al. (2023), the most prevalent kinds are found in soil and water habitats. Their presence is crucial for the breakdown of organic components such as cellulose and chitin. However, they support the maintenance of several biogeochemical cycles and aid in the replenishment of nutrient supplies in soil, which in turn involves humus production. *Streptomyces* is the type genus of the Streptomycetaceae family and the largest genus of actinobacteria. Secondary metabolites, including antibiotics, antihelminthic enzymes, herbicides, anticancer medications, vitamin B12 (growth factors), and immunological modulators are also produced by the *Streptomyces* (Mahajan and Balachandran 2012). The first antibiotic derived from *S. noursei* by Rachel Brown in the year 1949, was termed as "Nystatin". Numerous antibiotics as well as antifungal agents have been synthesized from different species of the genera that show diverse applications. The ability of soil-dwelling *Streptomyces* to thrive in harsh soil conditions and produce valuable antibiotics marks their importance commercially as well as industrially. Agriculture sector is the foremost user of this soil bacterium and various beneficial impacts produced by the genus was reported by the bioactive compounds produced by them (Fig. 3.1).

The present chapter seeks to explain the various bioactive compounds produced by *Streptomyces* sp. and their synthesis mechanisms. Further, their potential in agriculture, industrial, and medical sectors in the form of antimicrobial, antifungal, immunosuppressants, and biocontrol agents have been detailed in respective sections.

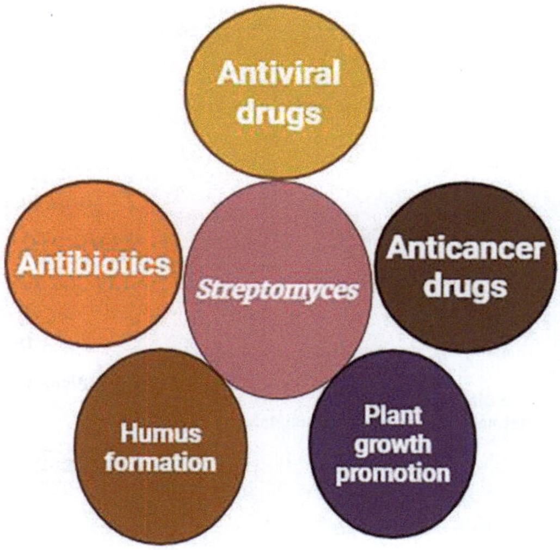

Fig. 3.1 Beneficial effects of *Streptomyces* in agriculture and human welfare

3.2 Bioactive Compounds and Their Synthesis

In common terms, bioactive compounds are defined as "an organic compound that provides detrimental and the beneficial effects to another organism in association" (Teodoro 2019). These compounds are polyketide derivatives and usually extracted from microorganisms, plants, and animals. Such compounds show beneficial effects in terms of

(a) Plant growth promotion
(b) Fungicides (for eukaryotes)
(c) Herbicides and pesticides
(d) Medical benefits such as anticancer, antioxidant, antiviral potentials, etc.

In *Streptomyces*, the production of secondary metabolites is regulated by the polyketide pathway (Risdian et al. 2019). The transition of simpler molecules into complex bioactive compounds involves multifunctional enzymes known as polyketide synthases (PKSs). Numerous enzymatic reactions use differential enzymes such as enoyl reductase (ER), acyltransferase (AT), ketoreductase (KR), ketosynthase (KS), and dehydratase (DH) (Shelest et al. 2015). The associated pathways for the synthesis of different metabolites use three types of PKSs, namely polyketide synthase 1, polyketide synthase 2, and polyketide synthase 3.

(A) *Polyketide synthase 1*

Also known as modular PKS as it involves multifunctional proteins. The enzyme complex is categorized into 15 domains, further classified into three modules. Each

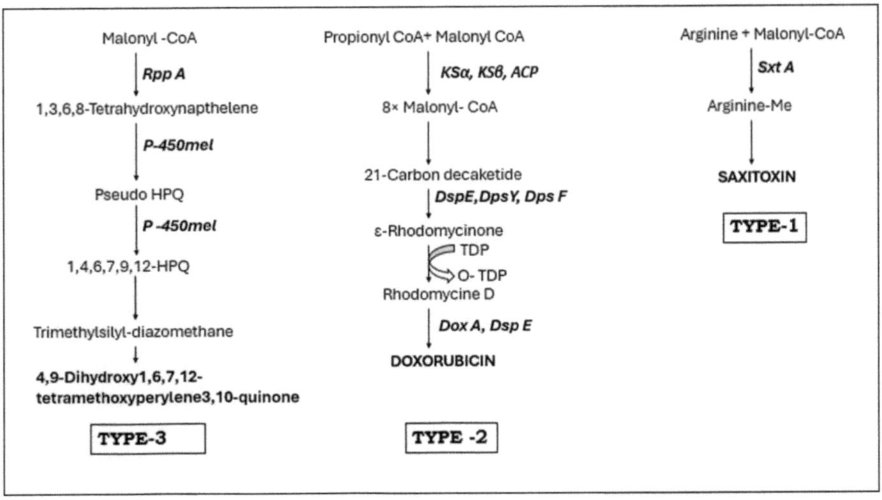

Fig. 3.2 Mode of action of polyketide synthases for antibiotic production (Funabashi et al. 2008; Zhang et al. 2017)

module contains the aforementioned crucial elements that work together to create the β-keto esterase intermediate. Furthermore, enoyl reductase (ER), dehydratase (DH), and β-keto esterase are the other domains that might be included in the module. The increasing polyketide chain is moved from one module to another during the polyketide production until a unique enzyme releases the finished molecule from the final module (Rokem et al. 2007). Macrocyclic polyketides of linear type are also synthesized using an enzyme complex such as tautomycin (Choi et al. 2007). Other antibiotics produced are rapamycin, methymycin, and spiramycin (Fig. 3.2) (Risdian et al. 2019).

(B) *Polyketide synthase 2*

Defined as type 2 polyketide producers, the enzymes also have monofunctional peptides that work in combination to produce aromatic compounds. It consists of acyl-carrier protein that functions as an anchor for singular polyketide chains. Type II PKSs have two ketosynthases units (KSα and KSβ) that cooperate to create poly-β-keto chains in addition to ACP. The KSβ functions as a factor that determines chain length, while the KSα unit catalyzes the condensation of the precursors. Known as "minimal PKS," these three main systems produce aromatic polyketides through repetitive processes. The poly-β-keto chain is converted into the aromatic compound core by the cooperation of various extra enzymes, including cyclases, aromatases, and ketoreductase. Additionally, oxygenases, as well as glycosyl and methyl transferases, carry out the post-tailoring processes (Fig. 3.2) (Zhang et al. 2017). Doxorubicin, mithramycin, glucocarcin, and similar aromatic compounds are chiefly produced by the enzymatic action (Yang et al. 2014; Fischer et al. 2003).

(C) *Polyketide synthase 3*

For the production of bioactive compounds using the synthase 3 enzyme, acyl-CoA is used as a substrate. PKSs such as RppA present in *S. griseus* are responsible for the synthesis of 1,3,6,8-tetrahydroxynapthalene (Funa et al. 2005). Ken 2, another PKS in *S. violaceoruber* played a key role in producing 3,5-dihydroxyphenylglycine, an initiator in the synthesis of Balhimycin and Vancomycin (Fig. 3.2) (Wenzel et al. 2008).

3.3 Economical Significance of *Streptomyces*: From Health to Field

Actinomycetes are the most prevalent sources of naturally occurring bioactive compounds that can be found in terrestrial as well as marine habitat (Zotchev 2012). The best producers of secondary metabolite are the marine actinomycetes. The single genus *Streptomyces* is the source of the majority of these substances, and it can be found on land and inside water as well (Prudhomme et al. 2008). Because of its ability to manufacture a variety of bioactive compounds that gives it a unique biological character, the members of the genus *Streptomyces* have greatly benefited humanity. More than 7000 compounds have been demonstrated to have significant clinical uses. It includes antioxidants, antimicrobials, immunosuppressants, anticancer, and antitumor agents. Bioprospection of *Streptomyces* also resulted in discovery of antibacterial compounds such as Chloramphenicol, Tetracycline, Amphotericin, and antifungals like Nystatin (Nisha et al. 2025).

3.3.1 Antibiotic Production

Secondary metabolites such as antibiotics, anticancer, antihelminthic enzymes, and their complex products including herbicides are produced by *Streptomyces*. With the discovery of Streptomycin (in 1943) and Streptothricin (in 1942), the history of *Streptomyces* began, and the scientists stepped up their search for additional antibiotics in the genus. One example is *Streptomyces thermoviola* that produces granaticin. It is a pigmented, pH sensitive antibiotic that works best at 45 °C (Procópio et al. 2012). Tetracyclines, macrolides, polyene, phenazine, anthraquinone, angucycline, non-polyene, benzoxazole, phenanthridine, heptadecaglycoside, lactone, and others are the antibiotics produced the species that inhibit the pathogenic growth. These antibiotics target the translation processes along with the development of cell wall and genetic makeup (Table 3.1).

3.3.2 Antifungal Potential

Numerous *Streptomyces* species have been effectively turned into commercial biofungicides, either by using their live spores and other bioactive metabolites

Table 3.1 Medically important bioactive compounds produced by *Streptomyces* spp.

Species	Antibiotics produced	References
Streptomyces clavuligerus	Cephamycin	Akintunde (2014)
Streptomyces venezuelae	Chloramphenicol	
Streptomyces aurofaciens	Tetracycline	
Streptomyces kanamycetus	Kanamycin	
Streptomyces spectabilis	Spectinomycin	
Streptomyces cinnamonesis	Monensin	
Streptomyces lavendulae	Mitomycin C	
Streptomyces griseus	Streptomycin	Nicault et al. (2021)
Streptomyces avermitilis	Avermectin	
Streptomyces cattleya	Fluorometaboltes	
Streptomyces parvulus	Actinomycin D	Rahman et al. CR1047(2011)
Streptomyces cheonanesis	2-Methyl-butyl-propyl phthalate and Diethyl phthalate	Mangamuri et al. (2016)
Antifungal compounds		
Streptomyces sp. ICBG 292	Pericidin-A, Nigericin	Ortega et al. (2019)
Streptomyces padanus PMS 702	Polyene fungalchromin	Shih et al. (2003)
Streptomyces sp.	Cyphomycin, Mycangimycin, Frontalamide A, Frontalamide B	Cheverette et al. (2019)
Streptomyces nodosus	Amphotericin	Akintunde (2014)
Streptomyces aurofaciens	Aurofacien	
S. griseus	Candicidin	
Streptomyces noursei	Oligomycin	
Streptomyces diastachromogens	Actinomycin D	
Streptomyces sp.	Daryamide	Fu et al. (2017)
Antiviral compounds		
Streptomyces sp. AM2504	Virantmycin B (Dengue virus)	Hirota-Takahata (2017)
Streptomyces fradiae VITMK2	9(10H)-Acridanone	Manimaran et al. (2018)
Streptomyces jiujiangensis NBERC-24992	Virantmycin	Liu et al. (2023a, b)
Streptomyces sp. SV21	Valinomycin	Wibowo et al. (2021)
Streptomyces sp. 1647	Pseudo-tetrapeptides	Sun et al. (2022)
Antitumor and anticancer compounds		
Streptomyces tunisiensis	Pigment extract	Ibrahim et al. (2023)
Streptomyces sp. NMF6	Bis(2-ethylhexyl) phthalate	Fahmy and Abdel-Tawab (2021)
Streptomyces antioxidans PU-12M	NA	Naseer et al. (2022)

(Blasticin S, Polyoxin, Kasugamycin, Mycostop R, and Rhizovit R) (Kim et al. 2011). Antifungals such as novonestmycin produced by *S. phytohabitans*, elairomycin B made by *Streptomyces* sp. BK 190, bafilomycin K made by *S. flavotricini* Y12-26 were assessed as novel biofungicides (Zhang et al. 2006). The primary antifungal agent was reveromycin, consisting of spiroacetal polyketide compounds made by *Streptomyces* sp. (Fremlin et al. 2011). Another carboxylic acid derivative, 1H-pyrrole-2-carboxylic acid (PCA), which was isolated from *S. griseus* H7602 by Nguyen et al. (2015), showed potent antifungal action against *Phytophthora capsici* in vitro. *Streptomyces* sp. A3265 yielded two antifungal compounds, i.e., guanidyl fungin A and methyl guanidyl fungin, which showed efficacy against a variety of plant pathogens (Minh et al. 2015). Likewise, the inhibitory effects of streptomycetal metabolites were studied against *Pythium aphanidermatum*, *Rhizoctonia solani*, and *Sclerotinia minor*, targeting fungal growth and oomycete formation (Table 3.1; Fig. 3.1) (Lyu 2017).

3.3.3 Antitumor and Anticancer Potential

Microorganisms, including different bacteria, fungi, and viruses are known for their ability to heal cancerous cells, opening a newer domain for their potent therapeutic applications. Different species of *Streptomyces* are now known for their antitumor effects, therefore reducing the costlier treatment options that is associated with better results. Brain tumor cells, known as Glioblastoma cells, were used to evaluate the tumor-suppressing potential of bacteria. Cell-growth and migration mediate through Akt/GSK-3β/β-catenin signaling pathway, explaining a diverging therapeutic mechanism. This mechanism targets the proliferation and apoptotic processes, leading to cell death (Liu et al. 2023a, b). Actinobacterial species *Streptomyces* sp. M12 extract was studied to regulate the apoptosis of U87_MG cells of brain tumor via activation of Caspase 8 and downregulation of Caspase 9 proteins, followed by induction of cytotoxic behavior (Table 3.1) (Rezaee et al. 2025).

3.3.4 Immunosuppressing Properties

Different species of *Streptomyces* are studied to produce powerful immunosuppressive agents. Rapamycin (Sirolimus) and FK-506 (Tacrolimus) are predominant immunosuppressors, which are known to inhibit T-cell activation and proliferation. The foremost mentioned immunosuppressant is produced by *S. hygroscopicus*, characterized as macrocyclic lactone that binds with FKBP-12, inducing T-cell activation and proliferation (Ganesh and Subathra Devi 2023). On the other hand, T-cell suppression occurred by FK-506 is seen as an outcome of calcineurin pathway and is produced by *S. tsukubaensis* (Martel et al. 2021). Clinical and medical importance of immunosuppressing bioactive compound Tacrolimus include 100 times more potent than the cyclosporine, which is a pharmaceutical drug. Higher potential with least side effects associated with usage explores its functionality in organ transplants such as liver, heart, and kidney (Pierce et al. 2021) as well as skin inflammation disorders (Chen and Gao 2024). However, Rapamycin, another bioactive compound produced by *Streptomyces*, has its major role in treating certain cancers, inflammation diseases (i.e., inflammatory bowel disease), and rheumatoid arthritis (Quan et al. 2022).

3.3.5 Antiparasitic and Antihelminthic Importance of Streptomyces

Certain species of focused actinobacteria impart antiparasitic properties. Metabolites such as Avermectins, Valinomycin, Stauosporine, Butenolide, and Leptomycin B are produced by different species of *Streptomyces* that depicts killing effects on human parasites such as *Leishmania major*, *Trypanosoma brucei*, and other agricultural and veterinarian pathogens. Derived compound from *S. canus* known as phenazine-1-carboxylic acid (PCA) showed anti-protozoal activity against *Toxoplasma gondii* at lethal dosage value of 55.5 μg/mL (LC_{50}) and *Plasmodium falciparum* (Pagmadulam et al. 2020). Another active metabolite, i.e., Venturicidin A, a macrolide, was found more vital against *T. brucei* and *L. donovani* (Hauser et al. 2024). Natural anti-kinetoplastid activities of indolocarbazole staurosporine (importantly 7-oxostaurosporine) have been reported, which was isolated from *S. sanyensis* and was explored as more potent against *Trypanosoma cruzi*, *Acanthamoeba castellanii*, *L. donovani*, and *L. amazonensis* (Cartuche et al. 2020).

Avermictin, a secondary metabolite produced by *S. avermitilis*, was found to possess an insecticidal property when ingested. This property was deduced because of activation of glutamate-gated chloride channels in host nerves and muscles, which disrupt pharyngeal function and locomotion (Sharma et al. 2021).

3.3.6 Antiviral Properties

Streptomyces bacteria possess broad-spectrum antiviral properties due to the secretion of bioactive compounds. These compounds inhibit the virus entry into host cells, interfering replication process and trigger apoptosis of pre-infected cells. Streptovaricins, a metabolite produced by *S. griseus* exhibit antiviral potential against sexually transmitted virus, Herpes Simplex, and vaccinia virus via disrupting membrane integrity (Saini et al. 2024). Other species *S. viridochromogenes* produces Avilamycin, which inhibit transcription process in retroviruses including HIV (Kumar et al. 2024). Azalomycin, a cyclic depsipeptide compound secreted by *S. hygroscopicus*, was studied to have a potent antiviral property against HSV-1 and 2 viruses (Kumar et al. 2024). Another species, *S. noursei*, secrete Nourseothricin, which interfere protein synthesis in adenoviruses, poxvirus, and herpesviruses during infective stage (Table 3.1) (Garcia et al. 2023). Chemical structures of some of the medically important compounds are given in Fig. 3.3.

A. Antibiotic compounds:

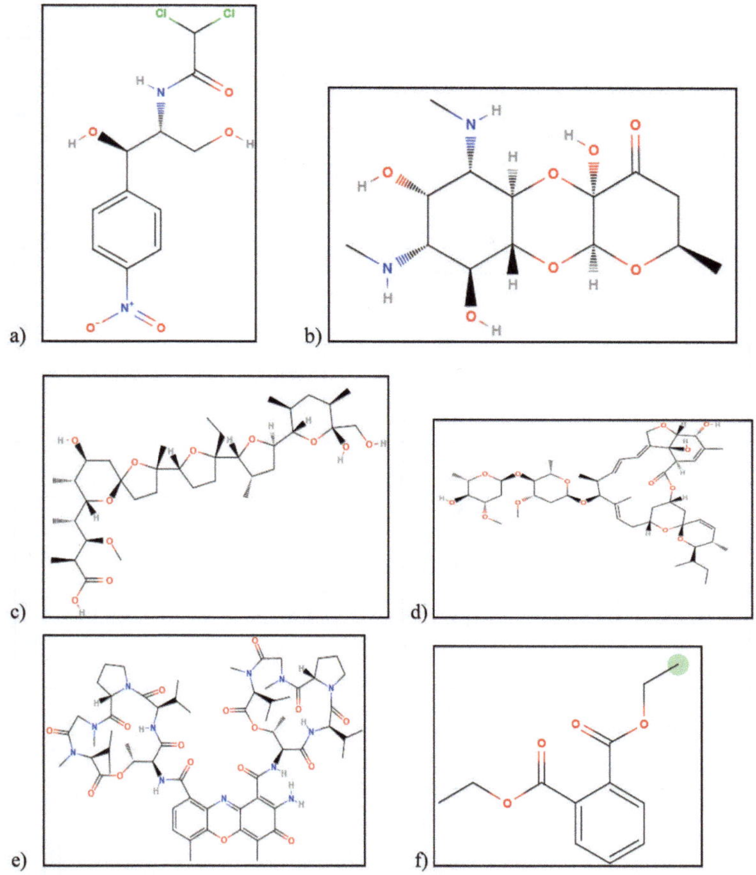

B. Antifungal compounds:

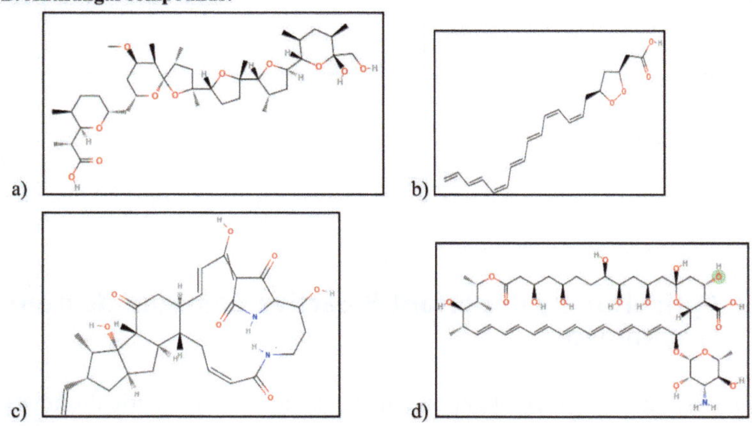

Fig. 3.3 Medicinally important compounds produced by *Streptomyces*: (**a**) antibiotic compounds (Cephamycin, Chloramphenicol, Tetracycline, Kanamycin, Spectinomycin, Monensin, Mitomycin C, Streptomycin, Avermectin, Actinomycin D); (**b**) antifungal compounds (Nigericin, Mycanganimycin, Frontalamide A, Frontalamide B, Amphotericin, Aureofacien, Candidicin, Oligomycin, Actinomycin D, Daryamide); (**c**) antiviral compounds (Virantmycin B, 9(10H)-Acridanone)

e) 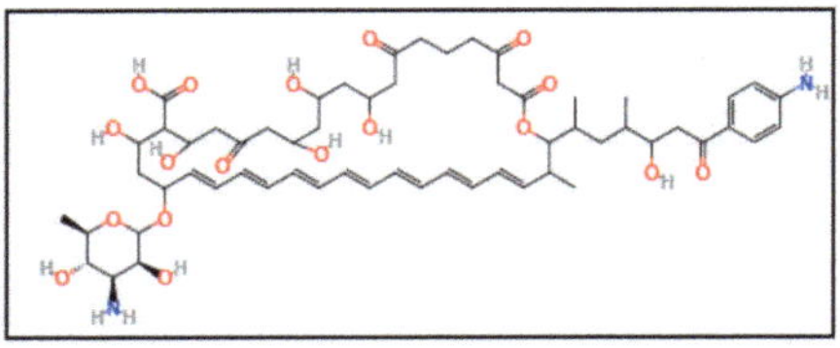

f) 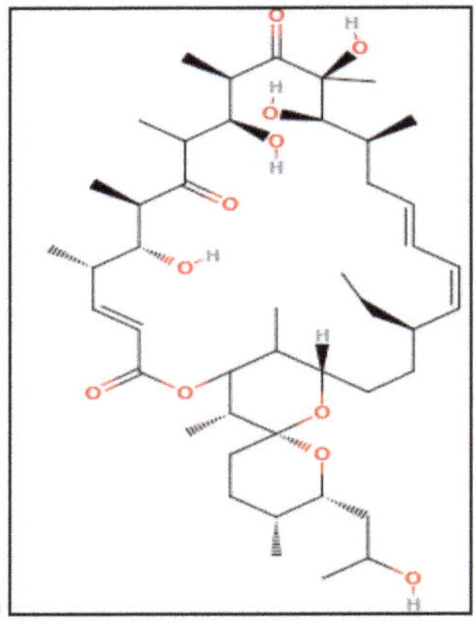

g) 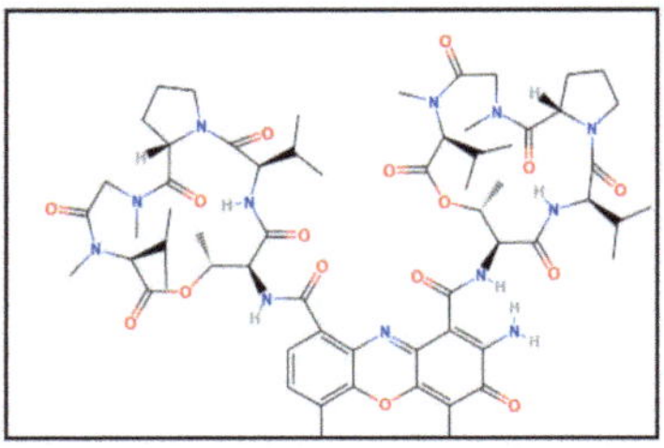

h) 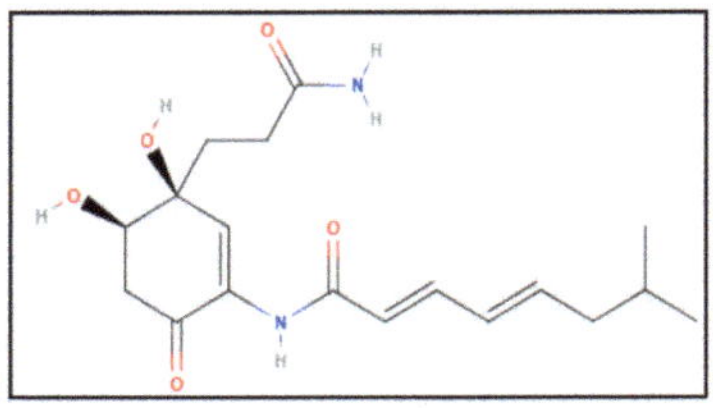

C. Antiviral compounds:

a) 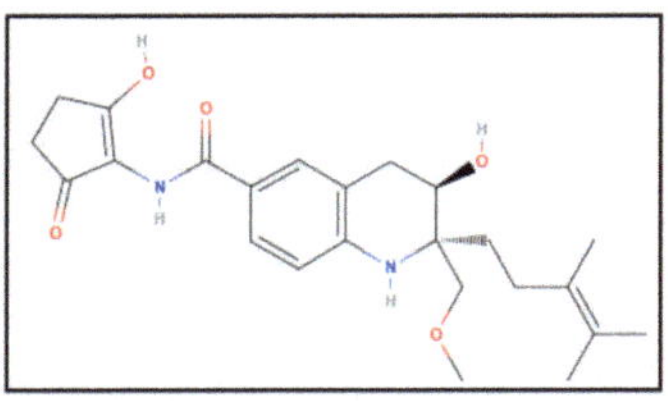

b) 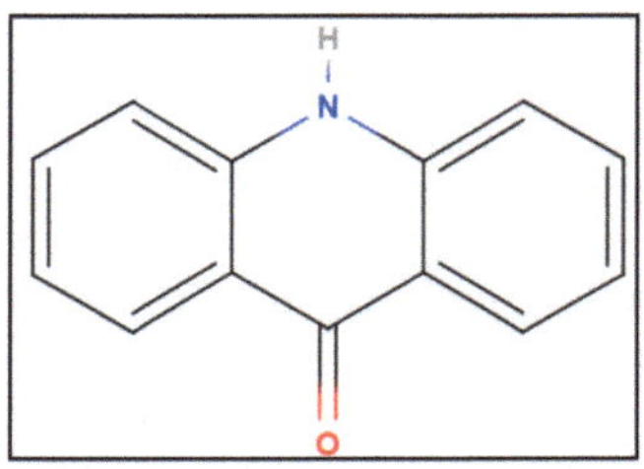

Fig. 3.3 (continued)

3.4 Agriculturally Important Bioactive Compounds from *Streptomyces*

Streptomyces have garnered attention among researchers and agriculturalists due to their potentiating effects as plant growth promoting bacteria (PGPB), biocontrol agents, antibiotic resistant inducers maintaining sustainability in environment. In recent years, due to their capacity to flourish in a variety of environments, they are emerging as a farmer's choice (Shivlata and Satyanarayana 2015). These bacteria are environmentally benign and extremely advantageous for crops.

According to de Jesus Sousa and Olivares (2016), the majority of PGPB stimulate plant growth by suppressing disease, improving nutrient uptake and producing phytohormones. However, the secretion of auxin, a plant hormone, also helps for root growth and development is elevated with the presence of *Streptomyces*. Nonetheless, these bacteria also aid in crop protection against phytopathogens via secretion of antimicrobial substances (Amaresan et al. 2018). The plant growth promoting activities of *Streptomyces* occurred via two ways, i.e., direct and indirect means.

3.4.1 Plant Growth Promotion Potential of Streptomyces Species

3.4.1.1 Direct Mechanisms

3.4.1.1.1 Nitrogen Fixation

Direct plant growth promotion requires nitrogen as a crucial component. Biological process termed as "nitrogen fixation" is necessary for converting the atmospheric nitrogen into the plant-utilizable forms. Nitrogen-fixing bacteria such as *Streptomyces* sp. are known to produce ammonia, nitrate ions, and trace elements that a plant can use for enhancing its growth and development via nitrogenase enzymes (Ahemed and Kibret 2014). *Streptomyces thermoautotrophicus* has an ability for fixing atmospheric nitrogen (Ribbe et al. 1997). Furthermore, sequential studies have reported the nitrogen fixing ability of the genus in recent years. Another species such as *S. chartreusis* WZS201 improved the nitrogen fixation in sugarcane by 9.16% after its inoculation in the rhizospheric region (Wang et al. 2017). Additionally, Dahal et al. (2017) reported the presence of *Streptomyces* in the arid moorland region of South Dakota.

3.4.1.1.2 Potassium Solubilization

Potassium is frequently used as fertilizer to maximize plant growth since it is an essential macronutrient. In addition, it is essential for photosynthesis, protein synthesis, as well as enzyme activation (Lin 2010). One study conducted by Bopin and Prajapati (2023) isolated *S. atacamensis* from the soils of Gujrat, India. The isolated strain has shown high potential for potassium solubilization compared to other rhizobacteria. However, the mechanism underlying the solubilizing potential of the genus has been due to the breakdown of rocks into powder forms by solid state fermentation, which liberates potassium from silicate minerals (Liu et al. 2016).

3.4.1.1.3 Production of Indole Acetic Acid (IAA)

The ubiquitous plant hormone known as indole-3-acetic acid (IAA) is a member of the auxin class. It stimulates cell division and elongation, which plays a significant part in plant growth and development. It has been observed that saprophytic actinomycetes produce zeatin, gibberellic acid, and indole acetic acid. These phytohormones are essential for plant–microbe interaction. *Streptomyces* isolates accounted for 85.3% of the isolates recovered in this investigation, and isolates were shown to produce a respectable amount of IAA. After inoculating mandarin seedlings with 12 actinomycete isolates, the seedlings' fresh shoot weight, fresh root weight, and shoot height all increased (Solans et al. 2011). *Streptomyces atrovirens* was identified from groundnut roots by Reddy et al. (2016), who also documented the plant growth and development brought on by the elevated amounts of IAA production. According to Lin and Xu (2013), tryptophan has been identified as a crucial precursor in the synthesis of IAA. Five distinct routes have been identified for the synthesis of IAA in microorganisms, with tryptophan serving as the primary beginning precursor (Glick 2012). *Streptomyces* species use indole-3-acetamide as one of these mechanisms to produce IAA in a tryptophan-dependent manner (Lin and Xu 2013). By boosting seed germination, root elongation, and plant dry weight, a number of *Streptomyces* species that produce IAA, including *Streptomyces violaceus*, *Streptomyces olivaceoviridis*, *S. scabies*, *S. griseus*, *S. rimosus*, *S. exfoliatus*, *S. coelicolor*, and *S. lividans*, have been shown to enhance plant growth (Goudjal et al. 2014).

3.4.1.1.4 Activity of ACC Deaminase Enzyme

The enzyme ACC deaminase, also known as 1-aminocyclopropane-1-carboxylase, is also involved in promoting plant growth. The bacterial endophytes absorb the ACC deaminase enzyme that is secreted from plant roots and cleave it into ammonia and α-ketoglutaric acid. As a result, the PGPB serves as a sink for the ACC enzyme, whose declining levels in plants indicate the drop in ethylene levels prior to the oxidation of ACC, which is carried out by ACC oxidase enzymes. Therefore, when exposed to unfavorable conditions such as high salinity, flooding, abiotic or biotic stresses such as nematodes, the ACC deaminase enzyme, which is present in plants, causes a decrease in hormones that inhibit plant growth, such as ethylene, which in turn promotes plant growth. According to Rashid et al. (2012), this enzyme is the primary component of the PGPB that stimulates plant growth. In addition, ACC deaminase activity was seen in flower stems when the cut portion of the stems absorbed ACC-PGPB, which delayed the senescence of the flowers (Ali et al. 2014).

3.4.1.2 Indirect Mechanisms

3.4.1.2.1 Antagonistic Potential

The inhibitory action displayed by two microorganisms is known as antagonism. Numerous rhizosphere-dwelling *Streptomycetes* are crucial in shielding plants from numerous plant diseases and promoting plant growth. Many actinomycetes species, especially those in the group of actinobacteria, are recognized to be antifungal biocontrol agents that prevent the growth of several plant–pathogenic fungus (Errakhi et al. 2007). Using in vitro tests, *S. griseus* H7602 demonstrated strong inhibitory activity against *P. capsici* (Nguyen et al. 2015).

Mycelial suppression against *R. solani*, *Pythium* sp., *Phytophthora* spp., *Alternaria alternate*, *F. oxysporum* f. sp. *lycopersici*, *F. oxysporum* f. sp. *radices-lycopersici*, and *F. solani* was demonstrated by *S. griseocarneusi* isolated from tomato rhizosphere soil (Sabaratnam and Traquair 2015). According to Mohandas et al. (2013), *Streptomyces* strains, i.e., *S. fradiae*, *S. avermitilis*, *S. cinnamonensis*, and *S. canus*, were isolated from the rhizosphere soil of guava plantations demonstrated possible antifungal activity against *F. solani* and *A. solani*. Martínez (2014) found that *Streptomyces* sp. exhibited broad antagonistic action against *Rhizoctonia* spp., *Colletotrichum* spp., *Fusarium* spp., and *A. niger* (Talubnak and Soytong 2010).

3.4.1.2.2 Siderophore Production

Iron-binding proteins called siderophores, which may bind ferric iron and distribute soil iron to the necessary bacteria, are produced by PGP microorganisms as a means of surviving in iron-stressed settings. Siderophore chemicals have the ability to reduce illness and stimulate plant growth. *Streptomyces rochei*, *S. carpinensis*, and *S. thermolilacinus* strains were able to produce siderophores on CAS agar medium, according to Jog et al. (2012). By reducing the amount of iron in the rhizosphere, *Streptomyces* species are known to produce hydroxamate-type siderophores, which prevent the formation of phytopathogens (Khamna et al. 2009). Enterobactin, a catecholate siderophore feature that was previously thought to be produced by the enterobacteriaceae group, was discovered to be synthesized by *S. tendae* (Fiedler et al. 2003). *Streptomyces* CMU-SK 126 isolated from *Curcuma monga* rhizospheric soil had high levels of siderophore, catechol, and hydroxamate types of synthesis, according to Khamna et al. (2009).

3.4.2 Streptomyces as Biocontrol Agents

Streptomyces strains exhibited biocontrol activity against plant pathogens on plate assays that are used as a traditional method (Passari et al. 2015). In an experiment, 29 *Streptomyces* strains were isolated from native plants of Algerian Sahara (*Euphorbia helioscopia*, *Astragallus armatus*, *Peganum harmala*, etc.). The strains were highly capable of inhibiting in vitro growth of the soil-borne pathogen and reduced the disease index in soil when tomato seeds were inoculated with these strains (Goudjal et al. 2014). Furthermore, treatment with several *Streptomyces* strains reduced wilt disease in chickpea (*Cicer arietinum*) caused by *Fusarium oxysporum* in soil by the assembly of several enzymes and metabolites, such as cellulases, proteases, and other metabolites that inhibited the pathogenic growth (Gopalakrishnan et al. 2016). Hence, Actinobacteria and its largest genus *Streptomyces* contributed to development of disease suppressive soils and are connected to disease limitation in soils infected with several fungal root pathogens such as *Rhizoctonia solani* and *F. oxysporum* (Cha et al. 2016).

The ability of *Streptomyces* to eradicate various phytopathogens is one of the characteristic features ensuring its biocontrol property. The major mechanisms involved are the enzymatic secretions, production of siderophores, and altering the membrane integrity via lipid peroxidation. *Rhizoctonia solani*, a phytopathogen causing tobacco target spot disease, was killed by the action of antifungal compounds released by *Streptomyces olivereticuli* in a dose-dependent manner (Zhong et al. 2023).

3.4.2.1 Bactericidal and Fungicidal Activity

Broad spectrum capability of *Streptomyces* sp. is often associated with their antimicrobial activities. This property of bacterial genus is foreseen in current era where problems like drug-resistance in bacteria are dominating with limited treatment options being effective. *Streptomycetes* are known for producing secondary metabolites, i.e., polyketides (aromatic ketides and macrolides), terpenoids, ribosomal and non-ribosomal peptides, alkaloids, phenazines, tetrazene derivatives, and so on, which are potent against diverse Gram-positive and Gram-negative bacteria (Ait Assou and El Hassouni 2025). A study conducted by Amimanan et al. (2024) showed antibacterial effects of culture supernatant and ethyl acetate extract of *Streptomyces* sp. PA5.6 against several drug-resistant pathogens such as *Staphylococcus. aureus*, *Enterococcus faecalis*, *E. coli*, *P. aeruginosa*, *A. baumannii*, and Methicillin Resistant *St. aureus* (MRSA). Likewise, antitubercular activity of *Streptomyces* sp. SbAr007 was seen against standard *Mycobacterium tuberculi* strains H37Ra and clinical strain H37Rv (Vadankula et al. 2024). Agricultural phytopathogens such as *St. aureus*, *Micrococcus luteus*, *Bacillus cereus*, *Pseudomonas fluorescents*, *Mycobacterium cyaneum*, and *E. coli* were also deteriorated by *S. scabiei* VKM As-305 when used (Teregulova et al. 2024).

Streptomycetes play a very important role in the decomposition of assorted biopolymers, such as lignin, cellulose, starch, xylan (hemicellulose), pectin (polygalacturonic acid), and chitin. The widespread ability of *Streptomycetes* to degrade chitin has attracted the eye of plant pathologists because it is a component of the fungal semipermeable membrane (Chater et al. 2010). For this reason, chitinolytic *Streptomyces* spp. strains are widely studied for fungal biocontrol. Chitinase and β-1,3-glucanase activity was investigated in *Paenibacillus* sp. 300 and *Streptomyces* sp. 385, which provided 71 and 64% reduction of wilt disease incidence in cucumber, respectively (Cao et al. 2019). *Streptomyces goshikiensis* YCXU, a strain producing catalase, β-1,3-glucanase, chitinase, urease, antifungal diffusible, and volatile organic compounds decreased 67% wilt of watermelon when vectored by a plant food. The treatment also reduced the pathogen population in soil by 89% and decreased the strain indicator enzymes and malondialdehyde content by 55–70 and 62%, respectively (Shen et al. 2016) (Table 3.2).

3.4.2.2 Pesticidal Activity

Applications of chemical insecticides and pesticides are mostly practiced across nations for decades. Common concerns associated with their use are accumulation of toxins and by-products in food and water, which in turn affect consumers health in negative way. Exposure of such chemicals to predator organisms results in killing of natural pest predators, declining the biodiversity variations. Biopesticides and secondary metabolites as a biocontrol agent have greater benefits as they are eco-friendly, selective, biodegradable, and nontoxic (Liu et al. 2024). With an LC50 value of 10–26 mg/L against *Mythimna separate*, bioactive compound derived from *Streptomyces diastatochromogenes* demonstrated notable insecticidal activity (Cao et al. 2019). In Pakistan, actinomycete isolates found in soil of Asteraceae

Table 3.2 *Streptomyces* as biocontrol agents

Species	Pathogen	References
Fungicidal activity		
Streptomyces luteogriseus HR40	*Verticillium nonalfaalfa*	Ghoreshizadeh et al. (2025)
Streptomyces flavofungini HR77		
Streptomyces violaceusniger SRA14	*Colletotrichum gloeosporioids*	Castañeda-Novoa et al. (2021)
Streptomyces phaeopurpeureus ExPro138	*Colletotrichum coccodes*	Palaniyandi et al. (2013)
Streptomyces scabies	*Solanum tuberosum*	Neeno-Eckwall (1999)
Streptomyces sp. TN258	*Pythium ultimum*	Sellem et al. (2017)
Streptomyces vinaceuscrappus	*Magnaporthe oryzae*	Ningathoujan et al. (2009)
Streptomyces philanthi RM-1-138		Boukaew and Prasertsan (2014)
Streptomyces globisporus		Li et al. (2011)
Streptomyces flavotricini		Khalil et al. (2014)
Streptomyces griseoviridis	*Ceratocystis radicicola*	Minuto et al. (2006)
Streptomyces albidoflavus	*Nocardia* sp.	Kavitha et al. (2010)
	Aspergillus sp.	Lotfy et al. (2019)
Bactericidal activity		
Streptomyces tenjimariensis ATCC 31603	*Bacillus subtilis* DSM 618	Gomaa et al. (2024)
Streptomyces thinghirensis MSA1	*Staphylococcus aureus, Salmonella typhi, Pseudomonas aeruginosa, E. coli*	Fathima et al. (2025)
S. thinghirensis EGDA65	*Candida albicans, B. subtilis, St. aureus, P. aeruginosa*	Hassanein et al. CR1024(2018)
Streptomyces rochei 3MKA	*Ralstonia solanacearum*	Elhalag et al. (2025)
Pesticidal activity		
Streptomyces gramineus	*Thrips palmi*	Kim et al. (2023)
Streptomyces catenula MY 12	*Galleria mellonella*	Soliman et al. (2021)
Streptomyces sp. SP5	*Zeugodacus cucurbitae*	Devi et al. (2023)
Streptomyces sp. KR0006	*Plutella xylostella*	Kim et al. (2022)
Streptomyces sp. SA61	*Trialeurodes vaporariorum*	Liu et al. (2024)

plants exhibited 100% lethal action against *Culex quinquefasciatus* larvae in their first and fourth instar stages (Tanvir et al. 2014). According to Kim et al. (2020), active compounds that were isolated from *Streptomyces* showed insecticidal and growth-regulating properties against *Plutella xylostella* and *Aedes albopictus*. Additionally, new metabolite from *Streptomyces* sp. CAI-155 was reported by Gopalakrishnan et al. (2016) to have outstanding larvicidal action against *Helicoverpa armigera* in greenhouse and laboratory setups. Active compounds from the genus, including Prassinons, Avermictins, Doramectin, Apinosad, and Nanchangmycin, have been identified to combat range of pest insects (Osman et al. 2019; Yu et al. 2020). *Streptomyces* sp. KSF103 showed a promising larvicidal activity against *Anopheles cracens*, *Aedes aegypti*, *Culex quinquefasciatus*, and *A. albopictus* (Amelia-Yap et al. 2023).

3.4.3 Production of Lytic Enzymes

β-Glucanases are a gaggle of glycosyl hydrolases that hydrolyzes β-glucan substrates and are broadly characterized in bacteria and fungi. Extracellular β-glucanases act as exo and endo hydrolases that are further categorized into differing kinds, supporting the substrate specificity. They have great ecological significance and extensive applications in various industries such as brewing, textile, food, and agriculture. Agricultural applications include pathogen and disease control, seed germination, placental nutrition, production of plant and fungal protoplasts, and so on, which are currently obtaining increased scientific attention for averting present and future agricultural crises (Edison and Sugathan 2018).

The lysing enzymes produced by *Streptomyces* sp. are widely utilized to combat a wide range of harmful fungus. Antagonistic effects of *Streptomyces* sp. EF-14 against *Phytophthora* sp. were studied despite the production of β-1,6-glucanase and β-1,3-glucanase (Bubici et al. 2019). Some *Streptomyces* strains (CAI-24, CAI-121, CAI-127, KAI-32, and KAI-90) exhibited significant biocontrol activity against *Fusarium oxysporum* f. sp. *cicero*, which causes wilt disease of chickpea by producing β-1,3-glucanase (Gopalakrishnan et al. 2013) (Table 3.2).

3.4.4 Use of Streptomyces as Biofertilizers

Exploring the use of bioorganic fertilizers as source of crop improvement and yield is a prerequisite in recent times. With the addition of microorganisms in manure, it enhanced efficiency, which makes it fit for use at larger scale. Microbes-containing compost is biologically important and responsible for seed germination, plant growth, disease control, and soil rehabilitation. Phosphate-solubilization, siderophore production, and biocontrol properties of *Streptomycetes* impact the quality of biofertilizers, hence broadens their utility. One strain of *Streptomyces*, i.e., *Streptomyces* sp. HN6, was studied to have good enzyme potential when used along with *S. thermophilus* and *Phanerochaete chrysosporium*. The microbial strains showed elevated activity of sucrase, catalase (164.3%), and urease (81.7%) enzyme when used in compost form on cowpea infected with *Fusarium* wilt (Wang et al. 2022).

Phosphate-solubilizing property of soil-dwelling microbes is essential at agriculture level. Flowering, crop maturity, and disease resistance are the foremost outcomes noted. Being a limiting nutrient in soil after nitrogen, proper P utilization, and consumption is required for plant health (Chouyia et al. 2022). Solubilization of tricalcium P from rocks and other inorganic matter of soil contributes to the fertility of soil (Boubekri et al. 2021). Strains, such as *Streptomyces lienomycini*, *Streptomyces ambofaciens*, *Streptomyces venezuela*, and *Streptomyces alboniger*, when isolated from different soil ecosystems, produce phytase enzymes that solubilize P. Another strain, *Streptomyces rishiriensis* 3AS4, isolated from rhizospheric soil of wheat, was able to utilize P, making its final concentration 400–200 mg/mL (Vargas Hoyos et al. 2021).

3.5 Conclusion and Future Prospectives

Producing food for the world's growing population while minimizing environmental damage and reliance on nonrenewable resources is agriculture's biggest problem of the twenty-first century. Utilizing microorganisms that promote plant growth to increase crop yields has become a viable and sustainable solution to this problem. Rhizobia and other Gram-negative bacteria are the most researched and technologically advanced plant growth-promoting microorganisms. However, members of the Gram-positive genus *Streptomyces* have been revealed to have advantageous relationships with plants, including those of agronomic significance, in addition to their well-known capacity to manufacture antibiotics. Their bioavailability in environment is crucial for agriculture, industrial, and medical purposes as well. Researches conducted in these areas have revealed the applications of bioactive metabolites produced by the consortia. Occurrence of antibiotic resistance among different bacterial genera and their prevalence across the generations is a global threat worldwide as suggested by the World Health Organization (WHO). Furthermore, long exposure to pharmaceutical medications against different health adversities results in prolonged side effects such as fatty liver, insulin resistance, weakness, nausea, and so on. Minimizing these health issues encompassing world has alarmed the use of bioactive compounds as alternative options. Improved accessibility and targeted delivery of *Streptomyces*-derived metabolites to cancer cells with reduced toxicity to healthy cells laid a new foundation stone in drug delivery systems. The microbial treasure trove of *Streptomyces*-derived secondary metabolites provides a glimmer of hope against the fatal illnesses amid the current and future difficulties. This highlights the need of future research to conduct and to collaborate among different agriculture, industrial, and health sectors.

References

Ahemad M, Kibret M (2014) Mechanisms and applications of plant growth promoting rhizobacteria: current perspective. J King saud Uni-sci 26(1):1–20

Akintunde OG (2014) Production of an antibiotic-like activity by Streptomyces sp. COUK1 under different growth conditions. Master's thesis, East Tennessee State University

Ali S, Charles TC, Glick BR (2014) Amelioration of high salinity stress damage by plant growth-promoting bacterial endophytes that contain ACC deaminase. Pl Physio Biochem 80:160–167

Ait Assou S, El Hassouni M (2025) New Streptomyces-derived antibacterial compounds targeting gram-positive bacteria: a systematic review. Sci World J 1:6659874

Amaresan N, Jayakumar V, Kumar K, Thajuddin N (2018) Biocontrol and plant growth-promoting ability of plant-associated bacteria from tomato (Lycopersicum esculentum) under field condition. Microb Pathog 136:103713

Amelia-Yap ZH, Low VL, Saeung A, Ng FL, Chen CD et al (2023) Insecticidal activities of Streptomyces sp. KSF103 ethyl acetate extract against medically important mosquitoes and non-target organisms. Sci Rep 13(1):4

Amimanan P, Ouancharee W, Muangsue N, Lasom S, Charoensri N (2024) Antibacterial compounds and draft genome sequence of Streptomyces sp. PA5.6 isolated from deciduous dipterocarp Forest soils, University of Phayao, Thailand. Trends Sci 21(10):8234. Available from: https://tis.wu.ac.th/index.php/tis/article/view/8234

Barka EA, Vatsa P, Sanchez L et al (2016) Taxonomy, physiology, and natural products of Actinobacteria. Microbiol Mol Biol Rev 80(1):1–43

Behal V (2000) Bioactive products from *Streptomyces*. Adv Appl Microbiol 47:113–157

Belin P, Le Du MH, Fielding A et al (2009) Identification and structural basis of the reaction catalyzed by the polyketide cyclase RemF from *Streptomyces* resistomycificus. J Biol Chem 284(43):28629–28638

Bentley SD, Chater KF, Cerdeño-Tárraga AM et al (2002) Complete genome sequence of the model actinomycete *Streptomyces* coelicolor A3(2). Nature 417(6885):141–147

Berdy J (2005) Bioactive microbial metabolites. J Antibiot (Tokyo) 58(1):1–26

Bérdy J (2012) Thoughts and facts about antibiotics: where we are now and where we are heading. J Antibiot (Tokyo) 65(8):385–395

Bharti D, Kumar A, Sharma A (2015) Antimicrobial activity of actinomycetes and its correlation with physiological characteristics. Int J Pharm Sci 7(1):564–572

Bibb MJ (2005) Regulation of secondary metabolism in streptomycetes. Curr Opin Microbiol 8(2): 208–215

Boubekri K, Soumare A, Mardad I, Lyamlouli K et al (2021) The screening of potassium-and phosphate-solubilizing actinobacteria and the assessment of their ability to promote wheat growth parameters. Microorganisms 9(3):470

Boukaew S, Prasertsan P (2014) Suppression of rice sheath blight disease using a heat stable culture filtrate from Streptomyces philanthi RM-1-138. Crop Prot 61:1–10

Bubici G, Kaushal M, Prigigallo MI et al (2019) Biological control agents against Fusarium wilt of banana. Front Microbiol 10:616

Cao H, Liu W, Chang Y, Chen H, Zhang Y, Xu W et al (2019) 5′-Epi-SPA-6952A,a new insecticidal 24-membered macrolide produced by Streptomyces diastatochromogenes SSPRC-11339. Nat Prod Res 33(5):659–664. https://doi.org/10.1080/14786419.2017.1405401. PMID: 29156970. Epub 20171120

Cao P, Li C, Wang H et al (2020) Community structures and antifungal activity of root-associated endophytic actinobacteria in cucumber and *Streptomyces* sp. HAAG3-15. Microorganisms 8(2): 236

Cartuche L, Sifaoui I, López-Arencibia A, Bethencourt-Estrella CJ et al (2020) Antikinetoplastid activity of indolocarbazoles from Streptomyces sanyensis. Biomolecules 10(4):657

Castañeda-Novoa CD, Vinchira-Villarraga DM et al (2021) Evaluation of the production of antifungal metabolites against Colletotrichum gloeosporioides in Streptomyces 5.1 by random mutagenesis. Acta Sci Biolog Sci 43

Cha JY, Han S, Hong HJ et al (2016) Microbial and biochemical basis of a Fusarium wilt-suppressive soil. ISME J 10(1):119–129

Challis GL (2008) Exploitation of the *Streptomyces* coelicolor A3(2) genome sequence for discovery of new natural products and biosynthetic pathways. J Ind Microbiol Biotechnol 35(6): 447–457

Chater KF (2006) *Streptomyces* inside-out: a new perspective on the bacteria that provide us with antibiotics. Philos Trans R Soc Lond Ser B Biol Sci 361(1469):761–768

Chater KF, Biró S, Lee KJ et al (2010) The complex extracellular biology of *Streptomyces*. FEMS Microbiol Rev 34(2):171–198

Chen J, Gao L (2024) Tacrolimus may play a role in dermatitis and radiation-induced skin injury through cellular senescence. Radiat Med Prot 5(3):161–164

Chen Y, Wendt-Pienkowski E, Shen B (2008) Identification of the gilvocarcin V gene cluster from *Streptomyces* griseoflavus. Mol BioSyst 4(11):1136–1146

Cheverette MG, Carlson CM, Ortega HE et al (2019) The antimicrobial potential of Streptomyces from insect microbiomes. Nat Commun 10:516

Choi SS, Hur YA, Sherman DH et al (2007) Isolation of the gene cluster for tautomycetin, a polyketide immunomodulator from *Streptomyces* sp. CK4412. Microbiology (Reading) 153(4): 1095–1102

Chouyia FE, Ventorino V, Pepe O (2022 Dec) Diversity, mechanisms and beneficial features of phosphate-solubilizing Streptomyces in sustainable agriculture: a review. Front Plant Sci 6(13): 1035358

Dahal B, NandaKafle G, Perkins L, Brözel VS (2017) Diversity of free-living nitrogen fixing Streptomyces in soils of the badlands of South Dakota. Microbiol Res 1(195):31–39

de Jesus Sousa JA, Olivares FL (2016) Plant growth promotion by streptomycetes: ecophysiology, mechanisms and applications. Chem Biol Technol Agric 3:24

Devi S, Verma J, Sohal SK, Manhas RK (2023) Insecticidal potential of endophytic Streptomyces sp. against Zeugodacus cucurbitae (Coquillett)(Diptera: Tephritidae) and biosafety evaluation. Toxicon 233:107246

Edison LK, Shiburaj S, Pradeep NS (2018) Microbial beta glucanase in agriculture. In: Advanced Biotechnology & Microbiology, vol 3. Apple Academic Press, pp 53–72

Elhalag KM, Hassan MG, Hamed AA, Elawady ME, Hassan A, Elmaaty SA (2025) Antibacterial efficacy of Streptomyces rochei metabolites against key bacterial diseases of potato. Indian Phytopathol 31:1–2

Errakhi R, Bouteau F, Lebrihi A, Barakate M (2007) Evidences of biological control capacities of Streptomyces spp. against Sclerotium rolfsii responsible for damping-off disease in sugar beet (Beta vulgaris L.). W J Microbiol Biotechn 23(11):1503–1509

Evangelista-Martínez Z (2014) Isolation and characterization of soil Streptomyces species as potential biological control agents against fungal plant pathogens. W J Microbiol Biotechn 30(5):1639–1647

Fahmy NM, Abdel-Tawab AM (2021) Isolation and characterization of marine sponge–associated Streptomyces sp. NMF6 strain producing secondary metabolite (s) possessing antimicrobial, antioxidant, anticancer, and antiviral activities. J Genet Eng Biotechnol 19(1):102

Fiedler HP (2003) Screening for bioactivity. Micro Div Biopros 7:324–335

Fathima SA, Arafath AY, Prathiviraj R, Hassan S et al. (2025) Bioactive fraction of Streptomyces thinghirensis MSA1 effectively inhibits biofilm forming clinically significant AMR pathogens. Curr Microbiol ;82(6):234

Fenical W, Jensen PR (2006) Developing a new resource for drug discovery: marine actinomycete bacteria. Nat Chem Biol 2(12):666–673

Fischer C, Lipata F, Rohr J (2003) Gene cluster of gilvocarcin V and implications for biosynthesis. J Am Chem Soc 125(26):7818–7819

Fremlin L, Farrugia M, Piggott AM, Khalil Z, Lacey E, Capon RJ (2011) Reveromycins revealed: new polyketide spiroketals from Australian marine-derived and terrestrial Streptomyces spp. a case of natural products vs. artifacts. Org Biomol Chem 9(4):1201–1211

Fu P, La S, MacMillan JB (2017) Daryamide analogues from a marine-derived Streptomyces species. J Nat Prod 80(4):1096–1101

Funa N, Funabashi M, Ohnishi Y et al (2005) Biosynthesis of melanin via oxidative aryl coupling in *Streptomyces* griseus. J Bacteriol 187(23):8149–8155

Funabashi M, Funa N, Horinouchi S (2008) Phenolic lipids synthesized by type III polyketide synthase confer penicillin resistance on Streptomyces griseus. J Biol Chem 283(20): 13983–13991

Ganesh SK, Subathra Devi C (2023) Molecular and therapeutic insights of rapamycin: a multi-faceted drug from Streptomyces hygroscopicus. Mol Biol Rep 50(4):3815–3833

Garcia VE, Pidatala V, Barcelos CA, Liu D et al (2023) Enhanced microbial production of protocatechuate from engineered sorghum using an integrated feedstock-to-product conversion technology. Green Chem 25(17):6797–6808

Ghoreshizadeh S, Calvo-Peña C, Ruiz-Muñoz M et al (2025) Isolation and characterization of pseudomonas sp. HX1, Streptomyces luteogriseus HR40, and Streptomyces flavofungini HR77 as promising biocontrol agents against Verticillium wilt in hops affected by Verticillium nonalfalfae. Horti 11(5):459

Gitari JM, Muraya MM, Onyango BO et al (2023) Characterization and antibacterial activity of soil actinomycetes from diverse land use systems in Meru South, Eastern Kenya. Actinomycetes 12: 13

Gokhale RS, Tsuji SY, Cane DE et al (1999) Dissecting and exploiting intermodular communication in polyketide synthases. Science 284(5413):482–485

Gomaa FA, Selim HM, Alshahrani MY, Aboshanab KM (2024) Central composite design for optimizing istamycin production by Streptomyces tenjimariensis. World J Microbiol Biotechnol 40(10):316

Gopalakrishnan S, Srinivas V, Vidya MS et al (2013) Plant growth-promoting activities of *Streptomyces* sp. in sorghum and rice. Springerplus 2:574

Gopalakrishnan S, Rajendran V, Arumugam S, Sharma HC, Vadlamudi S, Bhimineni RK et al (2016) Insecticidal activity of a novel fatty acid amide derivative from Streptomyces species against Helicoverpa armigera. Nat Prod Res 30(24):2760–2769. https://doi.org/10.1080/14786419.2016. 1154055PMID:26956775

Goudjal Y, Toumatia O, Yekkour A et al (2014) Biocontrol of Rhizoctonia solani damping-off and promotion of tomato growth by endophytic actinomycetes from Algerian Sahara. Microbiol Res 169(1):59–65

Hassanein NM, Shoala T, Gouda SA (2018) In vitro studies on biological control of Drechslera species causing brown spot disease in rice plants. PSM Microbiol 3(2):43–54

Hauser DA, Kaiser M, Mäser P, Albisetti A (2024) Venturicidin A affects the mitochondrial membrane potential and induces kDNA loss in Trypanosoma brucei. Antimicrob Agents Chemother 68(7):e01671–e01623

Hirota-Takahata Y, Kurosawa E, Ishimoto Y, Iwadate Y, Kizuka M, Chiba J et al (2017) Vestaines, novel vasoactive compounds, isolated from Streptomyces sp. SANK 63697. J Antibiot 70:179–186

Hopwood DA (2006) Soil to genomics: the *Streptomyces* chromosome. Annu Rev Genet 40:1–23

Hopwood DA (2007) Antibiotics: a hundred years of discovery. Microbiol Today 34:116

Hutchings MI, Truman AW, Wilkinson B (2019) Antibiotics: past, present and future. Curr Opin Microbiol 51:72–80

Ibrahim WM, Olama ZA, Abou-Elela GM, Ramadan HS, Hegazy GE, El Badan DE (2023) Exploring the antimicrobial, antiviral, antioxidant, and antitumor potentials of marine Streptomyces tunisiensis W4MT573222 pigment isolated from Abu-Qir sediments, Egypt. Micro Cell Fact 22(1):94

Igarashi Y (2004) Screening of novel bioactive compounds from plant-associated actinomycetes. J Antibiot (Tokyo) 57(1):1–10
Ikeda H, Shin-ya K, Omura S (2014) Genome mining of the *Streptomyces* avermitilis genome and development of genome-minimized hosts for secondary metabolism. Proc Jpn Acad Ser B Phys Biol Sci 90(7):238–252
Jog R, Nareshkumar G, Rajkumar S (2012) Plant growth promoting potential and soil enzyme production of the most abundant Streptomyces spp. from wheat rhizosphere. J Appl Microbiol 113(5):1154–1164
Kalita K, Joshi SR (2017) Study of growth and antimicrobial activity of endophytic *Streptomyces* from ethnomedicinal plants of Khasi Hills, India. Egypt J Basic Appl Sci 4(3):203–208
Kavitha A, Prabhakar P, Narasimhulu M, Vijayalakshmi M et al (2010) Isolation, characterization and biological evaluation of bioactive metabolites from Nocardia levis MK-VL_113. Microbiol Res 165(3):199–210
Khalil MS, Moubasher H, Hasan FF (2014) Biological control of rice blast disease by Streptomyces flavotricini. Res J Pharma Biol Chemical Sci 5(5):1453–1461
Khamna S, Yokota A, Peberdy JF, Lumyong S (2009) Antifungal activity of Streptomyces spp. isolated from rhizosphere of Thai medicinal plants. Inter J Integ Bio 6(3):143–147
Khamna S, Yokota A, Lumyong S (2009b) Actinomycetes isolated from medicinal plant rhizosphere soil: diversity and screening of antifungal compound, indole-3-acetic acid and siderophore production. World J Microbiol Biotechnol 25:649–655
Khosla C, Tang Y, Chen AY et al (2007) Structure and mechanism of the 6-deoxyerythronolide B synthase. Annu Rev Biochem 76:195–221
Kieser T, Bibb MJ, Buttner MJ et al (2000) Practical *Streptomyces* genetics. John Innes Foundation, Norwich
Kim JD, Han JW, Lee SC, Lee D, Hwang IC, Kim BS (2011) Disease control effect of strevertenes produced by Streptomyces psammoticus against tomato Fusarium wilt. J Agri Food Chem 59: 1893–1899
Kim JH, Choi JY, Park DH, Park D-J, Park MG, Kim SY et al (2020) Isolation and characterization of the insect growth regulatory substances from actinomycetes. Comp Biochem Physio Part C: Toxi Pharm 228:108651. https://doi.org/10.1016/j.cbpc.2019.108651. PMID: 31678310
Kim YS, Umurzokov M, Cho KM, Choi JS, Park KW (2022) Insecticidal characteristics and structural identification of the potential active compounds from Streptomyces sp. KR0006: strain improvement through mutagenesis. PLoS One 17(9):e0274766
Kim SH, Park DH, Choi JY, Wang M, Liu S, Je YH (2023) Characterization of insecticidal compound from Streptomyces gramineus against Thrips palmi. J Asia-Pac Ent 26(4):102166
Kumar PS, Duraipandiyan V, Ignacimuthu S (2014) Antimicrobial activity of *Streptomyces* sp. VPTS3-1 isolated from Southern Western Ghats. J Mycol Med 24(1):e1–e6
Kumar M, Kumar R, Kumar S, Jha MK, Tiwari SN, Gupta P (2024) Important antiviral properties of Streptomyces species compounds. Ind J Microbio Res 11(1):3–7
Laatsch H (2003) AntiBase 2003 – The natural compound identifier. Wiley-VCH
Laxminarayan R, Duse A, Wattal C et al (2013) Antibiotic resistance—the need for global solutions. Lancet Infect Dis 13(12):1057–1098
Li Q, Jiang Y, Ning P, Zheng L, Huang J, Li G et al (2011) Suppression of Magnaporthe oryzae by culture filtrates of Streptomyces globisporus JK-1. Biol Control 58(2):139–148
Lin L, Xu X (2013) Indole-3-acetic acid production by endophytic Streptomyces sp. En-1 isolated from medicinal plants. Curr Microbiol 67(2):209–217
Liu W, Du P, Zhao Z, Zhang L (2016) An adaptive weighting algorithm for interpolating the soil potassium content. Sci Rep 6(1):23889
Liu DN, Liu M, Zhang SS, Shang YF, Zhang WF, Song FH, Zhang HW, Du GH WYH (2023a) Chrysomycin a regulates proliferation and apoptosis of Neuroglioma cells via the Akt/GSK-3β signaling pathway in vivo and in vitro. Mar Drugs 21(6):329

Liu M, Ren M, Zhang Y, Wan Z, Wang Y, Wu Z, Wang K, Fang W, Yang X (2023b) Antiviral activity of benzoheterocyclic compounds from soil-derived Streptomyces jiujiangensis NBERC-24992. Molecules 28(2):878

Liu F, Wang N, Wang Y, Yu Z (2024) The insecticidal activity of secondary metabolites produced by Streptomyces sp. SA61 against Trialeurodes vaporariorum (Hemiptera: Aleyrodidae). Microorganisms 12(10):2031

Lotfy MM, Hassan HM, Mohammed R, Hetta M et al (2019) Chemical profiling and biological screening of some river Nile derived-microorganisms. Front Microbiol 12(10):787

Lyu A, Liu H, Che H, Yang L, Zhang J, Wu M et al (2017) Reveromycins a and b from Streptomyces sp. 3–10: antifungal activity against plant pathogenic fungi in vitro and in a strawberry food model system. Front Microbiol 8

Madigan MT, Martinko JM, Bender KS (2014) Brock biology of microorganisms, 14th edn. Pearson, Boston

Mahajan GB, Balachandran L (2012) Antibacterial agents from actinomycetes: a review. Front Biosci (Elite Ed) 4(1):240–253

Mangamuri U, Muvva V, Poda S, Naragani K, Munaganti RK, Chitturi B, Yenamandra V (2016) Bioactive metabolites produced by Streptomyces cheonanensis VUK-A from Coringa mangrove sediments: isolation, structure elucidation and bioactivity. 3 Biotech 6(1):–63

Manimaran M, Rajkumar T, Vimal S, Taju G, Majeed SA, Hameed AS, Kannabiran K (2018) Antiviral activity of 9 (10H)-Acridanone extracted from marine Streptomyces fradiae strain VITMK2 in Litopenaeus vannamei infected with white spot syndrome virus. Aquaculture 488: 66–73

Manivasagan P, Venkatesan J, Sivakumar K et al (2014) Marine actinobacterial metabolites: current status and future perspectives. Microbiol Res 169(4):262–278

Martel J, Chang SH, Wu CY, Peng HH, Hwang TL, Ko YF, Young JD, Ojcius DM (2021) Recent advances in the field of caloric restriction mimetics and anti-aging molecules. Ageing Res Rev 66:101240

Martínez-Burgo Y, Santos-Aberturas J, Vicente CM et al (2022) Discovery of novel secondary metabolites encoded in Actinobacteria by rational activation of biosynthetic gene clusters. Cell Chem Biol 29(12):1786–96.e5

Menzella HG, Reid R, Carney JR et al (2005) Combinatorial polyketide biosynthesis by de novo design and rearrangement of modular polyketide synthase genes. Nat Biotechnol 23(9): 1171–1176

Minh NV, Woo EE, Kim JY, Kim DW, Hwang BS, Lee YJ, Lee IK, Yun BS (2015) Antifungal substances from Streptomyces sp.A3265 antagonistic to plant pathogenic fungi. Mycobiology 43(3):333–338

Minuto A, Spadaro D, Garibaldi A, Gullino ML (2006) Control of soilborne pathogens of tomato using a commercial formulation of Streptomyces griseoviridis and solarization. Crop Prot 25(5): 468–475

Mohandas S, Poovarasan S, Panneerselvam P et al (2013) Guava (Psidium guajava L.) rhizosphere glomus mosseae spores harbor actinomycetes with growth promoting and antifungal attributes. Sci Horti 4(150):371–376

Naseer N, Fatima A, Cheema MT, Hasnain S, Sajid I (2022) Evaluation of antioxidant, antimicrobial and antitumor compounds production potential of Streptomyces antioxidans PU-12M. Res J Biotech 17:11

Nicault M, Zaiter A, Dumarcay S, Chaimbault P, Gelhaye E, Leblond P, Bontemps C (2021) Elicitation of antimicrobial active compounds by Streptomyces-fungus co-cultures. Microorganisms 9(1):178

Neeno-Eckwall EC, Schottel JL (1999) Occurrence of antibiotic resistance in the biological control of potato scab disease. Biol Control 16(2):199–208

Nguyen XH, Naing KW, Lee YS, Kim YH, Moon JH, Kim KY (2015) Antagonism of antifungal metabolites from Streptomyces griseus H7602 against Phytophthora capsici. J Basic Microbiol 55(1):45–53

Nisha SJ, Uma G, Sathishkumar R et al (2025) Optimization and characterization of bioactive secondary metabolites from *Streptomyces* sp. CMSTAAHL-4 isolated from mangrove sediment. BMC Microbiol 25(1):57

Ningthoujam DS, Sanasam S, Tamreihao K, Nimaichand S (2009) Antagonistic activities of local actinomycete isolates against rice fungal pathogens. Afr J Microbiol Res 3(11):737–742

Ochi K (2017) Insights into microbial cryptic gene activation and strain improvement: principle, application and technical aspects. J Antibiot (Tokyo) 70(1):25–40

Olano C, Lombo F, Méndez C et al (2008) Improving production of bioactive secondary metabolites in actinomycetes by metabolic engineering. Metab Eng 10(5):281–292

Olano C, Méndez C, Salas JA (2009) Antitumor compounds from marine actinomycetes. Mar Drugs 7(2):210–248

Omura S, Ikeda H, Ishikawa J et al (2001) Genome sequence of an industrial microorganism *Streptomyces* avermitilis: deducing the ability of producing secondary metabolites. Proc Natl Acad Sci USA 98(21):12215–12220

Ortega HE, Ferreira LL, Melo WG, Oliveira AL, Ramos Alvarenga RF, Lopes NP, Bugni TS, Andricopulo AD, Pupo MT (2019) Antifungal compounds from Streptomyces associated with attine ants also inhibit Leishmania donovani. PLoS Negl Trop Dis 13(8):e0007643

Osman ME, Abo Elnasr AA, Nawar MA, Hefnawy GA (2019) Myco-metabolites as biological control agents against the two-spotted spider mite, Tetranychus urticae Koch (Acari: Tetranychidae). Egyp J Biol Pest Cont 29(1):64

Pagmadulam B, Tserendulam D, Rentsenkhand T, Igarashi M, Sawa R, Nihei CI, Nishikawa Y (2020) Isolation and characterization of antiprotozoal compound-producing Streptomyces species from Mongolian soils. Par Int 74:101961

Passari AK, Mishra VK, Saikia R et al (2015) Isolation, abundance and phylogenetic affiliation of endophytic actinomycetes from medicinal plants and screening for antimicrobial potential. Front Microbiol 6:273

Pfeifer BA, Admiraal SJ, Gramajo H et al (2001) Biosynthesis of complex polyketides in a metabolically engineered strain of E. coli. Science 209(5509):1790–1792

Pierce DR, West-Thielke P, Hajjiri Z, Gaitonde S, Tzvetanov I, Benedetti E, Lichvar AB (2021) Clinical implications of tacrolimus time in therapeutic range and intrapatient variability in urban renal transplant recipients undergoing early corticosteroid withdrawal. Transplant Direct 7(6): e698

Procópio RE, Silva IR, Martins MK et al (2012) Antibiotics produced by *Streptomyces*. Braz J Infect Dis 16:466–471

Prudhomme J, McDaniel E, Ponts N et al (2008) Marine actinomycetes: a new source of compounds against the human malaria parasite. PLoS One 3:e2335

Quan Z, Yang Y, Zheng H, Zhan Y, Luo J, Ning Y, Fan S (2022) Clinical implications of the interaction between PD-1/PD-L1 and PI3K/AKT/mTOR pathway in progression and treatment of non-small cell lung cancer. J Cancer 13(13):3434

Rahman MA, Islam MZ, Islam MA (2011) Antibacterial activities of actinomycete isolates collected from soils of Rajshahi. Bangladesh Biotechnol Res Inter 2011(1):857925

Rashid S, Charles TC, Glick BR (2012) Isolation and characterization of new plant growth-promoting bacterial endophytes. Appl Soil Ecol 1(61):217–224

Rawlings BJ (1999) Biosynthesis of polyketides. Nat Prod Rep 16(4):425–484

Reddy KR, Jyothi G, Sowjanya C, Kusumanjali K et al (2016) Plant growth-promoting actinomycetes: mass production, delivery systems, and commercialization. In: Plant growth promoting Actinobacteria: a new avenue for enhancing the productivity and soil fertility of grain legumes. Singapore, Springer Singapore, pp 287–298

Rezaee H, Salehghamari E, Khamoushi A, Jalali H, Ardalani MA (2025) Investigating the anticancer effect of Streptomyces sp. M12 extract against glioblastoma U87_MG cell line. Cellular Microbiol 2025:7344471

Ribbe M, Gadkari D, Meyer O (1997) N2 fixation by Streptomyces thermoautotrophicus involves a molybdenum-Dinitrogenase and a manganese-superoxide oxidoreductase that couple

N2Reduction to the oxidation of superoxide produced from O2by a molybdenum-CO dehydrogenase. J Biol Chem 272(42):26627–26633
Ridley CP, Lee HY, Khosla C (2008) Evolution of polyketide synthases in bacteria. Proc Natl Acad Sci USA 105(12):4595–4600
Risdian C, Mozef T, Wink J (2019) Biosynthesis of polyketides in Streptomyces. Microorganisms 7(5):124
Rokem JS, Lantz AE, Nielsen J (2007) Systems biology of antibiotic production by microorganisms. Nat Prod Rep 24(6):1262–1287
Rudd BA (1984) Secondary metabolism and differentiation in *Streptomyces*: the advent of molecular genetics. Trends Biotechnol 2:40–44
Sabaratnam S, Traquair JA (2015) Mechanism of antagonism by Streptomyces griseocarneus (strain Di944) against fungal pathogens of greenhouse-grown tomato transplants. Can J Pl Path 37(2):197–211
Saini R, Ali MI, Pant M, Warghane A (2024) Current status of potential antiviral drugs derived from plant, marine, and microbial sources. Anti-Infect Agents 22(2):61–73
Samappito S, Lantz M, Khosla C et al (1998) Cloning and sequencing of the erythromycin gene cluster of Saccharopolyspora erythraea and heterologous expression in *Streptomyces* lividans. Biotechnol Prog 14(1):122–126
Sanchez S, Chavez A, Forero A et al (2010) Carbon source regulation of antibiotic production. J Antibiot (Tokyo) 63(8):442–459
Sellem I, Triki MA, Elleuch L, Cheffi M, Chakchouk A et al (2017) The use of newly isolated Streptomyces strain TN258 as potential biocontrol agent of potato tubers leak caused by Pythium ultimum. J Bas Microbiol 57(5):393–401
Sharma R, Magdum P, Jadhav T, Berge S, Shinde B (2021) Screening of secondary metabolites produced by Streptomyces species from a soil sample that can produce anti-Nematodal and antiprotozoal Avermectins. J Phar Res Int 33(6):90–98; Article no.JPRI.65484 ISSN: 2456-9119
Shelest E, Heimerl N, Fichtner M et al (2015) Multimodular type I polyketide synthases in algae evolve by module duplications and AT domain displacement. BMC Genomics 16:1–5
Shen T, Wang C, Yang H et al (2016) Identification, fermentation and biocontrol effects of *Streptomyces* hygroscopicus B04 on strawberry root rot. Appl Soil Ecol 103:36–43
Shih HD, Liu YC, Hsu FL, Mulabagal V, Dodda R, Huang JW (2003) Fungichromin: a substance from Streptomyces padanus with inhibitory effects on Rhizoctonia solani. J Agric Food Chem 51(1):95–99
Shivlata L, Satyanarayana T (2015) Agro-environmental sustainability. In: Actinobacteria in agricultural and environmental sustainability. Springer, Berlin/Heidelberg/Germany, pp 173–218
Singh LS, Mazumder S, Bora TC (2009) Optimisation of process parameters for growth and bioactive metabolite production by a *Streptomyces* sp. isolated from a pristine habitat in Assam. J Mycol Med 19(3):161–167
Solans M, Vobis G, Cassán F, Luna V, Wall LG (2011) Production of phytohormones by root-associated saprophytic actinomycetes isolated from the actinorhizal plant Ochetophila trinervis. W J Microbiol Biotechn 27(9):2195–2202
Soliman MA, Hamza AF, Zahran NF, Bassioni G (2021) Microbiological study and insecticidal potential of purified extract from Streptomyces sp. on the larvae of Galleria mellonella. J Plant Dis Prot 128(6):1565–1574
Staunton J, Weissman KJ (2001) Polyketide biosynthesis: a millennium review. Nat Prod Rep 18(4):380–416
Subramani R, Aalbersberg W (2012) Marine actinomycetes: an ongoing source of novel bioactive metabolites. Microbiol Res 167(10):571–580
Sun H, Li X, Chen M, Zhong M, Li Y et al (2022) Multi-omics-guided discovery of omicsynins produced by Streptomyces sp. 1647: pseudo-tetrapeptides active against influenza a viruses and coronavirus HCoV-229E. Engineering 16:176–186

Takano E (2006) Gamma-butyrolactones: *Streptomyces* signalling molecules regulating antibiotic production and differentiation. Curr Opin Microbiol 9(3):287–294

Talubnak C, Soytong K (2010) Biological control of vanilla anthracnose using Emericella nidulans. J Agr Tech 6:47–55

Tanaka Y, Omura S (1993) Metabolism and products of actinomycetes—an introduction. Actinomycetologica 7(1):1–12

Tanvir R, Sajid I, Hasnain S (2014) Larvicidal potential of Asteraceae family endophytic actinomycetes against Culex quinquefasciatus mosquito larvae. Nat Prod Res 28(22): 2048–2052. https://doi.org/10.1080/14786419.2014.919579

Teodoro AJ (2019) Bioactive compounds of food: their role in the prevention and treatment of diseases. Oxidative Med Cell Longev 11:3765986

Teregulova GA, Manucharova NA, Urazbakhtina NA, Zhemchuzhina NS, Yevtushenko LI, Stepanov AL (2024) Antimicrobial activity of specialized metabolites of soil chitinolytic streptomycetes. Moscow Univ Soil Sci Bull 79(1):47–55

Tiwari K, Gupta RK (2012) Rare actinomycetes: a potential storehouse for novel antibiotics. Crit Rev Biotechnol 32(2):108–132

Udwary DW, Zeigler L, Asolkar RN et al (2007) Genome sequencing reveals complex secondary metabolome in Salinispora tropica. Proc Natl Acad Sci USA 104(25):10376–10381

Vadankula GR, Rizvi A, Ali H, Khunjamayum R, Eedara VR, Nema V et al (2024) Secondary metabolites from a new antibiotic-producing Endophytic Streptomyces isolate inhibited pathogenic and multidrug-resistant mycobacterium tuberculosis strains. Trop Med Infect Dis 10(5): 117

Vargas Hoyos HA, Chiaramonte JB, Barbosa-Casteliani AG et al (2021) An actinobacterium strain from soil of cerrado promotes phosphorus solubilization and plant growth in soybean plants. Front Bioeng Biotechnol 22(9):579906

Wang F, Xu M, Li Q, Sattler I, Lin W (2010) P-Aminoacetophenonic acids produced by a mangrove endophyte Streptomyces sp.(strain HK10552). Molecules 15(4):2782–2790

Wang Z, Solanki MK, Pang F, Singh RK, Yang LT et al (2017) Identification and efficiency of a nitrogen-fixing endophytic actinobacterial strain from sugarcane. Sugar Tech 19(5):492–500

Wang T, Ahmad S, Yang L, Yan X, Zhang Y et al (2022) Preparation, biocontrol activity and growth promotion of biofertilizer containing Streptomyces aureoverticillatus HN6. Front Plant Sci 13:1090689

Wenzel SC, Bode HB, Kochems I et al (2008) A type I/type III PKS hybrid biosynthetic pathway for kendomycin. Chembiochem 9(16):2711–2721

Wibowo JT, Kellermann MY, Köck M et al (2021) Anti-infective and antiviral activity of valinomycin and its analogues from a sea cucumber-associated bacterium, Streptomyces sp. SV 21. Mar Drugs 19(2):81

Wietzorrek A, Bibb M (1997) A novel family of proteins that regulate antibiotic production in streptomycetes appears to contain an OmpR-like DNA-binding fold. Mol Microbiol 25(6): 1181–1184

Xu Z, Wang W, Cheng J et al (2010) A new clavulanic acid biosynthetic gene cluster from *Streptomyces* strain F6131. Arch Microbiol 192(10):853–860

Yang F, Teves SS, Kemp CJ (2014) doxorubicin, DNA torsion, and chromatin dynamics. Biochimica et Biophysica Acta (BBA) 1845(1):84–89

Yoon V, Nodwell JR (2014) Activating secondary metabolism with stress and chemicals. J Ind Microbiol Biotechnol 41(2):415–424

Yu B, Chen Z, Lu X, Huang Y et al (2020) Effects on soil microbial community after exposure to neonicotinoid insecticides thiamethoxam and dinotefuran. Sci Tot Env 10(725):138328

Zhang C, Albermann C, Fu X et al (2006) Biosynthetic engineering of aminocoumarin antibiotics and discovery of a new natural product. J Am Chem Soc 128(28):9634–9641

Zhang Z, Pan HX, Tang GL (2017) New insights into bacterial type II polyketide biosynthesis. F1000Res 6:172

Zhong J, Sui WW, Bai XY et al (2023) Characterization and biocontrol mechanism of *Streptomyces* olivoreticuli against Rhizoctonia solani. Pestic Biochem Physiol 197:105681

Zotchev SB (2012) Marine actinomycetes as an emerging resource for the drug development pipelines. J Biotechnol 158:168–175

Streptomyces as Mitigators of Biotic and Abiotic Stress

4

Sakshi Verma, Amandeep Singh, Nahida Rehman Mir, Arun Dev Singh, Kamini Devi, Shruti Kaushik, Sandeep Kumar, and Nitika Kapoor

Abstract

Rapid globalization has significantly increased the frequency of biotic and abiotic stresses, severely impacting plant growth, crop yields, and food quality, thereby posing risks to global food security. Soil microbial communities, particularly *Streptomyces*, a diverse group of Gram-positive filamentous bacteria, have emerged as promising eco-friendly biocontrol agents that enhance plant resilience through multiple mechanisms. This chapter explores the critical role of *Streptomycetes* in mitigating both biotic and abiotic stresses in plants. These bacteria promote plant growth by helping in nutrient uptake, disease suppression, and stress tolerance under drought, salinity, and heavy metal stress conditions through the production of variety of stress-protective metabolites such as phytohormones, vitamins, enzymes, and antimicrobial compounds. Although individual *Streptomyces* strains enhance plant growth, but utilizing microbial consortia presents a more comprehensive approach to abiotic stress management. The capacity of *Streptomyces* to suppress phytopathogens through the production of antibiotics, antifungal compounds, and enzymes is also explored. This chapter further delves into the mechanism of action underlying the protective interactions and induction of systemic resistance in plants, thereby helping to regulate the

S. Verma
Department of Zoology, Hans Raj Mahila MahaVidyalaya, Jalandhar, Punjab, India

A. Singh · N. R. Mir · A. D. Singh · K. Devi · N. Kapoor (✉)
Department of Botanical and Environmental Sciences, Guru Nanak Dev University, Amritsar, Punjab, India

S. Kaushik
ICAR-Central Soil Salinity Research Institute, Karnal, Haryana, India

S. Kumar
Department of Physics, DAV University Sarmastpur, Jalandhar, Punjab, India

T. Kaur et al. (eds.), *Streptomyces: Biological Candidates for Sustainable Agriculture*,
https://doi.org/10.1007/978-981-95-5803-2_4

rhizosphere microbiome. The synergistic interactions among different *Streptomyces* strains contribute to osmolyte production, ion homeostasis, and antioxidative defense, fostering plant resilience under harsh environmental conditions. Further, this chapter addresses the wide range of applications of *Streptomycetes* spp. in sustainable agriculture and associated challenges that must be overcome for successful field implementation.

Keywords

Streptomycetes · Phytopathogens · Abiotic stress · Sustainable agriculture · Plant resilience

4.1 Introduction

Climate change has led to variations in temperature, rainfall, and atmospheric conditions, exposing plants to harsh and extreme climates that negatively impact their morphological, developmental, cellular, and molecular processes (Chaudhry and Sidhu 2022). Besides, biotic stressors such as pathogenic bacteria, fungi, viruses, nematodes, and pests are recognized as major threat to plant health, and hence to the survival of natural ecosystems. Plant development and production are adversely affected by biotic and abiotic stressors, showing hindrance such as reduced photosynthesis, impeding nutrient uptake, and encouraging leaf senescence (Abbas et al. 2025; Khan et al. 2025; Jiang et al. 2025). To help plant sustaining these stresses and to stabilize food security, there has been a growing demand for eco-friendly and sustainable strategies to enhance plant resilience and productivity while minimizing the use of chemical fertilizers and pesticides. One among them is the application of *Streptomyces*. These filamentous Gram-positive bacteria have gained popularity as plant growth-promoting rhizobacteria (PGPR) as they enhance plant's ability to respond the adverse conditions (Quinn et al. 2020; Yang et al. 2023a). These *Streptomycetes* are well-known for their ability to produce a wide range of secondary metabolites, including antibiotics. These bacteria play a crucial role in soil ecology by breaking down complex organic materials. These beneficial microbes colonize the rhizosphere and root tissues of plants, where they form symbiotic associations that significantly contribute to plant growth and stress mitigation (Grover et al. 2016). In addition, *Streptomyces* are also known to release other beneficial substances such as auxins (e.g., indole-3-acetic acid, IAA) and siderophores. These compounds help plants by triggering systemic resistance, shaping the rhizosphere microbiome, and improving overall resilience. They also produce antibacterial agents and signaling molecules that support plant health and protect against harmful microbes (Myo et al. 2019).

Numerous evidences are available in the literature, elucidating the efficacy of *Streptomyces* as a stress mitigators. For example, a newly identified strain, *Streptomyces lasalocidi* JCM 3373^T, has been shown to reduce salt stress in *Glycine max* L. by releasing indole-3-carboxaldehyde (Lu et al. 2024). Inoculating plants with

Streptomyces species can also promote root growth by improving root length, root number, and overall biomass in crops such as *Eucalyptus* spp., wheat, sorghum, and chickpea (Pellegrini et al. 2022). Under biotic stress conditions, *Streptomyces* produce a broad spectrum of antibiotics, antifungal agents, and lytic enzymes that suppress phytopathogens. They also limit the growth of harmful microbes by competing for vital nutrients and occupying ecological niches (Manikandan et al. 2023). Therefore, leveraging *Streptomyces*-based bioinoculants offers a promising pathway toward sustainable agriculture, with the potential to enhance crop resilience, reduce chemical dependency, and support global food security. This chapter explores the role of *Streptomyces* spp. in alleviating various biotic and abiotic stressors in plants, highlighting the multifaceted mechanisms exerted by them. Here, we summarized the potential applications of various *Streptomyces* spp. as potent biofertilizers, biopesticides, and biostimulants and outlined the associated challenges that need to be addressed for using *Streptomyces* as potential environmentally friendly biostimulants and biocontrol agents.

4.2 *Streptomycetes*: Biology and Ecological Significance

Streptomycetes are Gram positive, filamentous bacteria, belonging to the phylum Actinobacteria. *Streptomycetes* are renowned for their complexity of life cycle and remarkable ability to produce secondary metabolites. The life cycle of *Streptomyces* commences with spore germination under favorable environmental conditions and sufficient nutrient availability. As germ tubes develop into syncytial vegetative or substrate mycelia, they establish networks of feeding hyphae for nutrient absorption. Upon nutrient depletion or the emergence of stress factors, programmed cell death of substrate mycelia induces cell differentiation and the formation of aerial hyphae (Flärdh and Buttner 2009; Hasani et al. 2014). These aerial hyphae are characterized by a hydrophobic fibrous coating, in contrast to the smooth, hydrophilic surface of feeding hyphae (Petrus and Claessen 2014). The growth of *Streptomyces* is marked by hyphal tip elongation and subapical branching, which are regulated by the apical protein complex DivIVA. Unlike rod-shaped bacteria, *Streptomyces* grow by generating hyphae at the cell pole. The last phase of the *Streptomyces* life cycle is marked by the conversion of the apical cells of aerial hyphae into chains of spores. This transformation involves coordinated, multiple cell divisions within the apical compartment undergoing differentiation (Willemse et al. 2011). The size of *Streptomyces* spores varies between 0.7 and 1.2 μm. The final stages of this life cycle are intricately linked to antibiotic production, which takes place alongside the programmed cell death of substrate mycelia (Claessen et al. 2014).

Streptomyces species are responsible for producing more than 70% of the 23,000 bioactive secondary metabolites generated by microorganisms. The creation of these secondary metabolites is associated with the formation of aerial hyphae when nutrients are limited. These substances demonstrate a range of biological activities, such as antimicrobial properties, toxicity to metazoans, hormone-like effects in microbes, and metal transport capabilities (Biarnes-Carrera et al. 2015; Morgenstern et al. 2015). Secondary metabolites produced by *Streptomycetes* help them to survive and thrive in crucial environmental conditions, therefore these are also referred to as stress metabolites (Lechevalier and Lechevalier 1970).

The physical traits of the *Streptomyces* genus that enable its widespread topographical dispersal include formation of spores under harsh conditions, production of antibiotics, and growth over a wide range of pH levels (>10) (Sultanpuram et al. 2015). These competent chemoorganotrophs metabolize various sources of carbon, including mono- and disaccharides, organic acids, polysaccharides, and other complex carbon sources (humic and fulvic acids) (Noda et al. 2015). They have also been found in different parts of plants, such as roots, tubers, stems, leaves, and seeds as endophytes. The ecological and physiological roles of these bacteria and their interactions with soil and plants are being explored, for appraising their potential for increasing plant development and defense (Hamedi and Mohammadipanah 2015).

Additionally, *Streptomycetes* promote plant growth and offer pest defense, making them potential bioinoculants for plant growth-promotion in sustainable farming (Sousa and Olivares 2016). Remarkably, while some *Streptomycetes* aid as biocontrol agents, others may help pathogenic fungi invading the roots by suppressing the plant defense system (Schrey and Tarkka 2008), highlighting the complex nature of *Streptomycete*–plant interactions. Moreover, the antibiotics produced by *Streptomycetes* can cause some competing bacteria to become antibiotic-resistant, which may affect the balance of the microbial communities. They play a significant role in the soil ecological system, particularly in plant–microbe interactions. These bacteria secrete proteins that are important for the successful colonization of the roots. Because of their ability to produce bioactive compounds and promote plant growth and defense, they are valuable agents for biocontrol and as bioinoculants in farming (Fyans et al. 2013).

4.3 Biotic and Abiotic Stresses in Agriculture and Environment

Land plants face various abiotic stresses, including drought, salinity, extreme temperatures, heavy metals, and nutrient deficiencies, as well as biotic stresses from bacteria, fungi, virus, weeds, insects, and nematodes. These factors negatively affect plant growth and productivity, and their impact is intensifying over time (Chaudhary et al. 2022; Mareri et al. 2022). The main culprit for these uncomplimentary environmental conditions is the steady rise in the average global temperature (Yang et al. 2023a). Also, it was reported that the global temperature is projected to rise by 1.5–2.0 °C by 2040, posing a serious threat to global food stability and security, potentially decreasing food production by 30–40% (Ray et al. 2015; Yang et al. 2023a). Heat stress reduces seed germination, photosynthesis, and yield, particularly during reproduction, causing petal cell dysfunction and another abnormalities (Liu et al. 2018). Likewise, drought is another major environmental stress and its severity depends on rainfall distribution, evaporative demand, and soil moisture retention. Climate change, marked by rising temperatures and CO_2 levels, exacerbates drought by altering rainfall patterns. Prolonged drought reduces soil water availability, leading to premature plant death. Under such stresses, plants respond by restricting shoot growth and lowering metabolic activity to conserve resources (Umar et al. 2021).

Another important environmental stress is the salinity stress, which is again a major threat to agriculture, reducing crop yields due to salt accumulation from natural processes or human activities such as weathering of rocks or irrigation with salt-rich water (He et al. 2018). It affects plants through osmotic stress (hindering water uptake) and ion toxicity (disrupting cellular functions), leading to reduced growth, impaired metabolism, and lower productivity (Fahad et al. 2017).

Heavy metal stress is a critical global research focus due to the increasing accumulation of toxic metals in agricultural soils as a consequence of industrialization, chemical fertilizers, and wastewater irrigation, harming the soil–plant system (Nievola et al. 2017). These metals enter plants through air, water, and the food chain, leading to bioaccumulation that poses environmental and health risks (Rauwane and Ntushelo 2020). Furthermore, plants face biotic stress from fungi, viruses, bacteria, nematodes, and insects, leading to diseases, infections, and yield loss (Fahad et al. 2017). Nematodes primarily affect roots, causing wilting and stunting. Viruses spread systemically, causing chlorosis and deformities, often transmitted by insect vectors. Insects damage plants by feeding and laying eggs, while fungi cause leaf spots, wilts, and cankers, either as necrotrophs (killing host cells) or biotrophs (living within host tissue) (Umar et al. 2021).

4.4 Role of *Streptomycetes* in Mitigating Abiotic Stress to Plants

Streptomyces is the well-researched genus of Gram-positive PGPB in context to its beneficial interactions with plants and their capacity to enhance plant growth (Sousa and Olivares 2016; Chouyia et al. 2022). This genus includes many species with high guanine–cytosine (G–C) concentration in their DNA, which can reach up to 75% of the genome. *Streptomyces* species have been documented to synthesize numerous physiologically active compounds, demonstrating plant growth-promoting activities (Devi et al. 2025a, b; Patel et al. 2025).

Plant growth-promoting rhizobacteria (PGPR) can mediate systemic resistance under abiotic stresses by initiating the production of phytohormones and volatile organic compounds (VOCs), causing callose deposition and enabling Ca^{2+} influx (Yu et al. 2022). Phytohormones such as auxins, cytokinins (CK), gibberellins, abscisic acid (ABA), and ethylene (via ACC deaminase) synthesis and exposure result in physiological modulations in plants that mitigate abiotic stresses, including drought and salinity (Yang et al. 2023a). Auxins produced by PGPR helps in stimulating root growth, thereby enhancing nutrient and water uptake, which is one of the mechanisms to combat drought stress. Further, PGPR contain an enzyme 1-aminocyclopropane-1-carboxylic acid (ACC) deaminase that acts by degrading ACC (the precursor of ethylene). Plants produce more ethylene during stress conditions that inhibit growth of roots. ACC deaminase lowers these ethylene levels, thereby supporting the better root development (del Carmen Orozco-Mosqueda et al. 2020). Furthermore, cytokinin biosynthesis, which acts against the reactive oxygen species (ROS), bursts and upregulates antioxidant enzyme activity that might help to alleviate the impacts of abiotic stresses in plants (Kumar and Verma 2018). Subsequently, VOCs such as N-acylhomoserine lactone and cyclo-dipeptides generated by PGPR are capable of regulating Na^+ homeostasis. This aids in countering the salt stress by plants (Numan et al. 2018). These compounds increase the induction of the high-affinity K^+ transporter (HKT1) in the branches while reducing it in roots. This decreases the entry of Na^+ into the roots and facilitates recirculation of Na^+ between branches and roots (Becquer 2022).

4.4.1 Plant Growth Promotion under Stress Conditions

Streptomyces species are deemed crucial in augmenting plant resilience to stress by modulating phytohormone levels, suppressing phytopathogens and triggering systemic tolerance processes (Sathya et al. 2017; Kumar and Verma 2018; Mondal et al. 2024). *Streptomyces* species enable their functioning by synthesizing extracellular polysaccharides and other osmoregulatory compounds that mitigate osmotic stress and prevent water loss, therefore maintaining internal homeostasis (Niu et al. 2022a, b; Bodade et al. 2024).

Streptomyces promotes stress tolerance primarily by hydrolyzing 1-aminocyclopropane-1-carboxylic acid (ACC), an ethylene precursor, by ACC deaminase, hence reducing ethylene levels and augmenting plant stress responses (Nascimento et al. 2016; Vejan et al. 2016; Kour et al. 2024). Considering ethylene's involvement in senescence, abscission, and stress signaling, its management is crucial for plant survival under abiotic stress (Patial et al. 2024). *Streptomyces* spp. have demonstrated their ability to enhance growth in crops such as wheat, maize, chickpea, and soybean under saline environment, underscoring their promise in sustainable agriculture (Table 4.1).

4.4.2 Osmolyte-Mediated Stress Resilience Induced by Streptomycetes in Plants

Plants respond to abiotic stress such as drought and salinity by adjusting their osmotic status via osmolytes production, which maintain cell turgor and metabolic processes of plants (Niu et al. 2022a, b). Many rhizosphere microorganisms such as soil-derived microbes and endophytes have been explored for their cooperative utilization in plants' defense against abiotic stresses. Among these, *Streptomycetes* are well-known for the production of osmolytes including proline, polysaccharides, and ectoine, which acts as compatible solutes to stabilize proteins and membranes by forming hydration shells around macromolecules, scavenge ROS and balance osmotic pressure within the plants (Singh and Dubey 2018; Niu et al. 2022a, b). These bacteria may synthesize osmolytes themselves or induce their production and

Table 4.1 *Streptomyces* species (rhizospheric or endophytic) exhibiting plant growth-promoting (PGP) activities

S. no.	*Streptomyces* species	Host plant	Applications/ implications	References
1.	*Streptomyces olivaceoviridis*	Wheat	Auxin, cytokinin, and gibberellin production	Aldesuquy et al. (1998)
2.	*Streptomyces lydicus*	Pea	Nodulation	Tokala et al. (2002)
3.	*Streptomyces hygroscopicus*	Kidney beans	Improved the growth of adventitious roots in hypocotyls	Igarashi et al. (2002)
4.	*Streptomyces filipinensis*	Tomato	Enhanced growth in plants	El-Tarabily (2008)
5.	*Streptomyces* spp.	*Eleagnus angustifolia* and *Alnus glutinosa*	Improved IAA and gibberellic acid content Gibberellic acid and the production of zeatin	Ghodhbane-Gtari et al. (2010)
6.	*Streptomyces rochei* and *S. Thermolilacinus*	Rhizosphere of wheat	Siderophore production and phosphate solubilization	Jog et al. (2012)
7.	*Streptomyces* spp.	Rice and sorghum	Helps in improving traits such as root biomass, total dry matter content, and grain yield Improvement in the agronomic traits of crops plants were observed	Gopalakrishnan et al. (2013)
8.	*Streptomyces* spp.	Soybean	Nutrient uptake to regulate plant growth	Nimnoi et al. (2014)
9.	*Streptomyces aurantiogriseus*	Rice	Production of IAA	Harikrishnan et al. (2014)
10.	*Streptomyces* spp.	*Terfezialeonis* Tul.	Enhanced the production of gibberellic acid, IAA, and siderophores were observed	Goudjal et al. (2016)
11.	*Streptomyces* sp. SNN289	*Vigna radiata* (mung bean)	Improved root/shoot growth and antagonistic to *Phytophthora*	Nonthakaew et al. (2022)
12.	*Streptomyces* sp. STD57	Wheat, rice, and tomato	Possess plant growth-promoting activity such as	He et al. (2024)

(continued)

Table 4.1 (continued)

S. no.	*Streptomyces* species	Host plant	Applications/ implications	References
			stimulation of root and shoot growth, which was determined to be associated with IAA production	
13.	*Streptomyces colwelliae* sp. nov.	Isolated from *Alnus glutinosa*	Traits to fix nitrogen, solubilize phosphate, and produce siderophore and ACC deaminase were reported	Nouioui et al. (2025)
14.	*Streptomyces* sp. A13	Isolated from the leaves of tea clone P312	Exhibited PGP traits, such as nitrogen fixation, ammonia production, indole-acetic acid (IAA) production, siderophore production, and ACC deaminase activity and prevent diseases	Devi et al. (2025a, b)
15.	*Streptomyces* sp. GD-4	Three grass species: *Lolium perenne* L., *Elymus dahuricus* Turcz., and *Elymus sibiricus* L.	Exhibited plant growth-promoting activities and possess specific stress resistance genes, including those responsible for the production of siderophores, trehalose, and cold- and heat-shock proteins, enabling it to survive in environments characterized by abiotic stress	Xu et al. (2025)

accumulation in plants through regulating plants' metabolic pathways (Jogawat 2019). Inoculation of rice with *Streptomyces albidoflavus* OsiLf-2 has been shown to enhance the production of osmolytes by inducing proline synthetic genes (*pro*A, *pro*B, *pro*C, and *Roc*D) and ectoine biosynthetic gene cluster. OsiLf-2 improved the ion (K^+/Na^+) balance in rice by exclusion of Na^+ ions and its partition. Furthermore, proline also acts as ROS scavenger, contributing to plant resistance to abiotic stress (Niu et al. 2022a, b).

4.4.3 Improvement of Nutrient Uptake and Soil Health

Plant growth-promoting rhizobacteria (PGPR) including *Streptomyces* are advantageous microorganisms residing in the rhizosphere that stimulate plant growth and development via direct and indirect methods. These encompass the nutrition cycle, metal mobilization, and phytohormone production. PGPR contribute to soil health by preserving microbial diversity and rhizospheric equilibrium, therefore augmenting soil fertility and bolstering plant resilience to abiotic and biotic challenges (Pramanik et al. 2021; Trivedi et al. 2021).Nitrogen, predominantly found in the atmosphere as inert N_2 and is inaccessible to plants in this state. Their availability depend on the assimilable forms such as nitrate (NO_3^-), ammonium (NH_4^+) and trace amounts of amino acids and peptides. Nitrogen fixation is extensively reported in several bacterial species, although there is very few knowledge concerning this mechanism in *Streptomyces*. Besides nitrogen fixation, *Streptomyces* species facilitate phosphorus solubilization by transforming insoluble soil phosphorus into useable forms that plants can assimilate (Alori et al. 2017; Chen et al. 2025a, b). They are essential for sustaining phosphate bioavailability, especially in low-phosphate environments. A two-component system governs this process, in which the sensor kinase PhoR phosphorylates a response regulator (PhoB or PhoP), thereby activating genes associated with phosphate scavenging and uptake, including hydrolases and high-affinity ABC transporters (Martín and Liras 2021; Pang et al. 2022). Furthermore, PGPR augments iron availability by transforming ferric iron (Fe^{3+}) into its more bioavailable ferrous form (Fe^{2+}). Siderophores produced by plant growth-promoting rhizobacteria, such as *Streptomyces*, chelate iron and enhance its transport across the cell membrane. In Gram-positive *Streptomyces*, siderophore-binding proteins (SBPs) and related lipoproteins engage with extracellular siderophores, prompting conformational alterations in the SBP-permease complex to facilitate iron translocation into the cytoplasm. Therefore, by solubilizing phosphorus, fixing atmospheric nitrogen, and producing siderophores for chelation of iron, *Streptomyces* not only enhance plant nutrition but also improve soil structure and microbial diversity, thereby enhancing overall soil health for plant growth (Wilson et al. 2016; Hesse et al. 2018; Rocha et al. 2025).

4.4.4 Enhancement of Stress Tolerance

Streptomyces species significantly contribute to lessen various abiotic stressors, through various physiological and biochemical mechanisms. These are primarily attributed to the ability of microbes to produce phytohormones (that regulate plant growth under stress conditions), siderophores (that improve iron acquisitions and mitigate nutrient deficiency), to fix atmospheric, and to solubilize phosphate (Gao et al. 2022; Nozari et al. 2022; Tian et al. 2024). These potential characteristics further help plants to improve their resilience against salinity, drought, and heavy metal toxicity. Notably, the primary factors associated with increased tolerance to abiotic stressors brought on by *Streptomyces* spp. include elevated enzymatic activity within plant tissues (such as roots, stems, and leaves), and the increased production of specific amino acids (Damodharan et al. 2018; Manullang and Chuang 2020; Niu et al. 2022a, b). Among these species, *Streptomyces pactum* has demonstrated strong heavy metal bioremediation potential by reducing the bioavailability of heavy metals (HMs) such as zinc, nickel, chromium, and cadmium in soil, thereby improving plant growth and soil health (Kaari et al. 2024; Patel et al. 2025). Table 4.2 summarizes various strains of *Streptomyces* studied for their role in combating the abiotic stress to plants along with the possible underlying mechanisms.

4.5 Role of *Streptomycetes* in Alleviating Biotic Stress to Plants

Streptomycetes contribute significantly to plant health and growth by alleviating biotic stress, caused by living organisms such as pathogens, insects, and competitors (Table 4.3) (Khan et al. 2023). They achieve this through various mechanisms, including the production of antibiotics and antifungal compounds, the induction of systemic resistance (ISR), and the competitive exclusion of pathogens (Manikandan et al. 2023). Table 4.4 summarizes the various strategies employed by *Streptomyces* spp. to suppress biotic stress in plants, along with their corresponding mechanisms of action.

4.5.1 Production of Antibiotics and Antifungal Compounds

The most notable characteristic of *Streptomyces* is its ability to produce bioactive secondary metabolites, including antifungals, antivirals, antitumorals, antihypertensives, immunosuppressants, and especially antibiotics that serve as natural deterrents to herbivores and pests (Procópio et al. 2012). Some of these metabolites can function as natural insecticides, helping to protect plants from herbivory by making them less palatable or toxic to pests (Ghanem et al. 2022). Additionally, certain species of *Streptomycetes* produce volatile compounds that repel insect pests, further reducing biotic stress on plants. Igarashi et al. (2003) reported Yatakemycin, a novel antifungal compound produced by *Streptomyces* sp. TP-A0356 and isolated from the soil sample of Yatake Highland, Japan. Likewise, in another study, three species of *Streptomyces* spp., i.e., *S. griseus* (MT210913 "DG5"), *S. rochei* (MN700192 "DG4"), and *S. sampsonii* (MN700191 "DG1" strains), showed antifungal activities by reducing population of plant pathogenic fungus *Sclerotinia sclerotiorum* in bean plant cultivation under in vitro and green house conditions. Furthermore, *Streptomyces* sp. increased root length, shoot length, chlorophyll content, and germination percentage as compared to the stressed plants (Ghanem et al. 2022). According to Al-husnan and Alkahtani (2016), five species of *Streptomyces* promote the production of antibiotic compounds. On the similar lines, another study conducted on tomato seedlings

Table 4.2 Different *Streptomyces* strains that help different plants cope up with abiotic stress, along with the suggested mechanisms of action for each

S. no.	Plant name	Abiotic/ biotic stress	*Streptomyces* strains	Effective mechanism of action	References
1.	Vine	Salinity	*Streptomyces rimosus* co-inoculated with the *Glomus mosseae*	Increased chlorophyll content, nutrient uptake by enhancing the production siderophores and process of phosphate solubilization	Karimi and Noori (2022)
2.	Corn	Salinity	*Streptomyces* sp. CLV95 and CLV79	Increase in growth parameters, upregulated activities of enzymatic antioxidants and contents of proline and other amino acids through the elevated production of indole and siderophores	Nozari et al. (2022)
3.	Soybean	Salinity	*Streptomyces paradoxus* D2–8	Increment in the growth parameters of stressed seedlings by improved levels of IAA, siderophores, organic acids, and glycine betaine clustering	Gao et al. (2022)
4.	Tomato	Salinity	*Streptomyces* sp. KLBMP5084	Improvement in the root and shoot lengths were observed under the potential activation of antioxidative enzymes as well as proline and soluble sugars	Gong et al. (2020)
5.	Corn	Salinity	*Streptomyces* sp. CLV97 and CLV179	Leaf length and the root length get enhanced under the production of ACC deaminase, siderophores, and indole compounds	Nozari et al. (2021)
6.	Tomato	Salinity	*Streptomyces* sp. SK68	Improved levels of dry biomass were found under the upregulated activity of enzymes such as cellulase, amylase, pectinase, and xylanase	Damodharan et al. (2018)
7.	*Puccinelliadistans*	Drought	*Streptomyces rochei* D74	Helps in improving the growth parameters such as fresh weight, fresh leaf weight, and total fresh weight. Also	Cao et al. (2023)

(continued)

Table 4.2 (continued)

S. no.	Plant name	Abiotic/ biotic stress	*Streptomyces* strains	Effective mechanism of action	References
				an increment in the proline content and sugar levels was observed under the upregulated activity of bacterial strains	
8.	Tomato	Drought	*Streptomyces* sp.	Five times increase in the content of fructose were observed by lowering in ethylene content, upregulation of enzymatic activity, ACC deaminase activity, and availability of nutrients	Abbasi et al. (2020)
9.	Rice	Drought	*Streptomyces albidoflavus* OsiLf-2	Survival rates of rice were enhanced and also an increase in the dry matter content was observed under the upregulated levels multiple osmolyte content	Niu et al. (2022a, b)
10.	Tomato	Aluminum (Al)	Co-inoculation of *Streptomyces corchorusii* CASL5 and *K. radicincitans* CABV2	A significant rise in the root and shoot length as well as dry and fresh weights were observed. Also a reduction in the Al accumulations were observed with the help of chelation via siderophores	Silambarasan et al. (2022)
11.	*Symphytum officinale*	Cadmium (Cd) and Zinc (Zn)	*Streptomyces pactum* Act12 and *Bacillus* consortium (*Bacillus licheniformis* and *B. subtilis* 1:1)	The metal content gets declined in shoots via siderophores production	Ali et al. (2021)
12.	Cowpea	Nickel (Ni)	*Streptomyces pactum* Act12	Growth parameters get enhanced by the upregulated activity of siderophores	Dimkpa et al. (2008)

Table 4.3 Different *Streptomyces* strains that help different plants cope up with biotic stress, along with the suggested mechanisms of action for each

S. no.	Biotic stress	*Streptomyces* strain	Target fungi	Compound produced and application as biocontrol agent	References
1.	Fungi	*Streptomyces bikiniensis*	*Magnaporthe oryzae*	Brusone disease gets eliminated by degrading the cell membrane and arresting the spore germination	Liu et al. (2022)
2.	Fungi	*Streptomyces albidoflavus*	*Fusarium oxysporum*	Helps in controlling the root rot disease by disrupting the stability of cell wall in fungi producing chitinases	Du et al. (2022)
3.	Fungi	*Streptomyces setonii*	*Ceratocystis fimbriata*	Brusone disease gets eliminated by volatile acids with the increase in the enzymatic antioxidants activity	Gong et al. (2022)
4.	Fungi	*Streptomyces salmonis*	*Colletotrichum gloeosporioides*	Production of volatile compounds leads in the hampered cell wall structures and prevention from anthracnose	Boukaew et al. (2021)
5.	Fungi	*Streptomyces malaysiensis*	*Colletotrichum gloeosporioides*	Hydrolytic enzymes synthesis leads in the suppression of anthracnose pathogenesis by the activation of apoptosis	Zhou et al. (2022)
6.	Fungi	*Streptomyces griseus*	*Sclerotinia sclerotiorum*	Mycelial growth gets arrested under white mold infections	Gebily et al. (2021)
7.	Fungi	*Streptomyces* sp.	*Fusarium oxysporum*	Volatile compounds were synthesized by the bacterial partners in order to suppress the fungal growth	Islamiati et al. (2022)
8.	Fungi	*Streptomyces yanglinensis*	*Aspergillus flavus*	Suppression in the mycelial growth and retardation in germ tube elongation were observed under aspergillosis	Lyu et al. (2020)

(continued)

Table 4.3 (continued)

S. no.	Biotic stress	*Streptomyces* strain	Target fungi	Compound produced and application as biocontrol agent	References
9.	Fungi	*Streptomyces alboflavus*	*Aspergillus flavus*	Shows a significant reduction in the conidial germination and sporulation process	Yang et al. (2019a, b)
10.	Insect	*Streptomyces* sp. KR0006	*P. xylostella*	Synthesis of antimycins A3a, A1a, and A8a was observed and further shows potential mortality against the insects	Kim et al. (2022)
11.	Nematode	*Streptomyces albogriseolus* ND41 and *S. fimicarius* D153	*M. incognita*	The bacteria-associated compounds show nematicidal effects	Hu et al. (2022)
12.	Nematode	*Streptomyces enissocaesilis* OM182843	*M. incognita*	Compounds such as 2,4-decadienal, chitinase, gelatinase, and protease were subjected to the cell wall degradation and death of the nematode cells	El-Akshar et al. (2022)

showed that *Streptomyces* sp. KLBMP5084 increases root length, shoot length, chlorophyll content, and plant stress tolerance against tomato early blight disease caused by *Alternaria solani* by promoting antifungal activities (Gong et al. 2020).

4.5.2 Induction of Systemic Resistance (ISR) in Plants

Streptomycetes can activate a plant's defense mechanisms, leading to a process known as induced systemic resistance (ISR) (Mun et al. 2020). This process enhances the plant's ability to withstand future infections (Taha et al. 2021). *Streptomyces* promotes the production of plant hormones, such as salicylic acids and jasmonic acids, which play crucial role in plant defense mechanisms (Kurth et al. 2014). *Streptomycetes* also stimulate the synthesis of antimicrobial compounds, pathogenesis-related proteins (PR-proteins), and other defense-related molecules (Chen et al. 2022). Multiple studies have indicated that *Streptomyces* species uplift plant systemic resistance. According to a study on potato cultivar infested by potato virus Y, application of *Streptomyces fradiae* reduced the pathogen population and enhanced the growth by increasing antioxidant enzymes activities, photosynthetic

Table 4.4 Strategies of *Streptomyces* spp. to suppress the biotic stress in plants

S. no.	*Streptomyces* species	Type of biotic stress	Host plant/system	Mode of application and concentration	Mechanism of action	References
1.	*Streptomyces* sp. GD-4	Bacterial wilt	*Elymus sibiricus* L.	Foliar application	Increase leaf, root dry weight, chlorophyll content, and active root absorption area as compared to control plants	Xu et al. (2025)
2.	*Streptomyces* spp.	Potato scab	*Solanum tuberosum*	Pellet culture	Increase plant height, root and shoot length, and chlorophyll content as compared to stressed plants	Liu et al. (2025)
3.	*Streptomyces pratensis* S10	*Fusarium graminearum*	*Triticum aestivum*	10^7, 10^8, 10^9 CFU/mL (Spore suspension)	Increase wheat height, weight, root activity, chlorophyll content, siderophore, 1-aminocyclopropane-1-carboxylic acid (ACC) deaminase, and indole-3-acetic acid (IAA) as compared to control plants	Tian et al. (2024)
4.	*Streptomyces* sp. UP1A-1	Bacterial wilt (*Ralstonia solanacearum*)	*Solanum lycopersicum* and *Solanum melongena*	2.5 kg/plot	Decrease disease incidence by increasing phosphate solubilization, nitrogen metabolism, and potassium metabolism and antioxidant enzyme activities as compared to stressed plants	Kaari et al. (2024)
5.	*Streptomyces* sp. AB-11	*Xanthomonas campestris*, *Pseudomonas syringae*	*Cicer arietinum*	Pellet culture	Enhance root length, fresh weight, germination %, relative water content (RWC), and total chlorophyll contents as compared to stressed plants	Thakur and Yadav (2023)

(continued)

Table 4.4 (continued)

S. no.	*Streptomyces* species	Type of biotic stress	Host plant/ system	Mode of application and concentration	Mechanism of action	References
6.	*Streptomyces corchorusii* TKR8, *Streptomyces corchorusii* JAS2, *Streptomyces misionensis* TBS5	Bacterial panicle blight disease (*Burkholderia glumae*)	*Oryza sativa*	Powder-based bioformulation	Reduce disease infection; increase grain yield and stress-related gene expression as compared to stressed plants	Ngalimat et al. (2023)
7.	*Streptomyces araujoniae* strains (TN11 and TN19)	Fusarium wilt	*Cicer arietinum*	Spore suspension	Increase antioxidant enzyme activities such as SOD, POD, and CAT and plant resistance against disease	Zeyad et al. (2022)
8.	*Streptomyces viridosporus* HH1 and *Rhizophagus irregularis*	Fusarium wilt	*Pisum sativum* L.	10^6 spore/mL (spore suspension)	Increase plant height, chlorophyll content, and grain yield by reducing disease contamination in pea plants	El-Sharkawy et al. (2022)
9.	*Streptomyces bellus* (SB) and *S. saprophyticus*	Root rot of sugar beet	*Beta vulgaris* L.	–	Increase root and shoot length, dry and fresh weight, and potassium and phosphate level in leaves of stressed plants as compared to control plants	Aallam et al. (2022)
10.	*Streptomyces fulvissimus Uts22*	Damping off disease (*Pythium aphanidermatum*)	*Cucumis sativus*	10^{10} CFU/mL	Reduce pathogen infestation; increase plant root surface area, number, and length compared to stressed plants	Saberi Riseh et al. (2022)
11.	*Streptomyces griseus* (CAI-24, CAI-121 and CAI-127), *Streptomyces africanus* (KAI-32) and	Fusarium wilt	*Cicer arietinum*	10 mL pot, 10^8 CFU/mL	Increase nodule weight, leaf area, root and shoot weight, biomass, seed weight, number and antioxidant enzymes activities such as SOD, CAT,	Ankati et al. (2021)

	Streptomyces coelicolor (KAI-90)				APOX, GPOX, glutathione reductase (GS-RH), and phenylalanine ammonia-lyase (PAL) as compared to stressed plants	
12.	*Streptomyces hydrogenans* strain DH-16	*Meloidogyne* sp. (root knot nematode)	*Solanum lycopersicum*	Pellet suspension	Enhance root and shoot length, weight, phenol and flavonoids content, antioxidant enzyme activities such as SOD, POD, CAT as compared to stressed plants	Sharma et al. (2020)
13.	*Streptomyces cellulosae*	Tobacco mosaic virus	*Solanum lycopersicum*	Foliar application; pellet suspension (2×10^7 CFU/mL)	Increase total phenolic contents, POD, and chitinase enzymes as compared to control plants	Abo-Zaid et al. (2020)
14.	*Streptomyces* sp. (strains IC10 and Y28)	*Fusarium oxysporum* f. sp. *lycopersici race 3* (FOL); *Fusarium* wilt	*Solanum lycopersicum*	10^6 CFU (pellet suspension)	Increase shoot length, fresh weight, dry weight, enzymatic antioxidant such as CAT and POD as compared to stressed plant and induced systemic resistance against FOL	Abbasi et al. (2019)
15.	*Streptomyces netropsis* DSM 40093 (SCF7), *S. ambofaciens* (SCF11) and *S. actuosus* (SCF20)	The potato virus Y	*Solanum tuberosum*	Foliar spray	Promotes the induction of systemic acquired resistance by decreasing the disease population in stressed plants	Nasr-Eldin et al. (2019)
16.	*Streptomyces* sp. AUR2 and AUR 4	*Botrytis* gray mold (BGM) disease	*Cicer arietinum*	10^9 CFU (pellet)	Increase seed number, weight, pod number, weight, biomass, and antioxidant enzyme activities such as SOD, POD, CAT, and PPO as compared to control plants	Vijayabharathi et al. (2018)

(continued)

Table 4.4 (continued)

S. no.	*Streptomyces* species	Type of biotic stress	Host plant/ system	Mode of application and concentration	Mechanism of action	References
17.	*Chitiniphilus* sp. MTN22 and *Streptomyces* sp. MTN14	Nematode stress (*Meloidogyne incognita*)	*Bacopa monnieri*	10^8 colony-forming units	Reduce H_2O_2 and MDA content; increase antioxidant enzyme activities such as SOD and free radical scavenging activities as compared to stressed plants	Gupta et al. (2017)
18.	*Streptomyces diastaticus* SP2, *Streptomyces fradiae* SP4, *Streptomyces olivochromogenes* SP5, *Streptomyces collinus* SP8, *Streptomyces ossamyceticus* SP10, and *Streptomyces griseus* SP12	Phytopathogen (*Sclerotium rolfsii*)	*Cicer arietinum*	10^8 CFU/mL (pellet culture)	Increase phenylalanine ammonia lyase polyphenol oxidase (PPO), superoxide dismutase, peroxidase, ascorbate peroxidase, and guaiacol peroxidase activities as compared to stressed plants	Singh and Gaur (2017)
19.	*Streptomyces rochei* SM3	*Sclerotinia sclerotiorum* infection	*Cicer arietinum* L.	Cell suspension (6.8×10^7 CFU/mL)	Increase PAL, total phenol content, proline, CAT activities as compared to stressed plants	Srivastava et al. (2015)

pigments content, and systemic resistance in potato plants (Mohammed et al. 2024). Similarly, another research indicated that tomato plant infected with yellow leaf curl disease, supplementation of *Streptomyces chromofuscus* strain RFS-23, promotes antioxidant enzymes activities such as catalase (CAT), peroxidase (POD), malonidialdehyde (MDA), superoxide dismutase (SOD), systemic resistance, and stress tolerance in stressed tomato plants (Chen et al. 2022). Likewise, in stressed cucumber plants, *Streptomyces* spp. LH 4 enhanced stress tolerance against white root disease (caused by *Sclerotinia sclerotiorum*) by increasing antioxidant enzyme activities and systematic resistance, and thereby promoted overall growth, development, and productivity of cucumber plants (Mun et al. 2020). Similarly, another research indicated that potato plants infested with tomato mosaic virus (TMV), a major pathogen in case of tomato plants, promotes crop damage and loss in crop yield and productivity. However, the application of *Streptomyces ovatisporus* LC597360 reduced the population of TM virus by increasing the antioxidant enzyme activities such as CAT, ascorbate peroxidase (APOX), polyphenol peroxidase (PPO), and systemic resistance, and therefore improved plant growth was observed in stressed plants (Taha et al. 2021).

4.5.3 Competitive Exclusion of Pathogens

Streptomyces spp. also demonstrated competitive exclusion of pathogens by occupying the ecological niche of harmful pathogens and prevents the establishment of these harmful microbes around plant roots (Hiltunen et al. 2009; Dave and Ingle 2022). Due to this competitive exclusion, *Streptomyces* reduce the activities of soil-borne pathogens, such as nematodes, bacteria, and fungi, thereby decreasing the incidence of pathogen infestation such as blight and root rot (Chater et al. 2010). According to Augustine et al. (2016), *Streptomyces rubrolavendulae* M56 strain isolated from the soil sediment of the Bay of Bengal promotes the complete exclusion of aquaculture pathogens of *Vibrio* under in vitro and in vivo conditions by reducing the biomass of the pathogen. A similar research indicated that the amendment of *Streptomyces* sp. S073 inhibited the growth and population of pathogenic *Vibrio parahaemolyticus* and promoted complete exclusion of pathogens (Yang et al. 2019a, b). Likewise, in another study, *Streptomyces* strain (346) decreased infection of common scab (a soil-borne disease of potato) by the pathogen exclusion process under greenhouse conditions (Liu et al. 1996; Hiltunen et al. 2009). Similarly, according to Kobayashi et al. (2012), field pot trials indicated that the amendment of *S. turgidiscabies* on potato plants decreased the common scab disease. In another study conducted on the rice plants, *Streptomyces* spp. act as a biocontrol agent to reduce bacterial panicle blight disease caused by a pathogen (*Burkholderia glumae*), which causes a significant decline in the yield and production of rice (Suárez-Moreno et al. 2019).

4.5.4 Secondary Metabolites and Their Role in Stress Mitigation

Streptomyces species are well known for their secondary metabolite biosynthetic gene clusters, which serve as a major source of potential bioactive chemicals. Advances in genome sequencing and bioinformatics technologies have significantly improved the identification and prediction of these biosynthetic gene clusters, as well as the chemical structures of their associated products (Kalkreuter et al. 2020). Genomic findings reveal that the majority of *Streptomyces* species harbor a multitude of secondary metabolite biosynthesis gene clusters than the actual variety of compounds isolated from these organisms to date (Pan et al. 2019; Donald et al. 2022; Hu et al. 2025). This suggests a significant untapped potential for *Streptomyces* to produce innovative novel compounds. Furthermore, research on marine *Streptomyces*, especially those isolated from deep-sea environments, has expanded our understanding of *Streptomyces* diversity and unveiled new avenues for the discovery of previously unknown bioactive compounds (Yang et al. 2022; Alam et al. 2022). *Streptomyces* species are especially known for producing macrocyclic tetralactones, including avermectins, milbemycin, piericidins, nanchangmycin, dianemycin, and meilingmycin, which exhibit significant potential against worms, mites, and insects, rendering them highly valuable for agricultural settings (Amelia-Yap et al. 2022). Moreover, it has been shown that *Streptomyces* A20 possesses the capacity to produce streptomycin D, E, and F, which play a protective role in rice against bacterial infections (Suárez-Moreno et al. 2019). Additionally, Actinomycin D, isolated from *Streptomyces* sp. NEAU-HV9, has demonstrated significant antagonistic activity against *Ralstonia solanacearum*, a significant pathogenic bacteria detrimental to tomatoes (Ling et al. 2020). Collectively, these findings underscore the pivotal function of *Streptomyces* in the biosynthesis of bioactive compounds with promising agricultural and medicinal applications.

4.6 Mechanism of Action

The mechanisms underlying the ability of *Streptomyces* sp. to reduce abiotic and biotic stress in plants include various genetic and biochemical pathways (Fig. 4.1). They produce variety of secondary metabolites, rendering them highly efficient in working both as biocontrol agents and as enhancers of abiotic stress tolerance (Yang et al. 2023b). The following are the possible mechanisms employed by *Streptomycetes* to alleviate the abiotic and biotic stress in plants.

The chemotactic movement of *Streptomyces* toward root exudates aids in the colonization of plant roots. The root exudates act as a source of energy and signaling compounds that influence microbial recruitment and community assembly. Phytohormones such as indole-3-acetic acid (IAA), cytokinins, gibberellins, and abscisic acid (ABA) produced by *Streptomyces* helps in regulating plant physiological responses under stress conditions (Yang et al. 2023b) (Fig. 4.1). These hormones influence root architecture, promote cell division and elongation, and improve water and nutrients absorption under adverse environmental conditions (Narsing Rao et al.

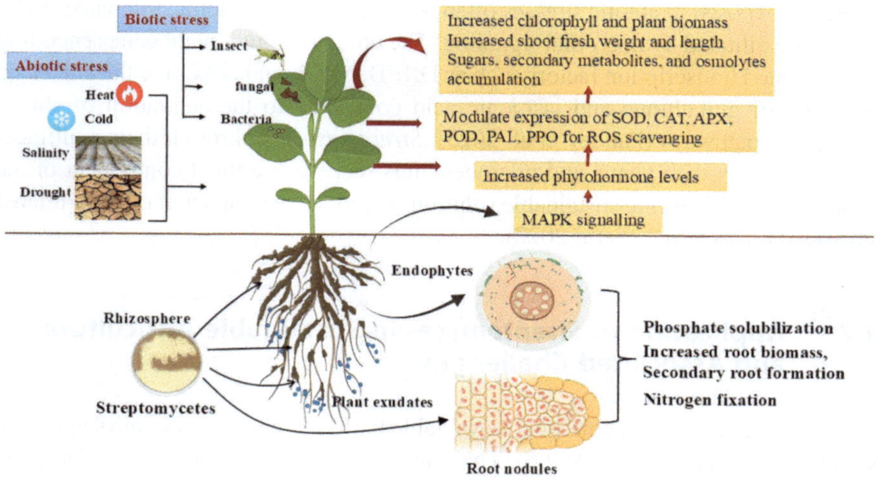

Fig. 4.1 The image illustrates the multifaceted role of *Streptomyces* in mitigating biotic and abiotic stresses in plants through interactions in the rhizosphere and root tissues. It highlights mechanisms such as phytohormone regulation, ROS scavenging enzyme activation, phosphate solubilization, nitrogen fixation, and MAPK signaling pathways that collectively enhance plant growth, biomass, and stress resilience

2022). The mechanisms behind these effects include increased phosphate solubilization, nitrogen fixation, and IAA production, which together improve nutrient uptake and stimulate root development (Pellegrini et al. 2022).

Furthermore, various antimicrobial compounds and enzymes such as chitinases and glucanases produced by *Streptomyces* sp. leads to biotic stress mitigation by helping in suppressing a broad range of phytopathogens either by depriving the microbes from essential nutrients or by degrading their cell walls (Grover et al. 2016). Additionally, *Streptomyces* enhances the accumulation of osmoprotectants such as proline, glycine betaine, and soluble sugars that maintain cellular homeostasis during osmotic stress. These bacteria also exhibit 1-aminocyclopropane-1-carboxylate (ACC) deaminase activity, which breaks down the ethylene precursor ACC into α-ketobutyrate and ammonia, thereby lowering stress-induced ethylene levels that typically inhibit plant growth (Munir et al. 2022). This enzymatic function is a key feature of PGPRs and plays a central role in alleviating the negative effects of abiotic stresses.

The synergistic interaction of *Streptomyces* with other soil microbes, particularly arbuscular mycorrhizal fungi (AMF), further amplifies their stress-mitigating effects. AMF form extensive mycelial networks that enhance phosphorus and nitrogen uptake, improve soil aggregation, and increase plant tolerance to drought, salinity, and nutrient deficiency (Liu et al. 2024). Further, microbial exopolysaccharides play an important role in improving soil structure, water retention, and protection of microbial cells from oxidative damage, which is crucial under extreme environmental conditions (Liu et al. 2024).

Plants exposed to abiotic stresses often activate induced systemic tolerance (IST), which is facilitated by microbial signaling. The upregulation of plant genes encoding stress-related transcription factors such as DREB, MYB, and NAC and the enhanced synthesis of glutathione and jasmonic acid contribute to the activation of stress-responsive pathways (Pandey et al. 2023). *Streptomycetes*, through their multifaceted interactions with plants and microbes, thus serve as a critical component of the plant holobiont, offering sustainable solutions to reduce the impact of climate-related stresses on agriculture (Sethi 2022).

4.7 Application of *Streptomyces* in Sustainable Agriculture and Associated Challenges

Streptomyces species are emerging as valuable biofertilizers and biocontrol agents in sustainable agriculture (Table 4.5). Their ability to solubilize phosphate, synthesize plant growth-promoting compounds such as indole acetic acid (IAA) and siderophores, and exhibit antimicrobial effects against plant pathogens makes them ideal candidates for use as biofertilizers, biopesticides, and biostimulants (Chouyia et al. 2022). *Streptomyces lydicus* M01 showed enhanced plant growth and mitigated foliar disease in cucumbers by altering the soil's microbial community and strengthening the plant's antioxidant defenses (Wang et al. 2020).

Despite their potential, the creation of commercial biofertilizers from *Streptomyces* faces several challenges. Although phosphate-solubilizing *Streptomyces* (PSS) has shown potential in both glasshouse and on-site experiments, no PSS-based biofertilizers have been commercially released (Chouyia et al. 2022). This lack of commercialization may be due to their variable effectiveness in real-world environments and the complex interactions among microbes in the soil. To overcome these challenges and maximize the benefits of *Streptomyces* in sustainable farming, more research is needed to improve its formulations and methods of application. Adding *Streptomyces* as a biofertilizer along with other various beneficial microbes may improve their effectiveness and enhance stability (Yadav and Kumar 2024). Additionally, studying the interactions between *Streptomyces*-based biofertilizers and other sustainable farming methods (such as push–pull technology) could lead to better and more efficient strategies for increasing crop yield and resilience while decreasing dependence on chemical fertilizers (Dzvene and Chiduza 2024).

4.8 Conclusion

Streptomyces, a prominent genus of soil actinobacteria, offers a promising nature-based solution to support plant health and resilience during abiotic and biotic stress. Drought and salinity are critical abiotic stresses that negatively impact agricultural output by disturbing water balance, ion homeostasis, and plant metabolism, frequently resulting in the formation of ROS. The remarkable ability of *Streptomyces* to produce various osmoprotectants, phytohormones, and stress-alleviating secondary

Table 4.5 Role of *Streptomyces* in sustainable agriculture: benefits and applications

S. no.	*Streptomyces* species	Type of abiotic stress	Host plant/system	Key findings	Applications/implications	References
1	*Streptomyces pactum* Act12 and *S. rochei* D74	Drought stress	*Triticum aestivum* L.	Induced activity of defense enzymes, increased the proline and soluble sugar content	Eco-friendly and sustainable agriculture	Shan et al. (2025)
2	*Streptomyces* strains (SP6C4::hyg), (KPB2::hyg and W1SF4::hyg), and (S8::yg	Drought and salinity stress	Tomato (cv. Heinz)	L-glutamic acid reduced the impact of abiotic stress by enhancing the *Streptomyces* population and modifying the composition of the microbial community	The role of L-glutamic acid in reshaping microbial communities and mitigating salinity stress may inform future strategies to engineer bacterial ecosystems or essential microbial characteristics that bolster host tolerance to abiotic stresses	Kim and Kwak (2023)
3	*Streptomyces iranensis*	Drought and salinity stress	*Arabidopsis* sp.	Pteridic acid produced by *Streptomycetes* sp. that improved plants' resistance to abiotic stress	It presents a valuable opportunity to harness the advantages of beneficial *Streptomyces* and their secondary metabolites in agriculture, aiming to reduce the adverse effects of climate change	Yang et al. (2023a)
4	*Streptomyces* sp. NRC21696	Metal polluted soil	*Brassica rapa* L., *Solanum tuberosum* L., *Capsicum annuum* L., and *Morus alba* L.	The release of PGP (zeatin, indole-acetic-acid, gibberellic acid, and ACC deaminase) and melanin by living cells were crucial methods developed by	*Streptomyces* sp. NRC21696 offers a cost-effective and eco-friendly solution for metal decontamination while also enhancing the growth and yield of maize cultivated	Al-Huqail and El-Bondkly (2022)

(continued)

Table 4.5 (continued)

S. no.	*Streptomyces* species	Type of abiotic stress	Host plant/system	Key findings	Applications/implications	References
				bacterial strains to reduce metal toxicity	in mineral-rich soils or irrigated with metal-laden waste	
5	*Streptomyces pactum* (Act12)	Metal polluted soil	*Triticum aestivum* L.	Act12 induced a decline in soil pH, enhanced the uptake of cadmium, copper, and zinc in both shoots and roots. Furthermore, Act12 facilitated an accumulation of lead (Pb) in the roots while reducing its translocation to the shoots. Additionally, it led to a decrease in the activity of antioxidant enzymes (peroxidase, superoxide dismutase, and catalase) along with a reduction in lipid peroxidation levels	Utilizing Act12 to boost the phytoextraction of metals presents an environmentally friendly strategy for the phytoremediation of metal-polluted mining soils, which can be effectively applied with nonedible plants or bioenergy crops	Ali et al. (2021)
6	PGP *Streptomyces* strains (IT25 and C-2012)	Drought stress	Tomato	*Streptomyces* isolates improved proline, leaf relative water content, malondialdehyde, hydrogen peroxide, ascorbate peroxidase activity, and total sugar levels, while they	The tomato yield was enhanced by 35% and 32% with the application of strains IT25 and C-2012, respectively	Abbasi et al. (2020)

				decreased catalase and guaiacol peroxidase activity		
7	*Streptomyces* sp.	Salt and heat stress	*Arabidopsis thaliana* L. and *Brassica chinensis* L.	Increased fresh weight, enhanced catalase activity, and reduced hydrogen peroxide and malondialdehyde levels. Triggered the accumulation of proteins in *Arabidopsis*, including those linked to nitrogen assimilation (GS1), carbohydrate metabolism (cFBPase), and the light-harvesting chlorophyll protein	The treatment of plants with *Streptomyces* sp. led to increased expression levels of GS1, cFBPase, and Lhcb1 gene proteins, which promoted growth in *Arabidopsis* and *brassica* sp.	Manullang and Chuang (2020)
8	*Streptomyces* sp. KLBMP5084	Salt stress	*Limonium sinense*	The increased fresh weight, longer shoots, and more fibrous roots suggest improved overall plant development. Biochemical changes in leaves and stems, including higher antioxidant enzyme activity, elevated soluble sugars, and increased proline levels, point to enhanced protective mechanisms against potential stressors	The strain *Streptomyces* sp. KLBMP5084 holds potential as a biological fertilizer, capable of boosting the growth of tomato seedlings under conditions of salinity stress	Gong et al. (2020)

(continued)

Table 4.5 (continued)

S. no.	*Streptomyces* species	Type of abiotic stress	Host plant/system	Key findings	Applications/implications	References
9	*Streptomyces variabilis* and *S. fradiae*	Salt stress	*Cucumis sativus* L. and *Salicornia europaea* L.	Increased levels of the major pronounced RuBisCO	Utilizing PGP streptomycetes to enhance the resilience of sugar crops that are of substantial economic significance	Tolba et al. (2019)
10	*Streptomyces pactum* (Act12)	Metal polluted soil	*Brassica juncea*	The findings revealed that trace element uptake (Zn, Cu, Pb, and Cd) by *B. juncea* shoots and roots was significantly higher in Feng County. In Tongguan samples, Cd and Cu absorption in shoots decreased but increased with higher Act12 doses	Application of Act12 to boost the phytoremediation of trace elements in soils affected by smelting and mining activities, along with examining the contribution of hyperaccumulator species, is recommended for future studies	Ali et al. (2017)

metabolites that facilitate improved nutrient absorption, plant growth, and protection of plant cells from ROS induced damage by activating stress–response pathways and enhancing systemic tolerance in plants. Moreover, they produce antifungal compounds, antibiotics, and various enzymes that suppress a wide range of plant pathogens. Production of siderophore such as substances further deprive the pathogens from essential nutrients thereby fosters a balanced microbial community. *Streptomyces* spp. demonstrate significant potential in current agricultural practices, such as their use as biofertilizers and for biological control. Harnessing *Streptomyces*-based bioinoculants aligns with the principles of sustainable agriculture, reducing reliance on chemical inputs while enhancing crop productivity. Despite their ecological versatility and metabolic richness, practical application of *Streptomyces* for sustainable agriculture faces challenges such as stability of formulations, inconsistent performance in field conditions, and complexity of interactions with soil microbiota and host plants. Further research into *Streptomyces* consortia-mediated defense mechanisms could lead to innovative biotechnological solutions for mitigating plant stress, thereby contributing to global food security and sustainable agricultural practices.

Use of Third-Party Material The authors confirm that no third-party materials were used in the preparation of this chapter.

References

Aallam Y, Dhiba D, El Rasafi T et al (2022) Growth promotion and protection against root rot of sugar beet (*Beta vulgaris* L.) by two rock phosphate and potassium solubilizing *Streptomyces* spp. under greenhouse conditions. Plant Soil 1:1–4

Abbas K, Javed M, Aslam S et al (2025) Co-application of potassium and thiourea for mitigating salinity stress in wheat seedlings. Sci Rep 15(1):14689

Abbasi S, Safaie N, Sadeghi A et al (2019) *Streptomyces* strains induce resistance to *Fusarium oxysporum* f. sp. *lycopersici* race 3 in tomato through different molecular mechanisms. Front Microbiol 10:1505

Abbasi S, Sadeghi A, Safaie N (2020) *Streptomyces* alleviate drought stress in tomato plants and modulate the expression of transcription factors ERF1 and WRKY70 genes. Sci Hortic 265: 109206

Abo-Zaid GA, Matar SM, Abdelkhalek A (2020) Induction of plant resistance against tobacco mosaic virus using the biocontrol agent *Streptomyces cellulosae* isolate Actino 48. Agronomy 10(11):1620

Alam K, Mazumder A, Sikdar S et al (2022) *Streptomyces*: the biofactory of secondary metabolites. Front Microbiol 13:968053

Aldesuquy HS, Mansour FA, Abo-Hamed SA (1998) Effect of the culture filtrates of *Streptomyces* on growth and productivity of wheat plants. Folia Microbiol 43:465–470

Al-Huqail AA, El-Bondkly AM (2022) Improvement of *Zea mays* L. growth parameters under chromium and arsenic stress by the heavy metal-resistant *Streptomyces* sp. NRC21696. Int J Environ Sci Technol 19(6):5301–5322

Al-husnan LA, Alkahtani MD (2016) Molecular identification of *Streptomyces* producing antibiotics and their antimicrobial activities. Ann Agric Sci 61(2):251–255

Ali A, Guo D, Mahar A et al (2017) Role of *Streptomyces pactum* in phytoremediation of trace elements by *Brassica juncea* in mine polluted soils. Ecotoxicol Environ Saf 144:387–395

Ali A, Guo D, Li Y et al (2021) *Streptomyces pactum* addition to contaminated mining soils improved soil quality and enhanced metals phytoextraction by wheat in a green remediation trial. Chemosphere 273:129692

Alori ET, Glick BR, Babalola OO (2017) Microbial phosphorus solubilization and its potential for use in sustainable agriculture. Front Microbiol 8:971

Amelia-Yap ZH, Azman AS, AbuBakar S et al (2022) *Streptomyces* derivatives as an insecticide: current perspectives, challenges and future research needs for mosquito control. Acta Trop 229: 106381

Ankati S, Srinivas V, Pratyusha S et al (2021) *Streptomyces* consortia-mediated plant defense against Fusarium wilt and plant growth-promotion in chickpea. Microb Pathog 157:104961

Augustine D, Jacob JC, Philip R (2016) Exclusion of *Vibrio* spp. by an antagonistic marine actinomycete *Streptomyces rubrolavendulae* M56. Aquac Res 47(9):2951–2960

Becquer CJ (2022) Rhizobacteria and their contribution to plant tolerance to drought and salinity. Cuban J Agric Sci 56(2):1–26

Biarnes-Carrera M, Breitling R, Takano E (2015) Butyrolactone signalling circuits for synthetic biology. Curr Opin Chem Biol 28:91–98

Bodade R, Kumar A, Sharma VK et al (2024) Recent advances in biotechnological applications of microbial secondary metabolites. Front Microbiol 15:1525146

Boukaew S, Cheirsilp B, Prasertsan P et al (2021) Antifungal effect of volatile organic compounds produced by *Streptomyces salmonis* PSRDC-09 against anthracnose pathogen *Colletotrichum gloeosporioides* PSU-03 in postharvest chili fruit. J Appl Microbiol 131(3):1452–1463

Cao S, Yang F, Zhang H et al (2023) Physiological and transcriptome profiling analyses reveal important roles of *Streptomyces rochei* D74 in improving drought tolerance of *Puccinellia distans* (Jacq.) Parl. Environ Exp Bot 207:105204

Chater KF, Biró S, Lee KJ et al (2010) The complex extracellular biology of *Streptomyces*. FEMS Microbiol Rev 34(2):171–198

Chaudhary P, Agri U, Chaudhary A et al (2022) Endophytes and their potential in biotic stress management and crop production. Front Microbiol 13:933017

Chaudhry S, Sidhu GP (2022) Climate change regulated abiotic stress mechanisms in plants: a comprehensive review. Plant Cell Rep 41(1):1–31

Chen D, Ali MN, Kamran M et al (2022) The *Streptomyces chromofuscus* strain RFS-23 induces systemic resistance and activates plant defense responses against tomato yellow leaf curl virus infection. Agronomy 12(10):2419

Chen Y, Gao Z, Yang Y et al (2025a) Revealing the existence of diverse strategies for phosphorus solubilization and acquisition in Plant-Growth Promoting *Streptomyces misionensis* SwB1. Microorganisms 13(2):378

Chen Z, Zhang P, Wang B et al (2025b) Harnessing the role of rhizo-bacteria to mitigate salinity stress in rice (*Orzya sativa*); focus on antioxidant defense system, photosynthesis response, and rhizosphere microbial diversity. Rhizosphere 33:101043

Chouyia FE, Ventorino V, Pepe O (2022) Diversity, mechanisms and beneficial features of phosphate-solubilizing *Streptomyces* in sustainable agriculture: a review. Front Plant Sci 13: 1035358

Claessen D, Rozen DE, Kuipers OP et al (2014) Bacterial solutions to multicellularity: a tale of biofilms, filaments and fruiting bodies. Nat Rev Microbiol 12(2):115–124

Damodharan K, Palaniyandi SA, Le B et al (2018) *Streptomyces* sp. strain SK68, isolated from peanut rhizosphere, promotes growth and alleviates salt stress in tomato (*Solanum lycopersicum* cv. Micro-Tom). J Microbiol 56:753–759

Dave A, Ingle S (2022) Potential of *Streptomyces* and its secondary metabolites for biocontrol of fungal plant pathogens. In: Sayyed RZ, Singh A, Ilyas N (eds) Antifungal metabolites of rhizobacteria for sustainable agriculture, vol 28. Springer International Publishing, Cham, pp 3–22

del Carmen Orozco-Mosqueda M, Glick BR, Santoyo G (2020) ACC deaminase in plant growth-promoting bacteria (PGPB): an efficient mechanism to counter salt stress in crops. Microbiol Res 235:126439

Devi CJ, Saikia K, Mazumdar R (2025a) Biocontrol and plant growth promotion potential of endophytic *Streptomyces* sp. a13. Curr Microbiol 82(2):64

Devi CJ, Saikia K, Mazumdar R et al (2025b) Identification, biocontrol and plant growth promotion potential of endophytic *Streptomyces* sp. a13. Curr Microbiol 82(2):64

Dimkpa C, Svatoš A, Merten D et al (2008) Hydroxamate siderophores produced by *Streptomyces acidiscabies* E13 bind nickel and promote growth in cowpea (*Vigna unguiculata* L.) under nickel stress. Can J Microbiol 54(3):163–172

Donald L, Pipite A, Subramani R et al (2022) *Streptomyces*: still the biggest producer of new natural secondary metabolites, a current perspective. Microbiol Res 13(3):418–465

Du Y, Wang T, Jiang J et al (2022) Biological control and plant growth promotion properties of *Streptomyces albidoflavus* St-220 isolated from *Salvia miltiorrhiza* rhizosphere. Front Plant Sci 13:976813

Dzvene AR, Chiduza C (2024) Application of biofertilizers for enhancing beneficial microbiomes in push–pull cropping systems: a review. Bacteria 3(4):271–286

El-Akshar EA, Elmeihy RM, Tewfike TA et al (2022) Endophytic *Streptomyces enissocaesilis* as a nematicidal and biostimulant agent. Egypt Acad J Biol Sci G Microbiol 14(2):123–133

El-Sharkawy HH, Rashad YM, Elazab NT (2022) Synergism between *Streptomyces viridosporus* HH1 and *Rhizophagus irregularis* effectively induces defense responses to fusarium wilt of pea and improves plant growth and yield. J Fungi 208(7):683

El-Tarabily KA (2008) Promotion of tomato (*Lycopersicon esculentum* Mill.) plant growth by rhizosphere competent 1-aminocyclopropane-1-carboxylic acid deaminase-producing treptomycete actinomycetes. Plant Soil 308:161–174

Fahad S, Bajwa AA, Nazir U et al (2017) Crop production under drought and heat stress: plant responses and management options. Front Plant Sci 8:1147

Flärdh K, Buttner MJ (2009) *Streptomyces* morphogenetics: dissecting differentiation in a filamentous bacterium. Nat Rev Microbiol 7(1):36–49

Fyans JK, Bignell D, Loria R et al (2013) The ESX/type VII secretion system modulates development, but not virulence, of the plant pathogen *Streptomyces scabies*. Mol Plant Pathol 14(2): 119–130

Gao Y, Han Y, Li X et al (2022) A salt-tolerant *Streptomyces paradoxus* D2-8 from rhizosphere soil of *Phragmites communis* augments soybean tolerance to soda saline-alkali stress. Pol J Microbiol 71(1):43

Gebily DA, Ghanem GA, Ragab MM et al (2021) Characterization and potential antifungal activities of three *Streptomyces* spp. as biocontrol agents against *Sclerotinia sclerotiorum* (Lib.) de Bary infecting green bean. Egypt J Bìol Pest Control 31:1–5

Ghanem GA, Gebily DA, Ragab MM et al (2022) Efficacy of antifungal substances of three *Streptomyces* spp. against different plant pathogenic fungi. Egypt J Biol Pest Control 32(1):112

Ghodhbane-Gtari F, Essoussi I, Chattaoui M et al (2010) Isolation and characterization of non-Frankia actinobacteria from root nodules of *Alnus glutinosa*, *Casuarina glauca* and *Elaeagnus angustifolia*. Symbiosis 50:51–57

Gong Y, Chen LJ, Pan SY et al (2020) Antifungal potential evaluation and alleviation of salt stress in tomato seedlings by a halotolerant plant growth-promoting actinomycete *Streptomyces* sp. KLBMP5084. Rhizosphere 16:100262

Gong Y, Liu JQ, Xu MJ et al (2022) Antifungal volatile organic compounds from *Streptomyces setonii* WY228 control black spot disease of sweet potato. Appl Environ Microbiol 88(6): e02317–21

Gopalakrishnan S, Srinivas V, Sree Vidya M et al (2013) Plant growth-promoting activities of *Streptomyces* spp. in sorghum and rice. Springerplus 2:1–8

Goudjal Y, Zamoum M, Sabaou N et al (2016) Potential of endophytic *Streptomyces* spp. for biocontrol of *Fusarium* root rot disease and growth promotion of tomato seedlings. Biocontrol Sci Technol 26(12):1691–1705

Grover M, Bodhankar S, Maheswari M et al (2016) Actinomycetes as mitigators of climate change and abiotic stress. In: Subramaniam G, Arumugam S, Rajendran V (eds) Plant growth promoting Actinobacteria. Springer, Singapore, pp 203–212

Gupta R, Singh A, Ajayakumar PV et al (2017) Microbial interference mitigates *Meloidogyne incognita* mediated oxidative stress and augments bacoside content in *Bacopa monnieri* L. Microbiol Res 199:67–78

Hamedi J, Mohammadipanah F (2015) Biotechnological application and taxonomical distribution of plant growth promoting actinobacteria. J Ind Microbiol Biotechnol 42(2):157–171

Harikrishnan H, Shanmugaiah V, Balasubramanian N (2014) Optimization for production of Indole acetic acid (IAA) by plant growth promoting *Streptomyces* sp VSMGT1014 isolated from rice rhizosphere. Int J Curr Microbiol Appl Sci 3(8):158–171

Hasani A, Kariminik A, Issazadeh K (2014) Streptomycetes: characteristics and their antimicrobial activities. Int J Adv Biol Biomed Res 2(1):63–75

He M, He CQ, Ding NZ (2018) Abiotic stresses: general defenses of land plants and chances for engineering multistress tolerance. Front Plant Sci 9:1771

He D, Gao C, Zhao S et al (2024) Antibacterial, herbicidal, and plant growth-promoting properties of *Streptomyces* sp. STD57 from the rhizosphere of *Adenophora stricta*. Microorganisms 12(11):2245

Hesse E, O'Brien S, Tromas N et al (2018) Ecological selection of siderophore-producing microbial taxa in response to heavy metal contamination. Ecollett 21(1):117–127

Hiltunen LH, Ojanperä T, Kortemaa H et al (2009) Interactions and biocontrol of pathogenic *Streptomyces* strains co-occurring in potato scab lesions. J Appl Microbiol 106(1):199–212

Hu Q, Yang M, Bo T et al (2022) Soluble macromolecules from two *Streptomyces* strains with potent nematicidal activity against *Meloidogyne incognita*. Rhizosphere 22:100529

Hu WJ, Deng LX, Huang YY et al (2025) Genome mining and metabolite profiling illuminate the taxonomy status and the cytotoxic activity of a mangrove-derived *Microbacterium alkaliflavum* sp. nov. BMC Microbiol 25(1):103

Igarashi Y, Iida T, Yoshida R et al (2002) Pteridic acids A and B, novel plant growth promoters with auxin-like activity from *Streptomyces hygroscopicus* TP-A0451. J Antibiot 55(8):764–767

Igarashi Y, Futamata K, Fujita T et al (2003) A novel antifungal antibiotic produced by *Streptomyces* sp. TP-A0356. J Antibiot 56(2):107–113

Islamiati ED, Widada J, Wahyuningsih TD et al (2022) Volatile organic compounds of *Streptomyces* sp. GMR22 inhibit growth of two plant pathogenic fungi. J Agric Nat Resour 56(5):899–908

Jiang J, Zhang Y, Liu J et al (2025) The regulatory roles of plant miRNAs in biotic stress responses. Biochem Biophys Res Commun 755:151568

Jog R, Nareshkumar G, Rajkumar S (2012) Plant growth promoting potential and soil enzyme production of the most abundant *Streptomyces* spp. from wheat rhizosphere. J Appl Microbiol 113(5):1154–1164

Jogawat A (2019) Osmolytes and their role in abiotic stress tolerance in plants. In: Molecular plant abiotic stress: biology and biotechnology. Wiley, pp 91–104

Kaari M, Manikkam R, Joseph J et al (2024) Integrated genomic and functional analysis of *Streptomyces* sp. UP1A-1 for bacterial wilt control and solanaceae yield increase. Gene Rep 37:102012

Kalkreuter E, Pan G, Cepeda AJ (2020) Targeting bacterial genomes for natural product discovery. Trends Pharmacol Sci 41(1):13–26

Karimi R, Noori A (2022) *Streptomyces rimosus* rhizobacteria and *Glomus mosseae* mycorrhizal fungus inoculation alleviate salinity stress in grapevine through morphophysiological changes and nutritional balance. Sci Hortic 305:111433

Khan S, Srivastava S, Karnwal A (2023) *Streptomyces* as a promising biological control agents for plant pathogens. Front Microbiol 14:1285543

Khan A, Liu D, Jie Z et al (2025) Pre-treatment of roots with melatonin triggers a key tolerance mechanism underlying cadmium stress in cotton seedlings. Ind Crops Prod 228:120870

Kim DR, Kwak YS (2023) Endophytic *Streptomyces* population induced by L-glutamic acid enhances plant resilience to abiotic stresses in tomato. Front Microbiol 14:1180538

Kim YS, Umurzokov M, Cho KM et al (2022) Insecticidal characteristics and structural identification of the potential active compounds from *Streptomyces* sp. KR0006: strain improvement through mutagenesis. PLoS One 17(9):e0274766

Kobayashi YO, Kobayashi A, Maeda M et al (2012) Isolation of antagonistic *Streptomyces* sp. against a potato scab pathogen from a field cultivated with wild oat. J Gen Plant Pathol 78:62–72

Kour D, Khan SS, Kour H et al (2024) ACC deaminase producing phytomicrobiomes for amelioration of abiotic stresses in plants for agricultural sustainability. J Plant Growth Regul 43(4): 963–985

Kumar A, Verma JP (2018) Does plant—microbe interaction confer stress tolerance in plants: a review? Microbiol Res 207:41–52

Kurth F, Mailänder S, Bönn M et al (2014) *Streptomyces*-induced resistance against oak powdery mildew involves host plant responses in defense, photosynthesis, and secondary metabolism pathways. Mol Plant-Microb Interact 27(9):891–900

Lechevalier MP, Lechevalier H (1970) Chemical composition as a criterion in the classification of aerobic actinomycetes. Int J Syst Evol Microbiol 20(4):435–443

Ling L, Han X, Li X et al (2020) A *Streptomyces* sp. NEAU-HV9: isolation, identification, and potential as a biocontrol agent against *Ralstonia solanacearum* of tomato plants. Microorganisms 8(3):351

Liu D, Anderson NA, Kinkel LL (1996) Selection and characterization of strains of *Streptomyces* suppressive to the potato scab pathogen. Can J Microbiol 42(5):487–502

Liu X, Zhou Y, Xiao J et al (2018) Effects of chilling on the structure, function and development of chloroplasts. Front Plant Sci 9:1715

Liu W, Wang J, Li S et al (2022) Genomic and biocontrol potential of the crude lipopeptide by *Streptomyces bikiniensis* HD-087 against *Magnaporthe oryzae*. Front Microbiol 13:888645

Liu H, Li J, Singh BK (2024) Harnessing co-evolutionary interactions between plants and *Streptomyces* to combat drought stress. Nat Plants 10(8):1159–1171

Liu H, Zhang Y, Zhang L et al (2025) Nano-selenium strengthens potato resistance to potato scab induced by *Streptomyces* spp., increases yield, and elevates tuber quality by influencing rhizosphere microbiomes. Front Plant Sci 16:1523174

Lu L, Liu N, Fan Z et al (2024) A novel PGPR strain, *Streptomyces lasalocidi* JCM 3373T, alleviates salt stress and shapes root architecture in soybean by secreting indole-3-carboxaldehyde. Plant Cell Environ 47(6):1941–1956

Lyu A, Yang L, Wu M et al (2020) High efficacy of the volatile organic compounds of *Streptomyces yanglinensis* 3-10 in suppression of *Aspergillus* contamination on peanut kernels. Front Microbiol 11:142

Manikandan A, Anandham R, Johnson I et al (2023) *Bacillus* and *Streptomyces* for management of biotic stresses in plants for sustainable agriculture. In: Chhabra S, Prasad R, Maddela NR, Tuteja N (eds) Plant microbiome for plant productivity and sustainable agriculture. Springer Nature Singapore, pp 263–288

Manullang W, Chuang HW (2020) *Streptomyces* sp. mitigates abiotic stress response and promotes plant growth. J Plant Prot Res 60(3):263–274

Mareri L, Parrotta L, Cai G (2022) Environmental stress and plants. Int J Mol Sci 23(10):5416

Martín JF, Liras P (2021) Molecular mechanisms of phosphate sensing, transport and signalling in *Streptomyces* and related actinobacteria. Int J Mol Sci 22(3):1129

Mohammed FA, Abu-Hussien SH, Dougdoug NKE et al (2024) *Streptomyces fradiae* mitigates the impact of Potato Virus Y by inducing systemic resistance in two Egyptian potato (*Solanum tuberosum* L.) cultivars. MicrobEcol 87(1):131

Mondal S, Acharya U, Mukherjee T et al (2024) Exploring the dynamics of ISR signaling in maize upon seed priming with plant growth promoting actinobacteria isolated from tea rhizosphere of Darjeeling. Arch Microbiol 206(6):282

Morgenstern A, Paetz C, Behrend A et al (2015) Divalent transition-metal-ion stress induces prodigiosin biosynthesis in *Streptomyces coelicolor* M145: formation of coeligiosins. Chem Eur J 21(16):6027–6032

Mun BG, Lee WH, Kang SM et al (2020) *Streptomyces* sp. LH 4 promotes plant growth and resistance against *Sclerotinia sclerotiorum* in cucumber via modulation of enzymatic and defense pathways. Plant Soil 448:87–103

Munir N, Hanif M, Abideen Z et al (2022) Mechanisms and strategies of plant microbiome interactions to mitigate abiotic stresses. Agronomy 12(9):2069

Myo EM, Ge B, Ma J et al (2019) Indole-3-acetic acid production by *Streptomyces fradiae* NKZ-259 and its formulation to enhance plant growth. BMC Microbiol 19:1–14

Narsing Rao MP, Lohmaneeratana K, Bunyoo C et al (2022) Actinobacteria–plant interactions in alleviating abiotic stress. Plants 11(21):2976

Nascimento FX, Rossi MJ, Glick BR (2016) Role of ACC deaminase in stress control of leguminous plants. In: Subramaniam G, Arumugam S, Rajendran V (eds) Plant growth promoting actinobacteria: a new avenue for enhancing the productivity and soil fertility of grain legumes. Springer Singapore, pp 179–192

Nasr-Eldin M, Messiha N, Othman B et al (2019) Induction of potato systemic resistance against the potato virus Y (PVY NTN), using crude filtrates of *Streptomyces* spp. under greenhouse conditions. Egypt J Biol Pest Control 29:1–11

Ngalimat MS, Hata EM, Zulperi D et al (2023) Induction of systemic resistance in rice plants against *Burkholderiaglumae* infection by bioformulation of *Streptomyces* spp. under greenhouse conditions. Biol Control 184:105286

Nievola CC, Carvalho CP, Carvalho V et al (2017) Rapid responses of plants to temperature changes. Temperature 4(4):371–405

Nimnoi P, Pongsilp N, Lumyong S (2014) Co-inoculation of soybean (*Glycine max*) with actinomycetes and *Bradyrhizobium japonicum* enhances plant growth, nitrogenase activity and plant nutrition. J Plant Nutr 37(3):432–446

Niu S, Gao Y, Zi H (2022a) The osmolyte-producing endophyte *Streptomyces albidoflavus* OsiLf-2 induces drought and salt tolerance in rice via a multi-level mechanism. Crop J 10(2):375–386

Niu Z, Yue Y, Su D et al (2022b) The characterization of *Streptomyces alfalfae* strain 11F and its effect on seed germination and growth promotion in switchgrass. Biomass Bioenergy 158: 106360

Noda S, Kawai Y, Tanaka T et al (2015) 4-Vinylphenol biosynthesis from cellulose as the sole carbon source using phenolic acid decarboxylase-and tyrosine ammonia lyase-expressing *Streptomyces lividans*. Bioresour Technol 180:59–65

Nonthakaew N, Panbangred W, Songnuan W (2022) Plant growth-promoting properties of *Streptomyces* spp. isolates and their impact on mung bean plantlets' rhizosphere microbiome. Front Microbiol 13:967415

Nouioui I, Gomez-Escribano JP, Pötter G et al (2025) A novel endophytic species, *Streptomyces colwelliae* sp. nov., isolated from root nodule of *Alnus glutinosa*. BMC Microbiol 25(1):577

Nozari RM, Ortolan F, Astarita LV et al (2021) *Streptomyces* spp. enhance vegetative growth of maize plants under saline stress. Braz J Microbiol 52(3):1371–1383

Nozari RM, Ramos LM, da Luz LA et al (2022) Halotolerant *Streptomyces* spp. induce salt tolerance in maize through systemic induction of the antioxidant system and accumulation of proline. Rhizosphere 24:100623

Numan M, Bashir S, Khan Y et al (2018) Plant growth promoting bacteria as an alternative strategy for salt tolerance in plants: a review. Microbiol Res 209:21–32

Pan R, Bai X, Chen J et al (2019) Exploring structural diversity of microbe secondary metabolites using OSMAC strategy: a literature review. Front Microbiol 10:294

Pandey P, Tripathi A, Dwivedi S et al (2023) Deciphering the mechanisms, hormonal signaling, and potential applications of endophytic microbes to mediate stress tolerance in medicinal plants. Front Plant Sci 14:1250020

Pang F, Solanki MK, Wang Z (2022) *Streptomyces* can be an excellent plant growth manager. World J Microbiol Biotechnol 38(11):193

Patel S, Naik L, Rai A et al (2025) Diversity of secondary metabolites from marine *Streptomyces* with potential anti-tubercular activity: a review. Arch Microbiol 207(3):1–39

Patial D, Yadav S, Kumar A (2024) Beneficial bacteria in regulating drought stress in plants. In: Chauhan PS, Bisht N, Agarwal R (eds) Microbial biostimulants for plant growth and abiotic stress amelioration. Elsevier, Amsterdam, pp 109–123

Pellegrini M, Djebaili R, Pagnani G et al (2022) Role of Actinomycetes in mitigating the impact of climate change: mechanisms of action and perspectives. In: Sayyed RZ, Uarrota VG (eds)

Secondary metabolites and volatiles of PGPR in plant-growth promotion. Springer, Berlin/Heidelberg, pp 153–171

Petrus ML, Claessen D (2014) Pivotal roles for *Streptomyces* cell surface polymers in morphological differentiation, attachment and mycelial architecture. Antonie Van Leeuwenhoek 106:127–139

Pramanik K, Banerjee S, Mukherjee D et al (2021) Beneficial role of plant growth-promoting rhizobacteria in bioremediation of heavy metal (loid)-contaminated agricultural fields. In: Hurst CJ (ed) Microbes: the foundation stone of the biosphere. Springer Nature Switzerland, pp 441–495

Procópio REDL, Silva IRD, Martins MK et al (2012) Antibiotics produced by *Streptomyces*. Braz J Infect Dis 16:466–471

Quinn GA, Banat AM, Abdelhameed AM et al (2020) *Streptomyces* from traditional medicine: sources of new innovations in antibiotic discovery. J Med Microbiol 69(8):1040–1048

Rauwane M, Ntushelo K (2020) Understanding biotic stress and hormone signalling in cassava (*Manihot esculenta*): potential for using hyphenated analytical techniques. Appl Sci 10(22): 8152

Ray DK, Gerber JS, MacDonald GK et al (2015) Climate variation explains a third of global crop yield variability. Nat Commun 6(1):5989

Rocha BM, Pinto E, Sousa E et al (2025) Targeting Siderophore biosynthesis to thwart microbial growth. Int J Mol Sci 26(8):3611

Saberi Riseh R, Moradi Pour M, Ait Barka E (2022) A novel route for double-layered encapsulation of *Streptomyces fulvissimus* Uts22 by alginate–Arabic gum for controlling of *Pythium aphanidermatum* in Cucumber. Agron 12(3):655

Sathya A, Vijayabharathi R, Gopalakrishnan S (2017) Plant growth-promoting actinobacteria: a new strategy for enhancing sustainable production and protection of grain legumes. 3 Biotech 7: 1–10

Schrey SD, Tarkka MT (2008) Friends and foes: *Streptomycetes* as modulators of plant disease and symbiosis. Antonie Van Leeuwenhoek 94:11–19

Sethi S (2022) Utilization of beneficial fungal strain/bacterial strains in climate-resilient agriculture. In: Kumar A, Panwar JS, Ferreira LFR (eds) Microbiome under changing climate. Woodhead Publishing, pp 313–331

Shan H, Wen H, Zhang J et al (2025) Synergistic role of *Streptomyces* composite inoculants in mitigating wheat drought stress under field conditions. Plants 14(3):366

Sharma N, Khanna K, Manhas RK et al (2020) Insights into the role of *Streptomyces hydrogenans* as the plant growth promoter, photosynthetic pigment enhancer and biocontrol agent against *Meloidogyne incognita* in *Solanum lycopersicum* seedlings. Plants 9(9):1109

Silambarasan S, Logeswari P, Sivaramakrishnan R (2022) Amelioration of aluminum phytotoxicity in *Solanum lycopersicum* by co-inoculation of plant growth promoting *Kosakonia radicincitans* strain CABV2 and *Streptomyces corchorusii* strain CASL5. Sci Total Environ 832:154935

Singh R, Dubey AK (2018) Diversity and applications of endophytic actinobacteria of plants in special and other ecological niches. Front Microbiol 9:1767

Singh SP, Gaur R (2017) Endophytic *Streptomyces* spp. underscore induction of defense regulatory genes and confers resistance against *Sclerotium rolfsii* in chickpea. Biol Control 104:44–56

Sousa JADJ, Olivares FL (2016) Plant growth promotion by *streptomycetes*: ecophysiology, mechanisms and applications. Chem Biol Technol Agric 3:1–12

Srivastava S, Patel JS, Singh HB et al (2015) *Streptomyces rochei* SM 3 induces stress tolerance in chickpea against *Sclerotinia sclerotiorum* and NaCl. J Phytopathol 163(7–8):583–592

Suárez-Moreno ZR, Vinchira-Villarraga DM, Vergara-Morales DI et al (2019) Plant-growth promotion and biocontrol properties of three *Streptomyces* spp. isolates to control bacterial rice pathogens. Front Microbiol 10:290

Sultanpuram VR, Mothe T, Mohammed F (2015) *Streptomyces alkalithermotolerans* sp. nov., a novel alkaliphilic and thermotolerant actinomycete isolated from a soda lake. Antonie Leeuwenhoek 107:337–344

Taha M, Ghaly M, Atwa H et al (2021) Evaluation of the effectiveness of soil *Streptomyces* isolates for induction of plant resistance against Tomato mosaic virus (ToMV). Curr Microbiol 78(8): 3032–3043

Thakur R, Yadav S (2023) Thermotolerant and halotolerant *Streptomyces* sp. isolated from *Ajuga parviflora* having biocontrol activity against *Pseudomonas syringae* and *Xanthomonas campestris* acts as a sustainable bioadditive in growth promotion of *Cicer arietinum*. Physiol Mol Plant Pathol 127:102059

Tian X, Hu L, Jia R et al (2024) *Streptomyces pratensis* S10 promotes wheat plant growth and induces resistance in wheat seedlings against *Fusarium graminearum*. J Fungi 10(8):578

Tokala RK, Strap JL, Jung CM et al (2002) Novel plant-microbe rhizosphere interaction involving *Streptomyces lydicus* WYEC108 and the pea plant (*Pisum sativum*). Appl Environ Microbiol 68(5):2161–2171

Tolba ST, Ibrahim M, Amer EA et al (2019) First insights into salt tolerance improvement of Stevia by plant growth-promoting *Streptomyces* species. Arch Microbiol 201:1295–1306

Trivedi S, Butani N, Patel H et al (2021) Microbial consortia and their application for the development of a sustainable environment. In: Microbial technology for sustainable environment, pp 191–202

Umar OB, Ranti LA, Abdulbaki AS et al (2021) Stresses in plants: biotic and abiotic. In: Ansari M (ed) Current trends in wheat research. IntechOpen

Vejan P, Abdullah R, Khadiran T et al (2016) Role of plant growth promoting rhizobacteria in agricultural sustainability—a review. Molecules 21(5):573

Vijayabharathi R, Gopalakrishnan S, Sathya A et al (2018) Deciphering the tri-dimensional effect of endophytic *Streptomyces* sp. on chickpea for plant growth promotion, helper effect with *Mesorhizobium ciceri* and host-plant resistance induction against Botrytis cinerea. Microb Pathog 122:98–107

Wang M, Xue J, Ma J et al (2020) *Streptomyces lydicus* M01 regulates soil microbial community and alleviates foliar disease caused by *Alternaria alternata* on cucumbers. Front Microbiol 11: 942

Willemse J, Borst JW, de Waal E et al (2011) Positive control of cell division: FtsZ is recruited by SsgB during sporulation of *Streptomyces*. Genes Dev 25(1):89–99

Wilson BR, Bogdan AR, Miyazawa M et al (2016) Siderophores in iron metabolism: from mechanism to therapy potential. Trends Mol Med 22(12):1077–1090

Xu W, Liu Y, Cheng Y et al (2025) Plant growth-promoting effect and complete genomic sequence analysis of the beneficial rhizosphere *Streptomyces* sp. GD-4 isolated from *Leymus secalinus*. Microorganisms 13(2):286

Yang M, Lu L, Pang J et al (2019a) Biocontrol activity of volatile organic compounds from *Streptomyces alboflavus* TD-1 against *Aspergillus flavus* growth and aflatoxin production. J Microbiol 57:396–404

Yang M, Zhang J, Liang Q et al (2019b) Antagonistic activity of marine *Streptomyces* sp. S073 on pathogenic *Vibrio parahaemolyticus*. Fish Sci 85:533–543

Yang M, Zhang W, Lv Z et al (2022) Evaluation of the inhibitory effects of Wuyiencin, a secondary metabolite of *Streptomyces albulus* CK-15, against *Sclerotinia sclerotiorum* in vitro. Plant Dis 106(1):156–164

Yang Z, Qiao Y, Konakalla NC et al (2023a) *Streptomyces* alleviate abiotic stress in plant by producing pteridic acids. Nat Commun 14(1):7398

Yang Z, Cao Y, Shi Y et al (2023b) Genetic and molecular exploration of maize environmental stress resilience: toward sustainable agriculture. Mol Plant 16(10):1496–1517

Yu Y, Gui Y, Li Z et al (2022) Induced systemic resistance for improving plant immunity by beneficial microbes. Plants 11(3):386

Zeyad MT, Tiwari P, Ansari WA et al (2022) Bio-priming with a consortium of *Streptomyces araujoniae* strains modulates defense response in chickpea against *Fusarium* wilt. Front Microbiol 13:998546

Zhou D, Jing T, Chen Y et al (2022) Biocontrol potential of a newly isolated *Streptomyces* sp. HSL-9B from mangrove forest on postharvest anthracnose of mango fruit caused by *Colletotrichum gloeosporioides*. Food Control 135:108836

Potential of Streptomycetes as Biocontrol and Plant Growth-Promoting Agents

5

Sahar Ahmed El-Shatoury and Mohamed Khaled Diab

Abstract

Streptomycetes are ubiquitous actinobacteria that play crucial roles in biocontrol, as well as in plant growth promotion, making them sustainable alternatives to chemical pesticides and inorganic fertilizers. This power of producing antimicrobial secondary metabolites, phytohormones, and bioactive compounds also plays a key role in disease suppression, solubilization of inorganic phosphates, and crop stress tolerance. In this chapter, we discuss *Streptomyces*-mediated biocontrol mechanisms such as the flow of antimicrobial metabolites, induced systemic resistance, and competition strategies that inhibit phytopathogens. Besides, it discusses the role of PGPR in the production of plant growth regulators such as auxins, cytokinins, and gibberellins and increasing the nutrient uptake, contributing to better cropping under different stressful climate conditions. It further elaborates on the incorporation of streptomycetes with organic farming and modern agriculture, its utilization in biofertilizers, and sustainable soil maintenance. Genomics, transcriptomics, proteomics, and metabolomics can double as quality tools for improvement and selection of bioagents toward the identification and characterization of novel biocontrol agents and better beneficial traits via genetic engineering. Field case studies, where commercially similar effectiveness to synthetic agrochemicals is found along with improved soil health and environmental impact. Despite their promise, however, they did face

S. A. El-Shatoury
Botany & Microbiology Department, Faculty of Science, Suez Canal University, Ismailia, Egypt
e-mail: sahar_hassan@science.suez.edu.eg

M. K. Diab (✉)
Pest Physiology Department, Plant Protection Research Institute, Agricultural Research Center, Giza, Egypt
e-mail: mohamed.diab_pgs@science.suez.edu.eg

T. Kaur et al. (eds.), *Streptomyces: Biological Candidates for Sustainable Agriculture*,
https://doi.org/10.1007/978-981-95-5803-2_5

regulatory challenges and commercialization hurdles, such as lengthy approval processes, stabilization of formulations, and barriers to adoption in the marketplace. Overcoming these barriers will need harmonized policies, biotechnological developments, and public–private partnerships to avoid dichotomous policy spaces that would hinder the use of streptomycetes-based agricultural products. This chapter emphasizes the potential of streptomycetes in biological control and plant growth promotion in facilitating environment-friendly agriculture for climate-resilient crops.

Keywords

Streptomyces · Biofertilizers; Biopesticides; Sustainable agriculture · Quorum sensing inhibition · Induced systemic resistance; Jasmonic acid signaling

Abbreviations

AMF	Arbuscular mycorrhizal fungi
ARGs	Antibiotic resistance genes
BGCs	Biosynthetic gene clusters
CAT	Catalase
EFSA	European food safety authority
EPA	Environmental protection agency
EPS	Exopolysaccharides
ERAs	Environmental risk assessments
ET	Ethylene
GAs	Gibberellins
HGT	Horizontal gene transfer
IAA	Indole-3-acetic acid
IPM	Integrated pest management
ISR	Inducing systemic resistance
JA	Jasmonic acid
MAPK	Mitogen-activated protein kinase
NP	Natural product
PGPB	Plant growth-promoting bacteria
PGPR	Plant growth-promoting rhizobacteria
POD	Peroxidase
PR	Pathogenesis-related
ROS	Reactive oxygen species
SA	Salicylic acid
SAR	Systemic acquired resistance
SOD	Superoxide dismutase
VOCs	Volatile organic compounds

5.1 Introduction

Streptomycetes are filamentous, Gram-positive bacteria belonging to the most predominant genus within the phylum Actinobacteria, characterized by their diverse metabolic capabilities and distribution in soil habitats (Mohammadipanah and Dehhaghi 2017). These microorganisms are instrumental in promoting soil health and stimulating plant growth, and act as natural antagonists to several plant pathogens (Al-Quwaie 2024). *Streptomyces* species are of particular importance owing to the diverse secondary metabolites they produce, such as antibiotics, antifungals, and plant growth-promoting substances, which are relevant to their ecosystem roles and agricultural uses (Olanrewaju and Babalola 2019; Nazari et al. 2023). The genus *Streptomyces* is one of the largest and best-characterized groups of actinobacteria and currently contains a record of over 500 known species (Gopalakrishnan et al. 2020). As soilborne microorganisms, these actinobacteria have a unique life cycle that includes sporulation, germination, mycelial growth, and the formation of aerial hyphae (Hamedi et al. 2017). In contrast to most bacteria, which reproduce by binary fission, streptomycetes grow in vermiform (branching filamentous) networks like fungi (Flärdh and Buttner 2009). The mycelial pattern of growth enables them to rapidly colonize particles of soil and roots of plants, allowing them to interact with other hosts and microbial communities more easily (Sousa and Olivares 2016). This ecological versatility can mainly be attributed to their secondary metabolism that allows them to outcompete other microorganisms by producing antibiotics, siderophores, and hydrolytic enzymes that degrade organic matter and inhibit plant pathogens (Khan et al. 2023), as summarized in Fig. 5.1.

Streptomyces species are also crucial for natural ecosystems because they degrade complex organic compounds, including cellulose, chitin, and lignin, as a part of nutrient cycling (Javed et al. 2021). They also play a vital role in soil fertility and the carbon cycle due to their ability to decompose organic matter (Jog et al. 2016). In addition, they form beneficial interactions with plants by colonizing either the rhizosphere (the soil region that is the direct vicinity with the roots of the plant) or forming an endophytic association within plant tissues (Vurukonda et al. 2018). In return, it promotes beneficial microbes to improve plant growth, enhance resistance to abiotic stress, and protect plants from pathogens (Pang et al. 2022). Among these, streptomycetes exhibit vital functional roles in an ecosystem, thriving in various activities, including biological control of plant diseases (Wang et al. 2024). As natural biopesticides, many *Streptomyces* spp. produce antibiotics capable of inhibiting the growth of fungal and bacterial phytopathogens (Dave and Ingle 2022). For example, the soil actinobacterium *Streptomyces griseus* produces the antibiotic streptomycin, which is widely used in agriculture to control bacterial infections in large crops (Wang et al. 2024). *Streptomyces hygroscopicus* produces antifungal rapamycin (Cai et al. 2023). Moreover, these microorganisms release hydrolytic enzymes such as chitinases, glucanases, and proteases that decompose the cell walls of pathogenic fungi and consequently lower the incidence of crop diseases (Fróes et al. 2012; Torres-Rodriguez et al. 2022).

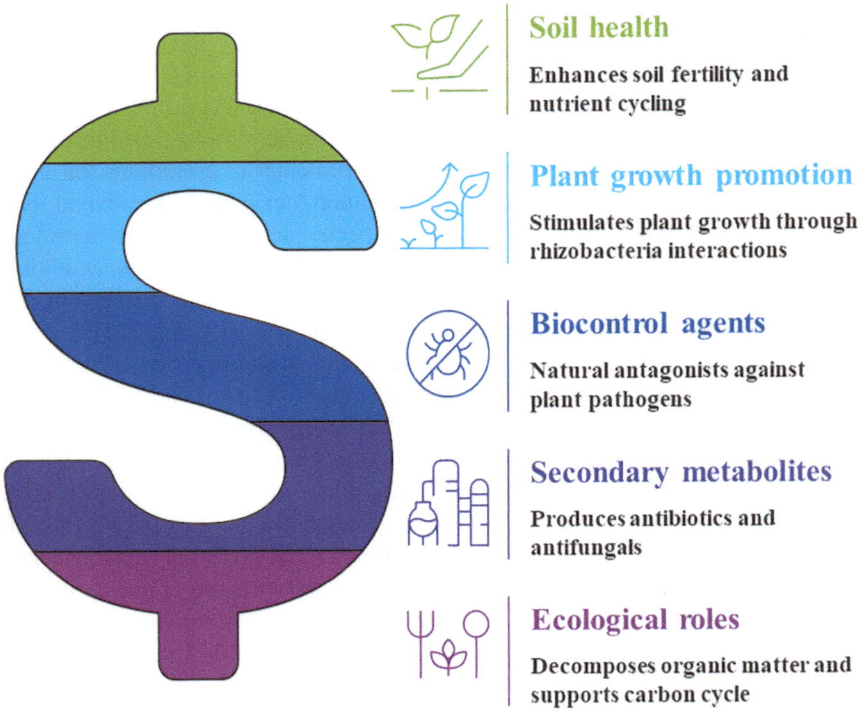

Fig. 5.1 The multifaceted role of streptomycetes

In addition to their antimicrobial activity, streptomycetes also act as plant growth-promoting rhizobacteria (PGPR) due to the production of phytohormones such as auxins, cytokinins, and gibberellins (Solans et al. 2011; Oyedoh et al. 2023b). For more about this, see why it seems to drive root hormones at the root tips (Worsley et al. 2021). For example, the plant pathogen *Streptomyces scabies* has been evaluated for its complex interactions with plants as a beneficial microbe and opportunistic pathogen (Lerat et al. 2009). Yet most streptomycetes are beneficial to plant health, further highlighting their significance for sustainable agriculture (Buzón-Durán et al. 2020). With their immense biotechnological potential, streptomycetes research has been rapidly growing in recent years, especially in the area of biocontrol, plant growth promotion, and soil microbiome engineering (Liu et al. 2024; Boukelloul et al. 2024). Recent progress in genome sequencing and genetic engineering is widening the avenues of using species of the genus *Streptomyces* in agriculture: scientists can enhance the functional features that improve crop productivity (Krysenko 2025). With the rising fears of chemical pesticide usage and ecological sustainability, the ecological functions of streptomycetes could provide routes for sustainable, biological-based solutions to modern critical agricultural issues (Kumar et al. 2022).

5.2 Mechanisms of Biocontrol by Streptomycetes

5.2.1 Antimicrobial Metabolite Production by Streptomycetes

Streptomycetes are well recognized for their biocontrol potential and produce a wide array of antimicrobial secondary metabolites, which are responsible for suppressing the plant pathogenic microorganisms (Dave and Ingle 2022). These Gram-positive, filamentous actinobacteria are the most prominent producers of antibiotics and bioactive compounds in nature, which extrapolated to over two-thirds of all known antibiotics derived from *Streptomyces* species, including streptomycin, tetracycline, and erythromycin (Diab et al. 2024). Through the ingestion of organic matter, they produce antifungal, antibacterial, and even antiviral compounds, offering a cost-effective and environmentally friendly alternative to chemical pesticides that minimize crop losses while supporting sustainable agriculture (Khan et al. 2023). *Streptomyces* also produce polyketides, non-ribosomal peptides, and alkaloids with antimicrobial activity against numerous pathogenic fungi, bacteria, and oomycetes (Wei et al. 2024; Ling et al. 2025), as outlined in Fig. 5.2. Examples are *S. griseus*, which can produce streptomycin, a powerful bactericide with activity against the *Erwinia*, *Pseudomonas*, and *Xanthomonas* species responsible for bacterial wilt and blight in crops (Al-Quwaie 2024). In the same way, antifungal metabolites, including rapamycin, produced by *S. hygroscopicus*, are reported to restrict fungal pathogen growth, with the same pathogens, including *Fusarium*, *Rhizoctonia*, and *Phytophthora*, known for causing root rots and damping-off diseases (Yun et al. 2023). In addition, streptomycetes release lipopeptides, siderophores, and hydrolytic enzymes to enhance their biocontrol efficiency (Chen et al. 2021; Wang et al. 2023a). Lipopeptides damage the membranes of fungal cells, and siderophores serve to chelate iron, restricting its availability to pathogenic microbes and curtailing their growth (Wang et al. 2023a; Wang and Li 2025). Hydrolytic enzymes, including chitinases, glucanases, and proteases, enzymatically disrupt the cell walls of filamentous fungal pathogens, serving as a physical blockade against infection and disease development in plants (Xu et al. 2019). Through direct antimicrobial activity and stimulation of plant immunity, they contribute significantly to soil microbial

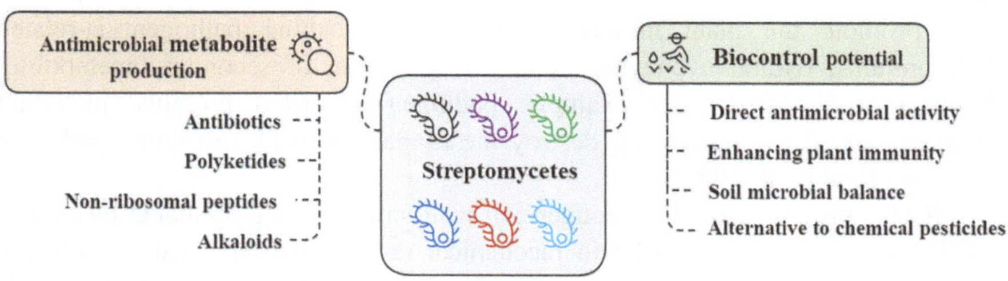

Fig. 5.2 Antimicrobial biocontrol by *Streptomyces*

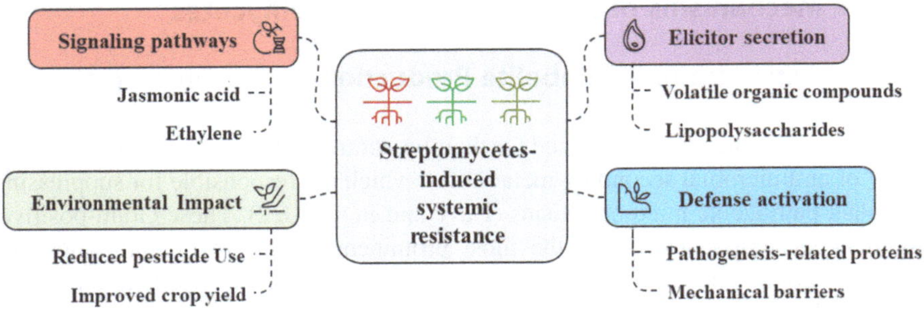

Fig. 5.3 Streptomycetes-induced systemic resistance in plants

balance and disease protection in crops (Nazari et al. 2023). Due to their innate capability to suppress pathogens and boost plant tolerance, they are crucial to biological control programs and a potential substitute for chemical crop protection strategies (LeBlanc 2022).

5.2.2 Streptomycetes-Induced Systemic Resistance in Plants

Streptomycetes are important for inducing systemic resistance (ISR), a plant defense response that generally increases resistance to a wide range of pathogens (Hata et al. 2021). ISR is a type of primed immunity that is stimulated by beneficial microbes such as members of the *Streptomyces* species that induce plant defense responses but do not cause disease (Yu et al. 2022), as categorized in Fig. 5.3. While pathogen-induced systemic acquired resistance (SAR) is invoked through salicylic acid (SA) signaling, ISR is predominantly mediated via jasmonic acid (JA) and ethylene (ET) pathways (Maithani et al. 2021). As signaling molecules, the plants induce those common defense responses to pathogens to react to rapid development (Roychowdhury et al. 2025). During the establishment of the rhizosphere or endophytic colonization, *Streptomyces* species are believed to secrete elicitors such as lipopolysaccharides and volatile organic compounds (VOCs), which interact with plant receptors to trigger ISR (Seth et al. 2023). The subsequent activation of defenses in the plant world is the requesters of defense genes, which in sequential order promote the plant immune system by depositing pathogenesis-related (PR) proteins, oxidative enzymes, and the formation of secondary metabolites (Sudisha et al. 2012). For example, pathogen-associated proteins, including chitinases and β-1,3-glucanases, destroy the fungal cell wall, restricting pathogen infection (Kaur et al. 2023).

ISR also primes plants for stronger cell wall strengthening mediated by callose and lignin deposition, resulting in mechanical defense barriers against pathogen ingress (Senthilraja 2016). Such studies have proven that *Streptomyces* spp. is prone not only to induce ISR but also to have the ability to modulate root exudates to create a beneficial microbiome that exerts further inhibition of pathogen growth

(Ebrahimi-Zarandi et al. 2022). For example, *Streptomyces* spp. have also been shown to elicit ISR in tomato plants in response to *Fusarium* wilt, leading to reduced disease symptoms (Kawicha et al. 2023). Streptomycetes bring a cost-effective solution in the way of ISR for crops, preventing their reliance on chemical pesticides and improving crop yield and productivity with lower impact on the environment (Nazari et al. 2023). This revealing of flowers of *Streptomyces* species in priming immunity is initiated to adopt their role as biocontrol, contributing to environmentally friendly disease management tactics in sustainable agriculture (Shrivastava and Kumar 2018).

5.2.3 Streptomycetes' Competition and Niche Exclusion Strategies

Streptomycetes have proven to be very effective biocontrol agents against plant pathogens, and it appears that exclusion by competition and niche are some of the mechanisms involved with their success as such agents in a soil/plant-associated environment (Johnson et al. 2025). They chiefly inhibit phytopathogens from colonizing and multiplying through rapid colonization, competition for resources, disrupting quorum sensing, and producing antimicrobial compounds (Pacios-Michelena et al. 2021; Khan et al. 2023). Another means by which *Streptomyces* species outcompete pathogenic microbes involves competition for limited resources with pathogenic microbes (Essarioui et al. 2017). They are fecund producers of hydrolytic enzymes (e.g., cellulases, chitinases, and proteases) that facilitate decomposition and effectively render complex organic matter source nutrients more bioavailable and starve the competing pathogenic microbes of needed nutrients (Javed et al. 2021). Iron sequestration through the production of siderophores also helps exclude some niches (Wang and Li 2025). Siderophores with stool-high-affinity binders are secreted by streptomycetes, which have an affinity for iron higher than many pathogens, to sequester iron from other microbes and to suppress the growth of competitors of microbes, mainly Gram-negative bacteria and fungal pathogens (Wang et al. 2014).

Additionally, *Streptomyces* species engage in quorum sensing interference to overthrow the communication network of pathogenic bacteria (Khan et al. 2023). They inhibit pathogens from orchestrating virulence factor production, biofilm development, and host infection by degrading or homologating quorum-sensing molecules (Sayın and Cihangir 2024). This interference also suppresses populations of pathogens, thus impairing their potential to effectively colonize the surfaces of plants or their rhizospheres (Cui et al. 2022). Streptomycetes also use hyperparasitism, a type of direct antagonism by which they colonize pathogenic fungal structures such as hyphae and spores and inhibit their growth and development directly (Sudha et al. 2022). This physical exclusion, combined with their production of antibiotics, volatile organic compounds (VOCs), and antifungal metabolites, prevents pathogens from thriving in such plant environments (Dave and Ingle 2022), as illustrated in Fig. 5.4. Streptomycetes find application in crop protection through the mechanisms of competition and niche exclusion that assist in establishing the balance of

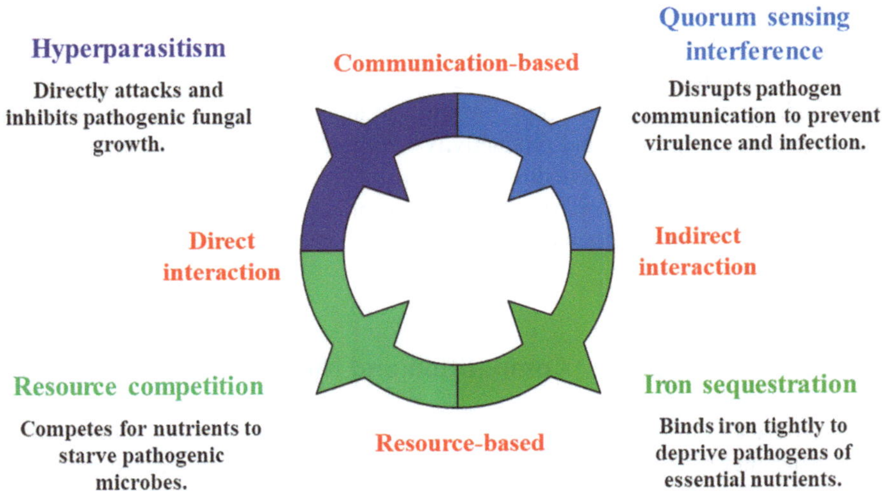

Fig. 5.4 Mechanisms of streptomycetes in biocontrol

microbial community in soil and disease suppression, thereby upholding the prospect of these organisms as eco-sustaining biocontrol agents in sustainable agricultural systems (Shepherdson et al. 2023).

5.3 Streptomycetes' Secondary Metabolites in Disease Suppression

5.3.1 Key Bioactive Compounds Produced by Streptomycetes

Streptomycetes represent one of the major sources of antibiotics and bioactive secondary metabolites and play an important role in the control of disease in plants (Khan et al. 2023). These Gram-positive filamentous actinobacteria generate an extensive diversity of antibacterial, antifungal, and antiparasitic agents, most of which are agriculturally and medically viable (Sousa and Olivares 2016; Abdellatif et al. 2024). *Streptomyces* species yield over two-thirds of all known antibiotics, highlighting their power in biocontrol mechanisms (Al-Quwaie 2024). One of the most famous antibiotics made by Streptomycetes is streptomycin (derived from *S. griseus*) (Pissowotzki et al. 1991). Aminoglycosides, which inhibit protein synthesis in bacteria, have been shown to be highly effective against plant pathogens such as *Erwiniaamylovora*, the cause of fire blight on apples and pears (He et al. 2024b). Tetracycline antibiotics are also key; they are produced by the soil actinobacterium *Streptomyces aureofaciens*, having a ribosome-wide action, stopping bacterial ribosome functions to prevent cellular proliferation (Palečková et al. 2006), as profiled in Fig. 5.5.

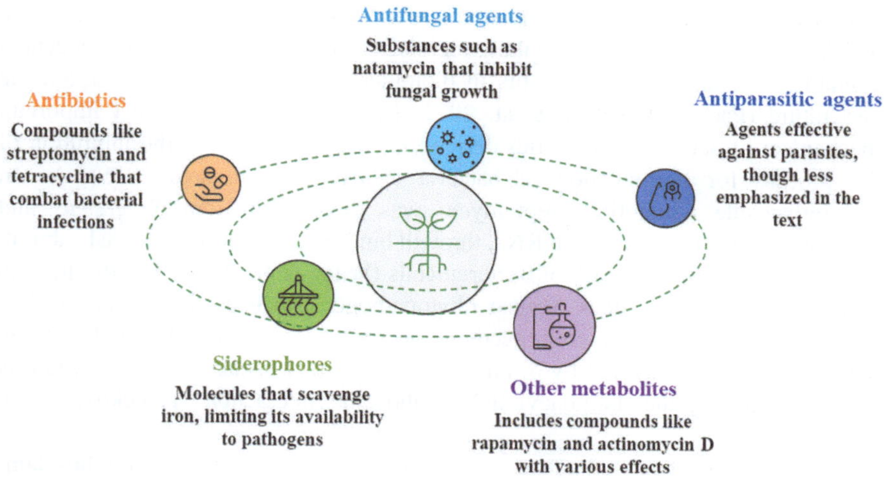

Fig. 5.5 Streptomycetes' role in disease control

Streptomycetes are also potent producers of antifungal compounds, in addition to their array of antibacterial compounds (Wang et al. 2024). Natamycin is a macrolide antifungal metabolite produced by *Streptomyces natalensis* and is extensively used in agriculture and the food industry to control neopterosporic diseases caused by fungi by specifically binding to ergosterol in fungal membranes, which causes exuded leak and cell death (Meena et al. 2021). Polyene antibiotics (nystatin and candicidin) have the same disrupting properties against fungi and, consequently, can suppress the development of pathogenic fungi/species (*Fusarium* and *Botrytis*) (Park et al. 2021). Other important secondary metabolites are rapamycin from *S. hygroscopicus*, which has antifungal and immunosuppressive effects (Ganesh and Subathra Devi 2023), and actinomycin D, which has highly potent antibacterial and anticancer activity (Kulkarni et al. 2017; Qureshi et al. 2021). Moreover, Streptomycetes also secrete siderophores to scavenge iron from the surrounding environment, thus restricting the availability of this essential nutrient to pathogens, thereby inhibiting their growth (Wang et al. 2014; Wang and Li 2025). Since Streptomycetes have the capacity to produce many different types of bioactive compounds, they are an important part of natural disease suppression and are therefore of great value for sustainable, environmentally friendly plant protection and agricultural biocontrol strategies (Vurukonda et al. 2018; Nazari et al. 2023).

5.3.2 Streptomycetes' Mode of Action Against Plant Pathogens

Streptomycetes exhibit broad-spectrum antimicrobial activity and especially produce an array of antibiotics and secondary metabolites that antagonize the pathogens of plants by multiple mechanisms such as cell wall synthesis inhibition, protein synthesis inhibition, DNA replication inhibition, and membrane destabilization

(Al-Quwaie 2024). They are potent antimicrobials capable of targeting both bacterial and fungal pathogens, thereby making members of the *Streptomyces* species genera valuable assistants and contributors in biological control strategies in sustainable agriculture (Pacios-Michelena et al. 2021; LeBlanc 2022). The most important mechanism of action of antibiotics derived from *Streptomyces* is the inhibition of the process for reproduction of bacteria (Sousa and Olivares 2016). As an aminoglycoside antibiotic, streptomycin was discovered from *S. griseus* and prevents correct translation of mRNA through binding of the 30S ribosomal subunit, ultimately inducing the death of bacterial cells (Wang et al. 2024). The mechanism serves as a powerful attack against phytopathogens such as fire blight *Erwinia amylovora*, a devastating crop bacterial necrosis (He et al. 2024b). Likewise, tetracycline that is secreted by *S. aureofaciens* inhibits bacterial protein synthesis through blocking the aminoacyl–tRNA–ribosome interaction (Palečková et al. 2006).

In the case of fungal pathogens, several Streptomycetes-derived antibiotics damage membrane integrity and sterol biosynthesis (Loboda et al. 2024). Derived from *S. natalensis*, natamycin attaches to ergosterol, which is an essential element of the cell membranes of fungi, resulting in increased permeability and subsequent cell death (Wang et al. 2013). Nystatin and candicidin, two other polyene antibiotics, have similar modes of action, targeting pathogenic fungi (e.g., *Fusarium* and *Botrytis*) through disruption of the fungal membrane architecture (Park et al. 2021; Ivanov et al. 2022). A different mode of action is based upon DNA intercalation and the inhibition of replication (Bhattacharjee 2022), as mentioned in Fig. 5.6.

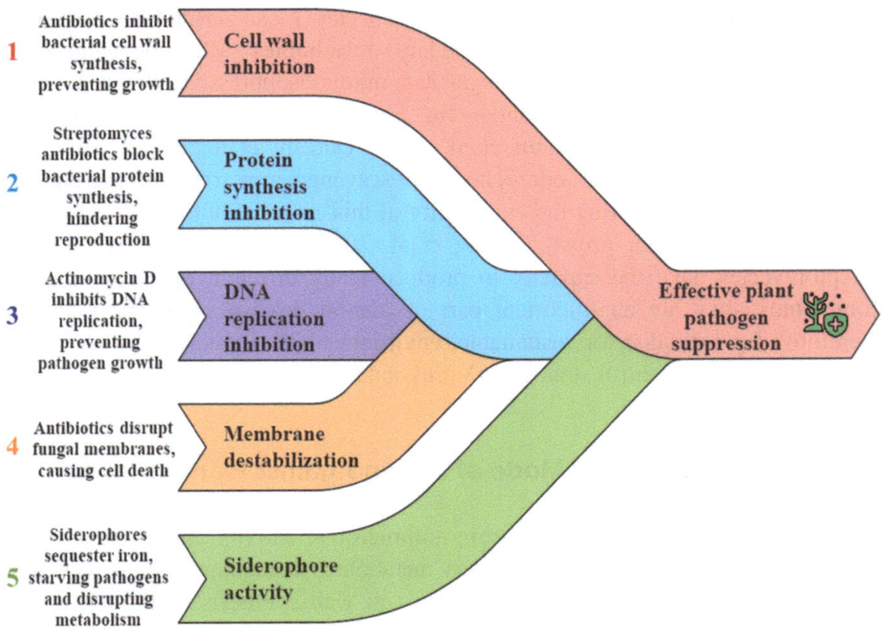

Fig. 5.6 Streptomycetes' multifaceted antimicrobial action

Streptomyces species produce actinomycin D, which attaches to DNA and prevents RNA synthesis and inhibits the growth of both bacteria and fungus (Gao et al. 2023). Siderophores are secreted by streptomycetes, which have a high affinity for binding to iron and sequestering it, starving the pathogens of this important nutrient and, in the process, disrupting their metabolism (Wang and Li 2025). Streptomycetes are becoming more and more appreciated for their roles in disease suppression and the future design of integrated systems for sustainable agriculture, as multiple antimicrobial strategies will provide a natural, green alternative to chemical pesticides and benefit plant protection (Khan et al. 2023).

5.3.3 Streptomycetes' Resistance Management and Sustainability

Streptomycetes are prominent indole-plant-associated microbes and prescribe strong disease-suppressing capability in agriculture due to the extensive utilization of antibiotics and secondary metabolites (Vurukonda et al. 2018). But pathogen resistance is a real concern that requires rational resistance management, as well as sustainable application methods (Pacios-Michelena et al. 2021). In contrast to synthetic pesticides, many of which result in the rapid mutational development of resistance in plant pathogens, streptomycetes-based biocontrol agents act through multiple mechanisms of action, rendering the evolution of resistance by a plant pathogen less likely (Beyari 2024). Actinobacteria produce complex collective products and metabolites that help reduce byproducts and improve the long-term efficacy of plant disease control strategies, which can lead to the development of less resistance by the pathogen (Oyedoh et al. 2023a). Combination treatment, the application of dual or multi-way antibiotics or bioactive metabolites from the *Streptomyces* species, including the new research, is another important way to manage its willful contamination and resistance (Al-Quwaie 2024). The concurrent production of streptomycin and tetracycline by some strains of Streptomycetes targets different bacterial processes, reducing the chances that a pathogen will develop resistance to both (Walsh et al. 2011). Finally, antibiotic-modifying enzymes and quorum-sensing inhibitors secreted by members of the genus *Streptomyces* also impede pathogen adaptation by preventing communication and downstream regulation (Xia et al. 2020), as classified in Fig. 5.7.

The integration of precision applications, improved formulations, and organic farming enhances sustainability in streptomycetes-mediated disease management (Buzón-Durán et al. 2020). Moreover, unlike chemical pesticides that tend to bioaccumulate in the environment, *Streptomyces*-derived antibiotics are biodegradable and, therefore, safer for nontarget organisms and soil microbiota (Shivlata and Satyanarayana 2017). Simultaneously, progress in biotechnological fermentation and genetic engineering processes has resulted in highly efficient streptomycetes that produce greater quantities of metabolites with minimal environmental footprints (Kumar et al. 2022). Farmers can achieve balance in pathogenesis by employing synergistic solutions such as crop rotation, biological soil amendments, and less pesticide by coupling these with streptomycetes-based biocontrol modalities (Beyari

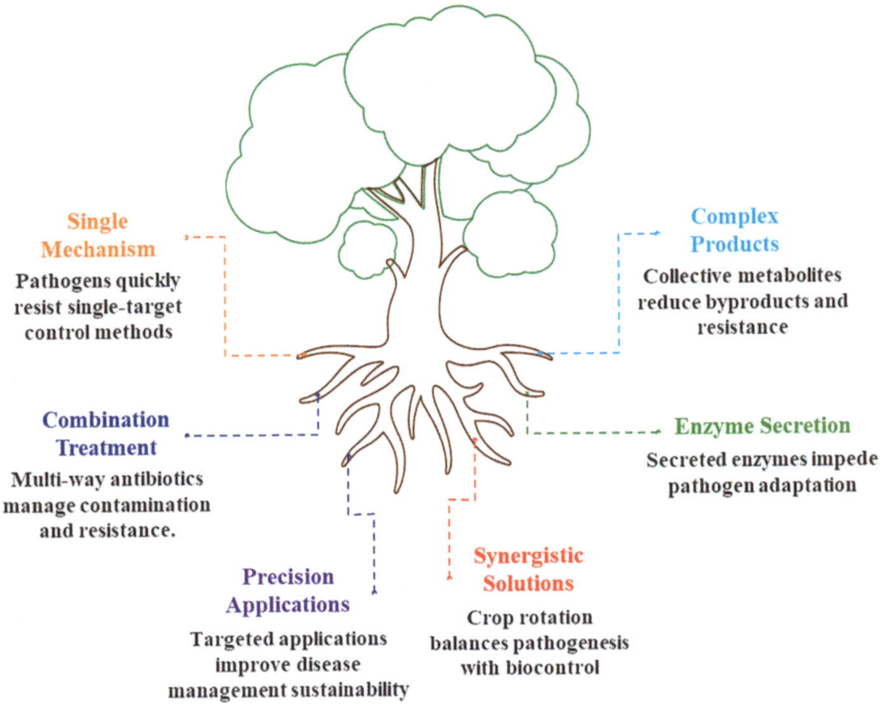

Fig. 5.7 Resistance management in streptomycetes-based disease control

2024). Thus, they secure streptomycete shelves of roots-derived secondary metabolites as effective tools in the extended plant disease mitigation and agroecological sustainability domains (Dave and Ingle 2022). Table 5.1 categorized some antibiotics that were derived from streptomycetes and applied in plant disease suppression.

5.4 Streptomycetes and Their Role in Inducing Systemic Resistance

5.4.1 Signaling Pathways Involved in Plant Defense Activation

Streptomycetes-induced systemic resistance (ISR) is widely recognized as dependent on direct activation of defense signaling pathways against a wide spectrum of pathogens (Hata et al. 2021), as presented in Fig. 5.8. In contrast to systemic acquired resistance (SAR), which is mediated by salicylic acid (SA) and generally induced by pathogen invasion, ISR is induced predominantly via jasmonic acid (JA) and ethylene (ET) signaling pathways (Maithani et al. 2021). These signaling molecules enable plants to implement priming mechanisms by which they can respond more rapidly and vigorously once attacked by a pathogen (Hönig et al.

Table 5.1 Antibiotics derived from streptomycetes that have been applied for plant disease suppression

Antibiotic/commercial products	Active ingredient	Producing *Streptomyces* strain	Targeted plant pathogen	Recommended dose	Mode of action	Target enzyme/protein or gene	References
Actinomycin D	Actinomycin D (0.6 mg/L = 0.48 μmol/L)	*Streptomyces* sp. NEAU-HV9	*Ralstonia solanacearum* (bacterial wilt of tomato)	10^9 CFU/mL of strain or 0.6–1.2 mg/L	Inhibits RNA synthesis; bacteriostatic	DNA-dependent RNA polymerase (general mechanism)	Ling et al. (2020)
Actinovate®	Streptomycin-like metabolites; lydicamycin or streptolydigin-like compound	*S. lydicus* WYEC108	*Pythium*, *Rhizoctonia*, *Fusarium*, *Botrytis*, downy mildew	30 g/100 L or 5 g/m^2	Antibiotic production, ISR	Chitinases, antifungals, and RNA polymerase inhibition	Pacios-Michelena et al. (2021 and Al-Quwaie (2024)
Action Iron®	Iron chelate plus actinomycete mix	*S. lydicus* WYEC108	Soilborne fungi	Soil/foliar application	Iron scavenging, microbial competition	Siderophore-associated pathways	Pacios-Michelena et al. (2021)
Albocycline	Macrolide (MIC: 5–6 μg/mL, LC_{50}: 10.57 μg/mL)	*S.* sp. OR6	*Verticillium dahliae* (olive)	Soil inoculation: ≥10^7 CFU/g or direct 5–6 μg/mL	Inhibits germination and mycelial growth	Possibly prolyl endopeptidase or peptidoglycan biosynthesis	Calvo-Peña et al. (2023)
Antimycin analog	N-formylantimycic acid methyl ester (MIC = 25–50 μg/mL	*S. melanogenes* YBS22	*Magnaporthe oryzae*	200 μg/mL crude extract; 5–50 μg/mL	Antibiosis, spore germination, and germ tube inhibition	Mitochondrial complex III inhibition (as with antimycins); non-dilactone mechanism	Yuan et al. (2019)

(continued)

Table 5.1 (continued)

Antibiotic/commercial products	Active ingredient	Producing *Streptomyces* strain	Targeted plant pathogen	Recommended dose	Mode of action	Target enzyme/protein or gene	References
Azalomycin B	Macrolide (isolated via bioactivity-guided separation, 12 mg yield)	*S.* sp. HAAG3-15 (related to *S. sporoclivatus*)	*Fusarium oxysporum* f. sp. *cucumerinum*	2 mL of 4–5 × 10^6 CFU/mL (greenhouse); extract ~1 mg/mL	Broad-spectrum macrolide; cell membrane interference; growth inhibition	Active site inferred via structure-based data	Cao et al. (2020)
Blasticidin-S	Blasticidin-S	*S. griseochromogenes*	*Pyricularia oryzae*	Foliar spray	Protein synthesis inhibitor	Inhibits peptidyl transferase on 50S ribosomal subunit	Al-Quwaie (2024)
CBQ-EA-2, CBQ-B-8	Secondary metabolites	*S.* sp. CBQ-EA-2,*S.* sp. CBQ-B-8	*Macrophomina phaseolina*, *R. solani*	10^8 CFU/mL (seed coating or foliar spray 1.5 × 10^6 CFU/mL biweekly)	Antibiosis, hydrolytic enzyme action (chitinases, glucanases)	Glucanases, chitinases	Díaz-Díaz et al. (2023)
Conprimycin	Thiopeptide	*S. globisporus* S4-7	*F. oxysporum* (Fusarium wilt of strawberry), *Botrytiscinerea* (gray mold)	Lab-based usage: 10 μL of 10^6 CFU/mL on filter disk	Inhibits fungal cell wall biosynthesis and RNA polymerase activity	Gene: *tsrD* (Cluster 17)	Kim and Kwak (2021)
C27 Wettable Powder	Gougerotin, Natamycin, Nystatin, Anisomycin (e.g., gougerotin)	*Streptomyces noursei* C27 (CGMCC NO.28268)	*B. cinerea*	WP: 50–200 μg/mL gougerotin; fermentation broth up to 40%	RNA/protein synthesis inhibition, membrane damage, volatile suppression	40 gene clusters: *PKS*, *NRPS*, *Gougerotin* (cluster 37), *Natamycin* (cluster 27), etc.	Lu et al. (2024)

Desferrioxamine B, Lydicamycin, Phenazine	Multiple (IAA: 17.54 μg/mL)	*S.* sp. NEAU-S7GS2	*S. sclerotiorum*, *R. solani*, *Exserohilum turcicum*, *Corynespora cassiicola*	10^7 CFU/mL	Cell wall degradation (glucanase, amylase), siderophores, phenazines	β-1,3-Glucanase, *lydicamycin*, *phz* genes, *des* genes	Liu et al. (2019)
Fungichromin	Polyene macrolide (EC_{50}: 3.07 mg/L; yield: 802 mg/L in MACC broth)	*S. padanus* PMS-702	*Rhizoctonia solani*	Soil sanitation and foliar spray: 100 dilutions of SMGC	Disrupts membrane and prevents infection structure formation	Disrupts fungal membranes (ergosterol binding), induces defense (*PR1a*, *PAL*, *POX*)	Yang et al. (2021)
Geosmin (volatile compounds)	Geosmin (45.74%), (+)-epi-bicyclosesquiphellandrene (2.88%), phenylethyl alcohol (1.88%)	*S. platensis* F-1	*R. solani*, *Sclerotinia sclerotiorum*, *B. cinerea*	50–75 g of wheat-seed culture per 20–50 cm desiccator; not applied directly	Volatile-mediated antifungal activity disrupts hyphal growth and infection structure	Mode inferred as membrane interference, metabolic disruption	Wan et al. (2008)
Grisin	Lantipeptide	*S. globisporus* SP6C4	*B. cinerea*	Lab-based usage: 10 μL of 10^6 CFU/mL	Inhibits fungal growth	Gene: *lanM* (Cluster 20)	Kim and Kwak (2021)
Imide substance	Imide compound (2.64 mg/mL in broth; 500–5000× dilution tested)	*S. atratus* PY-1 (CGMCC 7826)	*Phytophthora capsici*	Greenhouse: 10 mL imide solution per plant; 500 dilutions effective	Antibiosis via mycelial deformation and rupture; protein synthesis inhibition	Binds ribosomal 60S subunit	Zang et al. (2023)

(continued)

Table 5.1 (continued)

Antibiotic/commercial products	Active ingredient	Producing *Streptomyces* strain	Targeted plant pathogen	Recommended dose	Mode of action	Target enzyme/protein or gene	References
Isatropolone C	Isatropolone C (acetylated form; active at 25 μL MeOH extract per disk) + IAA (26 μg/mL)	*S.*A1RT	*S. scabies*, *S. turgidiscabies*, *S. stelliscabiei*, *S. europaeiscabiei* (potato common scab)	Soil drench, pot trial (1×10^6 CFU/mL); tested in greenhouse	Antibiosis (Isatropolone C), growth promotion (IAA)	Inferred antibiotic pathway via isatropolone-type polyketide	Sarwar et al. (2018)
JCK-8055 formulations	Aureothricin (24.33 μg/mL), Thiolutin (17.2 μg/mL); WP: 2 g FD broth or 1 g EtOAc extract per 10 g	*S.* sp. JCK-8055 (closely related to *S. luteireticuli*)	*Erwinia amylovora*, *R. solanacearum*	500 dilutions of WP or culture filtrate; 5–20 μg/mL for SYF	Antibiosis (DTP antibiotics); induced systemic resistance (PR1, PR2, PR3 genes)	*PR1*, *PR2*, *PR3*; systemic resistance pathways via SA and JA	Nguyen et al. (2024)
Kasugamycin (Kasumin®)	Kasugamycin	*S. kasugaensis*	*E. amylovora*, *Magnaporthe grisea*, *Xanthomonas* spp.	Foliar spray; 2 L/ha	RNA synthesis inhibition	30S ribosome (blocks mRNA binding)	Pacios-Michelena et al. (2021), Alam et al. (2022) and Al-Quwaie (2024)
Micro108®	Fermentation product	*S. lydicus* WYEC108	Broad-spectrum soil fungi	Soil drench	Enzyme production, competition	–	Pacios-Michelena et al. (2021)

Mycoshield®, Cuprimic	Streptomycin	*S. griseus*	*E. amylovora*, *Xanthomonas*, *Pseudomonas*	Spray (varies)	Protein synthesis inhibition	30S ribosomal subunit	Pacios-Michelena et al. (2021)
Mycostop®	Fermented cells form heptaene polyene (natamycin)	*S. griseoviridis* K61	*F. oxysporum*, *V. dahliae*, *Ceratocystis radicicola*	Varies (foliar/soil) by crop	Competition, antibiosis	Chitinases, glucanases, and ergosterol synthesis inhibition	Pacios-Michelena et al. (2021) and Al-Quwaie (2024)
n-Hexadecanoic acid, 9,12-octadecadienoic acid (Z,Z)-, bis (2-ethylhexyl) phthalate	n-Hexadecanoic acid, 9,12-octadecadienoic acid (Z,Z)-, bis(2-ethylhexyl) phthalate	*S. alfalfae* XN-04	*F. oxysporum* f. sp. *vasinfectum*	10^8 CFU/mL in soil; 50 μg/mL extract in vitro	Chitinase, glucanase, lipase, protease-mediated cell wall degradation; multiple antifungal metabolites; IAA, siderophores, phosphate solubilization	GH18, GH19 chitinases, ACC deaminase, *NRPS*, *PKS-I/III*, siderophore BGCs	Chen et al. (2021)
Niphimycin A and C	Niphimycin A and C (EC_{50}: 2.4–3.9 μg/mL)	*S.* sp. NEAU-ZSY13	*Ralstonia solanacearum*, *Bipolaris sorokiniana*	10^8 CFU/g soil (biofertilizer formulation)	Membrane disruption, antifungal and antibacterial activity	Polyketide-derived macrolide antibiotics	Wang et al. (2023b)

(continued)

Table 5.1 (continued)

Antibiotic/commercial products	Active ingredient	Producing *Streptomyces* strain	Targeted plant pathogen	Recommended dose	Mode of action	Target enzyme/protein or gene	References
Piericidin A	Polyketide (177.33 μg/L in broth)	*S.* sp. FXP04	*Phytophthora infestans* (potato late blight)	Tuber and soil treatment: 10^7 CFU/mL; Piericidin tested at 5 μg/mL	Interferences with NADH: ubiquinone oxidoreductase, disrupts respiration	Piericidin biosynthetic gene cluster (Cluster 1)	Fu et al. (2022)
Polyoxin D	Polyoxin D	*S. cacaoi* var. *asoensis*	*Botrytis*, *Rhizoctonia*, *Alternaria*, *Fusarium*	Foliar fungicide use; 0.5–1.5 g/L	Chitin biosynthesis inhibition	Chitin synthase	Pacios-Michelena et al. (2021) and Alam et al. (2022)
Quorum sensing inhibitors	Piericidin A, Glucopiericidin A (10–100 μM)	*S. xanthocidicus* KPP01532	*Erwinia carotovora* subsp. *atroseptica*	10 μL of 10^4 CFU/mL and 40 μL of compound (10–100 μM); paper disc on tuber slice	Prevents virulent gene expression	Downregulates *pelC*, *pehA*, *celV*, nip (via qRT-PCR)	Kang et al. (2016)
Reveromycin A and B	Reveromycin A (EC_{50}: 0.10–0.88 μg/mL at pH 4.5), Reveromycin B (less active; $EC_{50} > 1.15$ μg/mL)	*S.* sp. 3–10 (closely related to *S. yanglinensis*)	*B. cinerea*, *Mucor hiemails*, *Rhizopus stolonifer*, *S. sclerotiorum*	10–100 μg/mL (pre-inoculation dip); effective comparable to commercial fungicides	Inhibits spore germination and mycelial growth; more effective under acidic pH	Mitochondrial complex inhibition; acid-stable polyether mechanism	Lyu et al. (2017)

Rhizovit®	Fermentation product	*S. rimosus*	*Pythium*, *Phytophthora*, *Rhizoctonia*, *Alternaria*, *Botrytis*	Seed treatment or soil	Antifungal metabolites	Antibiotics (undecylprodigiosin, etc.)	Pacios-Michelena et al. (2021)
Streptomycin-based sprays (Dustret, AS-50, Paushak)	Streptomycin sulfate	*S. griseus*	*Xanthomonas*, *Pseudomonas*, *Erwinia*	100–200 ppm for foliar use	Protein synthesis inhibition	Binds to 16S rRNA in 30S ribosomal subunit	Pacios-Michelena et al. (2021) and Alam et al. (2022)
Streptothricin analogs	Streptothricin E acid, Streptothricin D, 12-carbamoylstreptothricin D; MIC: 0.31–10%	*S.* sp. JCK-6131	*E. amylovora*, *R. solanacearum*, *F. oxysporum*, *B. cinerea*	5-fold to 100-fold dilution of fermentation broth	Dual mechanism: antibiosis (streptothricins) and ISR activation (SA/JA pathways)	*PR1*, *PR3*, *PR5*, *PR12*; activation of SA and JA signaling; cell wall/ membrane damage	Le et al. (2021)
STR-1	4-oxo-4-[(1-phenylethyl) amino]but-2-enoic acid, 1,3,5-trimethylpyrazole, SMA-1	*S. graminearus* STR-1	*Magnaporthe oryzae*, *R. solani*, *Fusarium graminearum*, *Ustilaginoidea virens*, *Bipolaris maydis*	10% fermentation broth (in vitro); 1 mg/ mL for extract application	Inhibits conidial germination, protein synthesis, triggers plant immune response	*OsPR1a*, *OsPR10*, *OsKS4* expression induced; alkaloid/ macrolide activity	Shen et al. (2024)
SZF-179	Polyene-containing glycosylated macrolide (Fermentation broth)	*S. odonnellii* SZF-179	*A. alternata* (pear black spot)	1×10^8 CFU/mL; foliar spray	Disrupts mycelial growth, morphology; antifungal polyene action	Polyene macrolide	Zhang et al. (2023a)

(continued)

Table 5.1 (continued)

Antibiotic/commercial products	Active ingredient	Producing *Streptomyces* strain	Targeted plant pathogen	Recommended dose	Mode of action	Target enzyme/protein or gene	References
Tetranactin, Simeconazole, Wybutoxine, IAA, Phytosulfokine b, Fusicoccin H	Tetranactin, Simeconazole, Wybutoxine, IAA, Phytosulfokine b, Fusicoccin H	*S. albus* CINv1	*Colletotrichum acutatum*	10^6 CFU/mL weekly spray for 7 weeks	Antibiosis, ISR, plant hormone signaling, endophytic colonization	IAA production, chitinase, protease, cellulase, colonization of inter/intracellular spaces	Pupakdeepan et al. (2023)
Verpacamide A, 1-naphthylamine, L-tryptamine, Arglecin, Valindolmycin, N-acetylphenethylamine	Verpacamide A, 1-naphthylamine, L-tryptamine, Arglecin, Valindolmycin, N-acetylphenethylamine	*S.* sp. HU2014	*Rhizoctonia cerealis*, *R. solani*	Spore suspension (1 McFarland std); EF at 5 dilutions (control efficacy: 65.1%)	Antibiosis, membrane disruption, chitinase and protease activity, ISR	Chitinase activity, IAA, siderophore production, *PK* biosynthetic clusters	Zhu et al. (2023)
Volatile metabolites	VOCs: e.g., eugenol, 2,4-di-*tert*-butylphenol, indole, phthalates (concentration via 1:50, 1:100, 1:150 dilutions)	*S. sampsonii* DG1, *S. griseus* DG5, *Streptomyces rochei* DG4	*S. sclerotiorum*, *B. cinerea*, *R. solani*, *Macrophomina phaseolina*	15×10^6 CFU/mL; 1:50 dilution most effective; foliar spray × 4 times	VOC-induced mycelial inhibition, growth promotion via PGRs	Aromatic hydrocarbons, phenylacetic acid, gibberellic acid	Ghanem et al. (2022)
(Z)-13-docosenamide	(Z)-13-docosenamide	*S.* sp. CB-75 (99.93% identity to *S. spectabilis*)	*F. oxysporum* f. sp. *cubense*, *Colletotrichum* spp., *B. cinerea*, *Alternaria tenuissima*	100 mL of 50 diluted broth per pot; MIC = 0.781–50 μg/mL	Broad-spectrum antibiosis, membrane rupture, protein synthesis inhibition, and spore germination inhibition	*PKS-I*, *NRPS* biosynthetic genes; novel antifungal biosynthetic cluster	Chen et al. (2018)

2,3-Butanediol, Oxime derivatives	2.0 mg/mL culture filtrate: 2,3-Butanediol (52.2%), Oxime-methoxy-phenyl (33.6%), Butanoic acid (40.7%)	*S. rochei* DSM 41729	*F. graminearum* (FHB in triticale)	5 mL filtrate per pot weekly (1.9 mg/mL); 10^8 CFU/mL in vitro	VOC stabilization, antibiosis, induced systemic resistance	Acid phosphatase, β-glucosidase, ethylene suppression pathways	Buzón-Durán et al. (2025)
2,4-DTBP	2,4-Di-*tert*-butylphenol (main component)	*S.* sp. UT4A49	*Ralstonia solanacearum*	20 mL pot drench of 1×10^9 CFU/mL culture; 20 μL/disc for extract	Antibiosis via 2,4-DTBP; inhibits protein–ligand targets in silico	D-glucaratedehydratase, mannose-binding lectin (via docking)	Kaari et al. (2022)
5-Hydroxymethyl-2-furancarboxaldehyde	5-Hydroxymethyl-2-furancarboxaldehyde (52.75% of extract); MIC: 1.563–6.25 μg/mL	*S. violaceusniger* JBS5–6	*Fusarium oxysporum* f. sp. *cubense* (TR4), *Curvularia lunata*, *Colletotrichum gloeosporioides*, *B. cinerea*	100 mL diluted broth (50×) to roots of banana seedlings	Cell membrane disruption, inhibition of spore germination and mycelial growth, increased β-glucanase and chitinase	Upregulated chitinase, β-glucanase, *PKS-I* and *NRPS* gene clusters	Jing et al. (2020)
6-Amino-5-nitrosopyrimidine-2,4-diol	Nitrosopyrimidines, betaine, pyrazole/ piperazine derivatives	*S. amritsarensis* V31	*R. solani*, *F. oxysporum*, *Aspergillus flavus*, *A. alternata*, *Sitophilus oryzae*, *S. sclerotiorum*	2% metabolite spray (curative and preventive), diluted 10–1000	Cell wall disruption, multi-target antifungal metabolites	Nitrosopyrimidines, betaine, VOCs	Shahid et al. (2021)

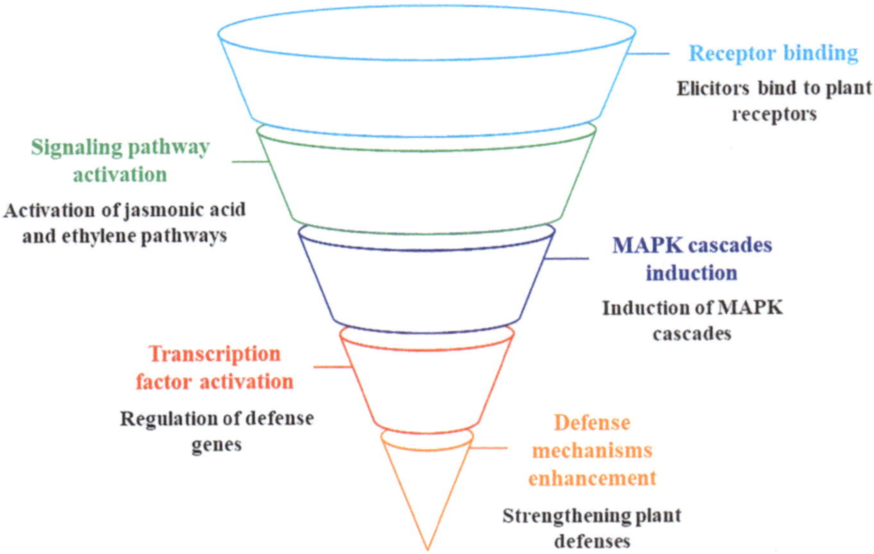

Fig. 5.8 Streptomycetes signaling pathways that induce plant defense mechanisms

2023). After colonizing the rhizosphere or endophytic tissues, *Streptomyces* species exude elicitors such as lipopolysaccharide, volatile organic compounds (VOCs), lipopeptides, and siderophores, which binds to receptors on plants to activate defense signaling cascades (Correa et al. 2022; Kaur et al. 2023). In turn, these elicitors induce MAPK cascades, leading to the activation of transcription factors that regulate the expression of defense-related genes (Gao et al. 2020).

A well-known example of the signaling pathways responsible for ISR activation is the jasmonic acid (JA) pathway, which is thought to be primarily active during systemic immunity and mediates increased expression of pathogenesis-related (PR) proteins, secondary metabolites, and mechanical barriers (e.g., callose and lignin deposition) (Kamle et al. 2020; Roychowdhury et al. 2025). It includes ethylene (ET) signaling careers regulation in plant stress responses involving plant cell wall strengthening, the activation of defense enzymes peroxidase, and polyphenol oxidase activity that reduce the pathogen penetration (Chaudhary et al. 2022). In addition, cross-talk with the salicylic acid (SA) pathway: while ISR is largely JA- and ET-dependent, previous studies also showed that *Streptomyces*-derived ISR could also have an antagonistic effect with systemic acquired resistance (SAR) pathways, resulting in a broader umbrella of defense (Thomas and Singh 2020; Yu et al. 2022). Inducing ISR through these signal transduction pathways contributes to the protection of plants by streptomycetes, leading to long-lasting, systemic resistance of plants, therefore minimizing their dependence on chemical pesticides and promoting sustainable biocontrol in modern agriculture (Qin et al. 2024).

5.4.2 Cross-Talk Between Plant Immune Responses and Streptomyces Activity

Streptomycetes interact with plants in a nonpathogenic manner by inducing systemic resistance (ISR), which arises from complex signaling cross-talk, where streptomycete-activated plant immunity is mediated by the interrelationship between the jasmonic acid (JA), ethylene (ET), and salicylic acid (SA) pathways (Gore and Patil 2025). This complex interaction controls plants immunity to many pathogens and the balance between growth and productivity (Liu et al. 2024). Unlike SAR, which is predominantly triggered by pathogen invasion and the accumulation of salicylic acid (SA), ISR is characterized by the mediation of JA and ET and induces in response to plant roots colonization with beneficial microbes, such as members of *Streptomyces* species (Maithani et al. 2021). After colonization, *Streptomyces* spp. produce elicitors including lipopeptides, siderophores, volatile organic compounds (VOCs), and polyketides that bind to the receptor proteins in the plants to activate downstream immune signaling (Vurukonda et al. 2018; Al-Quwaie 2024). This triggers mitogen-activated protein kinase (MAPK) cascade activation, resulting in the transcriptional induction of pathogenesis-related (PR) genes and antimicrobial compound production (Gao et al. 2020; Pandey et al. 2023). Although JA and ET are the major regulators of ISR, some evidence suggests the involvement of synergistic potential of *Streptomyces*-triggered resistance with SAR through SA-dependent pathways and it is mainly at the presence of combined biotic and abiotic stress (Thomas and Singh 2020).

Such cross-talk between ISR and SAR enables plants to fine-tune their immune responses according to environmental cues (Maithani et al. 2021; Gore and Patil 2025), as clarified in Fig. 5.9. JA/ET pathways are activated in response to necrotrophic pathogens (such as *Botrytiscinerea*), while responses to biotrophic pathogens (such as *Trichoderma longibrachiatum*) depend on SA (Yuan et al. 2019). Streptomycetes therefore optimize plant immunity, mediating basal resistance while preventing trade-offs between growth and defense to promote optimal plant development (Ku et al. 2024). Hence, the ISR mediated by *Streptomyces* offers an eco-friendly option for sustainable agriculture as it helps in promoting the disease resistance of plants by reducing the use of synthetic chemical pesticides by improving the microbiome of the soil (Khan et al. 2023). Table 5.2 summarized *Streptomyces* strains involved in plant defense activation.

5.5 Plant Growth Promotion by Streptomycetes

5.5.1 Phytohormone Production (Auxins, Cytokinins, Gibberellins)

The production of several phytohormones such as auxins, cytokinins, and gibberellins is attributed to one of the unique abilities of streptomycetes for plant growth promotion (Nazari et al. 2023). *Streptomyces* genus of actinobacteria is well known as biologically active compounds that might help by regulating some plant physiology, such as root architecture, shoot growth, seed germination, and plant

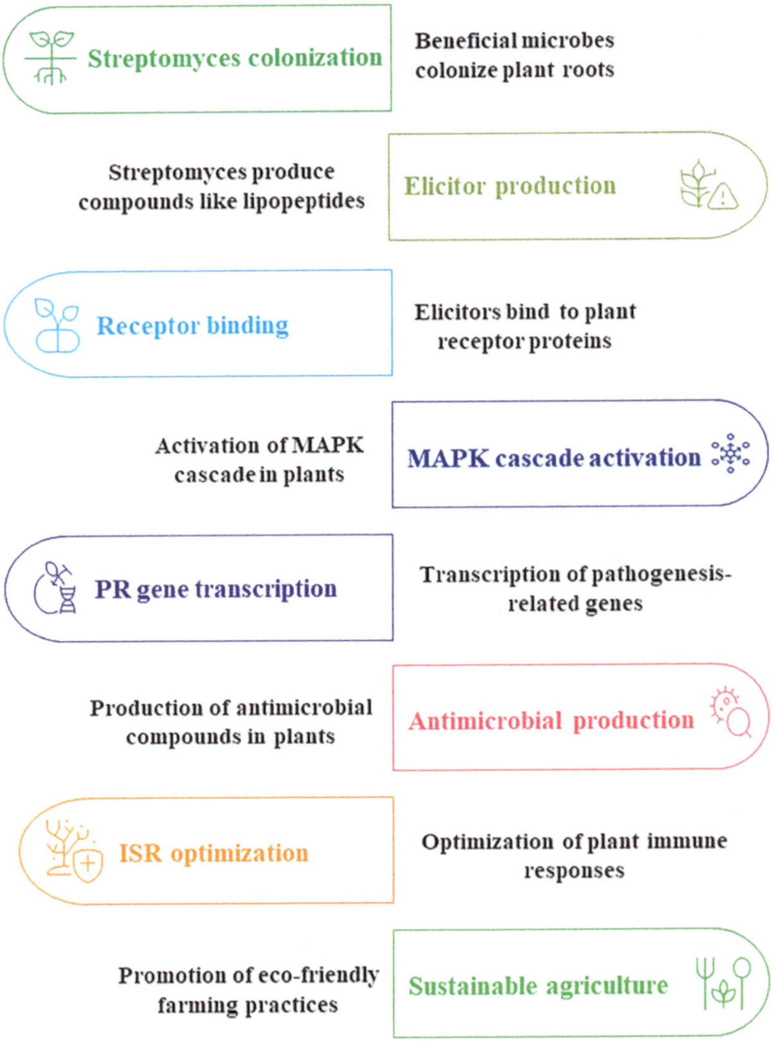

Fig. 5.9 Cross-talk between plant immunity and *Streptomyces* activity

vigor, which can serve as an ecologically friendly strategy for sustainable agriculture (Abdellatif et al. 2024). Auxin, specifically indole-3-acetic acid (IAA), is one of the essential phytohormones produced by this microbial genus (Myo et al. 2019). IAA enhances the elongation of the primary root, formation of lateral roots, and development of adventitious roots that increase water and nutrient uptake (Kaur and Manhas 2022). Streptomycetes, which synthesize IAA via tryptophan-dependent pathways, have been reported to enhance plant root growth through the release of IAA-like compounds, although strains such as *S. scabies* and *S. rochei* are not capable of producing IAA (Legault et al. 2011; Devi et al. 2022).

Table 5.2 *Streptomyces* strains involved in plant defense activation

Streptomyces strain	Habitat	Gene involved in plant immune response	GenBank accession number	Associated plant signaling pathway	References
Streptomyces albulus CK-15	Soil (China)	PR1, NPR1, PR4, ERF1, MYC2, COI1, PAL, POD, CAT, SOD, PPO, Chitinase, β-1,3-Glucanase	CGMCC No. 0703	SA and JA/ET pathways, ROS balance, and PR protein induction	Yang et al. (2023)
S. araujoniae TN11 and *S. araujoniae* TN19	Rhizosphere soil (India)	SOD, POD, CAT	NAIMCC-B-02868, NAIMCC-B-02870	ROS signaling and antioxidant enzyme system	Zeyad et al. (2022)
S. chromofuscus RFS-23	*Nicotiana benthamiana* rhizosphere	NbCIS, NbNCED, SOD, CAT, POD, MDA, CLH1, HEMA1, PORA	EU301837	SA and ABA pathways, ROS detox, ISR, and SAR	Chen et al. (2022)
S. corchorusii CG-G2	*Calotropis gigantea* rhizosphere (China)	Pinocembrin, *cis*-3,4-leucopelargonidin	KQ948396	Flavonoid biosynthesis	Li et al. (2024)
S. globisporus F8	Tomato rhizosphere (Iran)	PAL1, LOXB	MW857095	Phenylpropanoid SA and JA	Ebrahimi-Zarandi et al. (2021)
S. graminearus STR-1	Rice rhizosphere (China)	OsPR1a, OsPR10, OsKS4	PRJNA1047497	SA-dependent and diterpenoid biosynthesis	Shen et al. (2024)
S. griseoaurantiacus XQ-29	Pepper rhizosphere (China)	PR1, ERF1, LOX, PAL, CAT, SOD, POD, PPO	PP423018, CP146837–40	SA and JA pathways, antioxidant enzyme system, and ROS signaling	Qiu et al. (2024)
S. hydrogenans DH-16	*Solanum lycopersicum* rhizosphere	CAT, SOD, APOX, PPO, GuPOX, GST, flavonoids, glutathione, tocopherol	JX123130	ROS scavenging, antioxidant activation, and flavonoid pathway	Sharma et al. (2022)

(continued)

Table 5.2 (continued)

Streptomyces strain	Habitat	Gene involved in plant immune response	GenBank accession number	Associated plant signaling pathway	References
S. malaysiensis 8ZJF-21	Root of *Curculigo capitulata* (China)	MaPR1, MaPAL, Maβ-1,3-Glu, MaMAPK1, MaPPO, MaPOD, MaCAT, MaSOD	JAJQWY000000000, XM009388962, XM009403673, XM018826311, AF001523	SA, MAPK, and antioxidants	Zhang et al. (2022)
S. netropsis IMV Ac-5025	Chestnut soil (Ukraine)	Polyenes, IAA, zeatin, sterols as β-sitosterol, ergosterol, stigmasterol	–	Phytohormone mimicry; indirect induction of plant hormone-like molecules (auxin, cytokinin), possibly modulating plant immune gene expression	Loboda et al. (2024)
S. nigrescens NEAU-L66	Pumpkin rhizosphere (China)	PAL, POD, CAT, SOD	–	ROS signaling and antioxidant enzymes	Gu et al. (2025)
S. pratensis S10	Tomato leaf mold (China)	PR1.1, PR2, PR5, PR3, PR4a, PR9, PAL1, PDF1.2, CAT, SOD, POD, PAL, PPO	CP051486	SA and JA/ET pathways, ROS burst, and callose deposition	Tian et al. (2024)
S. praecox R7	Tomato rhizosphere (Iran)	PAL1, LOXB	MW857093	Phenylpropanoid SA and JA	Ebrahimi-Zarandi et al. (2021)
S. rochei A-1	Apple endophyte (China)	PR8, GLU, CHI, PAL, CAT, SOD, POD	CCTCC M2012008	SA-dependent, ROS signaling, and PR protein induction	Zhang et al. (2016)
S. scabiei EF-35 (pathogen)	Potato tuber pathogen (soil origin)	PAL, C4H, HCT	–	Phenylpropanoid pathway and SA-related phenolic biosynthesis	Isayenka and Beaudoin (2022)

S. setonii WY228	Endophyte of *Arctium lappa* L. (China)	IAA5, GH3, PYL6, ABF4, MYB75, and others involved in auxin, ethylene, ABA, and anthocyanin synthesis	–	Auxin, ethylene, ABA, and anthocyanin synthesis	Qin et al. (2024)
S. tauricus XF	Peony rhizospheric soil (China)	PR1, PR2, PR3, PR4, PR9, PAL (all Pathogenesis-Related genes)	OM541329 (16S rRNA)	SA and JA/ET pathways	Jia et al. (2023)
S. viridosporus HH1	Wheat rhizosphere (Egypt)	JERF3, CHI, GLU, POD, PPO, CAT	–	JA/ET pathway, ROS detox, and PR proteins	El-Sharkawy et al. (2023)
S. sp. DLS2013	Tomato stem (Italy)	PR1, PR3, JAZ25, LOX4, ARG, HCT, plus DEGs in SA, JA, ET, phenylpropanoid, and polyamine pathways	OR364037	SA, JA/ET, polyamine, phenylpropanoid, ROS burst	Cassanelli et al. (2025)
S. sp. JCK-6131	Suppressive soil (Korea)	PR1, PR3, PR5, PR12	MW911616	SA and JA/ET pathways, SAR, and ISR	Le et al. (2021)
S. sp. JCK-8055	*Capsicum annuum* root (Korea)	PR1, PR2, PR3	OP872729.1, gyrB, OP886867.1	SA pathway (SAR)	Villafañe et al. (2025)
S. sp. LH4	*Cucumber* rhizosphere (Korea)	SA, JA, β-1,3-glucanase, cellulase, protease, phenylalanine, leucine, tyrosine	KY684252	JA and SA pathways, amino acid-mediated signaling, and ROS scavenging	Mun et al. (2020)
S. sp. N2	*Oryza sativa* rhizosphere	PAL, PPO, SOD, CAT, POD, β-1,3-glucanase, chitinase	–	SA pathway, ROS detoxification, and PR protein induction	Zhang et al. (2020)
S. sp. N2A	Soybean rhizosphere (Argentina)	PR1, PR3, PR5, PDF1.2, CHS, NPR1, WRKY40, ERF, PAL, LOX7, CAT, ICS	RiPPs: albAI-III	SA and JA/ET pathways, ROS burst, and flavonoid biosynthesis	Villafañe et al. (2025)

(continued)

Table 5.2 (continued)

Streptomyces strain	Habitat	Gene involved in plant immune response	GenBank accession number	Associated plant signaling pathway	References
S. sp. PR69	Cuatro Ciénegas Basin (Mexico)	Ectoine, Melanin, Geosmin, SapB	GCA_026420845	Abiotic stress, osmoprotection, and SAR/ISR	López-Reyes et al. (2024)
S. sp. S-9	*Cajanus cajan* rhizosphere (India)	SOD, CAT, Proline, MDA, β-1,3-glucanase	MK158952	ROS detox, osmotic balance, and PR enzyme induction	Dave and Ingle (2021)
S. sp. SA51	Olive rhizosphere (Italy)	Indirect ISR activation via VOCs: 2-methylisoborneol, geosmin, etc.	–	Indirect SAR/ISR via VOCs; JA/SA cross-talk suspected	Pipponzi et al. (2025)

ABA abscisic acid signaling, *ABF4* abscisic acid-binding protein 4, *APOX* ascorbate peroxidase, *ARG* arginine, *C4H* cinnamate 4-hydroxylase, *CAT* catalase, *CHI* chitinase, *CHS* chalcone synthase, *CLH1* chlorophyllase 1, *COI1* coronatine-insensitive 1, *DEGs* differentially expressed genes, *ET* ethylene, *ERF* ethylene response factor, *ERF1* ethylene response factor 1, *GH3* Gretchen Hagen 3, *GLU* β-1,3-glucanase, *GST* glutathione S-transferases, *GuPOX* guaiacol peroxidase, *HCT* hydroxycinnamoyl-CoA, *HEMA1* hemic acid methyl esterase 1, *IAA* indole-3-acetic acid, *IAA5* indole-3-acetic acid 5, *ICS* isochorismate synthase, *ISR* inducing systemic resistance, *JA* jasmonic acid, *JAZ25* JAZ repressor protein 25, *JERF3* ethylene response factor (ERF) transcription factor, *LOX4* lipoxygenase 4, *LOX7* lipoxygenase 7, *LOXB* lipoxygenase 13 or lipoxygenase-related enzymes, *Maβ-1* specific receptor-like kinase (RLK), *MaCAT* methylated and carboxylated resistance induction associated with catalase, *MaMAPK1* methylated and carboxylated resistance induction associated with mitogen-activated protein kinase, *MaPAL* methylated and carboxylated resistance induction associated with phenylalanine ammonia-lyase, *MAPK* mitogen-activated protein kinase signaling, *MDA* malondialdehyde, *MaPOD* methylated and carboxylated resistance induction associated with peroxidase, *MaPPO* methylated and carboxylated resistance induction associated with polyphenol oxidase, *MaPR1* methylated and carboxylated resistance induction associated with pathogenesis-related protein 1, *MaSOD* methylated and carboxylated resistance induction associated with superoxide dismutase, *MYB75* transcription factor regulating the biosynthesis of anthocyanins and flavonoids, *MYC2* transcription factor regulates gene expression in the jasmonate (JA) signaling pathway, *NbCIS* nicotiana benthamiana cysteine synthase, *NbNCED* 9-*cis*-epoxycarotenoid dioxygenase gene family crucial for abscisic acid (ABA) biosynthesis, *NPR1* non-expressor of pathogenesis-related genes1, *OsKS4* gene in rice (*Oryza sativa*) involved in the biosynthesis of diterpenoid phytoalexins, *(OsPR1a and OsPR10)* pathogenesis-related (PR) protein genes in rice, *Oryza sativa*, *PAL* phenylalanine ammonia-lyase, *PAL1* phenylalanine ammonia-lyase 1, *PDF1.2* plant defensin 1.2 protein involved in plant defense against fungal pathogens, *POD* peroxidase, *PORA* protochlorophyllide oxidoreductase A, *PPO* polyphenol oxidase, *PR* pathogenesis-related, *(PR1, PR2, PR3, PR4, PR5, PR8, PR9, PR12)* pathogenesis-related (protein 1-12), *PR1.1* pathogenesis-related protein 1.1, *PR4a* pathogenesis-related alpha-protein, *PYL6* abscisic acid (ABA) receptor protein, *ROS* reactive oxygen species signaling, *SA* salicylic acid, *SapB* morphogenetic peptide, *SAR* systemic acquired resistance, *S. Streptomyces*, *SOD* superoxide dismutase, *VOCs* volatile organic compounds, *WRKY40* transcription factor regulating the expression of defense-related genes, *3-Glu* β-1,3-glucanase

Another class of growth-promoting phytohormones produced by streptomycetes are cytokinins, which primarily control cell division, shoot development, and delay of leaf senescence (Wu et al. 2021). The necrotic nature of the necrogenic tissue indicates active proliferation, and the hormones involved (cytokinins such as zeatin and isopentenyladenine) promote shoot proliferation, thus increasing biomass and photosynthetic efficiency (Prasad 2022). Reaction and stimulation of plant growth by streptomycetes via the modulation of cytokinin biosynthesis triggered by local nutrient availability (Oyedoh et al. 2023b). Gibberellins (GAs) are responsible for controlling the stem elongation, seed germination, and flowering and are biosynthesized by various species of streptomycetes (Keswani et al. 2022). The stimulating functions of these streptomycetes exist in the GA signal transduction pathways to promote the positive effects on plant height, seed viability, and the degree of stress (Abdellatif et al. 2024). *Streptomyces* spp. has been shown to produce bioactive gibberellins that induce shoot elongation and impact early seed germination in crops, e.g., rice and wheat (Nazari et al. 2023). Streptomycetes produce growth enhancers such as auxins, cytokinins, and gibberellins, which play a vital role in the growth and productivity of the host plants (Omar et al. 2022). Hence, due to the ability to endogenously modulate the levels of phytohormones, they can efficiently serve as an eco-friendly alternative to synthetic growth stimulators and, as such, can help promote sustainable agriculture as well as soil health management (Kumar et al. 2022).

5.5.2 Nutrient Solubilization and Uptake Enhancement

Streptomycetes are significant in plant growth promotion, as they can solubilize fundamental nutrients and boost nutrient uptake that effectively increases the health and yield of plants (Vurukonda et al. 2018). Soil–microbe interaction is one of the key processes that include filamentous soil-dwelling bacteria that interact with plant roots and promote the bioavailability of essential but immobile soil nutrients, such as P, N, K, and Fe (Basit et al. 2021; Chauhan et al. 2023). *Streptomyces* spp. also increases nutrient availability in alkaline or nutrient-poor soils by releasing organic acids and siderophores and secreting ions (Cui et al. 2022). One of the major mechanisms by which streptomycetes improve nutrient uptake is phosphate solubilization (Chouyia et al. 2022). *Streptomyces* spp. (e.g., *S. griseus*, *S. rochei*) as well as other genera have been shown to release organic acids (e.g., citric acid, gluconic acid, oxalic acid) capable of solubilizing insoluble phosphate sources (e.g., rock phosphate, tricalcium phosphate) to forms that can be taken up by plants (Al-Quwaie 2024). Increased phosphate availability helps to better root development and energy metabolism as well as photosynthesis, which enhances the positive growth of plants (Liu 2021).

Streptomyces spp. also has a role in biological nitrogen fixation as it develops symbiotic relationships with other nitrogen-fixing bacteria or through the direct interference of the uptake of nitrate and ammonium by the plant (Amaresan et al. 2018). Some strains encourage potassium and zinc solubilization, in which form those elements are released from mineral complexes and made metabolically available for plant uptake (Chouyia et al. 2022). Another essential component of plant nutrition is represented by the so-called siderophores, high-affinity iron-chelating compounds synthesized by several species of the genus *Streptomyces* that adsorb and transport iron to the roots of plants (Wang and Li 2025). In addition, they can reduce iron competition with plant pathogens and thus are important for disease resistance in plants (Zeng et al. 2018). Thus, streptomycetes play a sustainable agricultural role by reducing the use of inorganic fertilizers, ameliorating soil fertility, and promoting plant adaptation in nutrient-deficient systems via nutrient solubilization and uptake stimulation Abdellatif et al. 2024).

5.5.3 Stress Tolerance Improvement

Streptomycetes are accelerating the plant stress tolerance, and its role is gaining more attention, especially for environmental stressors due to climate change-like drought, salinity, high temperature, and soil desiccation (Liu et al. 2024). These filamentous bacteria support plants adopting to the harsh environment via different mechanisms such as antioxidant enzyme induction, osmolyte production, root architecture reformation, and stress-related gene activation (Abbasi et al. 2020; Wang et al. 2023c). One of the main ways through which *Streptomyces* species indoctrinate stress tolerance is via their improvement of volumetric water content and establishment of drought resistance (Yang et al. 2024). This is done by the production of exopolysaccharides (EPS), which improves soil structure and increases water-holding capacity (Morcillo and Manzanera 2021; Roychowdhury et al. 2025). Moreover, streptomycetes ensure root formation and branching for deeper, wider root systems to obtain water and nutritional units more effectively under dry conditions (Sousa and Olivares 2016).

Streptomycetes help plants under salinity stress by secreting osmoprotectants (e.g., proline, glycine betaine, and trehalose) that maintain cell structure and prevent dehydration (Niu et al. 2022). Moreover, it enhances antioxidant enzymes, such as superoxide dismutase (SOD), catalase (CAT), and peroxidase (POD), which defend plant cells against oxidative stress caused by high salinity and extreme temperatures (Nivedita et al. 2024). However, the severity of the impacts of climate change further complicates heavy metal toxicity and soil degradation, while *Streptomyces* species alleviate these consequences by sequestering toxic metals, regulating phytoremediation, and modulating nutrient cycling (Khanam et al. 2024). Some strains produce siderophores that chelate excess iron and consequently prevent oxidative damage, hence enhancing plant resilience (Wang et al. 2023c). Thus, streptomycetes provide an environmentally friendly biopesticide in climate-resilient agriculture by improving drought tolerance, salinity adaptation, and oxidative stress resistance (Nivedita et al. 2024). Their ability to modulate biotic stress and enhance soil health makes them indispensable in the fight against climate change and assures sustainable crop production and consequently global food security (Phour and Sindhu 2023).

5.6 Interactions of Streptomycetes with Plant Microbiomes

5.6.1 Synergies with Rhizosphere Microbes

The study of synergistic interactions of streptomycetes with rhizosphere microbes is a fundamental factor for plant microbiome assembly (Gates et al. 2023). The rhizosphere is the zone of soil surrounding plant roots where the concentration of soil bacteria, archaea, fungi, and other microorganisms is significantly higher than that in bulk soil, defined as the well-mixed mass of soil beyond the direct influence of roots and their secretion of organic substances (Prashar et al. 2014). The *Streptomyces* microbes enter symbiotic associations with other bacteria, fungi, and plant roots to improve soil fertility and plant resistance (Olanrewaju and Babalola 2019). This reflects one of the important synergies between streptomycetes and nitrogen-fixing bacteria, mainly *Rhizobium*, *Azospirillum*, and *Frankia* (Rashid et al. 2015). Apart from biological nitrogen fixation, streptomycetes accelerate this process by the production of plant growth-promoting metabolites and root architecture modulations (Wei et al. 2024). It prevents overfertilization by more than one organic nutrient and minimizes the use of synthetic at the same time (Kumar et al. 2022).

Streptomyces species interact with arbuscular mycorrhizal fungi (AMF) to facilitate phosphorus bioavailability and drought tolerance in plants due to the presence of nitrogen-fixing bacteria (Jin et al. 2024). Collectively, streptomycetes exert a beneficial role by producing antifungal metabolites that can suppress pathogenic fungi and encourage fungal root colonization of AMF (El-Sharkawy et al. 2023). This mutualistic association enhances the nutrition and stress tolerance of plants. Quorum sensing interactions with beneficial *Pseudomonas* and *Bacillus* species also occur and induce the modulation of root exudates that favors the establishment of a beneficial microbial community by streptomycetes (Ansari et al. 2023; Singh and Singh 2025). They create siderophores, antimicrobial compounds, and volatiles (VOCs) that inhibit pathogens but encourage the proliferation of microbes beneficial to the plant (Al-Quwaie 2024). Streptomycetes contribute to maintaining a balanced and disease-suppressive rhizosphere by inducing positive interactions between the microbial populations (pathogenic and beneficial) and thus promoting plant health and sustainable agriculture (Diab et al. 2024). Their ability to associate with other soil microbes also has led to their use in biocontrol, nutrient acquisition, and mitigation of environmental stress conditions (Sun et al. 2025). The synergistic interactions of streptomycetes in the rhizosphere were concluded, as represented in Fig. 5.10.

5.6.2 Impact on Soil Biodiversity and Ecosystem Stability

Streptomycetes are pivotal in the rhizosphere and bulk soil microbiome dynamics, influencing both the microbial community structure and the soil ecophysiology, and subsequently the soil environments, resulting in soil transformation (Liang et al. 2024). As filamentous Gram-positive bacteria, they play a crucial role in soil health

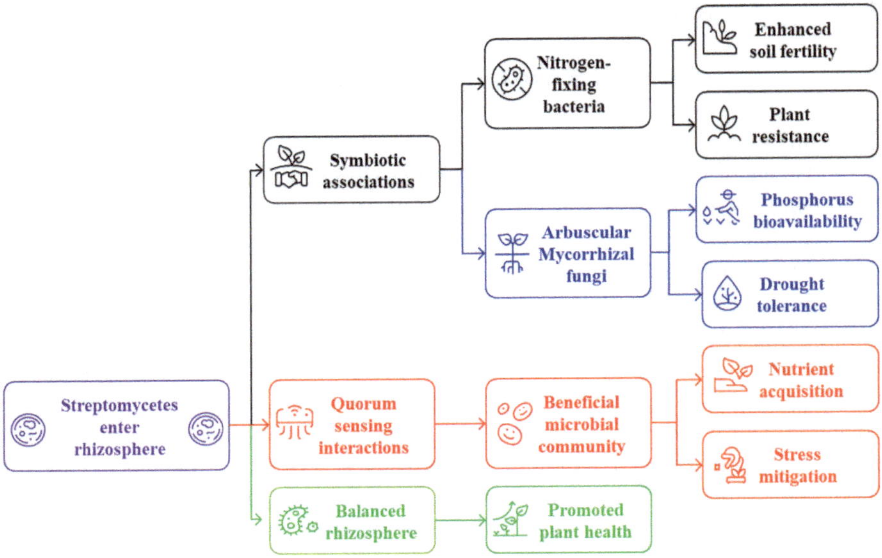

Fig. 5.10 Synergistic interactions of streptomycetes in the rhizosphere

through the decomposition of organic matter, nutrient cycling, pathogen suppression, and plant growth promotion, all of which affect the stability of soil ecosystems (Jog et al. 2016; Sousa and Olivares 2016). Enhancing microbial diversity by maintaining a balanced and competitive microbial population, *Streptomyces* species are known to enhance microbial diversity by generating a balanced and competitive microbial population (Shepherdson et al. 2023). This allows them to produce antibiotics, secondary metabolites, and volatile organic compounds (VOCs) to control the populations of pathogenic bacteria while also promoting the proliferation of health-associated bacteria and fungi (Baloch et al. 2024). Through regulation of deleterious microbes, they also play a role in the formation of a soil suppressive to disease, wherein particular plant pathogens fail to gain dominance (Khan et al. 2023).

Streptomycetes play an important role in soil organic matter decomposition, wherein they degrade complex macromolecules (polymers) such as cell wall polysaccharides (cellulose, hemicellulose, and pectin), lignocellulose, and chitin into smaller components that are then accessible to other microbes and plants (Javed et al. 2021). These are decomposed, thus contributing to carbon and nitrogen cycling transferring into soil, thus leading to long-term fertility and sustainability (Shanthi 2021). They also increase the availability of iron in the rhizosphere by the siderophore production, which is beneficial to other rhizosphere microbes and promotes an overall microbial balance (Cui et al. 2022). In addition, *Streptomyces* species affect plant–microbe interactions, enhancing beneficial symbioses between mycorrhizal fungi, nitrogen-fixing bacteria, and phosphate-solubilizing microbial species (Chouyia et al. 2022). This extensive network of microbes must interact and

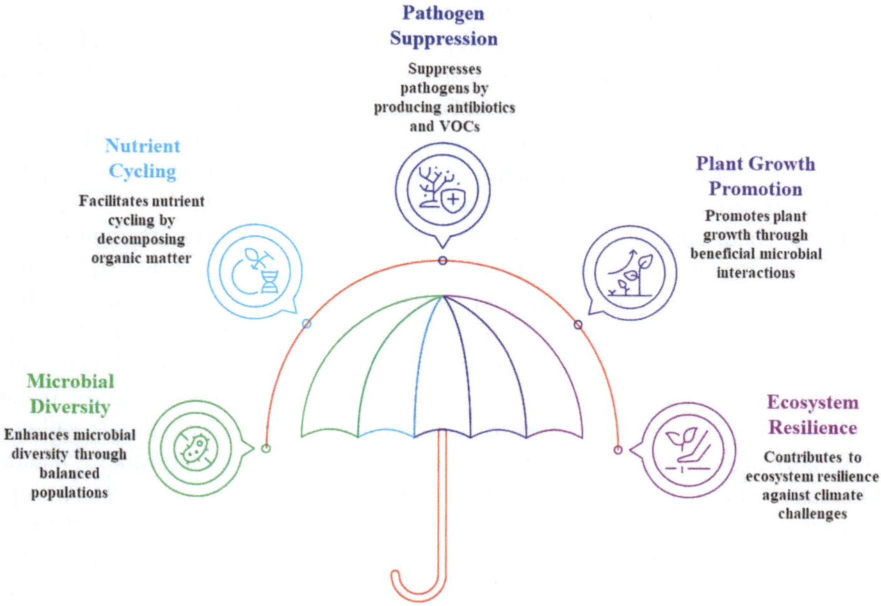

Fig. 5.11 Streptomycetes' role in soil ecosystems

cycle nutrients efficiently, leading to better plant health and ecosystem resilience against challenges posed by climate change, including drought, salinity, and soil erosion (Liu et al. 2024). Streptomycetes act on soil ecosystem stability due to microbial diversity, nutrient cycling, and pathogen suppression, contributing to sustainable agriculture, environmental restoration, and long-term soil health management (Jauri et al. 2016), as shown in Fig. 5.11.

5.7 Streptomycetes in Sustainable Agriculture

5.7.1 Streptomycetes in Organic Farming and Biofertilizers

Streptomycetes play a key role in sustainable agriculture, where they are increasingly being used in organic farming and biofertilizers (Buzón-Durán et al. 2020). As naturally occurring soil actinobacteria enhance nutrient cycling and plant growth, as well as promote disease suppression, they are thus good candidates for sustainable agriculture (Boubekri et al. 2022). As such, they can produce some plant growth-promoting compounds, bioactive metabolites, and other soil-enhancing enzymes, which also reduce dependence on the use of synthetic fertilizers and chemical pesticides, and that is more harmonious with organic farming principles (Kumar et al. 2022). Referring to organic farming, one of the major applications of *Streptomyces* species is in the form of biofertilizers, which are used to increase the soil fertility and availability of nutrients (Chouyia et al. 2022). Different strains of

Streptomyces are phosphate-providing strains; they dissolve insoluble phosphorus compounds and release phosphorus sources for the plants (Soumare et al. 2021). They can also directly promote biological nitrogen fixation or stimulate such activity of nitrogen-fixing in the big gun microbes such as *Rhizobium* and *Azospirillum* to reduce synthetic nitrogen fertilizer dependence (Ansabayeva et al. 2025). Moreover, the solubilization of potassium and iron by these microorganisms promotes plant nutrition in organic cropping systems as well (Kumar et al. 2022).

For beyond nutrient availability, *Streptomyces*-based biofertilizers contribute to soil health by improving microbial diversity and structure (Li et al. 2022). They secrete siderophores that chelate iron and help to maintain a balanced microbial ecosystem and to avoid overgrowth of pathogenic microbes (Wang and Li 2025). In addition, the synthesis of phytohormones (especially auxins, cytokinins, and gibberellins) enhances the growth of roots and increases the resistance of plants either to drought, salinity, or other stresses (Nazari et al. 2023). Even in agroecology, biocontrol agents based on *Streptomyces* are one of the most widely studied natural alternatives to chemical pesticides that crop producers can use to protect crops against soilborne pathogens such as *Fusarium*, *Rhizoctonia*, and *Pythium* (Al-Quwaie 2024). This work allows for prolonged biocontrol of diseases associated with ISR activation as well and does not disrupt biodiversity in the soil (Gore and Patil 2025). Streptomycetes provide a great weapon for organic agriculture by contributing to the adoption of ecological practices that avoid chemicals, sustainably increasing production, and protecting the soil and the environment (Ansabayeva et al. 2025).

5.7.2 Integration with Traditional and Modern Farming Practices

Beginning with their most straightforward morphological aspects, streptomycetes cross the gap between old and new farming by improving soil health, plant resistance, and sustainable crop production (Abdellatif et al. 2024). They are naturally occurring soil bacteria that promote nutrient cycling, biocontrol of pathogens, and growth promotion of plants, and therefore play as important roles in traditional organic farming systems as well as in modern agricultural biotechnology applications (Olanrewaju and Babalola 2019). In conventional agriculture, which uses soil fertility and organic amendments, Streptomycetes are regarded as key microbial agents for compost breakdown, soil structure development, and plant protection (Buzón-Durán et al. 2020). However, *Streptomyces* species contribute to the breakdown of organic matter into bioavailable nutrients that are then recycled and make many organic traditional agricultural systems more sustainable by reducing the need for external fertilizers (Javed et al. 2021). As these beneficial bacteria produce antimicrobials and outcompete soilborne pathogens, the reliance on chemical pesticides is reduced through a process known as natural disease suppression (Khan et al. 2023).

With the innovation, the progress in Streptomycetes-based biotechnological tools such as biofertilizers, biopesticides, and microbial inoculants utilized in modern agriculture (Pathma et al. 2021). Bioproducts based on *Streptomyces* are designed to synergize specifically with new breeding technologies to help elevate the production of crops, in addition to being well-compatible with advances in precision farming technology, controlled-release fertilizers, and microbial consortia applications to enable farmers to overcome yield degradation and improve efficiency without the negative feedback on the environment (Liu et al. 2024; Krysenko 2025). As genome sequencing and genetic engineering become more common, researchers are now isolating and optimizing streptomycetes strains to confer additional benefits such as increased stress tolerance and nitrogen fixation, as well as the ability for induced systemic resistance (ISR), more deeply embedding microbial solutions into the modern paradigm of precision agriculture (Wang et al. 2023c; Abdellatif et al. 2024). Streptomycetes represent an eco-friendly method of sustainable agriculture through the integration of traditional organic practices with advanced microbial technologies aimed at plant protection without chemical pesticides, improved nutrient absorption, and soil sustainability (Pacios-Michelena et al. 2021; Diab et al. 2024). Due to their functionality in different agricultural systems, they can play key roles in the transition to, and upscaling of, regenerative agriculture, climate-smart agriculture, and sustainable food systems (Oyedoh et al. 2023a).

5.8 Formulation and Application Methods for Agricultural Use

5.8.1 Carrier-Based Formulations for Field Application

Formulations based on *Streptomyces* are attracting greater attention in sustainable agriculture as they play a fundamental role in plant growth promotion and pathogenic biocontrol (Khan et al. 2023). These beneficial bacteria offer improved stability, shelf life, formulation, and delivery systems to crops and thus thrive through carrier-based formulation systems for improved viability, efficacy, and application in the field (Pacios-Michelena et al. 2021). Carrier-based formulations increase the viability of *Streptomyces*, promote their interaction with plant roots, and stimulate their biotic activity under varying soils and environmental conditions (Zamoum et al. 2017). Peat-based bioformulations are the most common carrier-based formulations of *Streptomyces* for application (Devi and Manhas 2023). As a good organic carrier of moisture and nutrients, peat helps to maintain *Streptomyces* viability. The approach is commonly used in seed coatings and soil additives to precursor seedling root colonization and growth (Al-Quwaie 2024). In a similar way, biochar-based carriers are successful carriers due to their high porosity, promoting microorganism persistence in the soil and improving soil structure and water retention (Mohamed et al. 2024).

Talc-based formulations are among the most used carriers and are stable, dry powders that are transportable and easy to apply in the field, in seed treatments, and in soil inoculation (Keswani et al. 2016). *Streptomyces* spp. talc formulations have been effectively applied to suppress *Fusarium*, *Rhizoctonia*, and *Phytophthora* infections, providing further resistance to plants (Devi and Manhas 2023). Granular formulations, generally based on bentonite, lignite, or vermiculite, are formulated for

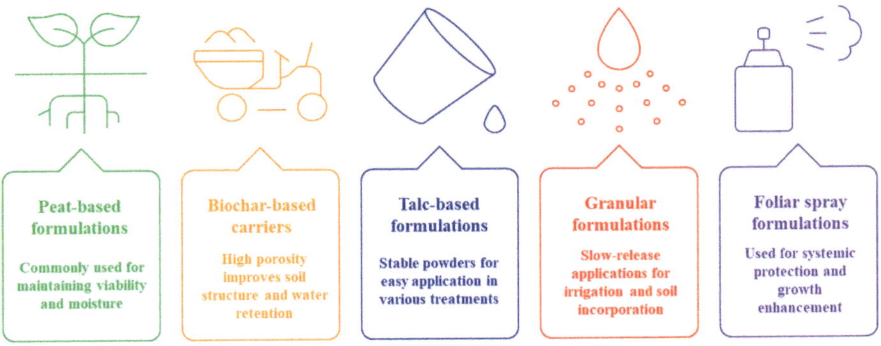

Fig. 5.12 Streptomycetes' carrier-based formulations

slow-release applications in irrigation systems (fertigation) or for direct incorporation into the soil (Narayanasamy 2013). Aside from a wide range of suspension concentrates and emulsions, *Streptomyces* formulations for application as foliar sprays, including drench and drip irrigation for the systemic protection of plants and for plant growth, are widely used (Reddy 2014). Carrier-based formulations of *Streptomyces* bioinoculants enable their use in different cropping systems and the minimization of chemical agrochemicals, providing a bioresource in sustainable and climate-resilient agriculture (Khan et al. 2024b), as introduced in Fig. 5.12.

5.8.2 Delivery Systems and Commercialization Strategies

Streptomyces-derived bioproducts to be commercialized and successfully applied in the field, and highly efficient delivery systems should guarantee microbial viability, convenience, and cost-efficiency (Reddy et al. 2016). Streptomycetes serve as an important bacterium in agriculture, and their overall impact can be boosted to its fullest by providing better formulation stability, a recommended shelf life, storage temperature, and compatibility with already existing practices (Khan et al. 2023). Recently, researchers and industry experts have been stressing these factors to increase streptomycetes' utility in agriculture (Jauri et al. 2016). Seed coating technology is one of the best delivery systems for *Streptomyces*-based formulations where the bacterial inoculants can be applied to seeds through polymeric coatings, bioadhesives, or film coating (Boukaew et al. 2023). The plant growth starts from germination and protects the roots from fungi such as *Fusarium*, *Rhizoctonia*, and so on, by colonizing the roots early (Vurukonda et al. 2018). Soil inoculation methods such as granular formulations, wettable powders, and suspension concentrates offer the possibility of direct application to soil by broadcasting, drip irrigation, or furrow application (Pathma et al. 2021). Granules are usually mixed with bentonite, lignite, or vermiculite to give slow-release action of microbes; these are also well-suited for long-term improvement of soil health (Keswani et al. 2016).

Formulations based on liquids (emulsions, bioactive suspension, etc.) are used as sprays in foliar applications, which help activate the immune system of the plants and inhibit pathogenic agents in the air (Pacios-Michelena et al. 2021). These formulations have ancillary benefits when integrated into integrated pest management (IPM) programs to have a lower dependency on chemical pesticides (Khan et al. 2023). On the other hand, for the commercial aspect, companies are working on methods such as industrial-scale fermentation and using modern drying techniques (i.e., freeze-drying, spray-drying), encapsulation technologies for improved shelf life, and ease of distribution (Reddy et al. 2016). Key approaches for advancing the global *Streptomyces*-based biofertilizer and biopesticide market include regulatory approvals and farmer training programs as well as public–private collaborations (Kumar et al. 2022), as charted in Fig. 5.13. Through the advancements in delivery systems and commercialization strategies, formulations based on *Streptomyces* are becoming increasingly important for sustainable agriculture, providing environmentally sound and effective alternatives to chemical-based inputs in farming (Wang et al. 2024).

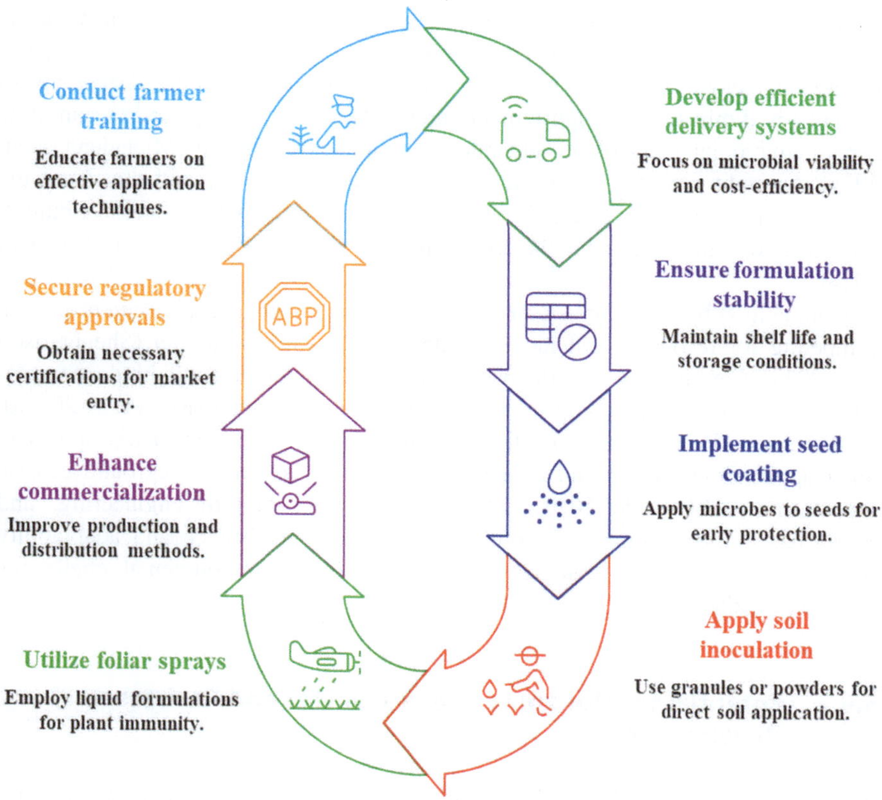

Fig. 5.13 *Streptomyces* commercialization cycle

5.9 Limitations and Challenges in *Streptomyces*-Based Biocontrol

5.9.1 Environmental Factors Influencing Efficacy

Although *Streptomyces*-based biocontrol as an eco-friendly alternative for chemical pesticides is promising, environmental factors can significantly affect their efficacy, thus confounding their field application (Nazari et al. 2023). The ability of *Streptomyces* strains to survive, colonize, and exert biocontrol activity in variable soil conditions can be influenced by their ability to cope with environmental pressures, such as composition of soil, temperature, moisture content, microbial competition, UV exposure, and thus limit the efficiency of *Streptomyces* strains in variable agricultural conditions (Shepherdson et al. 2023). Among these challenges, soil variability stands out, with distinct soil types affecting *Streptomyces* spore germination, root colonization, and bioactive compound production (Buzón-Durán et al. 2025). Soils high in organic matter tend to improve streptomycetes activity and compacted or low-nutrition soils may limit microbial movement and effectiveness (Javed et al. 2021). Similar to shifts in their pH, antibiotic biosynthesis is also stimulated by pH, as some strains of *Streptomyces* do not generate suitable bioactivity levels under unfavorable pH conditions (Zong et al. 2022). Other crucial environmental factors include temperature and moisture. For example, *Streptomyces* viability is diminished at high temperatures, whereas low temperatures can slow general microbial metabolism and rates of colonization (Pacios-Michelena et al. 2021). Drought conditions may also reduce water availability and thus limit the proliferation of streptomycetes, disallowing their adaptation, while excessive rainfall can wash away the inoculants, thereby decreasing their persistence in soil (Liu et al. 2024).

Competition by opportunistic microbes is another challenge, as *Streptomyces* are competing for resources and niches against native soil microbiota (Shepherdson et al. 2023). Certain fungi and rhizosphere-associated bacteria could hinder *Streptomyces* colonization or outcompete them in nutrient-poor habitats (Khan et al. 2024a). Furthermore, UV exposure can diminish the viability of *Streptomyces* spores in foliar applications, which can reduce their effectiveness for aboveground disease suppression (Al-Quwaie 2024). Microencapsulation, biofilm engineering, and carrier-based formulations to improve *Streptomyces* persistence and adaptability are important different strategies to overcome these environmental challenges (Guha et al. 2025).

5.9.2 Potential for Horizontal Gene Transfer and Resistance Development

Although *Streptomyces*-based biocontrol is an environment-friendly option compared to synthetic pesticides, the key concern remains horizontal gene transfer (HGT) and the development of resistance in the plant pathogens, and the residues may persist in other environmental microbial communities (LeBlanc 2022). The ability of *Streptomyces* species to separately express transiently acquired genes may contribute to the risk of unintended consequences in the soil microbiome, especially given their genetic plasticity and the ability to transfer genes via conjugation, transformation, and transduction (Krysenko 2025). Horizontal gene transfer is a major risk due to its ability to allow the dissemination of antibiotic resistance genes (ARGs) to pathogenic bacteria and other soil organisms (Ullah et al. 2025). Although *Streptomyces* strains produce antibiotics as a biocontrol, long-term exposure to these bioactive molecules may trigger pathogen adaptation and selection of resistant mutants (Ebrahimi-Zarandi et al. 2022). Moreover, some of the rhizosphere-associated bacteria can obtain resistance genes from the exchange of plasmids, which decreases the efficiency of *Streptomyces*-based biopesticides over time (Pathma et al. 2021).

Moreover, the indiscriminate application of *Streptomyces* inoculants in agriculture may unintentionally select for resistant microbial populations, potentially diminishing the long-term efficacy of biocontrol strategies (Shanthi 2021). Resistance development can be accelerated under suboptimal application conditions, such as low inoculant concentrations or inconsistent field applications, which allow pathogens to develop adaptive countermeasures (Wang et al. 2023c). Mitigation strategies, including the rotation of biocontrol agents with *Streptomyces* species and non-antibiotic-based biopesticides and precision application to minimize resistance risk, must be implemented to sustainably manage the risk associated with the use of *Streptomyces* species as biocontrol agents (Nazari et al. 2023). Genomic monitoring and microbial surveillance can also be used to identify the emergence of resistance genes and help researchers to create next-generation biocontrol formulations that can sustain suppression of pathogens without undue HGT risks (Krysenko 2025). Hence, by combating HGT concerns and resistance developments, *Streptomyces*-based biocontrol will continue to be a sustainable and long-lasting strategy for eco-friendly disease management with limited or no unwanted ecological side effects (Khan et al. 2023).

5.10 Advances in Genomics and Biotechnology of Streptomycetes

5.10.1 Genetic Engineering for Enhanced Biocontrol and Growth Promotion

Due to these distinctive biotechnological strategies utilized through these creatures, they have produced certain herbicides that gain high yields and are the most powerful, such as the horticulture herbicide (common for the streptomycetes strain), as well as plants grown through biocontrol and general plant promotion through new genomics, genetic control, and biotechnology methods used particularly by streptomycetes (Olanrewaju and Babalola 2019; He et al. 2024a; López-Reyes et al. 2024). For instance, researchers may find that CRISPR-Cas9, synthetic biology, metabolic engineering, and other advanced molecular techniques will enhance antibiotic production, biosynthesize secondary metabolite, and stress resistance in streptomycetes to improve their reliability and efficiency in agriculture (De Lima Júnior et al. 2023; Lee et al. 2024). Optimization of secondary metabolite pathways is a key strategy in genetic engineering (Alam et al. 2022). *Streptomyces* species produce a wide range of antimicrobial compounds that inhibit plant pathogens, but their biosynthesis is frequently governed by intricate genetic networks (Al-Quwaie 2024). Using genome editing and pathway refactoring can increase the yield of bioactive compounds, enhance antifungal and antibacterial efficacy, and expand the spectrum of biocontrol activity (Krysenko 2025). For instance, the antibiotic produced by the *S. griseus* strain was redesigned for streptomycin hyperfunction that also enhances the efficacy of fire blight and bacterial wilt (Wang et al. 2024).

Moreover, using genetic tools, plant growth-promoting traits in streptomycetes can be enhanced (Omar et al. 2022), as schematized in Fig. 5.14. Improving the net balance of these phytohormones, either through the introduction or upregulation of endogenous biosynthesis genes (e.g., for indole-3-acetic acid (IAA), cytokinins, and gibberellins) or through engineered strains, substantially increases crops' root development, nutrient uptake, and stress resistance (Pang et al. 2022; Nazari et al. 2023). Researchers have also altered nitrogen fixation and phosphate solubilization pathways to enhance soil fertility and to minimize the use of chemical fertilizers (Chouyia et al. 2022). Multi-omics approaches (genome sequencing, transcriptomics, and proteomics) can be utilized to accurately pinpoint regulatory elements and specific biosynthetic gene clusters associated with biocontrol and plant growth promotion (Chen et al. 2021; Zhang et al. 2024). Such findings enable the rational design of better *Streptomyces* strains with accurate stability, efficacy, and enzyme environmental safety (Das and Maity 2025). As there will be a growing demand to optimize streptomycetes via gene manipulation and biotechnology-based approaches to address next-generation sustainable agriculture, such developments will provide environmentally benign as well as highly effective alternatives to conventional pesticides as well as fertilizers (Beyari 2024).

5.10.2 Omics Approaches for Strain Improvement and Selection

Emerging omics technologies (genomics, transcriptomics, proteomics, and metabolomics) have considerably improved the potential for *Streptomyces* strain improvement and selection for biocontrol and plant growth-promoting traits (Sarethy and Saharan 2021; Das and Maity 2025). By enabling researchers to deconvolute complex biosynthetic pathways, select for desirable traits, and generate elite microbial strains, these state-of-the-art methods are paving the way for sustainable agriculture (Oyedoh et al. 2023a; Abdellatif et al. 2024). Biosynthetic gene clusters (BGCs) encoding the biosynthesis of antibiotics, secondary metabolites, and plant growth-promoting traits have been identified and characterized through genomics (Fatima et al. 2024). The successful sequencing of the entire genomes of

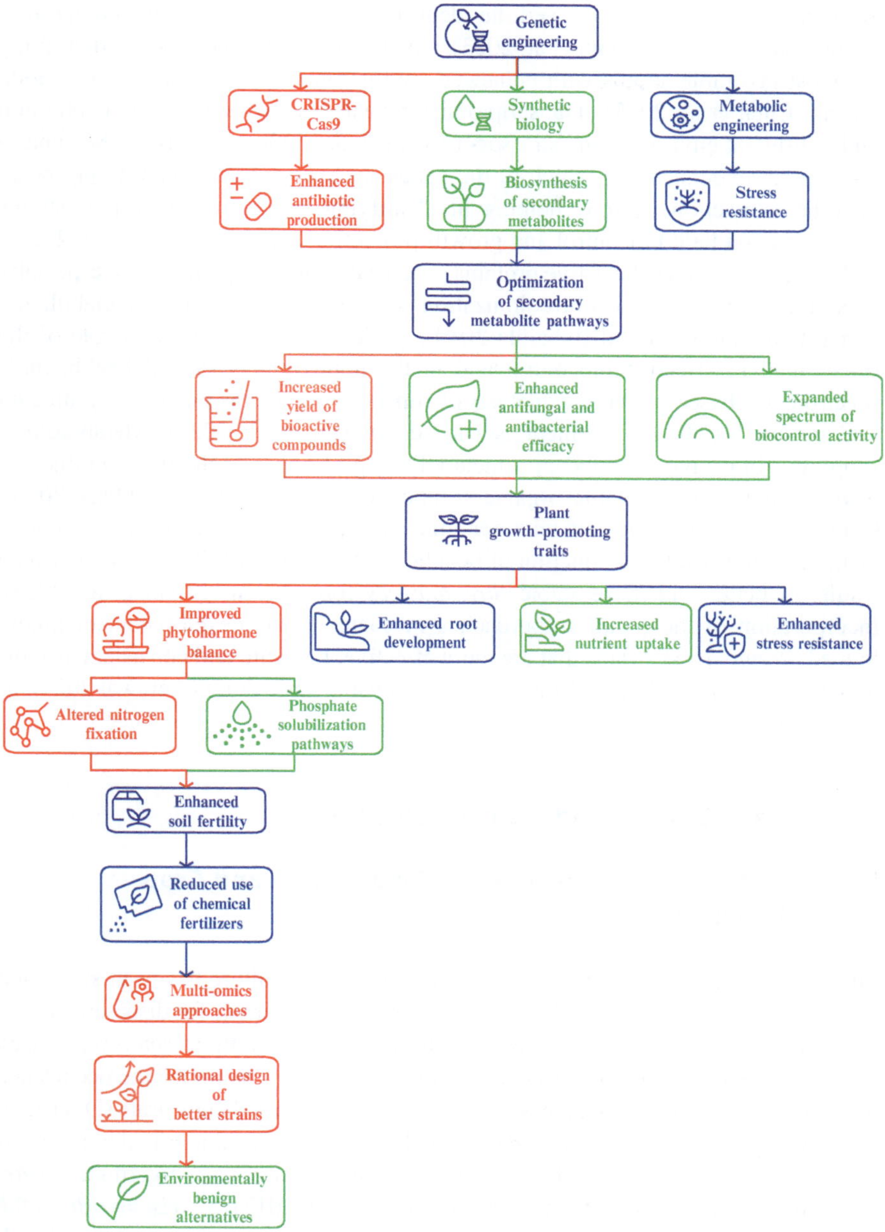

Fig. 5.14 Streptomycetes' genetic engineering for enhanced biocontrol and growth promotion

Streptomyces species has allowed the identification of novel bioactive compounds for biocontrol purposes (Zhu et al. 2023). Comparative genomics allows identifying the most favorable strains with enhanced antimicrobial and nutrient solubilization effects (Chen et al. 2025). Transcriptomics revealed the dynamic gene expression under different environmental stressors (Zhang et al. 2023b). Such RNA-Seq analysis helps in understanding which of the genes are expressed/signaled during host–microbe interaction, thereby helping in identifying suitable *Streptomyces* strains with improved field biocontrol and growth-promotion activities (Chen et al. 2021).

Proteomics studies the active proteins produced by the streptomycetes, especially those that play a role in root colonization, pathogen interaction, or metabolizing nutrients (Olanrewaju and Babalola 2019). To identify the own major role of the supernatant in effective biocontrol and in the improvement of soil health, mass spectrometry-based proteomic analysis of supernatant proteins playing a significant role in soil health improvement is carried out (Fu et al. 2022). Metabolomics supports bioactive secondary metabolite profiling, including antibiotics, phytohormones, and volatile organic compounds (VOCs) (Das and Maity 2025). Combining metabolomic data with genomic and transcriptomic analyses is a potent method to improve the production of beneficial metabolites of *Streptomyces*, which results in better and more stable *Streptomyces* formulations (Zhao et al. 2023). Incorporation of these omics approaches translates to the development of highly refined *Streptomyces* strains that are more effective, resilient, and consistent as tools in biological control, biofertilization, and sustainable agriculture (Agrawal 2024), as drafted in Fig. 5.15.

5.11 Case Studies and Field Applications

5.11.1 Success Stories in Disease Management and Crop Enhancement

The important role of streptomycetes in sustainable agriculture has been supported by a number of successful field applications, including their use for disease management and/or crop enhancement (Abdellatif et al. 2024). Both *Streptomyces*-based biocontrol agents and biofertilizers promote healthy crops, minimize the occurrence of a pathogen in crops, and increase yield; these case studies demonstrate (Devi et al. 2022; Wang et al. 2023b). An attractive and well-studied example is the ability of *Streptomyces griseorubens* E44G *Streptomyces griseoviridis* and to reduce *Fusarium* wilt in tomatoes (Minuto et al. 2006; Rashad et al. 2017). *S. griseoviridis* is the source of a commercial product, Mycostop, which has been widely used in North America and Europe for the suppression of soilborne fungal pathogens (e.g., *Fusarium oxysporum*, *Rhizoctonia solani*, and *Pythium* spp.) (Hansen et al. 2010). Mycostop applications resulted in 60–70% less disease in tomato plants compared with untreated controls, and plants receiving Mycostop also produced greater yields and healthier plants (Tsolakidou et al. 2024).

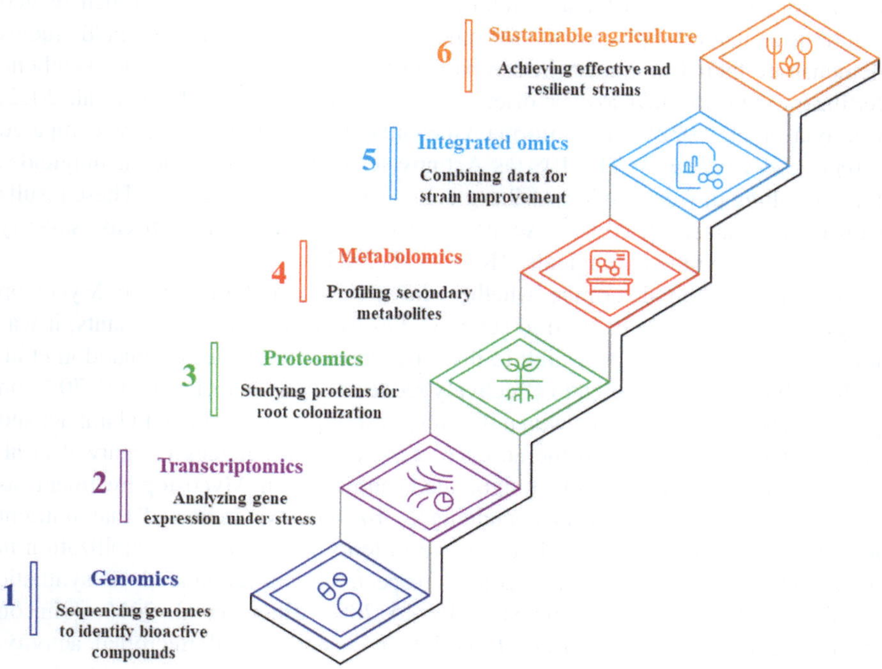

Fig. 5.15 Advancing *Streptomyces* for sustainable agriculture

An additional case study highlights the success of using Actinovate, a formulated product of *Streptomyces lydicus* WYEC108, in which the biocontrol agent was used to control powdery mildew and root rot diseases on grapevines, lettuce, and cucurbits (LeBlanc 2022; Khan et al. 2023). Moreover, *S. rochei* has also shown an increase in phosphate solubilization, alongside promoting root development in wheat and maize (Wei et al. 2024; Johnson et al. 2025). Field trials in India indicated that biomass formulations containing *S. rochei* improved wheat yield by 15–20% while decreasing reliance on chemical fertilizers (Kumar et al. 2022). These success stories emphasize the promise of streptomycetes-based solutions to provide potent disease control, increased resilience of our crops, and greener substitutes for synthetic pesticides and fertilizers in contemporary agriculture (Abdellatif et al. 2024). Increasingly important for sustainable and organic farming systems globally, they boast many field-tested applications (Oyedoh et al. 2023a).

5.11.2 Comparative Analysis with Chemical Pesticides and Synthetic Fertilizers

Due to their eco-friendly nature and effectiveness in disease suppression and plant growth promotion, streptomycetes-based bioproducts have been emerging and widely used in agriculture (Vurukonda et al. 2018; Oyedoh et al. 2023a). Healthy crops, combined with sustainable solutions, position *Streptomyces* and their related solutions as natural replacements for other products, and comparative field studies demonstrate their low impact in comparison to chemical pesticides and synthetic fertilizers with no yield loss or disease control differences (Boubekri et al. 2022; Chouyia et al. 2022). In California vineyards, for example, a study compared *Streptomyces lydicus* WYEC108 (as Actinovate) with the conventional fungicides for controlling powdery mildew (Zhang et al. 2011; Khan et al. 2023). These results showed that plants treated with Actinovate had a 65% reduction in disease severity and 75% with synthetic fungicides (Kaur et al. 2023).

In trials held to determine whether the commercial biofungicide Mycostop (*S. griseoviridis*) could be used to suppress *Fusarium* wilt in tomato plants, it was found that its efficacy was similar to that of chemical fungicides (Tsolakidou et al. 2024). Mycostop was shown to effectively reduce disease incidence by 60–70%, on par with chemical treatments, with a structure ensuring no environmental impact and no residual pollution due to the absence of synthetic fungicides (Aggarwal et al. 2016). Farmers noted better plant vigor and healthier soil in Mycostop treatments as well. Studies on biofertilizers containing *S. rochei* have indicated that nutrient management could be effectively enhanced in terms of phosphate solubilization in maize and wheat in India and Brazil, respectively, as compared to synthetic fertilizers, where yield was increased by 15–20% (Zhang et al. 2016; Johnson et al. 2025). Since *Streptomyces formulations* sustained soil microbial activity, they could possibly improve soil fertility in the long term, unlike chemical fertilizers (Kumar et al. 2022). Though streptomycetes-based bioproducts can sometimes perform slightly below or equal to synthetic agrochemicals in disease suppression, they provide many other benefits that favor their use, such as improving soil health, decreasing chemical residues, and longer sustainability, making streptomycetes-based bioproducts precious tools for modern sustainable agriculture (Buzón-Durán et al. 2020; Wang et al. 2024).

5.12 Future Perspectives and Research Directions

5.12.1 Prospects for Novel Strain Discovery

The prospect of streptomycetes application as a biocontrol agent and plant growth-promoting bacteria (PGPB) is a matter of the discovery of novel strains with improved biocontrol efficacy, stress tolerance, and plant growth promotion properties (Nazari et al. 2023). As climate change as well as soil degradation and recalcitrance of pathogens to synthetic pesticides become more menacing threats to the global agriculture system, the natural world is continuing to supply unexplored *Streptomyces* diversity that may open routes toward sustainable farming practices (Khan et al. 2023; Diab et al. 2024). This specific area of research is concerned with discovering new streptomycetes species in extreme environments (such as desert soils, deep-sea sediments, and Antarctic soils), exhibiting novel secondary metabolites and stress-resistance capabilities (Trenozhnikova et al. 2024). New antibiotics, antifungal agents, and bioactive compounds that were not previously present in existing biocontrol agents may be found in these strains (Johnson et al. 2025).

Since bacteria cannot be cultured, and *Streptomyces* strains are too rare to be collected in sufficient quantities to carry out complex investigations, breakthroughs in metagenomics and high-throughput sequencing have transformed the expedition of natural product (NP) biosynthetic gene clusters (BGCs) from soil-dwelling microorganisms (Al-Quwaie 2024; Schniete and Fernández-Martínez 2024). This approach, trained in genome mining and synthetic biology, uncovers silent/cryptic gene clusters to discover novel antimicrobial metabolites that may be exploited for their disease-suppressive potential (Lee et al. 2021). Strain improvement with genetic engineering and adaptive evolution is also an area of emerging research (De Lima Júnior et al. 2023). These techniques include CRISPR-Cas9 genome editing, metabolic engineering, and multi-omics approaches to engineer *Streptomyces* strains for improved antibiotic production, phytohormone biosynthesis, and stress tolerance (Agrawal 2024). Moreover, the bioprospecting of plant endophytic microbiomes has great potential for isolated strains of *Streptomyces* with better abilities for root colonization, which can increase the performance of this bacteria when applied in fields (Bhakat et al. 2025). A combined approach of genomics, synthetic biology, and microbial ecology has the potential to rewrite the future of streptomycetes research and will lead to the rational design of next-generation biocontrol agents and biofertilizers for eco-friendly, high-yield, and climate-resilient agriculture (Khan et al. 2023).

5.12.2 Regulatory Aspects and Commercialization Hurdles

Although streptomycetes have high potential as biocontrol agents and PGPR, the commercialization and approval of streptomycetes-based biocontrol agents and/or PGPR remain challenging (Beyari 2024). However, the adoption of *Streptomyces*-derived biofertilizers and biopesticides for agricultural use is largely constrained by stringent regulatory frameworks, prolonged approval processes, and market adoption barriers that prevent their widespread use, despite evidence of their efficacy in sustainable agriculture (Pathma et al. 2021). Foremost, one of the main regulatory hurdles is the need for comprehensive safety evaluations prior to market introduction (Ansari et al. 2023). Unlike synthetic chemicals, microbial bioproducts must also be rigorously evaluated through environmental risk assessments (ERAs) to determine their persistence, potential horizontal gene transfer (HGT), and inadvertent ecological consequences (Krysenko 2025). Regulatory authorities such as the U.-S. Environmental Protection Agency (EPA), the European Food Safety Authority (EFSA), and national agricultural agencies internationally require extensive toxicological, environmental, and efficacy studies, which can be lengthy and expensive (Wend et al. 2024).

Moreover, the absence of globally harmonized regulations makes it difficult to use *Streptomyces*-based products in various countries (Alam et al. 2022). There is often differentiation in data requirements, bioformulation standards, and safety protocols, which complicates how products enter the market and adds to the cost of development (Mawar et al. 2021). For chemical pesticides, testing protocols are already established, but microbial bioproducts need tailored regulatory pathways, adding another layer of complexity to commercialization (Agrawal 2024). There is a significant challenge in scaling production and maintaining strain stability and efficacy. These include, but are not limited to, large-scale fermentation, formulation development, and shelf-life stability, which continue to be technical hurdles for the widespread use of streptomycetes-based products (Boubekri et al. 2022; Vermelho et al. 2024). For a field application, companies need to optimize carrier-based formulations, storage conditions, and delivery systems (Pacios-Michelena et al. 2021). Public–private partnerships, efficiency of regulatory processes, and investment in microbial formulation technologies are needed to help overcome these barriers (Karuppiah et al. 2025). Commercialization of streptomycetes-based agro-solutions can be mostly fostered to find innovative alternatives with respect to synthetic agrochemicals by rationalizing approval pathways and enhancing microbial bioformulations for sustainable and eco-friendly farming (Priya et al. 2023; Oyedoh et al. 2023a; Beyari 2024).

5.13 Conclusion

In conclusion, the streptomycetes have become very hopeful biocontrol and plant growth-promoting agents in terms of sustainability and eco-friendly toward the betterment of modern agriculture. Antibiotics, secondary metabolites, and phytohormones emerging from these organisms push aside disease and supply energy to plants, thereby enhancing crops by means of nutrient solubilization and stress tolerance. The widening scope of biopesticide applications can be attributed to the advances in genomics, metabolomics, and genetic engineering, which have improved their biocontrol efficiency and agricultural applications, including the exploration of high trait performance novel strains. While they have great potential, challenges such as environmental variability, regulation, and commercialization will need to be accounted for to wave them to the mainstream. Going forward, strain improvement, large-scale formulation development, and regulatory streamlining efforts to ensure robust and consistent application in the field should be priorities for further research. The integration of *Streptomyces*-based solutions into traditional as well as modern farming practices will lead us a step further toward the development of a resilient, sustainable, and eco-friendly agriculture with a low dependence on synthetic agrochemicals, as recapped in Fig. 5.16.

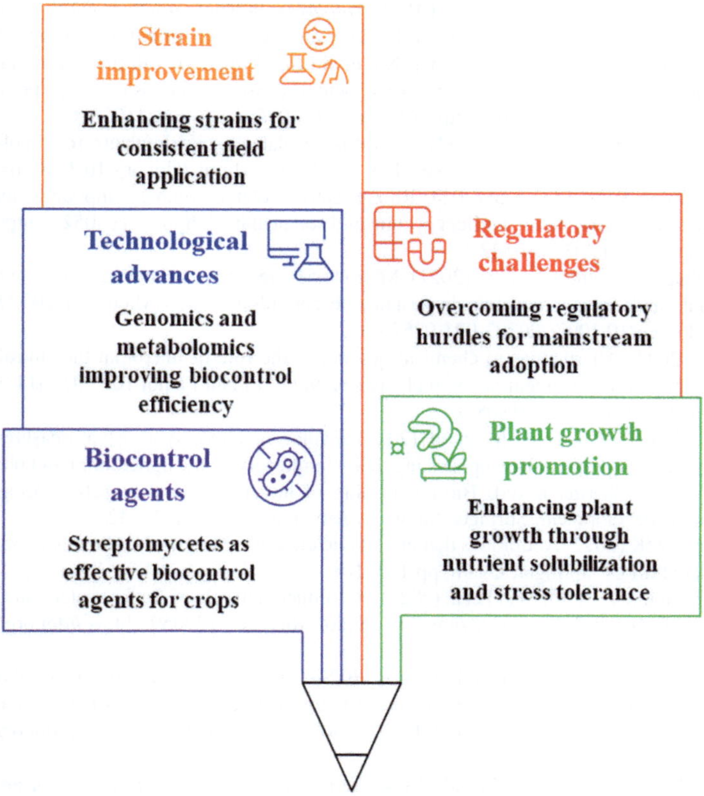

Fig. 5.16 Streptomycetes in modern agriculture

References

Abbasi S, Sadeghi A, Safaie N (2020) *Streptomyces* alleviate drought stress in tomato plants and modulate the expression of transcription factors ERF1 and WRKY70 genes. Sci Hortic 265: 109206. https://doi.org/10.1016/j.scienta.2020.109206

Abdellatif AAM, Gebily DAS, Elmaghraby MMK et al (2024) Transforming roles of Actinobacteria in sustainable agriculture: from soil health and plant productivity perspective. In: Kaur S, Dwibedi V, Sahu PK (eds) Metabolomics, proteomics and gene editing approaches in biofertilizer industry. Springer Nature Singapore, Singapore, pp 299–338

Aggarwal N, Thind SK, Sharma S (2016) Role of secondary metabolites of Actinomycetes in crop protection. In: Subramaniam G, Arumugam S, Rajendran V (eds) Plant growth promoting Actinobacteria. Springer Singapore, Singapore, pp 99–121

Agrawal R (2024) Emerging technologies for improvement of microbial strains using Omic sciences and measuring methods for maximum product formation. In: Textbook of industrial microbiology. Springer Nature Singapore, Singapore, pp 13–39

Alam K, Mazumder A, Sikdar S et al (2022) *Streptomyces*: the biofactory of secondary metabolites. Front Microbiol 13:968053. https://doi.org/10.3389/fmicb.2022.968053

Al-Quwaie DA (2024) The role of *Streptomyces* species in controlling plant diseases: a comprehensive review. Australas Plant Pathol 53:1–14. https://doi.org/10.1007/s13313-023-00959-z

Amaresan N, Kumar K, Naik JH et al (2018) *Streptomyces* in plant growth promotion. In: New and future developments in microbial biotechnology and bioengineering. Elsevier, pp 125–135

Ansabayeva A, Makhambetov M, Rebouh NY et al (2025) Plant growth-promoting microbes for resilient farming systems: mitigating environmental stressors and boosting crops productivity—a review. Horticulturae 11:260. https://doi.org/10.3390/horticulturae11030260

Ansari M, Devi BM, Sarkar A et al (2023) Microbial exudates as biostimulants: role in plant growth promotion and stress mitigation. J Xenobiot 13:572–603. https://doi.org/10.3390/jox13040037

Baloch FB, Zeng N, Gong H et al (2024) Rhizobacterial volatile organic compounds: implications for agricultural ecosystems' nutrient cycling and soil health. Heliyon 10:e40522. https://doi.org/10.1016/j.heliyon.2024.e40522

Basit A, Shah ST, Ullah I et al (2021) Microbe-assisted phytoremediation of environmental pollutants and energy recycling in sustainable agriculture. Arch Microbiol 203:5859–5885. https://doi.org/10.1007/s00203-021-02576-0

Beyari EA (2024) Alternatives to chemical pesticides: the role of microbial biocontrol agents in phytopathogen management: a comprehensive review. J Plant Pathol 107:291–314. https://doi.org/10.1007/s42161-024-01808-8

Bhakat A, Das P, Goswami M et al (2025) Exploring endophytes as a treasure house of bioprospecting compounds, emphasizing their significance in the agricultural and pharmaceutical fields. In: Sharma K (ed) Bio-prospecting of novel microbial bioactive compounds for sustainable development. Springer Nature Switzerland, Cham, pp 27–42

Bhattacharjee MK (2022) Antibiotics that inhibit nucleic acid synthesis. In: Chemistry of antibiotics and related drugs. Springer, Cham, pp 125–148

Boubekri K, Soumare A, Mardad I et al (2022) Multifunctional role of Actinobacteria in agricultural production sustainability: a review. Microbiol Res 261:127059. https://doi.org/10.1016/j.micres.2022.127059

Boukaew S, Mahasawat P, Petlamul W et al (2023) Application of antifungal metabolites from *Streptomyces philanthi* RL-1-178 for maize grain coating formulations and their efficacy as biofungicide during storage. World J Microbiol Biotechnol 39:157. https://doi.org/10.1007/s11274-023-03604-5

Boukelloul I, Aouar L, Cherb N et al (2024) Actinobacteria isolated from soils of arid Saharan regions display simultaneous antifungal and plant growth promoting activities. Curr Microbiol 81:327. https://doi.org/10.1007/s00284-024-03851-1

Buzón-Durán L, Pérez-Lebeña E, Martín-Gil J et al (2020) Applications of *Streptomyces* spp. enhanced compost in sustainable agriculture. In: Meghvansi MK, Varma A (eds) Biology of composts. Springer, Cham, pp 257–291

Buzón-Durán L, Molinuevo-Salces B, García-González MC et al (2025) Effect of *Streptomyces* spp. metabolites and the combination of biochar and compost on Fusarium graminearum inhibition, triticale growth, and soil properties. Sci Total Environ 981:179595. https://doi.org/10.1016/j.scitotenv.2025.179595

Cai L, Zhang H, Deng Y et al (2023) Antifungal activity of *Streptomyces hygroscopicus* JY-22 against Alternaria alternata and its potential application as a biopesticide to control tobacco brown spot. Agronomy 13:1944. https://doi.org/10.3390/agronomy13071944

Calvo-Peña C, Cobos R, Sánchez-López JM et al (2023) Albocycline is the main bioactive antifungal compound produced by *Streptomyces* sp. OR6 against Verticillium dahliae. Plants 12:3612. https://doi.org/10.3390/plants12203612

Cao P, Li C, Wang H et al (2020) Community structures and antifungal activity of root-associated Endophytic Actinobacteria in healthy and diseased cucumber plants and *Streptomyces* sp. HAAG3-15 as a promising biocontrol agent. Microorganisms 8:236. https://doi.org/10.3390/microorganisms8020236

Cassanelli S, Bellameche F, Caradonia F et al (2025) Foliar application of *Streptomyces* sp. DLS2013 induces transcriptional changes on tomato plants and confers resistance to *Pseudomonas syringae* pv. tomato. J Plant Dis Prot 132:19. https://doi.org/10.1007/s41348-024-01027-4

Chaudhary P, Agri U, Chaudhary A et al (2022) Endophytes and their potential in biotic stress management and crop production. Front Microbiol 13:933017. https://doi.org/10.3389/fmicb.2022.933017

Chauhan P, Sharma N, Tapwal A et al (2023) Soil microbiome: diversity, benefits and interactions with plants. Sustainability 15:14643. https://doi.org/10.3390/su151914643

Chen Y, Zhou D, Qi D et al (2018) Growth promotion and disease suppression ability of a *Streptomyces* sp. CB-75 from Banana Rhizosphere Soil. Front Microbiol 8:2704. https://doi.org/10.3389/fmicb.2017.02704

Chen J, Hu L, Chen N et al (2021) The biocontrol and plant growth-promoting properties of *Streptomyces alfalfae* XN-04 revealed by functional and genomic analysis. Front Microbiol 12:745766. https://doi.org/10.3389/fmicb.2021.745766

Chen D, Ali MNHA, Kamran M et al (2022) The *Streptomyces chromofuscus* strain RFS-23 induces systemic resistance and activates plant defense responses against tomato yellow leaf curl virus infection. Agronomy 12:2419. https://doi.org/10.3390/agronomy12102419

Chen Y, Gao Z, Yang Y et al (2025) Revealing the existence of diverse strategies for phosphorus solubilization and acquisition in plant-growth promoting *Streptomyces misionensis* SwB1. Microorganisms 13:378. https://doi.org/10.3390/microorganisms13020378

Chouyia FE, Ventorino V, Pepe O (2022) Diversity, mechanisms and beneficial features of phosphate-solubilizing *Streptomyces* in sustainable agriculture: a review. Front Plant Sci 13:1035358. https://doi.org/10.3389/fpls.2022.1035358

Correa PA, Nosheen A, Yasmin H, Ansari MJ (2022) Antifungal antibiotics biosynthesized by major PGPR. In: Sayyed RZ, Uarrota VG (eds) Secondary metabolites and volatiles of PGPR in plant-growth promotion. Springer, Cham, pp 199–247

Cui K, Xu T, Chen J et al (2022) Siderophores, a potential phosphate solubilizer from the endophyte *Streptomyces* sp. CoT10, improved phosphorus mobilization for host plant growth and rhizosphere modulation. J Clean Prod 367:133110. https://doi.org/10.1016/j.jclepro.2022.133110

Das S, Maity S (2025) Metabolomics for plant growth-promoting microbe profiling and designing biofertilizer. In: Kaur S, Dhiman S, Tripathi M (eds) Microbial metabolomics. Springer Nature Singapore, Singapore, pp 249–290

Dave A, Ingle S (2021) *Streptomyces* sp S-9 promotes plant growth and confers resistance in Pigeon pea (*Cajanus cajan*) against Fusarium wilt. 3 Biotech 11:459. https://doi.org/10.1007/s13205-021-02989-0

Dave A, Ingle S (2022) Potential of *Streptomyces* and its secondary metabolites for biocontrol of fungal plant pathogens. In: Sayyed RZ, Singh A, Ilyas N (eds) Antifungal metabolites of Rhizobacteria for sustainable agriculture. Springer, Cham, pp 3–22

De Lima Júnior AA, De Sousa EC, De Oliveira THB et al (2023) Genus *Streptomyces*: recent advances for biotechnological purposes. Biotechnol Appl Biochem 70:1504–1517. https://doi.org/10.1002/bab.2455

Devi S, Manhas RK (2023) Induction of systemic resistance in *Solanum lycopersicum* and *Capsicum annum* seedlings against Fusarium wilt by *Streptomyces* bioformulations. Environ Sci Pollut Res 30:109438–109452. https://doi.org/10.1007/s11356-023-29973-w

Devi S, Sharma M, Manhas RK (2022) Investigating the plant growth promoting and biocontrol potentiality of endophytic *Streptomyces* sp. SP5 against early blight in *Solanum lycopersicum* seedlings. BMC Microbiol 22:285. https://doi.org/10.1186/s12866-022-02695-8

Diab MK, Mead HM, Ahmad Khedr MM et al (2024) Actinomycetes are a natural resource for sustainable pest control and safeguarding agriculture. Arch Microbiol 206:268. https://doi.org/10.1007/s00203-024-03975-9

Díaz-Díaz M, Bernal-Cabrera A, Trapero A et al (2023) Biocontrol of root rot complex disease of *Phaseolus vulgaris* by *Streptomyces* sp. strains in the field. Crop Prot 165:106164. https://doi.org/10.1016/j.cropro.2022.106164

Ebrahimi-Zarandi M, Bonjar GHS, Riseh RS et al (2021) Exploring two *Streptomyces* species to control *Rhizoctonia solani* in tomato. Agronomy 11:1384. https://doi.org/10.3390/agronomy11071384

Ebrahimi-Zarandi M, Saberi Riseh R, Tarkka MT (2022) Actinobacteria as effective biocontrol agents against plant pathogens, an overview on their role in eliciting plant defense. Microorganisms 10:1739. https://doi.org/10.3390/microorganisms10091739

El-Sharkawy HHA, Rashad YM, Elazab NT (2023) Induction of multiple defense responses in wheat plants against stripe rust using mycorrhizal fungi and *Streptomyces viridosporus* HH1. BioControl 68:525–535. https://doi.org/10.1007/s10526-023-10207-4

Essarioui A, LeBlanc N, Kistler HC, Kinkel LL (2017) Plant community richness mediates inhibitory interactions and resource competition between *Streptomyces* and *Fusarium* populations in the rhizosphere. Microb Ecol 74:157–167. https://doi.org/10.1007/s00248-016-0907-5

Fatima A, Abbas M, Nawaz S et al (2024) Whole genome sequencing (WGS) and genome mining of *Streptomyces* sp. AFD10 for antibiotics and bioactive secondary metabolites biosynthetic gene clusters (BGCs). Gene Rep 37:102050. https://doi.org/10.1016/j.genrep.2024.102050

Flärdh K, Buttner MJ (2009) *Streptomyces* morphogenetics: dissecting differentiation in a filamentous bacterium. Nat Rev Microbiol 7:36–49. https://doi.org/10.1038/nrmicro1968

Fróes A, Macrae A, Rosa J et al (2012) Selection of a *Streptomyces* strain able to produce cell wall degrading enzymes and active against Sclerotinia sclerotiorum. J Microbiol 50:798–806. https://doi.org/10.1007/s12275-012-2060-2

Fu X, Liu S, Ru J et al (2022) Biological control of potato late blight by *Streptomyces* sp. FXP04 and potential role of secondary metabolites. Biol Control 169:104891. https://doi.org/10.1016/j.biocontrol.2022.104891

Ganesh SK, Subathra Devi C (2023) Molecular and therapeutic insights of rapamycin: a multi-faceted drug from *Streptomyces hygroscopicus*. Mol Biol Rep 50:3815–3833. https://doi.org/10.1007/s11033-023-08283-x

Gao Y, Zeng XD, Ren B et al (2020) Antagonistic activity against rice blast disease and elicitation of host-defence response capability of an endophytic *Streptomyces albidoflavus* OsiLf-2. Plant Pathol 69:259–271. https://doi.org/10.1111/ppa.13118

Gao L, Kumaravel K, Xiong Q et al (2023) Actinomycins produced by endophyte *Streptomyces* sp. GLL −9 from navel orange plant exhibit high antimicrobial effect against *Xanthomonas citri* susp. *citri* and *Penicillium italicum*. Pest Manag Sci 79:4679–4693. https://doi.org/10.1002/ps.7668

Gates AD, French AM, Demetros AA et al (2023) A *Streptomyces* consortium contributes distinct microbial interactions during *Arabidopsis thaliana* microbiome assembly. Phytobiomes J 7: 515–525. https://doi.org/10.1094/PBIOMES-11-22-0081-R

Ghanem GAM, Gebily DAS, Ragab MM et al (2022) Efficacy of antifungal substances of three *Streptomyces* spp. against different plant pathogenic fungi. Egypt J Biol Pest Control 32:112. https://doi.org/10.1186/s41938-022-00612-9

Gopalakrishnan S, Srinivas V, Prasanna SL (2020) Streptomyces. In: Beneficial microbes in agro-ecology. Elsevier, pp 55–71

Gore NS, Patil PS (2025) Deciphering microbial cross-talk for plant protection through induced resistance. In: Mitra D, De Los Santos Villalobos S, Rani A et al (eds) Bio-control agents for sustainable agriculture. Springer Nature Singapore, Singapore, pp 141–165

Gu J, Luo Y, Cui L et al (2025) Control of gummy stem blight in pumpkin using *Streptomyces nigrescens* NEAU-L66: insights into its biocontrol mechanisms. Physiol Mol Plant Pathol 138: 102695. https://doi.org/10.1016/j.pmpp.2025.102695

Guha AA, Harish S, Sendhilvel V et al (2025) Revamping the encapsulation of microbes and their metabolites to nanoscale: a strategy to enhance crop biocontrol. J Nanopart Res 27:117. https://doi.org/10.1007/s11051-025-06314-1

Hamedi J, Poorinmohammad N, Papiran R (2017) Growth and life cycle of Actinobacteria. In: Wink J, Mohammadipanah F, Hamedi J (eds) Biology and biotechnology of Actinobacteria. Springer, Cham, pp 29–50

Hansen VM, Winding A, Madsen AM (2010) Exposure to bioaerosols during the growth season of tomatoes in an organic greenhouse using Supresivit (*Trichoderma harzianum*) and Mycostop (*Streptomyces griseoviridis*). Appl Environ Microbiol 76:5874–5881. https://doi.org/10.1128/AEM.00446-10

Hata EM, Yusof MT, Zulperi D (2021) Induction of systemic resistance against bacterial leaf streak disease and growth promotion in rice plant by *Streptomyces shenzhenesis* TKSC3 and *Streptomyces* sp. SS8. Plant Pathol J 37:173–181. https://doi.org/10.5423/PPJ.OA.05.2020.0083

He D, Gao C, Zhao S et al (2024a) Antibacterial, herbicidal, and plant growth-promoting properties of *Streptomyces* sp. STD57 from the rhizosphere of *Adenophora stricta*. Microorganisms 12: 2245. https://doi.org/10.3390/microorganisms12112245

He L, Kim Y, Kim S et al (2024b) Unveiling antibiotic resistance: genome sequencing of streptomycin-resistant *Erwinia amylovora* isolate. Microorganisms 12:2494. https://doi.org/10.3390/microorganisms12122494

Hönig M, Roeber VM, Schmülling T, Cortleven A (2023) Chemical priming of plant defense responses to pathogen attacks. Front Plant Sci 14:1146577. https://doi.org/10.3389/fpls.2023.1146577

Isayenka I, Beaudoin N (2022) The *Streptomyces scabiei* pathogenicity factor Thaxtomin A induces the production of phenolic compounds in potato tubers. Plants 11:3216. https://doi.org/10.3390/plants11233216

Ivanov M, Ćirić A, Stojković D (2022) Emerging antifungal targets and strategies. Int J Mol Sci 23: 2756. https://doi.org/10.3390/ijms23052756

Jauri PV, Altier N, Kinkel LL (2016) *Streptomyces* for sustainability. In: Microbial models: from environmental to industrial sustainability. Springer Singapore, Singapore, pp 251–276

Javed Z, Tripathi GD, Mishra M, Dashora K (2021) Actinomycetes—the microbial machinery for the organic-cycling, plant growth, and sustainable soil health. Biocatal Agric Biotechnol 31: 101893. https://doi.org/10.1016/j.bcab.2020.101893

Jia R, Xiao K, Yu L et al (2023) A potential biocontrol agent *Streptomyces tauricus* XF for managing wheat stripe rust. Phytopathol Res 5:14. https://doi.org/10.1186/s42483-023-00168-y

Jin Z, Jiang F, Wang L et al (2024) Arbuscular mycorrhizal fungi and *Streptomyces*: brothers in arms to shape the structure and function of the hyphosphere microbiome in the early stage of interaction. Microbiome 12:83. https://doi.org/10.1186/s40168-024-01811-2

Jing T, Zhou D, Zhang M et al (2020) Newly isolated *Streptomyces* sp. JBS5-6 as a potential biocontrol agent to control banana Fusarium wilt: genome sequencing and secondary metabolite cluster profiles. Front Microbiol 11:602591. https://doi.org/10.3389/fmicb.2020.602591

Jog R, Nareshkumar G, Rajkumar S (2016) Enhancing soil health and plant growth promotion by actinomycetes. In: Subramaniam G, Arumugam S, Rajendran V (eds) Plant growth promoting Actinobacteria. Springer Singapore, Singapore, pp 33–45

Johnson I, Kavitha R, Karthikeyan M et al (2025) Harnessing biocontrol potential of *Streptomyces rochei* against *Pythium aphanidermatum*: efficacy and mechanisms. J Basic Microbiol 65: e2400531. https://doi.org/10.1002/jobm.202400531

Kaari M, Joseph J, Manikkam R et al (2022) Biological control of *Streptomyces* sp. UT4A49 to suppress tomato bacterial wilt disease and its metabolite profiling. J King Saud Univ Sci 34: 101688. https://doi.org/10.1016/j.jksus.2021.101688

Kamle M, Borah R, Bora H, et al (2020) Systemic acquired resistance (SAR) and induced systemic resistance (ISR): role and mechanism of action against phytopathogens. In: Hesham AE-L, Upadhyay RS, Sharma GD, et al. (eds) Fungal biotechnology and bioengineering. Springer, Cham, pp 457–470

Kang JE, Han JW, Jeon BJ, Kim BS (2016) Efficacies of quorum sensing inhibitors, piericidin A and glucopiericidin A, produced by *Streptomyces xanthocidicus* KPP01532 for the control of potato soft rot caused by Erwinia carotovora subsp. atroseptica. Microbiol Res 184:32–41. https://doi.org/10.1016/j.micres.2015.12.005

Karuppiah P, Thayalan KJ, Thangam GS, Eswaranpillai U (2025) Bio-control agents: a sustainable approach for enhancing soil nutrients use efficiency in farming. In: Mitra D, De Los Santos Villalobos S, Rani A et al (eds) Bio-control agents for sustainable agriculture. Springer Nature Singapore, Singapore, pp 209–260

Kaur T, Manhas RK (2022) Evaluation of ACC deaminase and indole acetic acid production by *Streptomyces hydrogenans* DH16 and its effect on plant growth promotion. Biocatal Agric Biotechnol 42:102321. https://doi.org/10.1016/j.bcab.2022.102321

Kaur T, Khanna K, Sharma S, Manhas RK (2023) Mechanistic insights into the role of actinobacteria as potential biocontrol candidates against fungal phytopathogens. J Basic Microbiol 63:1196–1218. https://doi.org/10.1002/jobm.202300027

Kawicha P, Nitayaros J, Saman P et al (2023) Evaluation of soil *Streptomyces* spp. for the biological control of Fusarium wilt disease and growth promotion in tomato and Banana. Plant Pathol J 39: 108–122. https://doi.org/10.5423/PPJ.OA.08.2022.0124

Keswani C, Bisen K, Singh V et al (2016) Formulation technology of biocontrol agents: present status and future prospects. In: Arora NK, Mehnaz S, Balestrini R (eds) Bioformulations: for sustainable agriculture. Springer India, New Delhi, pp 35–52

Keswani C, Singh SP, García-Estrada C et al (2022) Biosynthesis and beneficial effects of microbial gibberellins on crops for sustainable agriculture. J Appl Microbiol 132:1597–1615. https://doi.org/10.1111/jam.15348

Khan S, Srivastava S, Karnwal A, Malik T (2023) *Streptomyces* as a promising biological control agents for plant pathogens. Front Microbiol 14:1285543. https://doi.org/10.3389/fmicb.2023.1285543

Khan M, Khan T, Tabassum B, Hashim M (2024a) Microbiome-driven soil fertility: understanding symbiotic relationships. In: Parray JA (ed) Progress in soil microbiome research. Springer Nature Switzerland, Cham, pp 77–115

Khan MT, Aleinikovienė J, Butkevičienė L-M (2024b) Innovative organic fertilizers and cover crops: perspectives for sustainable agriculture in the era of climate change and organic agriculture. Agronomy 14:2871. https://doi.org/10.3390/agronomy14122871

Khanam M, Kabir MH, Akter M et al (2024) Role of microorganisms in soil health management. In: Rahman MM, Biswas JC, Meena RS (eds) Climate change and soil-water-plant nexus. Springer Nature Singapore, Singapore, pp 223–263

Kim D-R, Kwak Y-S (2021) A genome-wide analysis of antibiotic producing genes in *Streptomyces* globisporus SP6C4. Plant Pathol J 37:389–395. https://doi.org/10.5423/ppj.nt.03.2021.0047

Krysenko S (2025) Current approaches for genetic manipulation of *Streptomyces* spp.—key bacteria for biotechnology and environment. Biotech 14:3. https://doi.org/10.3390/biotech14010003

Ku Y, Liao Y, Chiou S et al (2024) From trade-off to synergy: microbial insights into enhancing plant growth and immunity. Plant Biotechnol J 22:2461–2471. https://doi.org/10.1111/pbi.14360

Kulkarni M, Gorthi S, Banerjee G, Chattopadhyay P (2017) Production, characterization and optimization of actinomycin D from *Streptomyces hydrogenans* IB310, a(n antagonistic bacterium against phytopathogens). Biocatal Agric Biotechnol 10:69–74. https://doi.org/10.1016/j.bcab.2017.02.009

Kumar M, Kumar P, Das P et al (2022) Proactive role of *Streptomyces* spp. in plant growth stimulation and management of chemical pesticides and fertilizers. Int J Environ Sci Technol 19:10457–10476. https://doi.org/10.1007/s13762-021-03473-1

Le KD, Kim J, Nguyen HT et al (2021) *Streptomyces* sp. JCK-6131 protects plants against bacterial and fungal diseases via two mechanisms. Front Plant Sci 12:726266. https://doi.org/10.3389/fpls.2021.726266

LeBlanc N (2022) Bacteria in the genus *Streptomyces* are effective biological control agents for management of fungal plant pathogens: a meta-analysis. BioControl 67:111–121. https://doi.org/10.1007/s10526-021-10123-5

Lee N, Hwang S, Kim W et al (2021) Systems and synthetic biology to elucidate secondary metabolite biosynthetic gene clusters encoded in *Streptomyces* genomes. Nat Prod Rep 38: 1330–1361. https://doi.org/10.1039/D0NP00071J

Lee Y, Hwang S, Kim W et al (2024) CRISPR-aided genome engineering for secondary metabolite biosynthesis in *Streptomyces*. J Ind Microbiol Biotechnol 51:kuae009. https://doi.org/10.1093/jimb/kuae009

Legault GS, Lerat S, Nicolas P, Beaulieu C (2011) Tryptophan regulates Thaxtomin A and Indole-3-acetic acid production in *Streptomyces scabiei* and modifies its interactions with radish seedlings. Phytopathology 101:1045–1051. https://doi.org/10.1094/PHYTO-03-11-0064

Lerat S, Simao-Beaunoir A, Beaulieu C (2009) Genetic and physiological determinants of *Streptomyces scabies* pathogenicity. Mol Plant Pathol 10:579–585. https://doi.org/10.1111/j.1364-3703.2009.00561.x

Li Y, Li H, Han X et al (2022) Actinobacterial biofertilizer improves the yields of different plants and alters the assembly processes of rhizosphere microbial communities. Appl Soil Ecol 171:104345. https://doi.org/10.1016/j.apsoil.2021.104345

Li X, Zhang L, Zhao Y et al (2024) Biocontrol potential of volatile organic compounds produced by *Streptomyces corchorusii* CG-G2 to strawberry anthracnose caused by Colletotrichum gloeosporioides. Food Chem 437:137938. https://doi.org/10.1016/j.foodchem.2023.137938

Liang J, Wei C, Song X et al (2024) Bacterial wilt affects the structure and assembly of microbial communities along the soil-root continuum. Environ Microbiome 19:6. https://doi.org/10.1186/s40793-024-00548-7

Ling L, Han X, Li X et al (2020) A *Streptomyces* sp. NEAU-HV9: isolation, identification, and potential as a biocontrol agent against *Ralstonia solanacearum* of tomato plants. Microorganisms 8:351. https://doi.org/10.3390/microorganisms8030351

Ling L, Zhao T, Zhang S et al (2025) Rhizobacterium *Streptomyces* sp. NEAU-383 as a potential biocontrol agent to control tomato bacterial wilt. Braz J Microbiol 56(2):1179–1189. https://doi.org/10.1007/s42770-025-01645-6

Liu D (2021) Root developmental responses to phosphorus nutrition. J Integr Plant Biol 63:1065–1090. https://doi.org/10.1111/jipb.13090

Liu D, Yan R, Fu Y et al (2019) Antifungal, plant growth-promoting, and genomic properties of an Endophytic Actinobacterium *Streptomyces* sp. NEAU-S7GS2. Front Microbiol 10:2077. https://doi.org/10.3389/fmicb.2019.02077

Liu H, Li J, Singh BK (2024) Harnessing co-evolutionary interactions between plants and *Streptomyces* to combat drought stress. Nat Plants 10:1159–1171. https://doi.org/10.1038/s41477-024-01749-1

Loboda M, Biliavska L, Iutynska G et al (2024) Natural products biosynthesis by *Streptomyces netropsis* IMV ac-5025 under exogenous sterol action. Antibiotics 13:146. https://doi.org/10.3390/antibiotics13020146

López-Reyes PK, De La Torre-Zavala S, Cortés-González MM et al (2024) Biological control of *Streptomyces* sp. PR69 against Phytophthora capsici and its growth-promoting effects on plants. Horticulturae 10:1365. https://doi.org/10.3390/horticulturae10121365

Lu Y, Song W, Wang J et al (2024) Biocontrol of Botrytis cinerea by *Streptomyces noursei* C27 and preliminary identification of antimicrobial metabolites. Biol Control 196:105561. https://doi.org/10.1016/j.biocontrol.2024.105561

Lyu A, Liu H, Che H et al (2017) Reveromycins A and B from *Streptomyces* sp. 3–10: antifungal activity against plant pathogenic fungi *in vitro* and in a strawberry food model system. Front Microbiol 8:550. https://doi.org/10.3389/fmicb.2017.00550

Maithani D, Singh H, Sharma A (2021) Stress alleviation in plants using SAR and ISR: current views on stress signaling network. In: Sharma A (ed) Microbes and signaling biomolecules against plant stress. Springer Singapore, Singapore, pp 7–36

Mawar R, Manjunatha BL, Kumar S (2021) Commercialization, diffusion and adoption of bioformulations for sustainable disease management in Indian arid agriculture: prospects and challenges. Circ Econ Sustain 1:1367–1385. https://doi.org/10.1007/s43615-021-00089-y

Meena M, Prajapati P, Ravichandran C, Sehrawat R (2021) Natamycin: a natural preservative for food applications—a review. Food Sci Biotechnol 30:1481–1496. https://doi.org/10.1007/s10068-021-00981-1

Minuto A, Spadaro D, Garibaldi A, Gullino ML (2006) Control of soilborne pathogens of tomato using a commercial formulation of *Streptomyces griseoviridis* and solarization. Crop Prot 25: 468–475. https://doi.org/10.1016/j.cropro.2005.08.001

Mohamed A, EL-Shatoury S, Aboulfotoh A et al (2024) Synergistic effects (adsorption and biodegradation) of *Streptomyces hydrogenans* immobilization on nano-reed biochar for further application in upflow anaerobic sludge blanket. RSC Adv 14:22828–22846. https://doi.org/10.1039/D4RA02864C

Mohammadipanah F, Dehhaghi M (2017) Classification and taxonomy of Actinobacteria. In: Wink J, Mohammadipanah F, Hamedi J (eds) Biology and biotechnology of Actinobacteria. Springer, Cham, pp 51–77

Morcillo R, Manzanera M (2021) The effects of plant-associated bacterial exopolysaccharides on plant abiotic stress tolerance. Meta 11:337. https://doi.org/10.3390/metabo11060337

Mun B-G, Lee W-H, Kang S-M et al (2020) *Streptomyces* sp. LH 4 promotes plant growth and resistance against Sclerotinia sclerotiorum in cucumber via modulation of enzymatic and defense pathways. Plant Soil 448:87–103. https://doi.org/10.1007/s11104-019-04411-4

Myo EM, Ge B, Ma J et al (2019) Indole-3-acetic acid production by *Streptomyces fradiae* NKZ-259 and its formulation to enhance plant growth. BMC Microbiol 19:155. https://doi.org/10.1186/s12866-019-1528-1

Narayanasamy P (2013) Development of formulations and commercialization of biological products. In: Biological management of diseases of crops. Springer Netherlands, Dordrecht, pp 129–187

Nazari MT, Schommer VA, Braun JCA et al (2023) Using *Streptomyces* spp. as plant growth promoters and biocontrol agents. Rhizosphere 27:100741. https://doi.org/10.1016/j.rhisph.2023.100741

Nguyen LTT, Park AR, Van Le V et al (2024) Exploration of a multifunctional biocontrol agent *Streptomyces* sp. JCK-8055 for the management of apple fire blight. Appl Microbiol Biotechnol 108:49. https://doi.org/10.1007/s00253-023-12874-w

Niu S, Gao Y, Zi H et al (2022) The osmolyte-producing endophyte *Streptomyces albidoflavus* OsiLf-2 induces drought and salt tolerance in rice via a multi-level mechanism. Crop J 10:375–386. https://doi.org/10.1016/j.cj.2021.06.008

Nivedita S, Behera SS, Behera PK et al (2024) Salt-resistant *Streptomyces* consortia promote growth of rice (Oryza sativa var. Swarna) alleviating salinity and drought stress tolerance by enhancing photosynthesis, antioxidant function, and proline content. The Microbe 4:100124. https://doi.org/10.1016/j.microb.2024.100124

Olanrewaju OS, Babalola OO (2019) *Streptomyces*: implications and interactions in plant growth promotion. Appl Microbiol Biotechnol 103:1179–1188. https://doi.org/10.1007/s00253-018-09577-y

Omar AF, Abdelmageed AHA, Al-Turki A et al (2022) Exploring the plant growth-promotion of four *Streptomyces* strains from rhizosphere soil to enhance cucumber growth and yield. Plants 11:3316. https://doi.org/10.3390/plants11233316

Oyedoh OP, Yang W, Dhanasekaran D et al (2023a) Sustainable agriculture: rare-Actinomycetes to the rescue. Agronomy 13:666. https://doi.org/10.3390/agronomy13030666

Oyedoh OP, Yang W, Dhanasekaran D et al (2023b) Rare rhizo-Actinomycetes: a new source of agroactive metabolites. Biotechnol Adv 67:108205. https://doi.org/10.1016/j.biotechadv.2023.108205

Pacios-Michelena S, Aguilar González CN, Alvarez-Perez OB et al (2021) Application of *Streptomyces* antimicrobial compounds for the control of phytopathogens. Front Sustain Food Syst 5: 696518. https://doi.org/10.3389/fsufs.2021.696518

Palečková P, Bobek J, Felsberg J, Mikulík K (2006) Activity of translation system and abundance of tmRNA during development of *Streptomyces* aureofaciens producing tetracycline. Folia Microbiol 51:517–524. https://doi.org/10.1007/BF02931615

Pandey P, Tripathi A, Dwivedi S et al (2023) Deciphering the mechanisms, hormonal signaling, and potential applications of endophytic microbes to mediate stress tolerance in medicinal plants. Front Plant Sci 14:1250020. https://doi.org/10.3389/fpls.2023.1250020

Pang F, Solanki MK, Wang Z (2022) *Streptomyces* can be an excellent plant growth manager. World J Microbiol Biotechnol 38:193. https://doi.org/10.1007/s11274-022-03380-8

Park H-S, Nah H-J, Kang S-H et al (2021) Screening and isolation of a novel polyene-producing *Streptomyces* strain inhibiting phytopathogenic fungi in the soil environment. Front Bioeng Biotechnol 9:692340. https://doi.org/10.3389/fbioe.2021.692340

Pathma J, Kennedy RK, Bhushan LS et al (2021) Microbial biofertilizers and biopesticides: nature's assets fostering sustainable agriculture. In: Prasad R, Kumar V, Singh J, Upadhyaya CP (eds) Recent developments in microbial technologies. Springer Nature Singapore, Singapore, pp 39–69

Phour M, Sindhu SS (2023) Soil salinity and climate change: microbiome-based strategies for mitigation of salt stress to sustainable agriculture. In: Parray JA (ed) Climate change and microbiome dynamics. Springer, Cham, pp 191–243

Pipponzi S, Licciardello F, Caradonia F, Stefani E (2025) Volatile organic compounds produced by *Streptomyces* sp. SA51 mitigate Fusarium oxysporum f.sp. lactucae infection and stimulate lettuce seedling growth. Eur J Plant Pathol 173:125–142. https://doi.org/10.1007/s10658-025-03054-9

Pissowotzki K, Mansouri K, Piepersberg W (1991) Genetics of streptomycin production in *Streptomyces griseus:* molecular structure and putative function of genes strELMB2N. Mol Gen Genet 231:113–123. https://doi.org/10.1007/BF00293829

Prasad R (2022) Cytokinin and its key role to enrich the plant nutrients and growth under adverse conditions-an update. Front Genet 13:883924. https://doi.org/10.3389/fgene.2022.883924

Prashar P, Kapoor N, Sachdeva S (2014) Rhizosphere: its structure, bacterial diversity and significance. Rev Environ Sci Biotechnol 13:63–77. https://doi.org/10.1007/s11157-013-9317-z

Priya AK, Alagumalai A, Balaji D, Song H (2023) Bio-based agricultural products: a sustainable alternative to agrochemicals for promoting a circular economy. RSC Sustainability 1:746–762. https://doi.org/10.1039/D3SU00075C

Pupakdeepan W, Termsung N, Ruangwong O-U, Kunasakdakul K (2023) Plant growth promoting and colonization of Endophytic *Streptomyces albus* CINv1 against strawberry anthracnose. Horticulturae 9:766. https://doi.org/10.3390/horticulturae9070766

Qin Y-Y, Gong Y, Kong S-Y et al (2024) Aerial signaling by plant-associated *Streptomyces* setonii WY228 regulates plant growth and enhances salt stress tolerance. Microbiol Res 286:127823. https://doi.org/10.1016/j.micres.2024.127823

Qiu ZL, Da Liu S, Li XG et al (2024) Identification and mechanism characterization of *Streptomyces griseoaurantiacus* XQ-29 with biocontrol ability against pepper southern blight caused by Sclerotium rolfsii. Pestic Biochem Physiol 202:105956. https://doi.org/10.1016/j.pestbp.2024.105956

Qureshi KA, Bholay AD, Rai PK et al (2021) Isolation, characterization, anti-MRSA evaluation, and in-silico multi-target anti-microbial validations of actinomycin X2 and actinomycin D produced by novel *Streptomyces smyrnaeus* UKAQ_23. Sci Rep 11:14539. https://doi.org/10.1038/s41598-021-93285-7

Rashad YM, Al-Askar AA, Ghoneem KM et al (2017) Chitinolytic *Streptomyces griseorubens* E44G enhances the biocontrol efficacy against Fusarium wilt disease of tomato. Phytoparasitica 45:227–237. https://doi.org/10.1007/s12600-017-0580-3

Rashid MH, Krehenbrink M, Akhtar MS (2015) Nitrogen-fixing plant-microbe symbioses. In: Lichtfouse E (ed) Sustainable agriculture reviews. Springer, Cham, pp 193–234

Reddy PP (2014) Mass production, formulation, delivery and commercialization. In: Plant growth promoting Rhizobacteria for horticultural crop protection. Springer India, New Delhi, pp 69–83

Reddy KRK, Jyothi G, Sowjanya C et al (2016) Plant growth-promoting Actinomycetes: mass production, delivery systems, and commercialization. In: Subramaniam G, Arumugam S, Rajendran V (eds) Plant growth promoting Actinobacteria. Springer Singapore, Singapore, pp 287–298

Roychowdhury R, Hada A, Biswas S et al (2025) Jasmonic acid (JA) in plant immune response: unravelling complex molecular mechanisms and networking of defence signalling against pathogens. J Plant Growth Regul 44:89–114. https://doi.org/10.1007/s00344-024-11264-4

Sarethy IP, Saharan A (2021) Genomics, proteomics and transcriptomics in the biological control of plant pathogens: a review. Indian Phytopathol 74:3–12. https://doi.org/10.1007/s42360-020-00302-2

Sarwar A, Latif Z, Zhang S et al (2018) Biological control of potato common scab with rare Isatropolone C compound produced by plant growth promoting *Streptomyces* A1RT. Front Microbiol 9:1126. https://doi.org/10.3389/fmicb.2018.01126

Sayın M, Cihangir N (2024) Antimicrobial, anti-quorum sensing, and Antibiofilm potentials of newly isolated *Streptomyces* spp. Proc Natl Acad Sci India Sect B Biol Sci 94:121–134. https://doi.org/10.1007/s40011-023-01515-x

Schniete JK, Fernández-Martínez LT (2024) Natural product discovery in soil actinomycetes: unlocking their potential within an ecological context. Curr Opin Microbiol 79:102487. https://doi.org/10.1016/j.mib.2024.102487

Senthilraja G (2016) Induction of systemic resistance in crop plants against plant pathogens by plant growth-promoting Actinomycetes. In: Subramaniam G, Arumugam S, Rajendran V (eds) Plant growth promoting Actinobacteria. Springer Singapore, Singapore, pp 193–202

Seth K, Vyas P, Deora S et al (2023) Understanding plant-plant growth-promoting rhizobacteria (PGPR) interactions for inducing plant defense. In: Plant-microbe interaction – Recent advances in molecular and biochemical approaches. Elsevier, pp 201–226

Shahid M, Singh BN, Verma S et al (2021) Bioactive antifungal metabolites produced by *Streptomyces amritsarensis* V31 help to control diverse phytopathogenic fungi. Braz J Microbiol 52: 1687–1699. https://doi.org/10.1007/s42770-021-00625-w

Shanthi V (2021) Actinomycetes: implications and prospects in sustainable agriculture. In: Biofertilizers, 1st edn. Wiley, pp 335–370

Sharma N, Manhas RK, Ohri P (2022) *Streptomyces* hydrogenans strain DH-16 alleviates negative impact of Meloidogyne incognita stress by modifying physio-biochemical attributes in Solanum lycopersicum plants. Sci Rep 12:15214. https://doi.org/10.1038/s41598-022-19636-0

Shen W, Liu R, Wang J et al (2024) Characterization of a broad-spectrum antifungal strain, *Streptomyces graminearus* STR-1, against Magnaporthe oryzae. Front Microbiol 15:1298781. https://doi.org/10.3389/fmicb.2024.1298781

Shepherdson EM, Baglio CR, Elliot MA (2023) *Streptomyces* behavior and competition in the natural environment. Curr Opin Microbiol 71:102257. https://doi.org/10.1016/j.mib.2022.102257

Shivlata L, Satyanarayana T (2017) Actinobacteria in agricultural and environmental sustainability. In: Singh JS, Seneviratne G (eds) Agro-environmental sustainability. Springer, Cham, pp 173–218

Shrivastava P, Kumar R (2018) Actinobacteria. In: New and future developments in microbial biotechnology and bioengineering. Elsevier, pp 79–91

Singh AA, Singh AK (2025) Role of bacterial quorum sensing in plant growth promotion. World J Microbiol Biotechnol 41:18. https://doi.org/10.1007/s11274-024-04232-3

Solans M, Vobis G, Cassán F et al (2011) Production of phytohormones by root-associated saprophytic actinomycetes isolated from the actinorhizal plant *Ochetophila trinervis*. World J Microbiol Biotechnol 27:2195–2202. https://doi.org/10.1007/s11274-011-0685-7

Soumare A, Boubekri K, Lyamlouli K et al (2021) Efficacy of phosphate solubilizing Actinobacteria to improve rock phosphate agronomic effectiveness and plant growth promotion. Rhizosphere 17:100284. https://doi.org/10.1016/j.rhisph.2020.100284

Sousa JADJ, Olivares FL (2016) Plant growth promotion by streptomycetes: ecophysiology, mechanisms and applications. Chem Biol Technol Agric 3:24. https://doi.org/10.1186/s40538-016-0073-5

Sudha A, Durgadevi D, Archana S et al (2022) Unraveling the tripartite interaction of volatile compounds of *Streptomyces rochei* with grain mold pathogens infecting sorghum. Front Microbiol 13:923360. https://doi.org/10.3389/fmicb.2022.923360

Sudisha J, Sharathchandra RG, Amruthesh KN et al (2012) Pathogenesis related proteins in plant defense response. In: Mérillon JM, Ramawat KG (eds) Plant defence: biological control. Springer Netherlands, Dordrecht, pp 379–403

Sun T, Liu H, Wang N et al (2025) Interactions with native microbial keystone taxa enhance the biocontrol efficiency of *Streptomyces*. Microbiome 13:126. https://doi.org/10.1186/s40168-025-02120-y

Thomas L, Singh I (2020) Microbe-mediated biotic stress signaling and resistance mechanisms in plants. In: Giri B, Sharma MP (eds) Plant stress biology. Springer Singapore, Singapore, pp 297–334

Tian X, Hu L, Jia R et al (2024) *Streptomyces* pratensis S10 promotes wheat plant growth and induces resistance in wheat seedlings against *Fusarium graminearum*. J Fungi 10:578. https://doi.org/10.3390/jof10080578

Torres-Rodriguez JA, Reyes-Pérez JJ, Quiñones-Aguilar EE, Hernandez-Montiel LG (2022) Actinomycete potential as biocontrol agent of phytopathogenic fungi: mechanisms, source, and applications. Plants 11:3201. https://doi.org/10.3390/plants11233201

Trenozhnikova LP, Baimakhanova GB, Baimakhanova BB et al (2024) Beyond traditional screening: unveiling antibiotic potentials of actinomycetes in extreme environments. Heliyon 10: e40371. https://doi.org/10.1016/j.heliyon.2024.e40371

Tsolakidou M-D, Demetriou G, Panagiotou S et al (2024) Efficacy of commercial biocontrol products for the management of Verticillium and Fusarium wilt in greenhouse tomatoes: impact on disease severity, fruit yield, and quality. Agriculture (Basel) 14:882. https://doi.org/10.3390/agriculture14060882

Ullah H, Hassan SHA, Yang Q et al (2025) Dynamic interaction of antibiotic resistance between plant microbiome and organic fertilizers: sources, dissemination, and health risks. World J Microbiol Biotechnol 41:4. https://doi.org/10.1007/s11274-024-04214-5

Vermelho AB, Moreira JV, Akamine IT et al (2024) Agricultural pest management: the role of microorganisms in biopesticides and soil bioremediation. Plants 13:2762. https://doi.org/10.3390/plants13192762

Villafañe DL, Maldonado RA, Rodríguez E, Chiesa MA (2025) Endophytic *Streptomyces* sp. N2A protects soybean against fungal diseases through two distinct mechanisms. BioControl 70:529–542. https://doi.org/10.1007/s10526-025-10320-6

Vurukonda SSKP, Giovanardi D, Stefani E (2018) Plant growth promoting and biocontrol activity of *Streptomyces* spp. as endophytes. Int J Mol Sci 19:952. https://doi.org/10.3390/ijms19040952

Walsh F, Ingenfeld A, Zampicolli M et al (2011) Real-time PCR methods for quantitative monitoring of streptomycin and tetracycline resistance genes in agricultural ecosystems. J Microbiol Methods 86:150–155. https://doi.org/10.1016/j.mimet.2011.04.011

Wan M, Li G, Zhang J et al (2008) Effect of volatile substances of *Streptomyces platensis* F-1 on control of plant fungal diseases. Biol Control 46:552–559. https://doi.org/10.1016/j.biocontrol.2008.05.015

Wang M, Li H (2025) Structure, function, and biosynthesis of Siderophores produced by *Streptomyces* species. J Agric Food Chem 73:4425–4439. https://doi.org/10.1021/acs.jafc.4c08231

Wang D, Yuan J, Gu S, Shi Q (2013) Influence of fungal elicitors on biosynthesis of natamycin by *Streptomyces natalensis* HW-2. Appl Microbiol Biotechnol 97:5527–5534. https://doi.org/10.1007/s00253-013-4786-0

Wang W, Qiu Z, Tan H, Cao L (2014) Siderophore production by actinobacteria. Biometals 27: 623–631. https://doi.org/10.1007/s10534-014-9739-2

Wang X, Du Z, Chen C et al (2023a) Antifungal effects and biocontrol potential of lipopeptide-producing *Streptomyces* against banana Fusarium wilt fungus *Fusarium oxysporum* f. sp. *cubense*. Front Microbiol 14:1177393. https://doi.org/10.3389/fmicb.2023.1177393

Wang Z, Gao C, Yang J et al (2023b) Endophytic *Streptomyces* sp. NEAU-ZSY13 from the leaf of *Perilla frutescens,* as a promising broad-spectrum biocontrol agent against soil-borne diseases. Front Microbiol 14:1243610. https://doi.org/10.3389/fmicb.2023.1243610

Wang Z, Solanki MK, Kumar A et al (2023c) Promoting plant resilience against stress by engineering root microenvironment with *Streptomyces* inoculants. Microbiol Res 277:127509. https://doi.org/10.1016/j.micres.2023.127509

Wang M, Li H, Li J et al (2024) *Streptomyces* strains and their metabolites for biocontrol of phytopathogens in agriculture. J Agric Food Chem 72:2077–2088. https://doi.org/10.1021/acs.jafc.3c08265

Wei M, Jiao M, Nie X et al (2024) Genomic and metabolomic profiling reveal *Streptomyces rochei* S32 contributes to plant growth by nitrogen fixation and production of bioactive substances. Plant Soil 501:343–360. https://doi.org/10.1007/s11104-024-06667-x

Wend K, Zorrilla L, Freimoser FM, Gallet A (2024) Microbial pesticides—challenges and future perspectives for testing and safety assessment with respect to human health. Environ Health 23: 49. https://doi.org/10.1186/s12940-024-01090-2

Worsley SF, Macey MC, Prudence SMM et al (2021) Investigating the role of root exudates in recruiting *Streptomyces* bacteria to the *Arabidopsis thaliana* microbiome. Front Mol Biosci 8: 686110. https://doi.org/10.3389/fmolb.2021.686110

Wu W, Du K, Kang X, Wei H (2021) The diverse roles of cytokinins in regulating leaf development. Hortic Res 8:118. https://doi.org/10.1038/s41438-021-00558-3

Xia H, Li X, Li Z et al (2020) The application of regulatory cascades in *Streptomyces*: yield enhancement and metabolite mining. Front Microbiol 11:406. https://doi.org/10.3389/fmicb.2020.00406

Xu T, Cao L, Zeng J et al (2019) The antifungal action mode of the rice endophyte *Streptomyces hygroscopicus* OsiSh-2 as a potential biocontrol agent against the rice blast pathogen. Pestic Biochem Physiol 160:58–69. https://doi.org/10.1016/j.pestbp.2019.06.015

Yang C-J, Huang T-P, Huang J-W (2021) Field sanitation and foliar application of *Streptomyces padanus* PMS-702 for the control of rice sheath blight. Plant Pathol J 37:57–71. https://doi.org/10.5423/ppj.oa.12.2020.0227

Yang M, Zhang W, Lv Z et al (2023) Induced defense response in soybean to *Sclerotinia sclerotiorum* using Wuyiencin from *Streptomyces albulus* CK-15. Plant Dis 107:107–115. https://doi.org/10.1094/pdis-03-22-0582-re

Yang B, Wen H, Wang S et al (2024) Enhancing drought resistance and yield of wheat through inoculation with *Streptomyces pactum* Act12 in drought field environments. Agronomy 14:692. https://doi.org/10.3390/agronomy14040692

Yu Y, Gui Y, Li Z et al (2022) Induced systemic resistance for improving plant immunity by beneficial microbes. Plants 11:386. https://doi.org/10.3390/plants11030386

Yuan M, Huang Y, Ge W et al (2019) Involvement of jasmonic acid, ethylene and salicylic acid signaling pathways behind the systemic resistance induced by *Trichoderma longibrachiatum* H9 in cucumber. BMC Genomics 20:144. https://doi.org/10.1186/s12864-019-5513-8

Yun T, Jing T, Zang X et al (2023) Antimicrobial mechanisms and secondary metabolite profiles of *Streptomyces hygroscopicus* subsp. *hygroscopicus* 5–4 against banana fusarium wilt disease using metabolomics. Front Microbiol 14:1159534. https://doi.org/10.3389/fmicb.2023.1159534

Zamoum M, Goudjal Y, Sabaou N et al (2017) Development of formulations based on *Streptomyces rochei* strain PTL2 spores for biocontrol of *Rhizoctonia solani* damping-off of tomato seedlings. Biocontrol Sci Technol 27:723–738. https://doi.org/10.1080/09583157.2017.1334257

Zang C, Kong T, Liang B et al (2023) Evaluation of imide substance from *Streptomyces atratus* PY-1 for the biocontrol of Phytophthora blight. Eur J Plant Pathol 165:725–734. https://doi.org/10.1007/s10658-023-02648-5

Zeng J, Xu T, Cao L et al (2018) The role of iron competition in the antagonistic action of the Rice endophyte *Streptomyces sporocinereus* OsiSh-2 against the pathogen *Magnaporthe oryzae*. Microb Ecol 76:1021–1029. https://doi.org/10.1007/s00248-018-1189-x

Zeyad MT, Tiwari P, Ansari WA et al (2022) Bio-priming with a consortium of *Streptomyces araujoniae* strains modulates defense response in chickpea against Fusarium wilt. Front Microbiol 13:998546. https://doi.org/10.3389/fmicb.2022.998546

Zhang S, Vallad GE, White TL, Huang C-H (2011) Evaluation of microbial products for management of powdery mildew on summer squash and cantaloupe in Florida. Plant Dis 95:461–468. https://doi.org/10.1094/PDIS-07-10-0521

Zhang Q, Yong D, Zhang Y et al (2016) *Streptomyces rochei* A-1 induces resistance and defense-related responses against *Botryosphaeria dothidea* in apple fruit during storage. Postharvest Biol Technol 115:30–37. https://doi.org/10.1016/j.postharvbio.2015.12.013

Zhang S, Yang Y, Wu Z, Li K (2020) Induced defense responses against *Rhizoctonia solani* in rice seedling by a novel antifungalmycin N2 from *Streptomyces* sp. N2. Australas Plant Pathol 49: 267–276. https://doi.org/10.1007/s13313-020-00703-x

Zhang L, Liu Z, Wang Y et al (2022) Biocontrol potential of endophytic *Streptomyces malaysiensis* 8ZJF-21 from medicinal plant against banana Fusarium wilt caused by *Fusarium oxysporum* f. sp. *cubense* tropical race 4. Front Plant Sci 13:874819. https://doi.org/10.3389/fpls.2022.874819

Zhang F, Wen S, Wang B et al (2023a) Biocontrol potential of *Streptomyces odonnellii* SZF-179 toward *Alternaria alternata* to control pear black spot disease. Int J Mol Sci 24:17515. https://doi.org/10.3390/ijms242417515

Zhang Y, Pan L, Zhang Y et al (2023b) Understanding the *Streptomyces albulus* response to low-pH stress at the interface of physiology and transcriptomics. Appl Microbiol Biotechnol 107:2611–2626. https://doi.org/10.1007/s00253-023-12449-9

Zhang Y, Cao X, Liu Q et al (2024) Multi-omics analysis of *Streptomyces djakartensis* strain MEPS155 reveal a molecular response strategy combating *Ceratocystis fimbriata* causing sweet potato black rot. Food Microbiol 122:104557. https://doi.org/10.1016/j.fm.2024.104557

Zhao Y, Sun C, Wang S et al (2023) Widely targeted metabolomic, transcriptomic, and metagenomic profiling reveal microbe–plant–metabolic reprogramming patterns mediated by *Streptomyces pactum* Act12 enhance the fruit quality of Capsicum annuum L. Food Res Int 166: 112587. https://doi.org/10.1016/j.foodres.2023.112587

Zhu H, Hu L, Hu H et al (2023) Identification of a novel *Streptomyces* sp. strain HU2014 showing growth promotion and biocontrol effect against *Rhizoctonia* spp. in wheat. Plant Dis 107:1139–1150. https://doi.org/10.1094/pdis-06-22-1493-re

Zong G, Fu J, Zhang P et al (2022) Use of elicitors to enhance or activate the antibiotic production in *Streptomyces*. Crit Rev Biotechnol 42:1260–1283. https://doi.org/10.1080/07388551.2021.1987856

Exploration of Streptomycetes as Biocontrol Agents Against Plant Pathogens

6

Mekala Jeyaram, Yasotha Jeyaram, Thirumurugan Ayyadurai, and Manikandan Gurusamy

Abstract

Plant diseases pose a major threat to the world's food security, necessitating the development of sustainable and ecologically friendly control strategies. *Streptomyces* spp. have gained great attention in sustainable agriculture as biological control agents due to their unique ability to synthesize diverse antibacterial compounds. The filamentous bacteria *Streptomyces*, and plant hosts, have developed sophisticated interactions where both organisms are able to furnish defence against disease, as well as obtain benefits from plant-derived nutrients. Multiple members of the genus *Streptomyces* show antagonistic physiological activity against phytopathogens. Through various mechanisms including competition, parasitism, antibiosis and stimulation of the plant to induce systemic resistance, *Streptomyces* are exceptional at the suppression of plant diseases. Furthermore, production of lytic enzymes creates the ability to degrade organic polymers, increasing overall biocontrol potential. The goal of this chapter is to summarize the genus as an anthropogenic source of antimicrobial compounds with interest given to their biocontrol potential in the suppression of plant diseases. By

M. Jeyaram
PG Department of Zoology, Arulmigu Palaniandavar Arts College for Women, Palani, Tamil Nadu, India

Y. Jeyaram
Department of Botany, Sri Kaliswari College (Autonomous), Sivakasi, Virudhunagar Dist, Tamil Nadu, India

T. Ayyadurai
Department of Botany, Bharathidasan University, Tiruchirappalli, Tamil Nadu, India

M. Gurusamy (✉)
AgriSETA Research Group, College of Human Sciences, School of Research and Graduate Studies, University of South Africa, Pretoria, South Africa

T. Kaur et al. (eds.), *Streptomyces: Biological Candidates for Sustainable Agriculture*,
https://doi.org/10.1007/978-981-95-5803-2_6

studying the bioactive compounds produced by these microorganisms, including other classes being discovered at this time, we will develop better awareness of their modes of action and identify new tactics to apply the biocontrol potential of *Streptomyces*. Ultimately, these ideas can be translated into sustainable or eco-friendly strategies for plant disease management.

Keywords

Plant diseases · Phytopathogens · Chemical compounds · *Streptomyces* sp. · Biocontrol agents

6.1 Introduction

Plant pathogens such as bacteria, fungi, viruses and nematodes are responsible for several plant diseases, including black spots on the leaves, decreased fruiting, death of the tissues (necrosis) and wilting. These pathogens can enter plants through wounds or injury and secrete enzymes or toxins to break down the plant cells which causes either damaged or dead cells. As reported by Ahmad et al. (2023), a review of 70 studies reviewed the employment of deep learning methods for plant disease identification and analysis. Deep learning methods have been utilized across multiple crops in the early season for plant disease diagnosis and management of the plant disease system. Additionally, many other researchers have also indicated that deep learning methods have been useful for plant disease diagnosis and management in different crops (Johannes et al. 2017; Syed-Ab-Rahman et al. 2022; Barbedo 2016).

Recent technologies such as gene cloning, rDNA technology, molecular markers and other techniques in biotechnology have reduced crop losses due to early detection of diseases (Kumar 2014). Though artificial pesticides drastically reduce the plant diseases, but their excessive and indiscriminate use has resulted in significant negative impact to humans, environment, soil fertility, etc. In contrast, biocontrol agents have shown potential by interacting with natural microbiota, acting as microbial inoculants and protecting host plants against pathogens (Tariq et al. 2020; Pal and Gardener 2006). For example, *Bacillus licheniformis* BL06 strain was reported to suppress the plant pathogen, *Phytophthora capsici*, in pepper plants (2020). Similarly, *Bacillus subtilis* 26DCryChS controlled the effects of *Stagonospora nodorum* Berk., a pathogen of wheat crop (Maksimov et al. 2020). *Pseudomonas fluorescens* controlled apple mucor rot disease (Wallace et al. 2018) while panicle blight disease in soybean was controlled by *Pseudomonas parafulva* JBCS1880 (Kakembo and Lee 2019). Rice blast disease, caused by *Magnaporthe oryzae* reduces the yield, but it has been managed with the biocontrol agent, *Pseudomonas putida* BP25, which inhibit the *M. oryzae* in rice (Ashajyothi et al. 2020). Along with these microbial agents, the streptomycetes, belonging to the phylum Actinobacteria, are also known to produce many secondary metabolites with antimicrobial, insecticidal and herbicidal properties and many researchers have reported their potential to use as promising biological control agents in agriculture (Awad et al. 2009; Law et al. 2017; Ser et al. 2018). This chapter focuses on an exploration of streptomycetes as biocontrol agents against plant pathogens. It will highlight limitation of chemical pesticides and will describe exploration of streptomycetes as biocontrol agents, classification, morphology and physiology of *Streptomyces*; mechanisms of biocontrol exhibited by streptomycetes against plant pathogens, applications, formulations; challenges of streptomycetes as biocontrol agents and future perspectives.

6.2 Effect of Plant Pathogens on Agriculture

In agriculture, pathogenic microorganisms causing plant diseases constitute a significant barrier and result in around 30–40% of crop damage (Gurusamy and Katerere 2026; Khan et al. 2023). According to Nazarov et al. (2020) plant diseases are classified as two types, infectious and non-infectious based on their causal organism. Non-infectious plant diseases don't spread to another plant while infectious plant diseases can be transferred from a sick plant to a wealthy plant.

High-density monocultures of soybean and wheat are affected by a multitude of pests and pathogens, which compromise their yields. For instance, *Phakopsora pachyrhizi* induced soybean rust and *Zymoseptoria tritici* induced wheat blotch, the two most devastating diseases affecting these crops, with documented yield reductions exceeding 50% during intense outbreaks (Fones et al. 2020; Goellner et al. 2010). According to Xu et al. (2021) maize is confronted with a variety of threats, such as fungal, bacterial, viral, nematodal infections and parasitic flora. Maize crops are affected by roughly 112 different diseases, with about 70 of these being seed-transmitted (Lv et al. 2020).

Leaf rust is the most common of the wheat rust diseases and is caused by *Puccinia triticina*. The pathogen exists in mild, damp environments and causes a reduction in kernel weight and grain per head. Although leaf rot causes variability in grain losses depending on time and location, but it has significant economic implications. Leaf rust is also challenging because of pathogen diversity, its creation of new virulence profiles and adaptations, granting it an advantage under different conditions (Moolhuijzen et al. 2022).

Rice crops are prone to numerous destructive diseases caused by bacteria, fungi or viruses, including blast, sheath blight, leaf blights, brown spot, sheath rot and bakanae disease (Ngalimat et al. 2021). Foliar diseases, that affect rice crops as well as those that impact the stem, roots or leaf sheaths, have been caused by *Gerlachia, Helminthosporium oryzae*, *Pyricularia*, *Rhizoctonia solani*, *Sclerotium, Xanthomonas* and other pathogens in several countries that grow rice (Chukwu et al. 2019). These diseases could potentially cause losses between 1% and 100% depending on growth conditions, varietal susceptibility and other factors (Sankaran et al. 2010). Late blight, the result of the fungal pathogen *Phytophthora infestans*, responsible for the catastrophic Irish potato famine of the 1840s, has always been one of the important diseases. Today, late blight still presents a notable challenge to potato producers worldwide, demonstrating new characteristics of having a stronger preference for tomato and potato plants (Kumbar et al. 2019).

6.3 Limitations of Chemical Pesticides

Chemical pesticides are synthetic substances that are used to control or eradicate insects, weeds, fungi and other agricultural pests. Depending upon their target pest, they can be divided into many types such as insecticides, herbicides, fungicides, bactericides and nematicides. Many researchers have discussed their limitations, including environmental contaminations, chronic health problems, pest resistance, bioaccumulation and biomagnification (Nicolopoulou-Stamati et al. 2016). These agrochemicals can accumulate in the body, causing skin damage and nutritional deficiencies (Alewu and Nosiri 2011; Pirsaheb et al. 2015). Other serious health effects associated with exposure to chemical pesticides include respiratory disorders, neurological and gastrointestinal problems, reproductive and dermatological issues, and endocrine disruption (Alewu and Nosiri 2011; Sanborn et al. 2007; Mnif et al. 2011; Mostafalou and Abdollahi 2013; Thakur et al. 2014). For example, carbamate pesticides such as aldicarb, carbofuran and ziram have been reported to cause possible reproductive disorders and endocrine-disrupting activities (Kniewald et al. 2000; Goad et al. 2004; Mnif et al. 2011; Jin et al. 2014).

6.4 Agriculturally Important Antimicrobial Compounds Against Plant Pathogens

More than a century has passed since the introduction of the antimicrobial agents. In parts of the world, animals and plants receive far more antimicrobial agents than humans (O'Neill 2015). Mellon et al. (2001) estimated that 29.5 million pounds of antimicrobials are given to animals, and 40,000 pounds are applied to plant agriculture every year. Economic growth tends to encourage the use of antimicrobial agents in agriculture. Some level of use is due to the economic and environmental benefits of helping plants thrive and produce food for human consumption. In agriculture, antimicrobials are primarily used to improve plant health and promote faster growth (Lekshmi et al. 2017).

Bacterial diseases in plants and crops are not as common as those caused by viruses and fungi. According to Vidaver (2001), in the United States, antimicrobials are sprayed on 53,000 hectares annually. Nevertheless, the application of antimicrobials like antibiotics for prophylactic treatment is restricted in both variety and availability due to limited effectiveness and environmental issues. The emergence of antimicrobial resistance in plants and crops is a contributing factor, having begun as early as the 1960s (Jones 1982). The situation deteriorated when checks on bacterial resistance were done only periodically. Nevertheless, no documented human health effects have been observed from the introduction of antimicrobials in plants over the past 60 years (Vidaver 2002).

Citrus fruits have a high water and nutrient content, which can make them very susceptible to infection by microbial pathogens like *Penicillium digitatum*, during the period between the time of harvest and consumption (Talibi et al. 2014). Mohammadi et al. (2017) reported that *Bacillus subtilis* and its metabolites have considerable antimicrobial activity against green mould pathogen, that infects citrus fruits. To control bacterial plant diseases like fire blight of pear and apple caused by *Erwinia amylovora*, antibiotics, like streptomycin, oxytetracycline, oxolinic acid and gentamicin are utilized. Antibiotic sprays applied in spring prevent the growth of pathogens on flowers and leaves prior to infection. It is used when the risk of disease and infection is elevated. Plants are only susceptible to antibiotics for under a week, and these drugs have no effect after the onset of infection. For more than five decades in the United States, antibiotics have been crucial for crop protection, with no reported negative consequences to human health or persistency through the environment (Stockwell and Duffy 2012).

6.5 Commercial Formulations for Phytopathogens

Plant diseases diminish crop yields and render them inadequate to meet demand, food security has emerged as the foremost global issue (Strange and Scott 2005). The occurrence of various pathogens in crops has serious impact on agricultural yields. To manage plant pathogens, various strategies are employed, such as utilizing resistant plant varieties and practicing crop rotation (Shuping and Eloff 2017). These pathogens are even controlled with chemical pesticides, which is the most widely used method. However, the overuse of these chemical pesticides has resulted in considerable environmental and health issues (Nicolopoulou-Stamati et al. 2016). As a result, methods that are environmentally friendly are currently being investigated to manage harm caused by plant pathogens. Natural antagonistic microbes, referred to as biocontrol agents, are being utilized. This approach is eco-friendly and, in certain instances, the sole option for managing these plant pathogens (Köhl et al. 2019). In addition to managing the plant diseases, these biocontrol agents also contribute to improved plant growth. Moreover, they enhance stress tolerance, builds resistance in plants and aids nutrient acquisition (Andersen et al. 2018).

Diverse microbial populations have been marketed and utilized as biocontrol agents across various industries. They serve either to combat phytopathogens or to manufacture different medications. Various biocontrol agents are commercially available, targeting specific plant pathogens. These include *Agrobacterium radiobacter* (Galtrol) for *Agrobacterium tumefaciens*, *Ampelomyces quisqualis* (AQ10) for powdery mildew, *Aspergillus flavus* AF36 (Alfa guard) for *A. flavus*, *Bacillus subtilis* (Serenade, RhizoPlus) for various pathogens, *Trichoderma harzianum* (Planter Box, Root Shield) for *Fusarium* and *Rhizoctonia*, and *Pseudomonas fluorescens* (Frostban) for bunch rot and fire blight. These biocontrol agents offer alternatives to chemical pesticides (Singh et al. 2021).

6.6 Role of Streptomycetes as Biocontrol Agents

Biocontrol can be established using streptomycetes, offering a viable alternative to chemical control. *Streptomyces* species produce numerous antimicrobial chemicals to reduce the growth of plant pathogens. For instance, 9(10H)-acridanone produced by *Streptomyces fradiae* strain VITMK2 inhibited white spot syndrome virus (Hayakawa et al. 2013). *Streptomyces rimosus* can reduce the effects of pesticide pollution (Khajezadeh et al. 2020). *Streptomyces yanglinensis* controlled contamination of feed and food products caused by *A. flavus* mycelium (Lyu et al. 2020). *Streptomyces griseoviridis* was used to control verticillium and black scorch disease in tomatoes and date palms caused by *Fusarium oxysporum* sp., *lycopersici* and *Ceratocystis radicicola*, respectively (Suleman et al. 2002; Minuto et al. 2006) (Table 6.1).

Table 6.1 Exploring streptomycetes as biocontrol agents

S. No.	*Streptomyces* species/strain	Bioactive compounds/ mechanism	Target pathogens/diseases	References
1.	*S. fradiae* VITMK2	9(10H)-Acridanone	White spot virus	Hayakawa et al. (2013)
2.	*Streptomyces rimosus*	Biodegradation	Reduces pesticide	Khajezadeh et al. (2020)
3.	*Streptomyces yanglinensis*	Antifungal compounds	Controls *A. flavus* (food/feed contamination)	Lyu et al. (2020)
4.	*S. griseoviridis* K61	Antifungal activity	Controls *C. radicicola* (black scorch in date palms) and *F. oxysporum* (tomato wilt)	Suleman et al. (2002), Minuto et al. (2006)
5.	*Streptomyces violaceusniger* YCED9	Antifungal compounds	Controls soil fungal diseases	Rey and Dumas (2017)
6.	*Streptomyces saraceticus* KH400	Antifungal compounds		Rey and Dumas (2017)
7.	*S. lydicus* WYEC108	Antifungal compounds		Rey and Dumas (2017)
8.	*S. griseoviridis* K61	Antifungal compounds		Rey and Dumas (2017)
9.	Commercial compounds from *Streptomyces*	Kasugamycin, Polyoxin D, Streptomycin	Fungicides and bactericides	Rey and Dumas (2017)

6.6.1 Mechanisms of Biocontrol Exhibited by Streptomycetes

6.6.1.1 Secondary Metabolite Production: Antibiotics, Antifungal and Other Compounds

Streptomyces-derived natural compounds are an essential source of priceless medications and medicinal substances. About 100,000 antibiotic compounds are produced by *Streptomyces*; all these naturally occurring bioactive substances have pharmacological or agrochemical potential (Harir et al. 2018; Abdel-Razek et al. 2020). Bioactive products from *Streptomyces* may have antihypertensive, immunosuppressive, antimicrobial, insecticidal, antitumor and plant growth-promoting properties. Some endophytic *Streptomyce*s also demonstrate biosurfactant properties (Pham et al. 2021; Sharma et al. 2021; Nguyen et al. 2020; Salwan and Sharma 2020). For biosurfactant synthesis by *Streptomyces* sp. DPUA1566 isolate, corn steep liquor was used as medium component (Santos et al. 2019).

The production of a broad range of bioactive substances such as anthelmintics like avermectin, exoenzymes like chitinase, volatile organic compounds like HCN and antimicrobials make *Streptomyces* species valuable biological control agents (Newitt et al. 2019). These diverse activities help them to overpower several phytopathogens like *Rhizoctonia solani, Gaeumannomyces graminis* var. *tritici, Magnaporthe oryzae* and *Fusarium* species, and *Alternaria* spp. (Dean et al. 2012; Law et al. 2017). For instance, Jung et al. 2013 reported reduction in *Fusarium*-induced seedling stunning by *Streptomyces* strain BN1. They also act as key agent in regulating wheat take-all and other soil-borne illness due to their spore-forming and saprotrophic features, which also supports their persistence in soil (Newitt et al. 2019).

Additionally, *Streptomyces* produced volatile organic compounds which hold the potential as biofumigants to control fungal agents such as *R. solani* (Chapelle et al. 2016). However, their efficacy in natural environments needs to be confirmed with further elaborated research.

6.6.1.2 Competition

Streptomyces is a well-known gram-positive bacterium which is well known for generating several secondary metabolites (Kinkel et al. 2012). These secondary metabolites are needed for arbitrating microbial competition. They can help to inhibit the growth of plant pathogens and to make disease-suppressive soils. On the other hand, niche partitioning may lower the condition for antagonistic interactions by decreasing competition, which can reduce the suppression of illness (Kinkel et al. 2011).

These secondary metabolites in *Streptomyces* also increase food intake to stop multiplication of soil-borne pathogens (bacteria, fungi or viruses), and can also use new growth and dispersal strategies to compete for environmental niches. They are strongly associated with plants and insects, often offering their host a form of defence through the production of antimicrobial compounds with activity against pathogens (Shepherdson et al. 2023). They have the capacity to produce a wide variety of antibiotics that can inhibit the growth of other microorganisms that can exist in the soil. For example, *Streptomyces coelicolor* produce an antibiotic called actinorhodin, which can control the production of other microorganisms and maintain supremacy in the soil (Park et al. 2017) (Table 6.2).

6.6.1.3 Mycoparasitism

Streptomyces species have been documented to possess mycoparasitic activity through the secretion of cell wall degrading enzymes such as chitinases, glucanases and proteases. Along with the antifungal secondary metabolites, these enzymes allow *Streptomyces* to limit the fungal development and assist in the inhibition of fungal disease in rhizosphere and soil conditions (van der Meij et al. 2017). These hydrolytic enzymes break down fungal hypahe by targeting cell walls, inhibiting their growth and existence. For instance, a biopesticide, Mycostop, from *Streptomyces griseoviridis* reduces wheat head blight due to *Fusarium* spp. (Lahdenpera et al. 1991) by producing hydrolytic enzymes.

Similarly, *Streptomyces venezuelae*, a potent producer of hydrolytic enzymes also holds the significant potential as biocontrol agent against fungal phytopathogens. Its ecological flexibility and possible function as a mycoparasite in complicated microbial communities is maintained by the secretion of chitinase (Shin et al. 2018) (Table 6.2).

6.6.1.4 Induced Systemic Resistance (ISR)

Streptomyces can induce systemic resistance to different pathogens via competitive exclusion and activated resistance mechanisms in the host (Ebrahimi-Zarandi et al. 2022). Induced systemic resistance (ISR) can lead to the accumulation of defence-related compounds, localized cell death and cell wall reinforcement (Viaene et al. 2016). Kurth et al. reported upregulation of pathogenesis-related proteins in oak trees in the presence of *Streptomyces* sp. AcH505 (Kurth et al. 2014) (Table 6.2).

6.7 Streptomycetes as Biocontrol Agents Against Different Plant Pathogens

In agriculture plant disease brought on by pathogenic microbes like bacteria, fungi, virus and nematodes cause a major impact all over the world. About 30–40% of the crop damage is due to these pathogenic microbes. To control these pathogens, antimicrobials and chemical pesticides are used but these chemicals may cause major negative impact on environment and human health. Hence it is essential to use natural alternatives, where *Streptomyces* have been emerged as potentially effective natural substitute to agrochemicals because of their ability to produce a variety of antimicrobial chemicals. Due to multitude of mechanisms exhibited by *Streptomyces* spp., they show promising potential to be used as biological control agents (Khan et al. 2023).

Table 6.2 Mechanisms of action by *Streptomyces* spp.

Mechanism	Key features/ Compounds/ Enzymes	Target pathogens/ Diseases	Examples/ Strains	References
Antibiosis	Exoenzymes—chitinase Anthelmintics—avermectin	*Rhizoctonia solani*, *Fusarium* spp., *Gaeumannomyces graminis*, *Magnaporthe oryzae*	*Streptomyces* BN1 (reduces *Fusarium* seedling stunting)	Newitt et al. (2019), Dean et al. (2012), Jung et al. (2013)
	Volatile organic compounds as biofumigants	*R. solani*	*Streptomyces* spp. (antifungal VOCs)	Chapelle et al. (2016)
	Symbiotic nitrogen fixation and mycorrhizal associations	Enhances soil health indirectly	*Streptomyces* spp. in mutualistic relationships	Gregor et al. (2003), Schrey and Tarkka (2008)
Competition	Secondary metabolites mediate microbial competition; niche partitioning reduces pathogen survival	Broad-spectrum soil pathogens	*Streptomyces coelicolor* (produces actinorhodin)	Kinkel et al. (2012), Park et al. (2017)
	Antibiotics inhibit phage replication and competitor growth	Soil bacteria/fungi	*Streptomyces* spp. (host-defence symbionts)	Shepherdson et al. (2023)
Mycoparasitism	Secretion of cell wall-degrading enzymes (chitinases, glucanases and proteases)	Fungal pathogens (*Fusarium* spp.)	*S. griseoviridis* (Mycostop® controls wheat head blight)	van der Meij et al. (2017), Lahdenpera et al. (1991)
	Chitinase degrades fungal/arthropod cell walls	Chitin-rich pathogens (e.g., fungi, insects)	*S. venezuelae* (extracellular chitinases)	Shin et al. (2018)
Induced Systemic Resistance (ISR)	Activation of host defence pathways	Broad-spectrum pathogens	*Streptomyces* sp. AcH505 (induces ISR in oak trees)	Kurth et al. (2014), Ebrahimi-Zarandi et al. (2022)
	Competitive exclusion and localized cell death	Soil-borne diseases	*Streptomyces* spp. (rhizosphere colonization)	Viaene et al. (2016)

6.7.1 Bacterial Pathogens

Bacterial plant diseases are a significant threat to global agriculture, impacting crop yield, quality and food security. Among the most destructive are bacterial wilt and bacterial blight, which affect a wide range of crops, including solanaceous vegetables, cereals and legumes.

6.7.1.1 Bacterial Wilt

It is caused by *Ralstonia solanacearum*, which mainly infects tomato, potato, eggplant, pepper and banana. The main symptoms of this disease are sudden wilting of foliage, browning of vascular tissue, bacterial ooze from the cut stems (Fall et al. 2015) (Table 6.3).

Table 6.3 Effect of *Streptomyces* species against bacterial diseases

S. No.	*Streptomyces* strains	Crop disease	References
1.	*Streptomyces* sp.	Bacterial blight against potato	Jin et al. (2023)
2.	*Streptomyces diastatochromogenes*	Bacterial soft rot in potato, carrot and cabbage	Doolotkeldieva et al. (2016)
3.	*Streptomyces humidus*	Bacterial canker in almond	Solanki et al. (2016)
4.	*Streptomyces hydrogenans IB310*	Soft rot, black rot and crowngall disease in cabbage and almond	Kulkarni et al. (2017)
5.	*Streptomyces hygroscopicus* var. *limoneus*	Bacterial blight	Fan et al. (2013)
6.	*Streptomyces* sp. *161a*	Bacterial panicle blightdisease in rice	Betancur et al. (2019)
7.	*Streptomyces* sp. *AN090126*	Tomato bacterial wilt	Le et al. (2022)
8.	*Streptomyces* sp. *PNM-9*	Bacterial panicle blightdisease in rice	Betancur et al. (2019)
9.	*Streptomyces* sp. *strain J145*	Black rot in cabbage	Solanki et al. (2016)
10.	*Streptomyces strain 22-4*	Tomato bacterial wilt	Wattana-Amorn et al. (2016)
11.	*Streptomyces* strain FJAT-31547	Bacterial wilt	Fall et al. (2015)
12.	*Streptomyces strain FJAT-31547*	Bacterial wilt of tomato	Zheng et al. (2019)
13.	*Streptomyces strain JJ45*	Black rot in cabbage	Muangham et al. (2015)
14.	*Streptomyces termitum ATC-2*	Bacterial blight in rice	Solanki et al. (2016)
15.	*Streptomyces* sp. *JCK-6131*	Tomato bacterial wilt	Le et al. (2021)

6.7.1.2 Bacterial Blight

It is caused by *Xanthomonas oryzae* pv. *oryzae* (rice) and *Pseudomonas syringae* (various crops). It mainly affects rice, beans, cotton and soybean. The symptoms include water-soaked lesions, leaf blight, wilting, cankers and dieback. Some of the *Streptomyces* species that act against these diseases are listed in Table 6.3 (Fan et al. 2013).

6.7.2 Fungal Pathogens

Fungal diseases are among the most widespread and economically damaging afflictions in agriculture. Fungal pathogens affect almost every crop species, leading to yield loss, reduced quality and post-harvest decay. Notable diseases such as powdery mildew and rust are prevalent globally and require robust management strategies (Table 6.4). Powdery mildew is a fungal disease caused by *Erysiphe*, *Oidium* and *Podosphaera*. It affects the grapevine, cereals, cucurbits and ornamentals. The symptoms of powdery mildew include white, powdery fungal growth on leaves, stems and sometimes fruits, chlorosis and premature leaf drop. Rust is also another important fungal disease caused by *Puccinia*, *Uromyces* and *Melampsora*. This fungal disease affects the wheat, beans, coffee, sunflower and ornamental plants with symptoms such as rust-coloured pustules on leaves and stems, defoliation, reduced vigour and yield (Kishimoto et al. 1996; Sheng et al. 2015). Some of the *Streptomyces* sp., *Streptoverticillium rimofaciens* and *S. hygroscopicus* are effective against powdery mildew and rust diseases.

6.7.3 Viral Pathogens: Control of Viral Diseases

Plant viruses are microscopic agents capable of causing significant damage to crops, affecting global food production, marketability and quality. Unlike fungal or bacterial pathogens, viruses cannot be managed by direct chemical treatments. Diseases

Table 6.4 Effect of *Streptomyces* species against fungal pathogens

S. No.	Disease	Causal agent (s)	*Streptomyces* strains	References
1.	Powdery mildew	*Erysiphe*, *Oidium* and *Podosphaera*	*Streptomyces* sp., *Streptomyces* sp. KNF 2047, *Streptoverticillium rimofaciens*	Schumacher and Hall (1982), Satomi et al. (1982) and Kishimoto et al. (1996)
2.	Rust	*Puccinia*, *Uromyces* *Melampsora*	*Streptomyces hygroscopicus*	Sheng et al. (2015)

Table 6.5 *Streptomyces* species effective against TMV

Disease	Virus name	Hosts	Key component	References
Tobacco Mosaic Virus (TMV)	*Tobacco mosaic virus*	*Streptomyces cellulosae*	Actino 48	Abo-Zaid et al. (2020)
		S. ahygroscopicus	ε-poly-L-lysine	Shirling and Gottlieb (1966)
		Streptomyces sp.	GS-93-23	Xiao et al. (2002)
		Streptomyces albulus	Anisomycin	Shu et al. (2024)

such as tobacco mosaic virus (TMV) and other viral infections rely heavily on preventive and vector management strategies. Understanding the nature of viral pathogens and their transmission, effective integrated control measures are essential for maintaining plant health. The TMV mainly affects tobacco, tomato, pepper and petunia. Some of the symptoms include mottling, leaf distortion, mosaic patterns and stunted growth. Table 6.5 presents the control of viral diseases by *Streptomyces* spp.

6.7.4 Nematode Pathogens: Control of Nematode Diseases

Nematodes are small roundworms that can only be seen under a microscope. They are commonly found in soil and plant tissues including those that are plant parasites. Plant-parasitic nematodes, including the root-knot nematodes (*Meloidogyne* spp.) can cause yield losses worldwide. Plant-parasitic nematodes invade plant roots and interfere with a plant's ability to absorb water and nutrients, and therefore, results in stunted or wilted growth along with decreased productivity. Their effective management requires an integrated approach using cultural, biological, chemical and genetic strategies.

6.7.4.1 Root-Knot Nematodes (RKN)

Causal agents: *Meloidogyne incognita*, *Meloidogyne javanica*, *Meloidogyne arenaria*

Hosts: Tomato, potato, pepper, okra, cucurbits, soybean

Symptoms: Galls or knots on roots, stunted growth, wilting, poor yield

Several *Streptomyces* species demonstrated significant nematicidal potential against root knot nematodes. Strains such as KPS-C004 (*Streptomyces galilaeus*), KPS-E004, KPS-A032, ST01 (*Streptomyces levis*), ST07 (*Streptomyces enissocaesilis*) and KA-1 (*Streptomyces nigrescens*) have shown effectiveness in nematode control (Nimnoi et al. 2017; Nimnoi and Ruanpanun 2020; Atif et al. 2023; Park et al. 2023) (Table 6.6).

Table 6.6 Nematode pathogens: control of nematode diseases

S. No.	*Streptomyces* species	Strain designation	References
1.	*S. galilaeus*	KPS-C004	Atif et al. (2023), Park et al. (2023)
2.	*S. levis*	ST01	Nimnoi and Ruanpanun (2020)
3.	*S. Enissocaesilis*	ST07	Nimnoi et al. (2017)
4.	*S. nigrescens*	KA-1	Atif et al. (2023)
5.	*Streptomyces plicatus*	ST10	Nimnoi and Ruanpanun (2020)
6.	*S. rochei*	ST15	Nimnoi et al. (2017)
7.	*Streptomyces sparsus*	T16	Park et al. (2023)
8.	Other strains	KPS-E004, KPS-A032	Atif et al. (2023)

6.8 Application and Formulation of Streptomycetes as Biocontrol Agents

6.8.1 Inoculum Production: Mass Production of Streptomycetes

The production of streptomycetes inoculum often involves growing the organism in a suitable medium under controlled conditions to produce a high concentration of viable cells or spores for subsequent use in other processes like fermentation. Common methods for growing or propagating inoculum in the laboratory include working with liquid media, solid media (such as sterilized rice), liquid spore suspensions and even dried spores. Growing streptomycetes on solid agar is an important first step in creating inoculum, but one of the first steps in the actual inoculum production is growing or propagating streptomycetes in liquid media, especially important for genetic modification, but also to produce secondary metabolites, and short- and long-term storage (Shepherd et al. 2010).

Streptomyces can alternatively be cultured on solid media, like sterilized rice grains, as a stable substrate for growth and sporulation in solid state fermentation process. This method can be useful for mass production of inoculum. The spores can be collected from a culture plate or sterilized rice and either re-suspended in a liquid medium or used alone as an inoculum. Solid substrates like sterile rice offer an effective low cost alternative substrate for high spore yields which are stable for storage and mass application (Soares et al. 2007).

6.8.2 Formulation: Development of Biocontrol Products

The development of biocontrol products in the form of powders or liquids involves designing of stable, effective, safe and easy-to-use formulation for enhanced activity of biocontrol agents. The formulation aims mainly to help with the product's handling and use in order to ensure that it remains viable and effective at the target site; to stabilize the biocontrol agents throughout their shelf life, transport and storage; and to protect the microbes from harmful environmental factors. While solid formulations can enhance the stability of biocontrol agents, liquid formulations are generally easier and cheaper to produce (Keswani et al. 2016). Besides these criteria, different categories of criteria are identified that may be considered when industries and scientific institutions start new programmes for the development of biological control.

6.8.2.1 Liquid Formulation

Various liquid formulations contain whole cultures or cell suspensions mixed with ingredients such as colourants, stabilizers, surfactants and biocontrol agents for enhanced dispersion properties. Two formulations are generally used: one that contains protective agents, mainly sugars (particularly trehalose) or polyols (like glycerol), and one that involves culturing growth cells in a sorbitol-modified medium and using isotonic solutions with trehalose for preservation.

6.8.2.2 Solid Formulation

Dried products have low viability because the drying process requires heat and dehydration. However, they are a feasible alternative due to reduced costs of production and ease of storage and transportation. This implies that developing a method for each isolate will require its own experimental performance. Teixidó et al. (2022) outline effective dehydration methods: freeze-drying, spray-drying, fluidized bed-drying, with the newest being fluidized bed-spray-drying.

6.8.3 Application Methods: Seed Treatment, Soil Drench, Foliar Spray

Biocontrol products, which utilize natural enemies to control pests, are applied in various ways depending on the agent to target the pest. These methods include soil inoculation, seed treatment, foliar application and releases into the environment.

6.8.3.1 Seed Treatment

Applying chemicals to seeds prior to planting is known as seed treatment. Its objective is to shield the seeds and upcoming plants from diseases and pests. Before being sown, seeds are treated with a physical, chemical or biological treatment. Products like insecticides, fungicides, nematicides and biocontrol agents are examples of agents. Treating seeds helps prevent illnesses and pests that are spread via seeds and soil. Additionally, it protects crops from early-season pests and illnesses. If these hazardous germs are not controlled quickly, they can be quite destructive. It can be controlled by using either dry formulations, wet treatments, and/or slurry treatments, with or without a method of seed dressing, seed coating and seed pelleting (Tsegaye et al. 2018).

6.8.3.2 Soil Drench

Soil drenching involves applying intact solutions of streptomycetes to the soil surrounding plants; this method can significantly reduce the disease inoculums, and effectively introduce and colonize these biocontrol agents. Alkaline soils present several physical and chemical properties correlating to the reduced plant growth and productivity; on the other hand, soil drenching could help remediate these issues and improve plant growth and productivity (Youssef et al. 2022). As soil drenching combines the application process of diluted products that are dispensed to the base of the plant with a slow impact, it takes time for the diluted chemicals to be absorbed by roots to disperse through the stem, limbs and into the leaves of the plant (Rao et al. 2020).

6.8.3.3 Foliar Spray

Foliar feeding is another method of delivering plant nutrients through leaf application, in which an aqueous solution is sprayed onto the leaves and plants absorb the nutrients/antimicrobial metabolites through the stomata and cuticles of the leaves. Foliar spray can be a useful practice if a plant is deficient in a nutrient (Youssef et al. 2022).

6.9 Challenges

The stability and consistency of biocontrol agents is vital for their efficacy, which includes their genetic stability and viability, as well as their ability to colonize and interact with the target plant. To maintain the microbial population and ensure that the intended biocontrol effect can be delivered against the target plant pathogen, biocontrol agents are routinely formulated with matching carrier materials. Microbial formulations may be developed in either granular or powdered formats depending on the carrier materials incorporated (Wong et al. 2019). Lastly, multiple control measures can be implemented to deploy less successful control measures that would not work on their own. Genetic modification of microbes and increasing mass production, formulation and application of microbes are other ways to enhance efficacy in regard to the antagonist (Spadaro and Gullino 2005).

The regulatory frameworks for registering and approving biocontrol products involve specific guidelines and procedures, often differing from conventional pesticide regulations (Helepciuc and Todor 2022). Since 2011, the European phytopharmaceutical regulation EC 1107/2009 has governed the limitations of biocontrol agents (BCAs). BCAs, including those of the biopesticide type, are in fact less intrusive methods of protecting plants worldwide. To start, it's important to highlight that the plant protection products (PPP) regulation does not contain the term "biocontrol" (Robin et al. 2022). Regulation (EU.) No 546/2011 opened the way to generalized regulations to assess and authorize plant protection products, and concurrently it enhanced pesticide registration laws for MBC registration and MBCP operator permission post-expansion of the pesticide package (Helepciuc and Todor 2022).

The examination of new *Streptomyces* strains includes going through the process of identifying and characterizing these strains, as well as understanding how these strains have evolved their unique metabolic abilities particularly in the area of producing metabolites including secondary metabolites such as antibiotics and plant growth promoting substances. New isolates of the bacteria from traditional medicine indicate that *Streptomyces* has likely been important for human health longer than we had previously identified. This new source of *Streptomyces* may help replenish the critically low supply of emergency antibiotics for our multi-drug resistance pathogen, while simultaneously yielding the structural diversity needed to generate a new class of antibiotics (Quinn et al. 2020). Two strains of *Streptomyces* that were isolated from the roots of *Tamarix tamarisk* were also classified as new species based on morphological, molecular and chemical properties. The two strains were determined to be suitable sources of biocontrol potential against the fungal pathogen, *Alternaria tenuissima*, where their generation of the biocontrol metabolites was identified and validated performatively through genome analysis, mass spectrometry assay and a biocontrol activity assay (Chen et al. 2024).

6.10 Conclusion and Future Prospects

The investigation of the use of streptomycetes as biocontrol agents has revealed their considerable potential in sustainable agriculture. These actinobacteria, which inhabit soils, excrete a broad range of secondary metabolites, some of which are strong antifungals, antibacterials and nematicides. The use of streptomycetes is congruent with the growing interest in crop protection methods that are based on biology and possess minimal negative environmental impact. The persistence of streptomycetes in soils, capability of sporulation, plus affinity for colonizing plant roots contributes to their efficacy and utility under field conditions. However, performance varies based on environmental variables and complex plant-microbe-pathogen interactions, and more studies will be needed to fully evaluate sustained use of these bacteria in agriculture. Future investigations should improve with respect to strain selection, understanding their molecular interactions and development of their formulation and delivery methods. Integration of streptomycetes as part of current pest management programmes will be essential to their successful commercialization, and hopefully occur with field assessments and regulatory permits. To conclude, *Streptomyces* are a resilient and an ecologically responsible alternative to pesticides in the path towards sustainable agriculture.

References

Abdel-Razek AS, El-Naggar ME, Allam A, Morsy OM, Othman SI (2020) Microbial natural products in drug discovery. Processes 8(4):470

Abo-Zaid GA, Matar SM, Abdelkhalek A (2020) Induction of plant resistance against tobacco mosaic virus using the biocontrol agent Streptomyces cellulosae isolate Actino 48. Agronomy 10(11):1620

Ahmad A, Saraswat D, El Gamal A (2023) A survey on using deep learning techniques for plant disease diagnosis and recommendations for development of appropriate tools. Smart Agric Technol 3:100083

Alewu B, Nosiri C (2011) Pesticides and human health. In: Pesticides in the modern world–effects of pesticides exposure. InTech, pp 231–250

Andersen HR, Tinggaard J, Grandjean P, Jensen TK, Dalgård C, Main KM (2018) Prenatal pesticide exposure associated with glycated haemoglobin and markers of metabolic dysfunction in adolescents. Environ Res 166:71–77

Ashajyothi M, Kumar A, Sheoran N, Ganesan P, Gogoi R, Subbaiyan GK, Bhattacharya R (2020) Black pepper (Piper nigrum L.) associated endophytic Pseudomonas putida BP25 alters root phenotype and induces defense in rice (Oryza sativa L.) against blast disease incited by Magnaporthe oryzae. Biol Control 143:104181

Atif AM, Elzamik FI, Mohamed GM, Al-Quwaie DA, Ashkan MF, Alqahtani FS, Abdelbasit HM (2023) Biological control of the root-knot nematode (Meloidogyne incognita) on eggplants with various chitinase-producing Streptomyces strains. Eur J Plant Pathol 167(3):371–394

Awad HM, El-Sahed K, El-Nakkadi A (2009) Isolation, screening and identification of newly isolated soil Streptomyces (Streptomyces sp. NRC-35) for b-lactamase inhibitor production. World Appl Sci J 7(5):637–646

Barbedo JGA (2016) A review on the main challenges in automatic plant disease identification based on visible range images. Biosyst Eng 144:52–60

Betancur LA, Forero AM, Romero-Otero A, Sepúlveda LY, Moreno-Sarmiento NC, Castellanos L, Ramos FA (2019) Cyclic tetrapeptides from the marine strain Streptomyces sp. PNM-161a with activity against rice and yam phytopathogens. J Antibiot 72(10):744–751

Chapelle E, Mendes R, Bakker PA, Raaijmakers JM (2016) Fungal invasion of the rhizosphere microbiome. ISME J 10(1):265–268

Chen YH, Zhang JX, Min Y, Liu Y, Wang JM, Bai LQ, Luo XX (2024) Two new strains of Streptomyces with metabolic potential for biological control of pear black spot disease. BMC Microbiol 24(1):1–14

Chukwu SC, Rafii MY, Ramlee SI, Ismail SI, Hasan MM, Oladosu YA et al (2019) Bacterial leaf blight resistance in rice: a review of conventional breeding to molecular approach. Mol Biol Rep 46(1):1519–1532

Dean R, Van Kan JA, Pretorius ZA, Hammond-Kosack KE, Di Pietro A, Spanu PD et al (2012) The top 10 fungal pathogens in molecular plant pathology. Mol Plant Pathol 13(4):414–430

Doolotkeldieva T, Bobusheva S, Suleymankisi A (2016) Biological control of Erwinia carotovora ssp. carotovora by Streptomyces species. Adv Microbiol 6(02):104

Ebrahimi-Zarandi M, Saberi Riseh R, Tarkka MT (2022) Actinobacteria as effective biocontrol agents against plant pathogens, an overview on their role in eliciting plant defense. Microorganisms 10(9):1739

Fall ML, Tremblay DM, Gobeil-Richard M, Couillard J, Rocheleau H, Van der Heyden H, Carisse O (2015) Infection efficiency of four Phytophthora infestans clonal lineages and DNA-based quantification of sporangia. PLoS One 10(8):e0136312

Fan Y, Yu Y, Jia X, Chen X, Shen Y (2013) Cloning, expression and medium optimization of validamycin glycosyltransferase from Streptomyces hygroscopicus var. jinggangensis for the biotransformation of validoxylamine A to produce validamycin A using free resting cells. Bioresour Technol 131:13–20

Fones HN, Bebber DP, Chaloner TM, Kay WT, Steinberg G, Gurr SJ (2020) Threats to global food security from emerging fungal and oomycete crop pathogens. Nat Food 1(6):332–342

Goad RT, Goad JT, Atieh BH, Gupta RC (2004) Carbofuran-induced endocrine disruption in adult male rats. Toxicol Mech Methods 14(4):233–239

Goellner K, Loehrer M, Langenbach C, Conrath UWE, Koch E, Schaffrath U (2010) Phakopsora pachyrhizi, the causal agent of Asian soybean rust. Mol Plant Pathol 11(2):169–177

Gregor AK, Klubek B, Varsa EC (2003) Identification and use of actinomycetes for enhanced nodulation of soybean co-inoculated with Bradyrhizobium japonicum. Can J Microbiol 49(8): 483–491

Gurusamy M, Katerere DR (2026) Harnessing rhizospheric interactions for root disease control in plants. In Secondary Metabolites in Stress and Disease Management (pp. 289–313). Academic Press. https://doi.org/10.1016/B978-0-443-33959-2.00006-X

Harir M, Bendif H, Bellahcene M, Fortas Z, Pogni R (2018) Streptomyces secondary metabolites. In: Basic biology and applications of actinobacteria. IntechOpen

Hayakawa Y, Kobayashi T, Izawa M (2013) Indanostatin, a new neuroprotective compound from Streptomyces sp. J Antibiot 66(12):731–733

Helepciuc FE, Todor A (2022) Improving the authorization of microbial biological control products (MBCP) in the European Union within the EU Green Deal framework. Agronomy 12(5):1218

Jin Y, Wang L, Chen G, Lin X, Miao W, Fu Z (2014) Exposure of mice to atrazine and its metabolite diaminochlorotriazine elicits oxidative stress and endocrine disruption. Environ Toxicol Pharmacol 37(2):782–790

Jin L, Feng S, Tang S, Dong P, Li Z (2023) Biological control of potato late blight with a combination of Streptomyces strains and biochar. Biol Control 183:105248

Johannes A, Picon A, Alvarez-Gila A, Echazarra J, Rodriguez-Vaamonde S, Navajas AD, Ortiz-Barredo A (2017) Automatic plant disease diagnosis using mobile capture devices, applied on a wheat use case. Comput Electron Agric 138:200–209

Jones AL (1982) Chemical control of phytopathogenic prokaryotes. In: Phytopathogenic prokaryotes. Academic, pp 399–414

Jung B, Park SY, Lee YW, Lee J (2013) Biological efficacy of Streptomyces sp. strain BN1 against the cereal head blight pathogen Fusarium graminearum. Plant Pathol J 29(1):52

Kakembo D, Lee YH (2019) Analysis of traits for biocontrol performance of Pseudomonas parafulva JBCS1880 against bacterial pustule in soybean plants. Biol Control 134:72–81

Keswani C, Bisen K, Singh V, Sarma BK, Singh HB (2016) Formulation technology of biocontrol agents: present status and future prospects. In: Bioformulations: for sustainable agriculture, pp 35–52

Khajezadeh M, Abbaszadeh-Goudarzi K, Pourghadamyari H, Kafilzadeh F (2020) A newly isolated Streptomyces rimosus strain capable of degrading deltamethrin as a pesticide in agricultural soil. J Basic Microbiol 60(5):435–443

Khan S, Srivastava S, Karnwal A, Malik T (2023) Streptomyces as a promising biological control agents for plant pathogens. Front Microbiol 14:1285543

Kinkel LL, Bakker MG, Schlatter DC (2011) A coevolutionary framework for managing disease-suppressive soils. Annu Rev Phytopathol 49:47–67

Kinkel LL, Schlatter DC, Bakker MG, Arenz BE (2012) Streptomyces competition and co-evolution in relation to plant disease suppression. Res Microbiol 163(8):490–499

Kishimoto K, Park YS, Okabe M, Akiyama S-I (1996) Effect of ferrous ion on mildiomycin production by Streptoverticillium rimofaciens. J Antibiot 49:770–774. https://doi.org/10.7164/antibiotics.49.770

Kniewald J, Jakominić M, Tomljenović A, Šimić B, Romac P, Vranešić Đ, Kniewald Z (2000) Disorders of male rat reproductive tract under the influence of atrazine. J Appl Toxicol Int J 20(1):61–68

Köhl J, Kolnaar R, Ravensberg WJ (2019) Mode of action of microbial biological control agents against plant diseases: relevance beyond efficacy. Front Plant Sci 10:845

Kulkarni M, Gorthi S, Banerjee G, Chattopadhyay P (2017) Production, characterization and optimization of actinomycin D from Streptomyces hydrogenans IB310, a(n antagonistic bacterium against phytopathogens). Biocatal Agric Biotechnol 10:69–74

Kumar S (2014) Plant disease management in India: advances and challenges. Afr J Agric Res 9(15):1207–1217

Kumbar B, Mahmood R, Nagesha SN, Nagaraja MS, Prashant DG, Kerima OZ et al (2019) Field application of Bacillus subtilis isolates for controlling late blight disease of potato caused by Phytophthora infestans. Biocatal Agric Biotechnol 22:101366

Kurth F, Mailänder S, Bönn M, Feldhahn L, Herrmann S, Große I et al (2014) Streptomyces induced resistance against oak powdery mildew involves host plant responses in defense, photosynthesis, and secondary metabolism pathways. Mol Plant-Microbe Interact 27(9): 891–900

Lahdenpera ML, Simon E, Uoti J (1991) Mycostop-a novel biofungicide based on Streptomyces bacteria. In: Developments in agricultural and managed forest ecology, vol 23. Elsevier, pp 258–263

Law JWF, Ser HL, Khan TM, Chuah LH, Pusparajah P, Chan KG et al (2017) The potential of Streptomyces as biocontrol agents against the rice blast fungus, Magnaporthe oryzae (Pyricularia oryzae). Front Microbiol 8:3

Le KD, Kim J, Nguyen HT, Yu NH, Park AR, Lee CW, Kim JC (2021) Streptomyces sp. JCK-6131 protects plants against bacterial and fungal diseases via two mechanisms. Front Plant Sci 12: 726266

Le KD, Yu NH, Park AR, Park DJ, Kim CJ, Kim JC (2022) Streptomyces sp. AN090126 as a biocontrol agent against bacterial and fungal plant diseases. Microorganisms 10(4):791

Lekshmi M, Ammini P, Kumar S, Varela MF (2017) The food production environment and the development of antimicrobial resistance in human pathogens of animal origin. Microorganisms 5(1):11

Lv M, Zhou G, He M, Chen A, Zhang W, Hu Y (2020) Maize leaf disease identification based on feature enhancement and DMS-robust alexnet. IEEE Access 8:57952–57966

Lyu A, Yang L, Wu M, Zhang J, Li G (2020) High efficacy of the volatile organic compounds of Streptomyces yanglinensis 3-10 in suppression of Aspergillus contamination on peanut kernels. Front Microbiol 11:142

Maksimov IV, Blagova DK, Veselova SV, Sorokan AV, Burkhanova GF, Cherepanova EA et al (2020) Recombinant Bacillus subtilis 26DCryChS line with gene Btcry1Ia encoding Cry1Ia toxin from Bacillus thuringiensis promotes integrated wheat defense against pathogen Stagonospora nodorum Berk. and greenbug Schizaphis graminum Rond. Biol Control 144: 104242

Mellon MG, Benbrook C, Benbrook KL (2001) Hogging it: estimates of antimicrobial abuse in livestock. Union of Concerned Scientists

Minuto A, Spadaro D, Garibaldi A, Gullino ML (2006) Control of soilborne pathogens of tomato using a commercial formulation of Streptomyces griseoviridis and solarization. Crop Prot 25(5): 468–475

Mnif W, Hassine AIH, Bouaziz A, Bartegi A, Thomas O, Roig B (2011) Effect of endocrine disruptor pesticides: a review. Int J Environ Res Public Health 8(6):2265–2303

Mohammadi P, Tozlu E, KoTan R, KoTan MŞ (2017) Potential of some bacteria for biological control of postharvest citrus green Mould caused by Penicillium digitatum. Plant Prot Sci 53(3): 134–143

Moolhuijzen PM, See PT, Shi G, Powell HR, Cockram J, Jørgensen LN et al (2022) A global pangenome for the wheat fungal pathogen Pyrenophora tritici-repentis and prediction of effector protein structural homology. Microb Genom 8(10):000872

Mostafalou S, Abdollahi M (2013) Pesticides and human chronic diseases: evidences, mechanisms, and perspectives. Toxicol Appl Pharmacol 268(2):157–177

Muangham S, Pathom-Aree W, Duangmal K (2015) Melanogenic actinomycetes from rhizosphere soil—antagonistic activity against Xanthomonas oryzae and plant-growth-promoting traits. Can J Microbiol 61(2):164–170

Nazarov PA, Baleev DN, Ivanova MI, Sokolova LM, Karakozova MV (2020) Infectious plant diseases: etiology, current status, problems and prospects in plant protection. Acta Nat 12(3):46

Newitt JT, Prudence SM, Hutchings MI, Worsley SF (2019) Biocontrol of cereal crop diseases using streptomycetes. Pathogens 8(2):78

Ngalimat MS, Mohd Hata E, Zulperi D, Ismail SI, Ismail MR, Mohd Zainudin NAI et al (2021) Plant growth-promoting bacteria as an emerging tool to manage bacterial rice pathogens. Microorganisms 9(4):682
Nguyen CT, Dhakal D, Pham VTT, Nguyen HT, Sohng JK (2020) Recent advances in strategies for activation and discovery/characterization of cryptic biosynthetic gene clusters in Streptomyces. Microorganisms 8(4):616
Nicolopoulou-Stamati P, Maipas S, Kotampasi C, Stamatis P, Hens L (2016) Chemical pesticides and human health: the urgent need for a new concept in agriculture. Front Public Health 4:148
Nimnoi P, Ruanpanun P (2020) Suppression of root-knot nematode and plant growth promotion of chili (Capsicum flutescens L.) using co-inoculation of Streptomyces spp. Biol Control 145: 104244
Nimnoi P, Pongsilp N, Ruanpanun P (2017) Monitoring the efficiency of Streptomyces galilaeus strain KPS-C004 against root knot disease and the promotion of plant growth in the plant-parasitic nematode infested soils. Biol Control 114:158–166
O'Neill J (2015) Antimicrobials in agriculture and the environment: reducing unnecessary use and waste. In: The review on antimicrobial resistance. The Review on Antimicrobial Resistance, London
Pal KK, Gardener BM (2006) Biological control of plant pathogens. Plant Health Instruct 2(5): 1117–1142
Park HB, Park JS, Lee SI, Shin B, Oh DC, Kwon HC (2017) Gordonic acid, a polyketide glycoside derived from bacterial Coculture of Streptomyces and Gordonia species. J Nat Prod 80:2542–2546
Park EJ, Jang HJ, Park JY, Yang YK, Kim MJ, Park CS et al (2023) Efficacy evaluation of Streptomyces nigrescens KA-1 against the root-knot nematode Meloidogyne incognita. Biol Control 179:105150
Pham VTT, Nguyen CT, Dhakal D, Nguyen HT, Kim TS, Sohng JK (2021) Recent advances in the heterologous biosynthesis of natural products from Streptomyces. Appl Sci 11(4):1851
Pirsaheb M, Limoee M, Namdari F, Khamutian R (2015) Organochlorine pesticides residue in breast milk: a systematic review. Med J Islam Repub Iran 29:228
Quinn GA, Banat AM, Abdelhameed AM, Banat IM (2020) Streptomyces from traditional medicine: sources of new innovations in antibiotic discovery. J Med Microbiol 69(8):1040–1048
Rao HY, Mohana NC, Satish S (2020) Biocommercial aspects of microbial endophytes for sustainable agriculture. In: Microbial endophytes. Woodhead Publishing, pp 323–347
Rey T, Dumas B (2017) Plenty is no plague: Streptomyces symbiosis with crops. Trends Plant Sci 22(1):30–37
Robin D, Merlet L, Marchand P (2022) Regulatory aspects of biocontrol. In: Biocontrol of plant disease: recent advances and prospects in plant protection, pp 1–18
Salwan R, Sharma V (2020) Bioactive compounds of Streptomyces: biosynthesis to applications. Stud Nat Prod Chem 64:467–491
Sanborn M, Kerr KJ, Sanin LH, Cole DC, Bassil KL, Vakil C (2007) Non-cancer health effects of pesticides: systematic review and implications for family doctors. Can Fam Physician 53(10): 1712–1720
Sankaran S, Mishra A, Ehsani R, Davis C (2010) A review of advanced techniques for detecting plant diseases. Comput Electron Agric 72(1):1–13
Santos EF, Teixeira MFS, Converti A, Porto ALF, Sarubbo LA (2019) Production of a new lipoprotein biosurfactant by Streptomiyces sp. DPUA1566 isolated from lichens collected in the Brazilian Amazon using agroindustry wastes. Biocatal Agric Biotechnol 17:142–150
Satomi T, Kusakabe H, Nakamura G, Nishio T, Uramoto M, Isono K (1982) Neopeptins A and B, new antifungal antibiotics. Agric Biol Chem 46:2621–2623
Schrey SD, Tarkka MT (2008) Friends and foes: streptomycetes as modulators of plant disease and symbiosis. Antonie Van Leeuwenhoek 94:11–19
Schumacher DP, Hall SS (1982) An efficient stereospecific total synthesis of (.+−.)-anisomycin and related new synthetic antibiotics. J Am Chem Soc 104:6076–6080

Ser HL, Ab Mutalib NS, Yin WF, Goh BH, Lee LH, Chan KG (2018) Genome sequence of Streptomyces antioxidants MUSC 164T isolated from mangrove forest. Prog Microbes Mol Biol 1(1). https://doi.org/10.36877/pmmb.a0000001
Sharma V, Kaur R, Salwan R (2021) Streptomyces: host for refactoring of diverse bioactive secondary metabolites. 3 Biotech 11(7):340
Sheng Y, Lam PW, Shahab S, Santosa DA, Proteau PJ, Zabriskie TM et al (2015) Identification of elaiophylin skeletal variants from the Indonesian Streptomyces sp. ICBB 9297. J Nat Prod 78: 2768–2775
Shepherd MD, Kharel MK, Bosserman MA, Rohr J (2010) Laboratory maintenance of Streptomyces species. Curr Protoc Microbiol 18(1):10E-1
Shepherdson EM, Baglio CR, Elliot MA (2023) Streptomyces behavior and competition in the natural environment. Curr Opin Microbiol 71:102257
Shin D, Byun WS, Moon K, Kwon Y, Bae M, Um S et al (2018) Coculture of marine Streptomyces sp. with Bacillus sp. produces a new piperazic acid-bearing cyclic peptide. Front Chem 6:498
Shirling ET, Gottlieb D (1966) Methods for characterization of Streptomyces species. Int J Syst Evol Microbiol 16(3):313–340
Shu J, Cao K, Fei C, Dai H, Li Y, Cao Y, Wu Y (2024) Antiviral mechanisms of Anisomycin produced by Streptomyces albulus SN40 on potato virus Y. J Agric Food Chem 72(7): 3506–3519
Shuping DS, Eloff JN (2017) The use of plants to protect plants and food against fungal pathogens: A review. Afr J Tradit Complement Altern Med 14(4):120–127
Singh S, Kumar V, Datta S, Singh S, Dhanjal DS, Dhaka V et al (2021) Commercial production and formulation of microbial biocontrol agents. In: Food security and plant disease management. Woodhead Publishing, pp 241–256
Soares ACF, Sousa CDS, Garrido MDS, Perez JO (2007) Production of streptomycete inoculum in sterilized rice. Sci Agric 64:641–644
Solanki MK, Malviya MK, Wang Z (2016) Actinomycetes bio-inoculants: A modern prospectus for plant disease management. In: Plant growth promoting Actinobacteria: a new avenue for enhancing the productivity and soil fertility of grain legumes. Springer Singapore, Singapore, pp 63–81
Spadaro D, Gullino ML (2005) Improving the efficacy of biocontrol agents against soilborne pathogens. Crop Prot 24(7):601–613
Stockwell VO, Duffy B (2012) Use of antibiotics in plant agriculture. Rev Sci Tech 31(1):199–210
Strange RN, Scott PR (2005) Plant disease: a threat to global food security. Annu Rev Phytopathol 43(1):83–116
Suleman P, Al-Musallam A, Menezes CA (2002) The effect of biofungicide Mycostop on Ceratocystis radicicola, the causal agent of black scorch on date palm. BioControl 47:207–216
Syed-Ab-Rahman SF, Hesamian MH, Prasad M (2022) Citrus disease detection and classification using end-to-end anchor-based deep learning model. Appl Intell 52(1):927–938
Talibi I, Boubaker H, Boudyach EH, Ait Ben Aoumar A (2014) Alternative methods for the control of postharvest citrus diseases. J Appl Microbiol 117(1):1–17
Tariq M, Khan A, Asif M, Khan F, Ansari T, Shariq M, Siddiqui MA (2020) Biological control: a sustainable and practical approach for plant disease management. Acta Agric Scand Sect B—Soil Plant Sci 70(6):507–524
Teixidó N, Usall J, Torres R (2022) Insight into a successful development of biocontrol agents: production, formulation, packaging, and shelf life as key aspects. Horticulturae 8(4):305
Thakur DS, Khot R, Joshi PP, Pandharipande M, Nagpure K (2014) Glyphosate poisoning with acute pulmonary edema. Toxicol Int 21(3):328
Tsegaye Z, Assefa F, Tefera G, Alemu T, Gizaw B, Abatenh E (2018) Concept, principle and application of biological control and their role in sustainable plant diseases management strategies. Int J Res Stud Biosci 6(4):18–34
Van der Meij A, Worsley SF, Hutchings MI, van Wezel GP (2017) Chemical ecology of antibiotic production by actinomycetes. FEMS Microbiol Rev 41(3):392–416

Viaene T, Langendries S, Beirinckx S, Maes M, Goormachtig S (2016) Streptomyces as a plant's best friend? FEMS Microbiol Ecol 92(8):fiw119

Vidaver AK (2001) Horticultural and other uses of antibiotics. In: International congress and symposium Series-Royal Society of Medicine, vol 247. Royal Society of Medicine Press, London, pp 125–130; 1999

Vidaver AK (2002) Uses of antimicrobials in plant agriculture. Clin Infect Dis 34(Supplement_3): S107–S110

Wallace RL, Hirkala DL, Nelson LM (2018) Efficacy of Pseudomonas fluorescens for control of Mucor rot of apple during commercial storage and potential modes of action. Can J Microbiol 64(6):420–431

Wattana-Amorn P, Charoenwongsa W, Williams C, Crump MP, Apichaisataienchote B (2016) Antibacterial activity of cyclo(L -Pro- L -Tyr) and cyclo(D -Pro- L -Tyr) from Streptomyces sp. strain 22-4 against phytopathogenic bacteria. Nat Prod Res 30:1980–1983

Wong CKF, Saidi NB, Vadamalai G, Teh CY, Zulperi D (2019) Effect of bioformulations on the biocontrol efficacy, microbial viability and storage stability of a consortium of biocontrol agents against Fusarium wilt of banana. J Appl Microbiol 127(2):544–555

Xiao K, Kinkel LL, Samac DA (2002) Biological control of Phytophthora root rots on alfalfa and soybean with Streptomyces. Biol Control 23(3):285–295

Xu Y, Zhao B, Zhai Y, Chen Q, Zhou Y (2021) Maize diseases identification method based on multi-scale convolutional global pooling neural network. IEEE Access 9:27959–27970

Youssef SM, El-Serafy RS, Ghanem KZ, Elhakem A, Abdel Aal AA (2022) Foliar spray or soil drench: microalgae application impacts on soil microbiology, morpho-physiological and biochemical responses, oil and fatty acid profiles of chia plants under alkaline stress. Biology 11(12):1844

Zheng X, Wang J, Chen Z, Zhang H, Wang Z, Zhu Y, Liu B (2019) A Streptomyces sp. strain: isolation, identification, and potential as a biocontrol agent against soilborne diseases of tomato plants. Biol Control 136:104004

Exploration of Streptomycetes as Nematicidal and Herbicidal Agents

7

Ranjithkumar Murugesan, Mareeswari Kesavan, Salmiya Marimuthu, Manikandan Gurusamy, and Tendai Makwikwi

Abstract

Over 70% of the natural substances used in agriculture and medicine today are produced by the *Streptomyces* genus. Due to their extensive secondary metabolism, these species are highly sought as microbial inoculants. In terms of plant health, *Streptomyces* species can exhibit desirable functional characteristics by producing specialized metabolites that enhance nutrient uptake, promote plant growth, alleviate abiotic stress, promote resistance, and prevent the invasion and establishment of pathogens or pests through herbicidal and nematicidal properties. *Streptomyces* species can produce plant-protective compounds like enzymes, secondary metabolites, and volatile organic compounds, indicating promise as herbicidal and nematicidal agents. They can additionally stimulate immunity in plant systems to respond quickly to an infection of pathogens. Biocontrol methods using streptomycetes can provide cheaper, effective, and environmentally safe pest control, thereby reducing the hazards of chemical pesticides. Research on streptomycetes as nematicidal and herbicidal agents has key implications for agricultural sustainability and inspires the possibility of

R. Murugesan · M. Kesavan · S. Marimuthu
PG Department of Microbiology, Syed Ammal Arts and Science College, Kootampuli, Ramanathapuram, Tamil Nadu, India

M. Gurusamy
AgriSETA Research Group, College of Human Sciences, Office of Graduate Studies and Research, University of South Africa, Pretoria, South Africa

T. Makwikwi (✉)
Department of Laboratory, Diagnostic and Investigative Sciences, Medical Microbiology Unit, Faculty of Medicine and Health Sciences, University of Zimbabwe, Avondale, Harare, Zimbabwe

Mitochondria Research Laboratory, Biomedical and Molecular Metabolism Research (BioMMet), North-West University, Potchefstroom, South Africa

T. Kaur et al. (eds.), *Streptomyces: Biological Candidates for Sustainable Agriculture*,
https://doi.org/10.1007/978-981-95-5803-2_7

developing holistic pest management strategies. This chapter describes nematicidal and herbicidal aspects of streptomycetes and serves to discuss their potential as environmentally friendly alternatives to synthetic pesticides.

Keywords

Streptomyces · Biocontrol agent · Nematicidal · Herbicidal · Sustainable agriculture

7.1 Introduction

Streptomyces is the largest genus of Actinobacteria in the Streptomycetaceae family (Hong et al. 2009). According to Lee et al. (2005), there are more than 500 species of *Streptomyces* bacteria known to science. *Micromonospora*, *Mycobacterium*, *Actinomadura*, *Saccharopolyspora*, *Microbispora*, *Nonomurea*, *Frankia*, *Actinoplanes*, *Verrucosispora*, and *Nocardia* are the next most common genera among actinobacteria, after *Streptomyces* (Martínez-Hidalgo et al. 2014). According to Sathya et al. (2017), *Streptomyces* makes up half of the entire population of soil actinobacteria. Waksman and Henrici (1943) proposed the classification of *Streptomyces*, which was first based on morphology (colour of hyphae and spores), chemotype, whole-cell sugars, fatty acid, and phospholipid profiles, and cell wall composition (peptidoglycan type) (Kroppenstedt et al. 1990). Later, phenotypic and genotypic traits were used (Anderson and Wellington 2001). The intergeneric connections within the Streptomycetaceae family were previously resolved by the taxonomic systems using phenotypic features. Nonetheless, *Streptomyces* genus delimitation has advanced significantly, thanks to its genotypic features (Stackebrandt et al. 1997). For example, *Streptomyces* and *Streptoverticillium* were once thought to belong to separate genera, but they were later reclassified as synonyms of *Streptomyces* using immunodiffusion and 16S and 23S rRNA studies (Stackebrandt and Woese 1981; Logan 1994).

According to Smaoui et al. (2011), *Streptomyces* species are non-motile, catalase-positive, convert nitrates to nitrites, and break down L-L-tyrosine, adenine, esculin, casein, gelatin, hypoxanthine, and starch. The main component of the cell wall peptidoglycan is L-L-diaminopimelic acid (L-DAP). With nine isoprene units as the main isoprenolog, they have either hexa- or octahydrogenated menaquinones, no mycolic acids, and complex polar lipid patterns that usually contain diphosphatidyl glycerol, phosphatidyl ethanolamine, phosphatidyl inositol, and phosphatidyl inositol mannosides (Cummins and Harris 1958).

7.1.1 Importance of Biological Control Agents in Agriculture

In agriculture, plant diseases brought on by pathogenic microbes provide a significant challenge, causing between 30% and 40% of crop damage. Traditional controls for these microbes, for example chemical pesticides and antimicrobials, have been shown to negatively affect human health and the environment. They also encourage the development of pathogen resistance. Therefore, it is now critical to consider natural eco-friendly and safe alternatives (Khan et al. 2023).

The *Streptomyces* genus of Gram-positive bacteria has been extensively investigated as a naturally derived alternative to synthetic chemicals due to its ability to synthesize various antimicrobial and naturalized metabolites. Research efforts are currently underway across multiple continents to discover new bioactive products from these bacteria. It is even reported that bioactive compounds from *Streptomyces* sp. can help preserve marketable quality in plants by substantially decreasing the number of potentially harmful bacteria and fungi that could be biotic threats to plants and crops. As such, *Streptomyces* sp. has the potential to be used as microbial biocontrol/bioagent and biopesticide agents to reduce plant diseases. Bioactive compounds that are observed from this genus could be the basis of future agricultural research on how to utilize this natural source for biopesticides (Khan et al. 2023).

Recent research has suggested that certain species of *Streptomyces* may exhibit biocontrol activity in managing bacterial and fungal pathogens that could be potentially detrimental to plants (Suárez-Moreno et al. 2019). Numerous research studies have shown that streptomycetes can effectively manage several phytopathogenic fungi, such as *Magnaporthe oryzae*, *Pyrrhoderma noxium*, *Phytophthora capsici*, *Rhizoctonia solani*, *Puccinia triticina*, *Pythium aphanidermatum*, *Fusarium verticillioides*, *Botrytis cinerea*, and *Pythium ultimum* (Panchalingam et al. 2022). Studies have also shown that *Streptomyces* has considerable promise as a producer of secondary metabolites such as antibiotics, growth promoters, and herbicides (Harir et al. 2018). The genus also produces 66.67% of all antibiotics that occur naturally and produces 70% of the antibiotics that are currently used therapeutically for human diseases (Sivalingam et al. 2019).

The dynamic *Streptomyces* genus, appear to be a readily available natural source to develop new natural disease control tactics. Efforts to discover natural products as biocontrol agents for plant protection have recently taken hold globally. Streptomycetes can control plant pathogens by the following mechanisms: they produce siderophores (Vijayabharathi et al. 2015), synthesize plant growth substrates (Goudjal et al. 2013), produce antibiotics (Couillerot et al. 2013), produce volatile compounds (Jones and Elliot 2017), and compete for nutrients. Their primary biocontrol mechanism is attributed to the tremendous quantities of antibiotics, volatile compounds, and metabolites they create that enhance their antipathogenic activity. For example, different strains of *Streptomyces violaceusniger* YH27A, *S. violaceusniger* YCED-9 produce chitinase enzyme, antibiotic geldanamycin, and the antifungal nigericin (Gherbawy et al. 2012; Shrivastava and Kumar 2018). These biocontrol products are presumed to be derived from hyphal growth induced by nutrient deficiency (Al-Askar et al. 2015; Errakhi et al. 2016; Shekh 2017). Many *Streptomyces* species, such as *Streptomyces lividans*, *Streptomyces plicatus*, *Streptomyces humidus*, *Streptomyces scabies*, *Streptomyces violaceusniger*, *Streptomyces aureofaciens*, *Streptomyces hygroscopicus*, *Streptomyces olivaceoviridis*, *Streptomyces lydicus*, *Streptomyces avermitilis*, and *Streptomyces roseoflavus*, have been recognized for their capacity to produce effective bioactive fungicidal compounds for use against a range of pathogenic fungi.

Streptomyces spp., a significant group of soil-associated bacteria, enhance the functioning of the microbial community and contribute to plant and soil health in a range of activities, especially under nematode-induced stress (Olanrewaju and Babalola 2019). These bacteria produce secondary metabolites, including enzymes, phytohormones, and antibiotics, as well as other products that stimulate plant growth (Abd-Alla et al. 2013; Gupta et al. 2019; Inbar et al. 2005). Interestingly, the dual function of using *Streptomyces* as a biocontrol agent is that it combines nematicidal activity with plant growth-promoting activity (Vurukonda et al. 2018). These bacteria certainly require additional research to fully realize the potential of these beneficial bacteria in sustainable agriculture.

Weeds are an important threat to crop production, resulting in significant losses in yield. The habitual applications of synthetic herbicides produce environmental issues such as pollution, herbicide residues, mammalian hazards, and herbicide-resistant weeds (Yasuor et al. 2008; Vasilakoglou et al. 2000). As the concern for environmental protection increases, there is a pressing need to develop safe, non-toxic, and environment friendly bioherbicides to lessen chemical inputs into systems (Lee et al. 2013). Additionally, *Streptomyces* species have been shown to have herbicidal activity, which makes them a target group of microorganisms for evaluating herbicidal potential (Nakajima et al. 1991).

This chapter focuses on the biocontrol of plant parasitic nematodes and weeds using streptomycetes. It highlights the importance of biological control agents in agriculture, streptomycetes, and their bioactive metabolites as nematicidal and herbicidal agents, and the challenges and future directions to explore streptomycetes as nematicidal and herbicidal agents in agriculture.

7.2 Streptomycetes as Nematicidal Agents

The incredible adaptability and longevity of nematodes can be linked to their widespread habitats, which can be terrestrial, aquatic, or marine. Nematodes, classified as Ecdysozoa, can moult their cuticle. The phylum Nematoda is expansive and fascinating, with around 30,000 known species (Hugot et al. 2001). The phylum Nematoda is organized into three principal monophyletic groups: Chromadoria (which consists of nematodes found across different habitats), Dorylaimia (which contains numerous parasitic trichinellids and mermithids), and Enoplia (which is mostly marine).

With more than 4100 species known to exist, plant-parasitic nematodes (PPN) represent a serious financial risk to agricultural crop production (Decraemer and Hunt 2006). Global crop loss from these nematodes is estimated to be \$80--118 billion annually (Nicol et al. 2011). Remarkably, 15% of all nematode species that found to attack plant roots, obstruct the uptake of nutrients and water, and eventually lower crop yields, quality, and agronomic performance. Nematodes that infect plants, invertebrates, and fungi are members of the order Tylenchida, which is regarded as a significant category of agricultural pests (Sasser and Freckman 1987). To lessen the negative effects of these nematodes on the world's food output, effective management techniques are essential.

The capacity to create persistent feeding sites inside plant hosts enables sedentary plant-parasitic nematodes to effectively get nutrition from host plants while finishing their life cycles. The growth of sustained feeding structures is made possible by the intricate host cell modification, which gives immobile nematodes an edge over migratory species. Although more than 4000 plant-parasitic nematode genera have been identified, only a small number of them result in notable financial losses. According to a US assessment, *Heterodera, Hoplolaimus*, *Meloidogyne*, *Pratylenchus, Rotylenchulus*, and *Xiphinema* are the main genera of phytoparasitic nematodes that cause agricultural losses (Koenning et al. 1999).

7.2.1 Root-Knot Nematodes

Root-knot nematodes (*Meloidogyne* spp.), which include more than 100 species, are regarded as one of the worst problems in plant pathology. According to Postnikova et al. (2015), *Meloidogyne javanica*, *Meloidogyne arenaria*, *Meloidogyne hapla*, and *Meloidogyne incognita* are particularly dangerous for agricultural crop production. By pushing their stylets and releasing enzymes that break down cell walls, these nematodes enter plant cells and go through root cortex cells to target undifferentiated procambium cells in the vascular cylinder. The worms' dorsal gland releases proteinaceous secretions as the infection worsens, encouraging the growth of large cells, which act as the main feeding site (Davis et al. 2000). Crop damage is eventually caused by this intricate relationship between the nematode and plant cells.

Nematode-induced endoreduplication without cytokinesis produces the multinucleated large/giant cells that form the permanent feeding site of the root-knot nematodes within the plant root. These enormous cells produce cellular ingrowths that improve the xylem's ability (due to increased surface area) to absorb nutrients for the neamtode's benefit. J2 larvae after hatching from eggs invade roots. Inside the roots they feed through large cells and then move through the phases of their life cycle: J2 larvae moult into J3, then J4, and finally adulthood stages. The genus name *Meloidogyne* comes from Greek words meaning "apple shaped" and "female," referring to the apple-shaped form of a mature female worm. Adult females lay hundreds of eggs into a gelatinous matrix on root surfaces which are then discharged into the soil. In the soils eggs hatch to complete the life cycle. Stunted growth, wilting, root galling, and irregular root development are common symptoms of PPN infected plants (Caillaud et al. 2008).

7.2.2 Cyst Nematodes

In terms of economic significance, cyst nematodes (*Heterodera* and *Globodera* spp.) are second only to root-knot nematodes as a major agricultural concern. J2 larvae infect host roots as part of their life cycle, and then mature into adults inside host tissues. The eggs of cyst nematodes remain inside the dead female's body which hardens into a cyst. They remain there until ideal conditions stimulate their hatching. By using oesophageal gland secretions released by the stylet, these nematodes cause syncytia, which are specialized feeding structures, in infected roots (Hewezi and Baum 2013). Notable species include the soybean cyst nematode (*Heterodera glycines*), the potato cyst nematodes (*Globodera rostochiensis* and *Globodera pallida*), and the cereal cyst nematodes (*Heterodera avenae* and *H. filipjevi*). According to Wrather and Mitchum (2010), *H. glycines* is responsible for an estimated significant loss of $1.286 billion in the United States. While *Heterodera schachtii* can result in 30% output losses in sugar beets and other crops in the Chenopodiaceae family, *G. pallida* can cause 50–80% yield losses in densely infested potato areas.

7.2.3 Lesion Nematodes

With about 70 species worldwide and a host range of almost 400 plant species, root-lesion nematodes (*Pratylenchus* spp.) are among the most destructive agricultural worms (Davis and MacGuidwin 2000). Among these lesion nematodes, *Pratylenchus thornei* significantly reduces wheat yields up to 85% in Australia, 70% in Israel, 50% in Oregon, and 37% in Mexico (Smiley 2015). These migratory nematodes can penetrate vascular tissues and feed on the root cortex, causing necrosis, lesion development, and secondary infections (Jones and Fosu-Nyarko 2014; Smiley 2015). Chlorosis, decreased leaf quantity, and lesser leaf size are examples of aboveground symptoms. Using morphometric and genetic analyses, recent research has discovered new species in Korea, including *Pratylenchus kumamotoensis* and *Pratylenchus pseudocoffeae* (Kim et al. 2016).

7.2.4 Burrowing Nematodes

A migratory plant-parasitic nematode of major economic significance, the burrowing worm *Radopholus similis* is classified as a quarantine pest worldwide (Smith and Charles 1998). Particularly in banana-growing countries, *R. similis*, which has over 250 host plants, significantly reduces agricultural yields in crops like coffee, citrus, pepper, and bananas (Sarah et al. 1999). New methods are necessary since controlling *R. similis* is difficult. According to research, plant-mediated RNA interference can effectively suppress the expression of *Radopholus* Calreticulin (Rs-CRT) and increase resistance in transgenic tomato plants (Li et al. 2015). According to Hölscher et al. (2014), phenylphenalenones that are isolated from

Musa spp. have also demonstrated anti-nematode properties by preventing movement of *R. similis*. These studies described potential approaches for controlling *R. similis* infestations.

Several *Streptomyces* species have also been identified for their nematicidal properties. *Streptomyces spectabilis* and *Streptomyces avermitilis* produce spectinabilin and avermectins, respectively, and are widely used as effective nematicides. Additionally, *Streptomyces* sp. AE170020 produces aureothin, which suppress nematode growth and reproduction. Furthermore, *Streptomyces yatensis* KRA-28 has also demonstrated nematicidal activity against *Meloidogyne* species (Kang et al. 2022) (Table 7.1).

7.2.5 Mechanisms of Nematicidal Activity in Streptomycetes

As environmental and human health concerns have grown, synthetic nematicides have become less common, thus, underscoring the need for novel, effective, and low-toxicity biological alternatives (Chen and Song 2021). An eco-friendly way to manage plant-parasitic nematodes and enhance soil ecology is using biological control agents (Wang and Yang 2002). As already mentioned, *Streptomyces*, a crucial antibiotic producing genus of Actinobacteria, is a promising useful resource to explore for biological nematicides. These bacteria produce secondary metabolites that exhibit a range of biological activities, including antagonsim against fungi and nematodes (Nair et al. 1995; Zhang et al. 2017; Liu et al. 2019). Noteworthy examples of nematicidal compounds having nematicidal potential include nikkomycin, nanchangmycin, and avermectin obtained from *S. tendae*, *S. nanchangensis*, and *S. avermitilis*, respectively.

Streptomyces species can employ more than one mechanistic strategies for their nematicidal actions. Many nematicidal metabolites from streptomycetes show direct toxic effect on nematodes and kill them by interfering with their cellular functions. Some of the metabolites demonstrate nematicidal activity by impeding the reproduction of nematodes by blocking egg hatching or larval development. A few strains of *Streptomyces* may also induce systemic resistance (ISR) in plants and provide improved resistance against nematode infections. Furthermore, some species produce enzymes or volatile compounds that deter nematodes from feeding, or disrupt their behaviour (Ma et al. 2017).

7.2.6 Fermentation and Formulation of Streptomycetes-Based Nematicides

Streptomyces strains can be cultivated on ISP2 agar plates for 14 days at room temperature. To extract the active metabolites from agar cultures, ethyl acetate/methanol/acetic acid solution can be used. Afterwards, DMSO (dimethysufoxide) dissolved crude extracts are used to examine their impact on worms.

Table 7.1 Nematicidal potential of some *Streptomyces* spp.

S. No.	*Streptomyces* species	Bioactive compounds	Nematicidal activity	References
1.	*Streptomyces avermitilis* G8-17	Avermectin	Nematicidal activity	Yang et al. (2013)
2.	*S. avermitilis* G8-17	Actinomycins	Nematicidal activity	Huang et al. (2015)
3.	*S. avermitilis* NEAU1069	Macrocyclic lactones	Nematicidal activity	Wang et al. (2010)
4.	*Streptomyces bingchenggensis* BCJ60	Milbemycin	Nematicidal activity	Li et al. (2021)
5.	*Streptomyces microflavus* neau3	Nemadectin	Nematicidal activity	Yang et al. (2013)
6.	*Streptomyces nanchangensis*	Nanchangmycin	Nematicidal activity	Nair et al. (1995), Liu et al. (2019) and Zhang et al. (2017)
7.	*Streptomyces tendae*	Nikkomycin	Nematicidal activity	Nair et al. (1995), Liu et al. (2019) and Zhang et al. (2017)
8.	*Streptomyces albireticuli* HA10002	Fungichromin B	Nematicidal activity	Zeng et al. (2013)
9.	*Streptomyces antibioticus* strain M7	Actinomycins V, X_2 and D	*Meloidogyne incognita*	Sharma et al. (2019)
10.	*S. avermitilis*	Avermectins	Widely used as commercial nematicides	Khalil (2013)
11.	*S. avermitilis*	Avermectin B1a and Avermectin B1b	*M. incognita, M. arenaria* and *M. javenica*	Burg et al. (1979)
12.	*Streptomyces* sp.	Fervenulin	Active against *Meloidogyne* spp.	Ruanpanun et al. (2011a), Burg et al. (1979)
13.	*Streptomyces* sp.	Not specified (crude extracts)	Nematicidal activity	Park et al. (2002), Yang et al. (2013), Jang et al. (2015) and Kaur et al. (2016)
14.	*Streptomyces spectabilis*	Spectinabilin	Nematicidal activity	Park et al. (2020)
15.	*Streptomyces* sp. 680560	Teleocidin B4	Nematicidal activity	Kang et al. (2021)
16.	*Streptomyces* sp. AE170020	Aureothin, Alloaureothin	Suppresses nematode growth and reproduction	Kang et al. (2022)
17.	*Streptomyces* sp. AE170020	Aureothin and Alloaureothin	Nematicidal activity	Liu et al. (2019)

(continued)

Table 7.1 (continued)

S. No.	*Streptomyces* species	Bioactive compounds	Nematicidal activity	References
18.	*Streptomyces* sp. AN091965	Spectinabilin	Nematicidal activity	Liu et al. (2019)
19.	*Streptomyces* sp. CMU-JT005	Carbazomycins D and F	Nematicidal activity	Ruanpanun et al. (2011b)
20.	*Streptomyces* sp. CMU-MH021	Fervenulin	Nematicidal activity	Ruanpanun et al. (2011a)
21.	*Streptomyces* sp. P294	Angumycinones	Nematicidal activity	Su et al. (2016)
22.	*Streptomyces yatensis* KRA-28	Not specified (crude extracts)	Demonstrated activity against *Meloidogyne* species (root-knot nematodes)	Park et al. (2020)

For experimental setup, a 96-well plate with 0.1 g/L of extract and M9 buffer is filled with about 50 nematodes. As a control, a 1% DMSO solution is employed. Following an incubation period of 24 h at 20 °C, the dead and living worms are counted using an inverted microscope. To determine mortality, the number of dead worms is divided by the total number of initial worms. Worms which move freely or respond to plate shaking are considered as viable.

7.3 Weeds and Their Types

Weeds are undesirable or out-of-place plants that frequently hinder the development of crops or livestock. Common lamb quarters (*Chenopodium album*), crab grasses (*Digitaria* spp.), and pigweeds (*Amaranthus* spp.) are a few examples. The most expensive type of agricultural pest is weeds, which raise production costs globally and result in large output losses. Indeed, they represent a more significant economic risk than nematodes, crop diseases, insect pests, or larger grazers such as deer, birds, and rodents (Altieri 1995; Sullivan 2003).

Weeds have a negative impact on both the amount and quality of agricultural products, significantly reducing farmers' profits (Zimdahl and Basinger 2024). Weeds can release allelochemicals that inhibit germination of seeds and/or growth of crops, and they compete with crops for important factors like water, light, nutrients, and space (Riaz Marral et al. 2020; Farooq et al. 2020a). This competition of weeds and allelochemicals can have a large impact on crop production and overall agricultural production.

The large number of seeds that weeds produce also contributes to the seed bank in the soil and provides a path for new infestations (Farooq et al. 2019, 2021; Önen et al. 2018). Effective weed management is important to avoid crop losses and reduce the weed seed bank (Chauhan et al. 2017). Weeds need to be managed to be productive within agriculture and to provide for sustainable agriculture.

Therefore, effective weed management is crucial for reducing financial losses and promoting sustainable agriculture.

7.3.1 Herbicidal Activity in Streptomycetes

The herbicidal activities of *Streptomyces* can be attributed to: secondary metabolites, such as antibiotics, toxins, and/or volatile organic compounds (VOCs); metabolites acting as photosynthesis inhibitors that disrupt the movement of electrons in the chloroplast, and ultimately prevents the production of energy. In contrast, other metabolites work by disturbing the cell membrane, which leads to the leakage of electrolytes and ultimately leads to cell death. Additionally, certain *Streptomyces* species produce substances that inhibit essential plant enzymes and disrupt plant growth or defence mechanisms. While some *Streptomyces* strains promote plant growth by producing phytohormones like IAA, enhancing root development and nutrient uptake, others act as biocontrol agents, competing with pathogens by producing antimicrobial compounds that indirectly benefit crops by reducing weed competition (Nakajima et al. 1991).

Streptomyces sp. KRA18-249 produces metabolites with broad-spectrum herbicidal activity, showing both photosynthesis-inhibiting effects and mechanisms between paraquat and glyphosate. Another example is *Streptomyces scopuliridis*, whose metabolites disrupt cell membranes, potentially serving as a contact herbicide (Won et al. 2016).

7.3.2 Fermentation and Formulation of Streptomycetes-Based Herbicides

In Lee et al. (2003) experimental setup, isolates of *Streptomyces* were cultivated in M3 liquid medium and incubated for 1 week at 27 °C. After collecting the bacterial biomass, the supernatant was sprayed onto the foliage of *Digitaria ciliaris* in a greenhouse. Seven days following treatment, herbicidal activity was assessed, and phytotoxicity was graded according to morphological and physiological signs. With three replicates of 20 plants each, the scoring system went from 0% (no effect) to 100% (total mortality) (Lee et al. 2003).

7.4 Bioactive Compounds from Streptomycetes as Nematicidal and Herbicidal Agents

Streptomycetes as nematicides and herbicides work through diverse mechanisms, mainly the release of secondary metabolites, which modulate specific biological processes in nematodes and plants. The compounds with nematicidal activity typically target nematode physiology, while herbicides affect plant growth and development (Alam et al. 2022).

7.4.1 Active Compounds of Streptomycetes with Nematicidal Properties

Research has demonstrated that *Streptomyces* species have nematicidal activity against a variety of nematodes in agricultural systems (Kaur et al. 2016; Sharma et al. 2019). *Streptomyces* isolates produce secondary metabolites with nematicidal activity that affect all nematode's physiological processes and lead to paralysis, immobilization, and death. In addition, studies reported that the secondary metabolites produced by these isolates can induce systemic resistance to nematodes in plants and thus decrease the plants' susceptibility to nematodes (Mhatre et al. 2024). However, in studies involving *Streptomyces* isolates, researchers commonly screened the nematicidal potential of the organisms in the laboratory using nematode growth or reproduction assays (Kaur et al. 2016). Further, field trials have confirmed that formulations based on *Streptomyces* can reduce nematode populations, as well as promote plant health (Burtseva et al. 2019). For example, the *Streptomyces* spp. strain DT10 isolated from an ancient moss exhibited complete nematicidal activity against *C. elegans*, attributed to the compound spectinabilin with an IC50 of 2.948 μg/mL (Sun et al. 2023). Spectinabilin could serve as a candidate for a novel biological nematicide due to its unique mode of action.

The reports documenting the nematicidal potential of the genus *Streptomyces* include in vitro and in vivo (from glasshouse and field) experimental setups involving different agronomic crops. The potential to manage plant-parasitic nematodes using crude extracts and purified agents from *Streptomyces* spp. has been confirmed in many studies (Samac and Kinkel 2001; Park et al. 2002; Yang et al. 2013; Jang et al. 2015; Kaur et al. 2016) (See Fig. 7.1). Compounds such as avermectin and fervenulin also exhibited broad-spectrum nematicide activity against plant-parasitic nematodes such as *Meloidogyne* spp. (Burg et al. 1979; Ruanpanun et al. 2011).

7.4.2 Active Compounds from Streptomycetes with Herbicidal Properties

Streptomyces and other soil microbes produce helpful secondary metabolites with herbicidal activities and can be exploited as potential biocontrol agents for weed control (Nakajima et al. 1991; Christy et al. 1993). Commercialized bioherbicides such as glufosinate ammonium and bialaphos were obtained from *Streptomyces hygroscopicus* (Ha et al. 2017) and *S. hygroscopicus* SF1293 strains, respectively (Xu et al. 2009). Nakajima et al. (1991) and Won et al. (2016) reported that metabolites from *Streptomyces* species such as nigericin, hydantocidin, and geldanamycin (*Streptomyces hygroscopicus*); blasticidin S (*Streptomyces griseochromogenes*); and herbicidins A and B (*Streptomyces sahnonensis* and *Streptomyces copperiridis*) also display herbicidal activity. (Table 7.2).

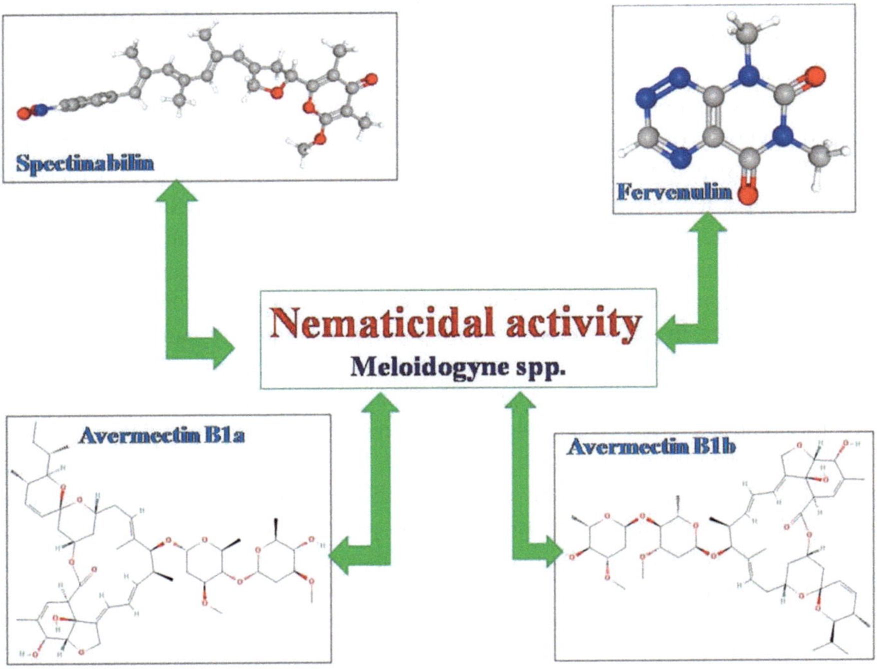

Fig. 7.1 Active compounds from streptomycetes with nematicidal properties

Table 7.2 Herbicidal compounds produced by *Streptomyces* species

S. No.	*Streptomyces* species	Bioactive compounds	References
1.	*Streptomyces saganonensis and Streptomyces scopuliridis*	Herbicidins A & B	Nakajima et al. (1991)
2.	*S. hygroscopicus* SF1293	Bialaphos	Xu et al. (2009)
3.	*Streptomyces griseochromogenes*	Blasticidin S	Won et al. (2016)
4.	*S. hygroscopicus*	Nigericin, hydantocidin, geldanamycin	Won et al. (2016)
5.	*S. hygroscopicus*	Glufosinate ammonium	Ha et al. (2017)

7.5 Challenges and Future Directions of *Streptomycetes* as Nematocidal and Herbicidal

Streptomyces are promising plant disease suppressive agents, but there are a number of issues that arise. One of the concerns is the selection of the right strain that can suppress plant diseases while not harming the beneficial bacteria. Strains of *Streptomyces* have been shown to have a wide range of efficacy (Kasuga et al. 2017). If

Streptomyces are going to function, we need to ensure specificity to protect against disrupting the biological environment in soils. Establishing ideal conditions in the dynamic nature of agriculture can also affect that specificity, given that varied factors such as temperature, moisture, and soil pH can affect the efficacy of *Streptomyces* (Le et al. 2021). Additionally, *Streptomyces* will interact and compete with the local microbial community for resources and niche space, which could affect their establishment and sustainability (Nguyen et al. 2015; Solanki et al. 2016; Le et al. 2021).

Using biocontrol agent products based on *Streptomyces* may present some challenges. Muangham et al. (2015) and Khan et al. (2023) indicated that these challenges include the requirement for multiple applications (as *Streptomyces* are limited in their persistence in soil), and the increased risk of pathogen resistance, which they often pose based on an increase in resistance over time. In addition, regulatory approval for Streptomyces-derived products will be costly and time-consuming, and the application process will be prohibitively expensive and too complex to ensure consistency in product performance (Sun et al. 2023). In addition, integrating *Streptomyces*-based biocontrol into the agricultural system could be difficult and consequently, farmers will require appropriate education and support to successfully implement it into their management paradigm (Khan et al. 2023).

Given the variability in effectiveness for different plant types and seasonality, *Streptomyces*-based biocontrol agents may not have the same reliability as chemical products (Zheng et al. 2019). Additionally, consumer perception and acceptance are another critical factor since consumers may be unaware of or cautions about biological control (Wu et al. 2015). Though *Streptomyces*-based biocontrol has promise as an alternative for plant disease management, we need to address these challenges through coordinated research and development, and a collaborative effort between researchers, policymakers, and farmers.

7.6 Conclusion

Streptomyces species continue to be a great source of bioactive compounds with nematicidal and herbicidal activity. These microorganisms produce a variety of secondary metabolites that could provide nematode and weed protection to plants, which offers an environmentally sustainable alternative to chemical pesticides. These bioactive compounds can alter nematode physiology, induce systemic resistance in plants, and alter plant growth and development, and they could be included in pest management strategies because they can increase yield, decrease pollution, and improve alternative ecologically sustainable agricultural systems. To capitalize on these microorganisms for ecologically sustainable production systems, further research is needed to develop and evaluate nematicides and herbicides based on *Streptomyces*.

References

Abd-Alla MH, El-Sayed ESA, Rasmey AHM (2013) Indole-3-acetic acid (IAA) production by *Streptomyces* atrovirens isolated from rhizospheric soil in Egypt. J Biol Earth Sci 3(2):182–193

Alam K, Mazumder A, Sikdar S, Zhao YM, Hao J, Song C, Wang Y, Sarkar R, Islam S, Zhang Y, Li A (2022) Streptomyces: The biofactory of secondary metabolites. Frontiers in microbiology. 13:968053.

Al-Askar AA, Baka ZA, Rashad YM, Ghoneem KM, Abdulkhair WM, Hafez EE, Shabana YM (2015) Evaluation of *Streptomyces* griseorubens E44G for the biocontrol of Fusarium oxysporum f. sp. lycopersici: ultrastructural and cytochemical investigations. Ann Microbiol 65:1815–1824

Altieri M (1995) Agroecology: the science of sustainable agriculture, 2nd edn. Westview Press, Boulder, pp 283–305

Anderson AS, Wellington EM (2001) The taxonomy of *Streptomyces* and related genera. Int J Syst Evol Microbiol 51(3):797–814

Burg, R. W., Miller, B. M., Baker, E. E., Birnbaum, J., Currie, S. A., Hartman, R., …, Ōmura, S. (1979). Avermectins, new family of potent anthelmintic agents: producing organism and fermentation. Antimicrob Agents Chemother, 15(3), 361–367

Burtseva S, Iurcu-Straistaru E, Birsa M, Sasanelli N (2019) Research on streptomycetes from moldovian soils with antifungal, nematicidal and phytostimulating activity on tomato. Microbiol J 81(1):9–21

Caillaud, M. C., Dubreuil, G., Quentin, M., Perfus-Barbeoch, L., Lecomte, P., de Almeida Engler, J., …, Favery, B. (2008). Root-knot nematodes manipulate plant cell functions during a compatible interaction. J Plant Physiol, 165(1), 104–113

Chauhan BS, Matloob A, Mahajan G, Aslam F, Florentine SK, Jha P (2017) Emerging challenges and opportunities for education and research in weed science. Front Plant Sci 8:1537

Chen JX, Song BA (2021) Natural nematicidal active compounds: recent research progress and outlook. J Integr Agric 20(8):2015–2031

Christy, A. L., Herbst, K. A., Kostka, S. J., Mullen, J. P., Carlson, P. S., Duke, S. O., …, Plimmer, J. R. (1993). Synergizing weed biocontrol agents with chemical herbicides. Pest control with enhanced environmental safety. American Chemical Society 87–100

Couillerot, O., Vatsa, P., Loqman, S., Ouhdouch, Y., Jane, H., Renault, J. H., …, Barka, E. A. (2013). Biocontrol and biofertilizer activities of the *Streptomyces* anulatus S37: an endophytic actinomycete with biocontrol and plant-growth promoting activities. IOBC-WPRS Bull, 86, 271–276

Cummins C, Harris H (1958) Studies on the cell-Wall composition and taxonomy of Actinomycetales and related groups. J Gen Microbiol 18:173–189

Davis E, MacGuidwin E (2000) Lesion nematode disease. The Plant Health Instructor. https://www.apsnet.org/edcenter/Pages/default.aspx

Davis EL, Hussey RS, Baum TJ, Bakker J, Schots A, Rosso MN, Abad P (2000) Nematode parasitism genes. Annu Rev Phytopathol 38(1):365–396

Decraemer W, Hunt DJ (2006) Structure and classification. In: Plant Nematol, vol 1, pp 3–32

Errakhi R, Bouteau F, Barakate M, Lebrihi A (2016) Isolation and characterization of antibiotics produced by *Streptomyces* J-2 and their role in biocontrol of plant diseases, especially grey mould. In: Biocontrol of major grapevine diseases: leading research. CABI, Wallingford, pp 76–83

Farooq S, Onen H, Ozaslan C, Baskin CC, Gunal H (2019) Seed germination niche for common ragweed (Ambrosia artemisiifolia L.) populations naturalized in Turkey. S Afr J Bot 123:361–371

Farooq M, Nawaz A, Saharawat YS, Reeves T, Siddique K (2020a) Crop and cropping systems management practices and benefits in Conservation Agriculture systems. In: Advances in conservation agriculture: volume 2: practice and benefits. Burleigh Dodds Science Publishing Limited, Cambridge

Farooq, S., Onen, H., Tad, S., Ozaslan, C., Mahmoud, S. F., Brestic, M., …, El-Shehawi, A. M. (2021). The influence of environmental factors on seed germination of Polygonum perfoliatum L.: implications for management. Agronomy, 11(6), 1123

Gherbawy Y, Elhariry H, Altalhi A, El-Deeb B, Khiralla G (2012) Molecular screening of *Streptomyces* isolates for antifungal activity and family 19 chitinase enzymes. J Microbiol 50: 459–468

Goudjal Y, Toumatia O, Sabaou N, Barakate M, Mathieu F, Zitouni A (2013) Endophytic actinomycetes from spontaneous plants of Algerian Sahara: indole-3-acetic acid production and tomato plants growth promoting activity. World J Microbiol Biotechnol 29:1821–1829

Gupta R, Singh A, Srivastava M, Shanker K, Pandey R (2019) Plant-microbe interactions endorse growth by uplifting microbial community structure of Bacopa monnieri rhizosphere under nematode stress. Microbiol Res 218:87–96

Ha, S., Lee, K. J., Lee, S. I., Gwak, H. J., Lee, J. H., Kim, T. W., …, Park, H. W. (2017). Optimization of Herbicidin a production in submerged culture of *Streptomyces* scopuliridis M40. J Microbiol Biotechnol, 27(5), 947–955

Harir M, Bendif H, Bellahcene M, Fortas Z, Pogni R (2018) *Streptomyces* secondary metabolites. In: Basic biology and applications of actinobacteria. IntechOpen

Hewezi T, Baum TJ (2013) Manipulation of plant cells by cyst and root-knot nematode effectors. Mol Plant-Microbe Interact 26(1):9–16

Hölscher, D., Dhakshinamoorthy, S., Alexandrov, T., Becker, M., Bretschneider, T., Buerkert, A., …, Swennen, R. L. (2014). Phenalenone-type phytoalexins mediate resistance of banana plants (Musa spp.) to the burrowing nematode Radopholussimilis. Proc Natl Acad Sci USA, 111(1), 105–110,

Hong, K., Gao, A. H., Xie, Q. Y., Gao, H., Zhuang, L., Lin, H. P., …, Ruan, J. S. (2009). Actinomycetes for marine drug discovery isolated from mangrove soils and plants in China. Mar Drugs, 7(1), 24–44

Huang, J., Chen, A. L., Zhang, H., Yu, Z., Li, M. H., Li, N., …, Zheng, Y. G. (2015). Gene replacement for the generation of designed novel avermectin derivatives with enhanced acaricidal and nematicidal activities. Appl Environ Microbiol, 81(16), 5326–5334

Hugot JP, Baujard P, Morand S (2001) Biodiversity in helminths and nematodes as a field of study: an overview. Nematology 3(3):199–208

Inbar E, Green SJ, Hadar Y, Minz D (2005) Competing factors of compost concentration and proximity to root affect the distribution of streptomycetes. Microb Ecol 50:73–81

Jang, J. Y., Choi, Y. H., Joo, Y. J., Kim, H., Choi, G. J., Jang, K. S., …, Kim, J. C. (2015). Characterization of *Streptomyces* netropsis showing a nematicidal activity against Meloidogyne incognita. Res Plant Dis, 21(2), 50–57

Jones SE, Elliot MA (2017) *Streptomyces* exploration: competition, volatile communication and new bacterial behaviours. Trends Microbiol 25(7):522–531

Jones MGK, Fosu-Nyarko J (2014) Molecular biology of root lesion nematodes (Pratylenchus spp.) and their interaction with host plants. Ann Appl Biol 164(2):163–181

Kang, M. K., Kim, M. H., Liu, M. J., Jin, C. Z., Park, S. H., Lee, J. M., …, Kim, C. J. (2021). Nematicidal activity of teleocidin B4 isolated from *Streptomyces* sp. against pine wood nematode, Bursaphelenchus xylophilus. Pest Manag Sci, 77(4), 1607–1615

Kang, M. K., Kim, J. H., Liu, M. J., Jin, C. Z., Park, D. J., Kim, J., …, Son, K. H. (2022). New discovery on the nematode activity of aureothin and alloaureothin isolated from endophytic bacteria *Streptomyces* sp. AE170020 Sci Rep, 12(1), 3947

Kasuga, K., Sasaki, A., Matsuo, T., Yamamoto, C., Minato, Y., Kuwahara, N., …, Kojima, I. (2017). Heterologous production of kasugamycin, an aminoglycoside antibiotic from *Streptomyces* kasugaensis, in *Streptomyces* lividans and Rhodococcus erythropolis L-88 by constitutive expression of the biosynthetic gene cluster. Appl Microbiol Biotechnol, 101, 4259–4268

Kaur T, Jasrotia S, Ohri P, Manhas RK (2016) Evaluation of in vitro and in vivo nematicidal potential of a multifunctional streptomycete, *Streptomyces* hydrogenans strain DH16 against Meloidogyne incognita. Microbiol Res 192:247–252

Khalil MS (2013) Abamectin and azadirachtin as eco-friendly promising biorational tools in integrated nematodes management programs. J Plant Pathol Microbiol 4(4):1–7

Khan S, Srivastava S, Karnwal A, Malik T (2023) *Streptomyces* as a promising biological control agents for plant pathogens. Front Microbiol 14:1285543

Kim D, Chun JY, Lee KY (2016) Morphometric and molecular characterization of populations of Pratylenchus kumamotoensis and P. pseudocoffeae (Nematoda, Pratylenchidae) newly recorded in Korea. ZooKeys 600:1

Koenning SR, Overstreet C, Noling JW, Donald PA, Becker JO, Fortnum BA (1999) Survey of crop losses in response to phytoparasitic nematodes in the United States for 1994. J Nematol 31(4S):587

Kroppenstedt RM, Stackebrandt E, Goodfellow M (1990) Taxonomic revision of the actinomycete genera Actinomadura and Microtetraspora. Syst Appl Microbiol 13(2):148–160

Le KD, Kim J, Nguyen HT, Yu NH, Park AR, Lee CW, Kim JC (2021) *Streptomyces* sp. JCK-6131 protects plants against bacterial and fungal diseases via two mechanisms. Front Plant Sci 12: 726266

Lee HB, Kim CJ, Kim JS, Hong KS, Cho KY (2003) A bleaching herbicidal activity of methoxyhygromycin (MHM) produced by an actinomycete strain *Streptomyces* sp. 8E-12. Lett Appl Microbiol 36(6):387–391

Lee J, Jung H, Hwang B (2005) *Streptomyces* koyangensis sp. nov., a novel actinomycete that produces 4-phenyl-3-butenoic acid. Int J Syst Evol Microbiol 55:257–262

Lee, B., Kim, J. D., Kim, Y. S., Ko, Y. K., Yon, G. H., Kim, C. J., …, Choi, J. S. (2013). Identification of *Streptomyces*scopuliridis KR-001 and its herbicidal characteristics. Weed Turfgrass Sci, 2(1), 38–46

Li, Y., Wang, K., Xie, H., Wang, Y. T., Wang, D. W., Xu, C. L., …, Wang, D. S. (2015). A nematode calreticulin, Rs-CRT, is a key effector in reproduction and pathogenicity of Radopholussimilis. PLoS One, 10(6), e0129351

Li, J. S., Qi, H., Zhang, S. Y., Xiang, W. S., Zhang, L. Q., Wang, H. Y., …, Wang, J. D. (2021). Two new milbemycin derivatives from a genetically engineered strain *Streptomyces* bingchenggensis. J Asian Nat Prod Res, 23(7), 660–665

Liu MJ, Hwang BS, Jin CZ, Li WJ, Park DJ, Seo ST, Kim CJ (2019) Screening, isolation and evaluation of a nematicidal compound from actinomycetes against the pine wood nematode, Bursaphelenchus xylophilus. *Pest Manag Sci* 75(6):1585–1593

Logan NA (1994) Bacterial systematics. Blackwell Scientific Publications, Oxford/London/ Edinburgh/Boston, p 1e269

Ma YY, Li YL, Lai HX, Guo Q, Xue QH (2017) Effects of two strains of Streptomyces on root-zone microbes and nematodes for biocontrol of root-knot nematode disease in tomato. Applied Soil Ecology. 112:34–41.

Martínez-Hidalgo P, Olivares J, Delgado A, Bedmar E, Martínez-Molina E (2014) Endophytic Micromonospora from Medicago sativa are apparently not able to fix atmospheric nitrogen. Soil Biol Biochem 74:201–203

Mhatre, P. H., Thorat, Y. E., Manimaran, B., Divya, K. L., Bairwa, A., Chavan, S. N., …, Karthik, C. (2024). Nematicidal activity of secondary metabolites from soil microbes. Sustainable management of nematodes in agriculture, Vol. 2: role of microbes-assisted strategies, 297-324

Muangham S, Pathom-Aree W, Duangmal K (2015) Melanogenic actinomycetes from rhizosphere soil—antagonistic activity against Xanthomonas oryzae and plant-growth-promoting traits. Can J Microbiol 61(2):164–170

Nair MG, Chandra A, Thorogod DL, Davis RMG (1995) Nematicidal and mosquitocidal aromatic nitro compounds produced by *Streptomyces* spp. cabidigitallibrary 43:361–365

Nakajima, M., Itoi, K., Takamatsu, Y., Kinoshita, T., Okazaki, T., Kawakubo, K., …, Haneishi, T. (1991). Hydantocidin: a new compound with herbicidal activity from *Streptomyces hygroscopicus*. J Antibiot, 44(3), 293–300

Nguyen XH, Naing KW, Lee YS, Kim YH, Moon JH, Kim KY (2015) Antagonism of antifungal metabolites from *Streptomyces* griseus H7602 against Phytophthora capsici. J Basic Microbiol 55(1):45–53

Nicol J, Turner D, Coyne L, den Nijs L, Hockland S, Maafi Z (2011) Current nematode threats to world agriculture. In: Jones J, Gheysen G, Fenoll C (eds) Genomics and molecular genetics of plant-nematode interactions. Springer Science Business Media, Berlin, pp 21–43

Olanrewaju OS, Babalola OO (2019) *Streptomyces*: implications and interactions in plant growth promotion. Appl Microbiol Biotechnol 103:1179–1188

Önen H, Farooq S, Tad S, Özaslan C, Gunal H, Chauhan BS (2018) The influence of environmental factors on germination of Burcucumber (Sicyos angulatus) seeds: implications for range expansion and management. Weed Sci 66(4):494–501

Panchalingam, H., Ashfield-Crook, N., Naik, V., Frenken, R., Foster, K., Tomlin, R., …, Kurtböke, D. İ. (2022). Testing the biocontrol ability of a Trichoderma-Streptomycetes consortium against Pyrrhodermanoxium (Corner) LW Zhou and YC Dai in soil. J Fungi, 9(1), 67

Park JO, El-Tarabily KA, Ghisalberti EL, Sivasithamparam K (2002) Pathogenesis of Streptoverticillium albireticuli on Caenorhabditis elegans and its antagonism to soil-borne fungal pathogens. Lett Appl Microbiol 35(5):361–365

Park EJ, Jang HJ, Park CS, Lee SJ, Lee S, Kim KH, Yun BS, Lee SW, Rho MC (2020) Evaluation of nematicidal activity of Streptomyces yatensis KRA-28 against Meloidogyne incognita. Journal of Microbiology and Biotechnology 30(5):700

Postnikova OA, Hult M, Shao J, Skantar A, Nemchinov LG (2015) Transcriptome analysis of resistant and susceptible alfalfa cultivars infected with root-knot nematode Meloidogyne incognita. PLoS One 10(2):e0118269

Riaz Marral MW, Khan MB, Ahmad F, Farooq S, Hussain M (2020) The influence of transgenic (Bt) and non-transgenic (non-Bt) cotton mulches on weed dynamics, soil properties and productivity of different winter crops. PLoS One 15(9):e0238716

Ruanpanun P, Dame ZT, Laatsch H, Lumyong S (2011a) 3-Methoxy-2-methyl-carbazole-1, 4-quinone, carbazomycins D and F from *Streptomyces* sp. CMU-JT005. FEMS Microbiol Lett 322(1):77–81

Ruanpanun P, Laatsch H, Tangchitsomkid N, Lumyong S (2011b) Nematicidal activity of fervenulin isolated from a nematicidal actinomycete, *Streptomyces* sp. CMU-MH021, on Meloidogyne incognita. World J Microbiol Biotechnol 27(6):1373–1380

Samac DA, Kinkel LL (2001) Suppression of the root-lesion nematode (Pratylenchus penetrans) in alfalfa (Medicago sativa) by *Streptomyces* spp. Plant Soil 235:35–44

Sarah J, Gowen S, De Waele D, Tessera M, Quimio A (1999) Nematode pathogens. In: Jones D (ed) Diseases of Banana, Abacá and Ensete. CABI Publishing, Wallingford, pp 295–303

Sasser JN, Freckman DW (1987) A world perspective on nematology: the role of the society. In: Veech J, Dickson D (eds) Vistas on nematology. Society of Nematologists, Inc, Hyattsville, pp 7–14

Sathya A, Vijayabharathi R, Gopalakrishnan S (2017) Plant growth-promoting actinobacteria: a new strategy for enhancing sustainable production and protection of grain legumes. 3 Biotech 7: 1–10

Sharma M, Jasrotia S, Ohri P, Manhas RK (2019) Nematicidal potential of *Streptomyces antibioticus* strain M7 against *Meloidogyne incognita*. AMB Express 9:1–8

Shekh MAN (2017) Isolation and characterization of antibacterial metabolites from Streptomyces species. Doctoral dissertation, Daffodil International University

Shrivastava P, Kumar R (2018) Actinobacteria: eco-friendly candidates for control of plant diseases in a sustainable manner. In: New and future developments in microbial biotechnology and bioengineering. Elsevier, pp 79–91

Sivalingam P, Hong K, Pote J, Prabakar K (2019) Extreme environment *Streptomyces*: potential sources for new antibacterial and anticancer drug leads? Int J Microbiol 2019(1):5283948

Smaoui S, Mathieu F, Fguira LFB, Merlina G, Mellouli L (2011) Taxonomy and antimicrobial activities of a new *Streptomyces* sp. TN17 isolated in the soil from an oasis in Tunis. Arch Biol Sci 63(4):1047–1056

Smiley RW (2015) Root-lesion nematodes. In: Biology and management in Pacific northwest wheat cropping systems, PNW extension bulletin 617. Oregon State University, Corvallis

Smith, I., Charles, L. (1998). Distribution maps of quarantine pests for Europe. Wallingford: CABI International. pp. 1–78

Solanki MK, Malviya MK, Wang Z (2016) Actinomycetes bio-inoculants: a modern prospectus for plant disease management. In: Plant growth promoting actinobacteria*:* a new avenue for enhancing the productivity and soil fertility of grain legumes, pp 63–81

Stackebrandt E, Woese CR (1981) Towards a phylogeny of the actinomycetes and related organisms. Curr Microbiol 5:197–202

Stackebrandt E, Rainey FA, Ward-Rainey NL (1997) Proposal for a new hierarchic classification system, Actinobacteria classis nov. Int J Syst Evol Microbiol 47(2):479–491

Su H, Shao H, Zhang K, Li G (2016) Antibacterial metabolites from the actinomycete *Streptomyces* sp. P294. J Microbiol 54(2):131–135

Suárez-Moreno, Z. R., Vinchira-Villarraga, D. M., Vergara-Morales, D. I., Castellanos, L., Ramos, F. A., Guarnaccia, C., …, Moreno-Sarmiento, N. (2019). Plant-growth promotion and biocontrol properties of three *Streptomyces* spp. isolates to control bacterial rice pathogens. Front Microbiol, 10, 290

Sullivan P (2003) Principles of sustainable weed management for croplands

Sun, Y., Xie, J., Tang, L., Odiba, A. S., Chen, Y., Fang, W., …, Wang, B. (2023). Isolation, identification and molecular mechanism analysis of the nematicidal compound spectinabilin from newly isolated *Streptomyces* sp. DT10. Molecules, 28(11), 4365

Vasilakoglou IB, Eleftherohorinos IG, Dhima KV (2000) Propanil-resistant barnyard grass (Echinochloa crus-galli) biotypes found in Greece. Weed Technol 14(3):524–529

Vijayabharathi R, Sathya A, Gopalakrishnan S (2015) Plant growth-promoting microbes from herbal vermicompost. In: Plant-growth-promoting rhizobacteria (PGPR) and medicinal plants, pp 71–88

Vurukonda SSKP, Giovanardi D, Stefani E (2018) Plant growth promoting and biocontrol activity of *Streptomyces* spp. as endophytes. Int J Mol Sci 19(4):952

Waksman SA, Henrici AT (1943) The nomenclature and classification of the actinomycetes. J Bacteriol 46(4):337–341

Wang LF, Yang BJ, Dao LC (2002) A review of biological control of root-knot nematodes. J Nanjing For Univ 26:64–68

Wang XJ, Wang M, Wang JD, Jiang L, Wang JJ, Xiang WS (2010) Isolation and identification of novel macrocyclic lactones from *Streptomyces* avermitilis NEAU1069 with acaricidal and nematocidal activity. J Agric Food Chem 58(5):2710–2714

Won OJ, Kim YT, Choi JS, Oh TK, Shinogi Y, Park KW (2016) Herbicidal activity and mode of action of *Streptomyces*scopuliridis metabolites. J Fac Agric Kyushu Univ 61:47–51

Wrather A, Mitchum M (2010) Soybean cyst nematodes: diagnosis and Management. Available from: http://extension.missouri.edu/p/g4450. Accessed 23 Mar 2017

Wu Y, Yuan J, E Y, Raza W, Shen Q, Huang Q (2015) Effects of volatile organic compounds from *Streptomyces* albulus NJZJSA2 on growth of two fungal pathogens. J Basic Microbiol 55(9): 1104–1117

Xu W, Tao L, Gu X, Shen X, Yuan S (2009) Herbicidal activity of the metabolite SPRI-70014 from *Streptomyces* griseolus. Weed Sci 57(5):547–553

Yang LY, Wang JD, Zhang J, Xue CY, Zhang H, Wang XJ, Xiang WS (2013) New nemadectin congeners with acaricidal and nematocidal activity from *Streptomyces* microflavus neau3 Y-3. Bioorg Med Chem Lett 23(20):5710–5713

Yasuor H, Ten Brook PL, Tjeerdema RS, Fischer AJ (2008) Responses to clomazone and 5-ketoclomazone by Echinochloa phyllopogon resistant to multiple herbicides in Californian rice fields. Pest Manag Sci Former Pestic Sci 64(10):1031–1039

Zeng Q, Huang H, Zhu J, Fang Z, Sun Q, Bao S (2013) A new nematicidal compound produced by *Streptomyces* albogriseolus HA10002. Antonie Van Leeuwenhoek *103*(5):1107–1111

Zhang, Z., Wu, X., Song, R., Zhang, J., Wang, H., Zhu, J., …, Shen, Y. (2017). Ansavaricins F–I, new DNA topoisomerase inhibitors produced by *Streptomyces* sp. S012. RSC Adv, 7(24), 14857–14867

Zheng X, Wang J, Chen Z, Zhang H, Wang Z, Zhu Y, Liu BA (2019) Streptomyces sp. strain: Isolation, identification, and potential as a biocontrol agent against soilborne diseases of tomato plants. Biological control. 136:104004.

Zimdahl RL, Basinger NT (2024) Fundamentals of weed science. Academic/Elsevier, Cambridge, MA

Streptomyces as Emerging Biocontrol Agents: Mechanistic Insights into Nematicidal and Herbicidal Action

8

Deepak Kumar, Roohi Sharma, Vikram, Alisha Sharma, Anchita Singh, Suhani Kiran Dutta, Diksha, Rajesh Kumari Manhas, and Puja Ohri

Abstract

Plant-parasitic nematodes (PPNs) represent a major threat to global crop yields, leading to a predictable annual loss of $157 billion in the agricultural sector. Further, weeds can negatively impact crop development and productivity by creating a competition with crops for nutrients and water. Nowadays, a range of potentially hazardous and poisonous chemical compounds are employed to reduce the amount of agricultural damage caused by plant-parasitic nematodes and weeds. As concerns over the risks of chemical pesticides grow, there is an increasing concern in using microbial methods to control PPNs and weeds, with biological nematicides and herbicides becoming a key part of eco-friendly management strategies. A natural alternative that has been well studied is the *Streptomyces* species of Gram-positive bacteria, which may yield a variety of antipathogenic chemicals, including organic compounds and metabolites. Researchers from all around the world employ a variety of techniques and strategies to identify novel bioactive substances in these bacteria. Bioactive chemicals have been shown to be effective in reducing the nematodes and weeds population that seriously endanger plants and crops. Therefore, the current

D. Kumar
Department of Zoology, Guru Nanak Dev University, Amritsar, Punjab, India

PG Department of Zoology, Khalsa College, Amritsar, Punjab, India

R. Sharma · Vikram · A. Sharma · A. Singh · S. K. Dutta · Diksha · P. Ohri (✉)
Department of Zoology, Guru Nanak Dev University, Amritsar, Punjab, India

R. K. Manhas
Department of Microbiology, Guru Nanak Dev University, Amritsar, Punjab, India

T. Kaur et al. (eds.), *Streptomyces: Biological Candidates for Sustainable Agriculture*,
https://doi.org/10.1007/978-981-95-5803-2_8

chapter summarizes the negative impacts of nematodes and weeds on plant growth and development and also explores role of streptomyces as nematicidal and herbicidal agents, possible mechanisms and challenges for using streptomyces as biocontrol agent.

Keywords

Streptomyces · Plant parasitic nematodes · Weeds · Biological nematicides · Bioherbicides

8.1 Introduction

Modern agriculture faces growing challenges due to the increasing prevalence of biotic stressors such as plant-parasitic nematodes and weeds, both of which significantly compromise crop productivity and quality (Sharma et al. 2023; Parven et al. 2025). The conventional use of synthetic agrochemicals, such as herbicides and nematicides, has provided temporary fixes but has had detrimental long-term effects as well, such as contaminating the environment, harming non-target creatures, and creating resistant populations of weeds and pests (Das et al. 2024; Kumar et al. 2023). A global movement for environmentally benign and sustainable agricultural inputs, especially the investigation of microbial-based biocontrol agents, has been spurred by these worries.

Among the diverse microbial taxa being investigated, *Streptomyces*, a genus within the phylum Actinobacteria stand out due to their remarkable capacity to synthesize a vast array of secondary metabolites. These spore-forming, filamentous bacteria are mostly found in soil conditions, where they are essential for the natural suppression of illness and the breakdown of organic materials (Pang et al. 2022; Sharma et al. 2022). More than two-thirds of therapeutically and agriculturally significant antibiotics are identified to be produced by streptomyces; however, new research has broadened their known bioactivity range to include nematicidal, herbicidal, antifungal, insecticidal, and plant growth-promoting activities.

In the context of nematode management, streptomyces have shown promising activity against economically significant species such as *Meloidogyne* spp. (root-knot nematodes), *Heterodera* spp. (cyst nematodes), and *Pratylenchus* spp. (lesion nematodes) (Marin-Bruzos et al. 2021; Sharma et al. 2022). Avermectins, actinomycins, and *Streptomyces*-derived enzymes like chitinases and proteases are among the bioactive substances produced by these bacteria that can break down the nematode's cuticle or interfere with its development and reproduction. Additionally, it has been shown that certain *Streptomycees* species efficiently colonize the rhizosphere, increasing their potential as persistent and long-lasting soil-dwelling biocontrol agents that provide nematode protection (Thakur and Singh 2018; Balla et al. 2021). Various studies have reported sustainable nematode management using *Streptomyces* species. Further, synthesis of those chemicals which are toxic to plants and disturb plant signaling, growth, and development have also been reported in specific streptomyces. Such *Streptomyces* species can be employed for eco-friendly management of weed species. For example, cycloheximide, a herbicide active compound is synthesized by *Streptomyces* sp. strain KR0005 which acts by inhibiting photosynthesis and integrity of cell membrane (Umurzokov et al. 2023). Similarly, *Streptomyces* strain KRA17-580 was reported to synthesize a cinnoline-4-carboxamide and cinnoline-4-carboxylic acid which exhibited phytotoxic activity against *Digitaria ciliaris* (Kim et al. 2020).

This dual nature of *Streptomyces* of acting as nematicidal and herbicidal agents attracts the researchers for employing this microbe in integrated pest management. Further, potential of *Streptomyces* to establish and grow in diverse soil environments, compatibility with other microbes and specificity to the target make it more appealing for researchers working on the sustainable pest management (Agale et al. 2018; Pacios-Michelena et al. 2021; Khan et al. 2023a, 2023b). Commercialization of certain *Streptomyces*-based products has been started but performance of microbe in laboratory conditions, strict regulatory process, and safety issues restrict the effective extrapolation of the lab studies to the field (Butt et al. 2024). However, use of advanced techniques and supportive regulatory processes can be employed to overcome these challenges (Chater 2016). Therefore, this chapter highlights use of *Streptomyces* as nematicidal and herbicidal agent and its underlying mechanism for a sustainable agriculture.

8.2 Nematodes-Induced Plant Damage

Belonging to the phylum Nematoda, small microscopic nematodes found globally in a range of habitats cause significant yield losses (Kumar et al. 2025a, 2025b). About 14.6% of crop losses in tropical and subtropical regions and an additional 8.8% in industrialized nations are attributed to plant-parasitic nematodes (PPNs). Every year, these pests cause agricultural losses worth over $358 billion worldwide (Khanal and Land 2023; Kamaraju et al. 2024). The interactions between RKN and plants have been the subject of numerous studies. According to these studies, the growth, development, and photosynthesis of the infected crop plants drastically decreased, which decreased crop yield (Sharma et al. 2023). Okra shoots and root length, plant height, stem diameter, number of leaves, and leaf area were drastically decreased as a result of *Meloidogyne javanica* (Ali et al. 2023). Cowpea infected with *Meloidogyne incognita* showed a further decline in plant growth and photosynthetic pigments, such as carotenoid concentration and chlorophyll a and b (Tauseef et al. 2021). Plant oxidative stress is caused by nematode infection, which modifies normal solute transport, causes electrolyte leakage, and increases the production of reactive oxygen species (Raza et al. 2022; Kumar et al. 2023). As byproducts of regular metabolism, reactive oxygen species (ROS) can be generated at harmful levels in a number of cellular systems, such as mitochondria, chloroplasts, peroxisomes, and apoplasts, in reaction to stress (Mansoor et al. 2022). When ROS levels increase beyond the threshold, cellular and organellar membranes experience increased lipid peroxidation, which impacts regular cellular activity (Aureliano et al. 2023). By producing lipid-derived radicals, which have the ability to interact with proteins and deoxyribonucleic acid (DNA) and cause damage, lipid peroxidation exacerbates oxidative stress. The level of lipid peroxidation has often been used as an indicator of ROS-mediated cell membrane degradation during nematode stress (Kaushal et al. 2023; Sharma et al. 2023). Elevated levels of oxidative stress in terms of increased content of malondialdehyde (MDA) and hydrogen peroxide (H_2O_2) has been reported in nematode-infected plants (Sharma et al. 2022; Kumar et al. 2025a;). Further, nematode infestation increases activity of various antioxidant including ascorbate peroxidase (APOX), catalase, (CAT), guaiacol peroxidase (GuPOX), polyphenol oxidase (PPO), glutathione-s-transferase (GST), and superoxide dismutase (SOD). Studies have reported that tomato plants infected with *M. incognita* exhibited the increased activities of these enzymes (Kavari et al. 2022; Kumar et al. 2025b). Similarly, another study reported chilli and tomato plants exhibited enhanced activities of PPO, peroxidase (POD), and SOD upon infection with *M. incognita* (Chandrawat et al. 2020). Table 8.1 enlists some effects of different nematode on various plants.

8.3 *Streptomyces* as Nematicidal Agent

Streptomyces species have been reported to synthesize numerous compounds that act as antimicrobials, enzymes, and secondary metabolites targeting PPNs make them suitable candidate to be used as nematicidal agents for sustainable agriculture (Sharma et al. 2019, 2022). By interfering with nematode growth, reproduction, or cuticle integrity, these substances can eventually lower the population of nematodes in the soil. Significantly, avermectins, which are produced by *S. avermitilis* and other species, are very efficient against different species of root-knot nematodes (*Meloidogyne* spp.). *Streptomyces* are attractive options for sustainable management of nematodes in agriculture because of their natural origin, specificity, and environmental safety. Further, in agriculture, use of *Streptomyces* have been reported to reduce negative effects induced by plant parasitic nemaotdes.

8.3.1 Streptomyces-Induced Nematode Mortality

Nematode mortality caused by *Streptomyces* is a complex biological phenomenon that is important for sustainable pest control. A genus of actinobacteria found in soil ecosystems, *Streptomyces* is widely known for generating a variety of secondary metabolites having nematicidal, antifungal, and antibacterial qualities (Aggarwal et al. 2016; Vurukonda et al. 2018). Certain strains of *Streptomyces* use both direct and indirect methods to cause death when controlling nematodes. Avermectins, a class of macrocyclic lactones generated from *Streptomyces avermitilis*, are one of the main techniques. They target the nematode nervous system by interfering with chloride ion channels, causing paralysis and ultimately death (Cerna-Chávez et al. 2024). Other bioactive molecules that *Streptomyces* make include lipopeptides,

Table 8.1 Nematode-induced damage to host plants

Nematode	Plant species	Effect on host plant	References
Meloidogyne incognita	Tomato	Reduced plant growth and induced oxidative stress	Kumar et al. (2025a, 2025b)
M. incognita	Pepper	Reduced chlorophyll a and b and total chlorophyll	Ghareeb et al. (2024)
M. incognita	Tomato	Enhanced level of H_2O_2	Ji et al. (2024)
M. enterolobii	Tomato	Increased levels of ascorbic acid and proline	Sikandar et al. (2024)
M. incognita	Tomato	Increased activities of APOX, SOD, CAT, and POD	Yang et al. (2024)
M. incognita	Carrot	Decrease chlorophyll and carotenoid content	Ahamad et al. (2023a)
M. incognita	Tomato	Reduced number of leaves and flowers and content of chlorophyll a and b and total chlorophyll	Ahamad et al. (2023b)
M. javanica	Okra	Reduced shoot and root length, plant height, stem diameter, number of leaves and leaf area	Ali et al. (2023)
M. incognita	Cucumber	Boosted the activities of SOD, POD, CAT, and PPO	Nofal and Attia (2023)
M. incognita	Tomato	Enhanced $H_2O_2$and MDA content	Kaushal et al. (2023)
M. incognita	Tomato	Enhanced contents of glutathione, phenol, flavonoid and proline	Kaushal et al. (2023)
M. graminicola	Rice	Increased activities of APOX, SOD, CAT, and POD	Malviya et al. (2023)
M. javanica	Gilo	Reduced plant height, shoot dry matter, root fresh weight, transpiration and photosynthetic rate and stomatal conductance	Santos et al. (2023)
M. javanica	Soybean	Increased concentration of H_2O_2 and MDA	Santos et al. (2023)
M. incognita	Chilli	Increased levels of proline and phenol	Ujjwal et al. (2023)
M. javanica	Chickpea	Augmented the activities of POD and PPO	Aljuboori et al. (2022)
M. javanica	Tomato	Increased content of proline	Bakr et al. (2022)
M. javanica	Tomato	Increased H_2O_2 concentration	Behzadian et al. (2022)

indole derivatives, and volatile organic compounds (VOCs), which can interfere with nematode development or function as poisons (Pacios-Michelena et al. 2021). In addition to using chemical warfare, these bacteria release extracellular enzymes such as cellulases, chitinases, and proteases that break down nematode cuticles and eggshells, impairing their structural integrity and decreasing their capacity to infect host plants (Lopes et al. 2021). By causing systemic resistance in plants, *Streptomyces* can also have an indirect impact on nematode populations by strengthening the plant's defenses against nematode invasion. Additionally, *Streptomyces* colonizing plant roots can produce a protective *rhizosphere* microbiome that is unsuitable for nematode survival. Nematode viability, reproduction, and infectivity are significantly decreased as a result of these combined effects (Khan et al. 2023a, b; Sharma et al. 2022). *Streptomyces* are becoming more widely acknowledged as effective biocontrol agents in integrated pest management programs due to continuous developments in microbial ecology, molecular biology, and bioformulation technology. They provide a practical and environmentally friendly alternative for synthetic nematicides in agricultural systems. For examples, guava plants treated with *S. rochei* GA exhibited a 75% increase in plant growth and 79.1% reduction in egg masses, hence inhibiting *M. enterolobii* parasitism (Mani et al. 2025a, b). Similarly, another study reported that *Streptomyces* sp. MR-14 induced 83% mortality on *M. incognita* juveniles and reduced number of galls in tomato seedlings (Sharma et al. 2021). Table 8.2 summarizes effect of *Streptomyces* on different nematodes.

8.3.2 Streptomyces-Induced Mitigation of Nematode Stress

The genus *Streptomyces*, a prominent group within actinobacteria, has been extensively investigated for its potential as a biological control agent against root-knot nematodes (RKN), particularly those of the genus *Meloidogyne* (Yoon et al. 2012; Zeng et al. 2013; Jin et al. 2017; Sharma et al. 2019; Meidani et al. 2020). Several species within this genus are recognized for producing nematicidal metabolites effective against juvenile stages of these nematodes. Notably, *S. cacaoi* (Yoon et al. 2012), *S. albogriseolus* (Zeng et al. 2013), *S. rubrogriseus* (Jin et al. 2017), and *S. antibioticus* (Sharma et al. 2019) have demonstrated such properties. Culture filtrates from *S. rubrogriseus* and *S. cacaoi* have been shown to inhibit the hatching of *M. incognita* eggs (Jin et al. 2017). Thus, *Streptomyces* demonstrated their potential to reduce nematode-induced damage (Table 8.3) along with improved morphological and photosynthetic parameters in nematode-stressed plants. Furthermore, nematode-induced oxidative damage is also reduced by *Streptomyces* via modulating antioxidant defense system of plants (Sharma et al. 2022).

Morphological Parameters

Various morphological parameters including root length, shoot length, root weight, shoot weight, and number of leaves have been reported to be negatively impacted by nematode infestation (de Oliveira Costa et al. 2025; Kumar et al. 2025a, b). However, treatment of seeds, seedlings, and plants with *Streptomyces* has been reported to mitigate these negative effects of nematode infestation on plant morphology. For instance, a study conducted by Sharma et al. (2022) reported that soil drenching with culture cells, supernatant, and crude extract of *Streptomyces* resulted in reduced nematode infestation in tomato plants, and improved morphological parameters as

Table 8.2 Nematocidal potential of *Streptomyces* spp. against different nematodes

Nematode	*Streptomyces* used	Effect of *Streptomyces* on nematodes	References
Meloidogyne enterolobii	*Streptomyces* spp.	Induced juvenile mortality (100%) and reduced hatching (100%)	Mani et al. (2025a)
M. enterolobii	*S. rochei*	100% efficiency in reducing nematode egg viability and inducing mortality	Mani et al. (2025b)
M. incognita	*S. aquilus*	The death rate for juveniles at 12 and 24 h was 70.86 and 85.58%	Xu et al. (2025)
M. incognita	*S. violascens*	99% immobility of second-stage juveniles (J2s)	Azlay et al. (2024)
M. javanica	*Streptomyces* sp.	Decreased in the quantity of eggs and galls	Mogollón-Ortiz et al. (2024)
M. incognita	*S. avermitilis*	Induced 100% mortality in 1 day	Radwan et al. (2024)
M. javanica	*Actinobacteria*	Substantial drops in the quantity of galls and eggs	Xing et al. (2024)
M. incognita	*Streptomyces* sp.	Substantially decreased egg hatching	Atif et al. (2023)
M. incognita	*Streptomyces* sp. DT10	The death rate was approximately 40%	Sun et al. (2023)
M. incognita	*Streptomyces albogriseolus* ND41	82.13% of juveniles died after 12 h	Hu et al. (2022)
M. incognita	*Streptomyces fimicarius* D153	After 12 h, the juvenile mortality rate was 86.96%	Hu et al. (2022)
Bursaphelenchus xylophilus	*Streptomyces* sp.	Prevented the hatching of eggs	Kang et al. (2021)
Pratylenchus penetrans	*Streptomyces fulvissimus* A12, *Streptomyces venezuelae* A30 and *Streptomyces annulatus* A34	Decreased nematode population on onion roots	Marin-Bruzos et al. (2021)
M. incognita	*Streptomyces* sp. MR-14	83% juvenile mortality	Sharma et al. (2021)
M. incognita	*Streptomyces hydrogenans*	Significantly reduced number of galls (19.9%)	Sharma et al. (2020)
M. incognita	*S. antibioticus* M7	97.0–99.0% mortality of juveniles	Sharma et al. (2019)

Table 8.3 Plant growth-promoting effect of *Streptomyces* during nematode stress

Plant species	Nematode	*Streptomyces* used as mitigator	Effect of *Streptomyces* on plant	Reference
Solanum melongena	*M. incognita*	*Streptomyces* spp.	Improved growth parameters Reduced nematode numbers	Abo-Elyousr et al. (2023)
Picea abies	*Bursaphelenchus xylophilus*	*S. spectabilis* KCTC9218^{t}	Increased survival rate of pine trees from 10 to 80% when treated	Lee et al. (2022)
S. lycopersicum	*M. incognita*	*S. hydrogenans* DH-16	Elevated chlorophyll a, b, total chlorophyll and carotenoids and activity of enzymatic antioxidants	Sharma et al. (2022)
S. lycopersicum	*M. incognita*	*S. antibioticus* M7	Reduced root galling enhanced plant growth parameters and photosynthetic pigments	Singh et al. (2022)
Allium cepa	*Pratylenchus penetrans*	*S. fulvissimus A12, S. venezuelae A30 and S. annulatus A34*	Enhanced root length and weight	Marin-Bruzos et al. (2021)
S. lycopersicum	*M. incognita*	*Streptomyces* sp. MR-14	Improved vigor of seedlings, enhanced root and shoot length and reduced oxidative stress	Sharma et al. (2021)
S. lycopersicum	*M. incognita*	*S. hydrogenans* DH-16	Enhanced shoot and root length, increased biomass, and reduced gall formation	Sharma et al. (2020)

(continued)

Table 8.3 (continued)

Plant species	Nematode	*Streptomyces* used as mitigator	Effect of *Streptomyces* on plant	Reference
S. lycopersicum	*M. incognita*	*S. antibioticus*	Enhanced plant dry weight, fresh weight, and length	Sharma et al. (2019)
Bacopa monnieri	*M. incognita*	*Streptomyces* sp. MTN14	Reduced oxidative stress in plants	Gupta et al. (2017)
Sorghum bicolor	Not specified	*Streptomyces* spp.	Increased plant height, leaf area, root, and shoot weight Enhanced panicle and seed weight	Alekhya and Gopalakrishnan (2016)
Capsicum annuum	*M. incognita*	*S. yatensis* KRA-28	Improved shoot length and dry weight Reduced egg masses in roots	Jang et al. (2015)
Actinidia deliciosa	*Meloidogyne* sp.	*Streptomyces* spp.	Enhanced plant growth Reduced root galling	Jaizme-Vega et al. (2006)

revealed by increased root and shoot length and root and shoot weight of treated plants when assessed against untreated nematode inoculated plants. Similarly, another study reported that after 12 h, the nematicidal activity of culture filtrates of *Streptomyces albogriseolus* ND41 and *S. fimicarius* D153 against *Meloidogyne incognita* was 82.13 and 86.96%, respectively. In addition to this, increased plant height, fresh shoot weight, and fresh root weight as compared to untreated nematode-infected tomato plants grown in pots were observed (Lima et al. 2017).

Photosynthetic Pigments

In addition to the morphological parameters, nematode infestation also affects photosynthetic pigment and photosynthetic efficiency and leads to the reduction in chlorophyll a, b, total chlorophyll, and carotenoid content along with altered stomatal conductance and transpiration rate (Kumar et al. 2025a, b). However, studies have reported that testament of nematode-infested plants with *Streptomyces* might alleviate these negative impacts of nematode and may improve the levels of photosynthetic pigments. For instance, a study conducted by Sharma et al. (2020) demonstrated that the nematode-tainted tomato plant seedlings when treated with supernatant and extract of *Streptomyces hydrogenans* displayed increased content of chlorophyll a, chlorophyll b, total chlorophyll, and carotenoids as compared to the seedlings that were untreated but inoculated with nematodes.

Oxidative Stress and Antioxidative Defense System

Oxidative stress in plants lead to the modulation of activities of antioxidative enzymes like APOX, GupOX, SOD, POD, PPO, and CAT (Manullang and Chuang 2020; Khan et al. 2023a, b; Ali and Ohri 2023). Use of *Streptomyces* species have also been reported to mitigate oxidative stress and strengthen the antioxidant defense system during stress conditions. Markers of oxidative stress were significantly decreased following microbial inoculations (Sharma et al. 2020). For instance, *Streptomyces canus* (GLY-P2) reduced the amounts of hydrogen peroxide, superoxide radicals, and malonaldehyde in leaves. Additionally, GLY-P2 increased the amounts of ascorbate and glutathione, increased the transcript levels of copper/zinc SOD, manganese SOD, catalase, and APOX in leaves (Wu et al. 2019). Ascorbate peroxidase, GuPOX, POD, and SOD all significantly increased when chickpeas were treated with *Sclerotium rolfsii*, which promotes the least amount of lipid peroxidation under pathogenic stress (Singh and Gaur 2017). *S. canus* (GLY-P2) reduced the amounts of H_2O_2, superoxide radicals, and MDA in leaves. Additionally, GLY-P2 increased the amounts of ascorbate and glutathione, increased the transcript levels of copper/zinc SOD, manganese SOD, CAT, and APOX in leaves and improved the activities of SOD, CAT, APOX, monodehydroascorbate reductase, dehydroascorbate reductase, and glutathione reductase (Wu et al. 2019). The culture cells, supernatant, and extract of *S. hydrogenans* strain DH-16 significantly reduced the levels of stress markers (MDA and H_2O_2) in infected plants and increased the activities of enzymatic antioxidants (CAT, SOD, APOX, GuPOX, GST PPO) and non-enzymatic antioxidants (glutathione, tocopherol) in *M. incognita*-infected tomato plants (Sharma et al. 2022). Tomato seedlings afflicted with nematodes had higher levels of MDA and H_2O_2 content than control seedlings, according to oxidative burst profiling. However, when comparing supernatant-pretreated seedlings to nematode-inoculated seedlings, a notable decrease of 20% in MDA and 31% in H_2O_2 was noted. In another study carried out by Sharma et al. (2021), MDA and H_2O_2 concentration in extract-treated nematode-inoculated seedlings decreased by 18 and 14%, respectively. According to another study, *Bacopa monnieri* treated with dual microorganisms (*Streptomyces* sp. MTN14 and *Chitiniphilus* sp. MTN22) infected with the pathogen initially produced more hydrogen peroxide (H_2O_2), but this production eventually decreased compared to control plants inoculated with *M. incognita*. The same treatment also had the greatest levels of superoxide dismutase and free radical scavenging activity, which were directly correlated with the lowest levels of lipid peroxidation and root gall development in *Bacopa monnieri* under biotic stress (Gupta et al. 2017).

The integration of *Streptomyces* spp. as a biocontrol agent offers a promising strategy to alleviate the detrimental effects of *nematode* infestation on plant morphology and photosynthesis. Its application not only restores but enhances growth parameters and photosynthetic efficiency, underscoring its potential in sustainable agricultural practices.

8.4 Plant Growth-Promoting Mechanisms Exhibited by *Streptomyces*

The production of secondary metabolites linked to the promotion of plant development and biocontrol potential makes the genus *Streptomyces* potentially useful in a number of agricultural applications. We go over a few of them as below.

8.4.1 Chemotaxis

The interaction between plant parasitic nematodes and *Streptomyces* is significantly mediated by chemotaxis, the movement of organism in response to chemical stimulus (Mani et al. 2024). Various studies have demonstrated that *Streptomyces* spp. produce wide range of secondary metabolites/volatile organic compounds (Guo et al. 2020). Nematodes use chemical signals such as pyrazines, alkenes, and other non-volatile chemicals to find their host plants. Some of these chemicals attract nematodes to the vicinity of *Streptomyces* whereas certain metabolites are nematocidal in nature resulting in killing of nematodes upon ingestion or when in close contact (Tran et al. 2017). Different variables such as species of nematode, developmental stage, environmental conditions affect chemotactic response of plant-parasitic nematodes towards *Streptomyces* spp. and their metabolites (Wang et al. 2021). Along with the production of nematocidal compounds, *Streptomyces* spp. also engage in other different defense strategies to enhance plant's resistance to nematode invasion such as producing antibiotics and enzymes, and outcompeting harmful microorganism by releasing siderophores (Yoon et al. 2012) (Fig. 8.1). Wang et al. (2019) performed an in-vitro experiment in which they evaluated the effect on plant parasitic nematode *M. incognita* of volatile secondary metabolites released by *Streptomyces plicatus*. One of the volatile compounds emitted by *S. plitacus*, namely dibenzofuran acts as potent attractant to nematode whereas another volatile compound acts as repellent to nematode population.

8.4.2 Plant Growth Promotion

8.4.2.1 Phytohormones Production

Numerous phytohormones, including auxins, gibberellins, ethylene, cytokinin-like chemicals, salicylic acid, jasmonic acid, abscisic acid, and gibberellic acid, have been found to be produced by *Streptomyces* spp. (Rashad et al. 2015; Passari et al. 2016). These hormones are essential for fostering plant growth and assisting plants in responding to biotic and abiotic stressors (Fu et al. 2022). This genus leads the production of the phytohormone indole-3-acetic acid (IAA), which is a member of the auxin group (Narayana et al. 2007). *Streptomyces* spp. produce IAA via the indole-3-acetamide pathway from L-tryptophan (Fitriani et al. 2022). Apart from the advantages of auxins, IAA can relax the cell walls of plant roots and promote the release of root exudates that aid in plant growth (Pang et al. 2022). In addition to the

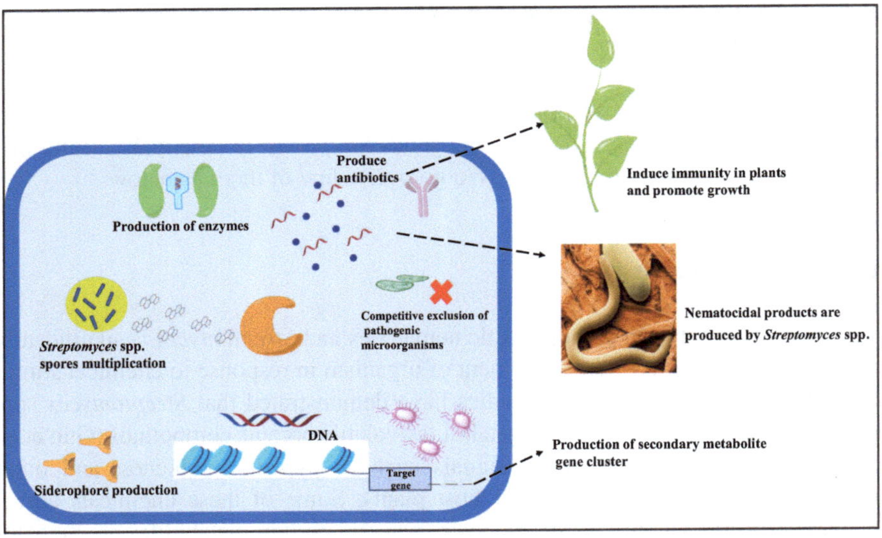

Fig. 8.1 Molecular mechanisms and interaction of *Streptomyces* spp. with plant parasitic nematodes

Fig. 8.2 General mechanism of *Streptomyces* spp. in inducing plant growth

phytohormones seen in Fig. 8.2, certain rhizobacteria that promote plant growth can also create gibberellic acid and abscisic acid, which are similarly effective in accelerating the plant's natural defenses and encouraging growth (Tran et al. 2021). *Streptomyces* synthesize phytohormones to promote nutrient cycling to remedy nutrient imbalance that occurs under stressful conditions (Nazari et al. 2023). One significant mechanism by which certain *Streptomyces* spp. increase plant tolerance to biotic and abiotic stress conditions is the ability of ACCD (1-aminocyclopropane-1-carboxylate deaminase) to hydrolyze

1-aminocyclopropane-1-carboxylate (ACC, the precursor of ethylene) to α-keto butyric acid and ammonia (Alkaabi 2022). Therefore, bacteria that produce ACCD promote plant development by reducing the levels of ethylene (Kumar et al. 2022). By generating secondary metabolites like plant hormones, certain enzymes, antibiotics, volatile organic compounds, stimulating defense system of plants and encouraging certain beneficial microbes, *Streptomyces* spp. have been described as a promising agents to improve the quality and yield of multiple cultivars (Abbasi et al. 2022; Li et al. 2022).

8.4.2.2 Siderophores Production

Plant growth-promoting bacteria, including *Streptomyces* spp., produce siderophores, a group of low molecular weight secondary metabolites (Cui et al. 2022). These metabolites, as their main characteristic, have high affinity with iron particles and their transport to the interior of cells (Rehan et al. 2022). The siderophores are mostly associated with Fe^{+3} particles which are subsequently reduced to Fe^{+2} particles. This component is useful to plants for a variety of processes, including the creation of phytohormones and other primary and secondary metabolites. Fe can also support the enzymes that repair nutrients (Nozari et al. 2021). Consequently, as iron is a necessary element that takes part in catalysis, DNA and RNA creation, and photosynthesis, releasing siderophores to extract it offers several benefits for agricultural output (Armin et al. 2021). These substances have a variety of uses for crops in agriculture. Chelated iron siderophores help with a variety of biological processes by delivering this element into plant and microbe cells via transporters in the cell membrane (Khan et al. 2018). Additionally, the excretion of these metabolites enhances competition for soil iron ions, which hinders the development of harmful soil microbes and pests (Saha et al. 2016). In addition, siderophores' ability to chelate metals promotes the production of phytohormones, which in turn accelerates plant development (Dimkpa et al. 2008). *Streptomyces* spp. can also contribute to the phosphates solubilization by releasing siderophores (Chouyia et al. 2022). Phosphorus is excellently immobilized by iron ions in acidic soils. Therefore, these secondary metabolites after release can form bond with iron, releasing phosphorus. Because *Streptomyces* spp. produce siderophores, which lengthen and increase the mass of plant roots and shoots, they are effective for agricultural purposes and boost crop productivity (Dimkpa et al. 2009).

8.4.2.3 Bioavailability of Nutrients

Certain *Streptomyces* species have the capacity to increase the bioavailability of nutrient by solubilizing nitrogen and phosphorus. One macronutrient that is necessary for the growth and development of plants is nitrogen. The plant's ability to absorb nitrogen allows for the synthesis of proteins and amino acids, which have a direct impact on photosynthesis and raise biomass (El-Serafy and El-Sheshtawy 2020). Consequently, the accessibility of this chemical, that restricts agricultural production, depends on the activity of N-fixing bacteria (Kaur and Manhas 2022). Studies have reported the solubilization of nitrogen by *Streptomyces* species using the mechanism of acetylene reduction (Sarwar et al. 2019), their nitrogenase activities (Sambangi and Gopalakrishnan 2023), and enhancing the incorporation of $^{15}N_2$ isotope into their biomass and environment that is ammonia free (Dahal et al. 2017).

Another macronutrient that plants need in large quantities is phosphorus and is required for molecules like deoxyribonucleic acid (DNA), ribonucleic acid (RNA) and phospholipids of plant cells. It is also necessary for photosynthesis, which enhances flower formation, seed production, crop maturity, and resistance to plant diseases (Bechtaoui et al. 2021; Khan et al. 2023a, b). Additionally, phosphorus promotes root development and supports plant development throughout its life cycle (Chouyia et al. 2022). Despite having a significant phosphorus reserve, plants cannot consume much of it, and a significant amount of phosphate that is added to soils as fertilizer is quickly immobilized, resulting in low P-use efficiency and perhaps surplus phosphorus (Kochian 2012; Yadav et al. 2015). An economical and environmentally sustainable method of using native soil P is by solubilization and microbial mineralization, which reduces the need for chemical fertilizers and offers advantages for the environment and the economy. Many *Streptomyces* species that have been isolated from rhizosphere soils have been identified as possible phosphorus solubilizers (Hamdali et al. 2010). These species can solubilize various phosphorus sources; multiple studies have proven that *Streptomyces* may solubilize the tricalcium phosphorus form. Specifically, it was demonstrated that several *Streptomyces* species isolated from tomato and wheat rhizospheres could generate large levels of phosphorus, ranging from 70.36 μg/mL to 1916.12 ± 10.35 μg/mL (Jog et al. 2012, 2014; Chouyia et al. 2022). Additionally, a number of secondary metabolites are released by *Streptomyces* which may aid in solubilization of phosphate, nitrogen, and other nutrients (Liu et al. 2016; Dahal et al. 2017; Chouyia et al. 2022; Suriyachadkun et al. 2022) and suggest that biofertilizers produced using these bacteria can be employed as an effective strategy to enhance plant growth and yield by improving bioavailability of nutrients to the plants.

8.5 Weeds and Crop Plants

Due to their competition for vital resources like light, water, nutrients, and space, weeds are one of the main biotic stressors in agriculture and can seriously harm cultivated plants. Their aggressive colonization and quick development frequently result in stunted growth, decreased crop vitality, and poorer yields (Kostina-Bednarz et al. 2023; Behera et al. 2024). Allelopathic substances released by certain plant species also prevent neighboring crops' seeds from germinating and their roots from growing. Weeds can also act as different hosts for diseases and pests, which make crop losses much worse. Weed infestation can result in significant financial losses and decreased agricultural system efficiency if it is not adequately controlled (Kumar et al. 2021; Demis 2024). For instance, a study by Amare et al. (2014) reported that crop rotations and weed management significantly impact wheat production, and the absence of any weed control strategy significantly reduced wheat growth and yield. Chemical herbicides are frequently used in agricultural areas to decrease physical work for weed removal, and increase crop yields. Herbicides are effective, yet they have a number of detrimental impacts on biodiversity, human health, and the environment (Parven et al. 2025). Non-target plants and aquatic life may be harmed by overuse or incorrect application, which can contaminate soil and water supplies. Additionally, herbicides may cause resistant weed species to emerge, which will eventually make them more difficult to manage (Green and Owen 2011). Furthermore, human health concerns such as respiratory disorders, skin irritation, hormone disturbance, and even an increased risk of cancer have been connected to exposure to certain herbicides (Ádám et al. 2024; Gupta and Gupta 2025). These issues draw attention to the necessity of using herbicides responsibly and encouraging the use of more sustainable, alternative weed control techniques.

8.6 *Streptomyces* as Herbicidal Agent

Weeds significantly impact global crop production by competing with crops for vital resources such as water, nutrients, sunlight, and space which ultimately reduce both yield and quality. Globally, weeds are estimated to cause up to a 34% loss in crop yields (Abouziena et al. 2014). Although chemical herbicides have been widely used to control weeds, their adverse effects such as residues in food, environmental contamination, and water pollution have raised concerns. This has increased the demand for eco-friendly alternatives, including microbial-based herbicides (Priya et al. 2023). Among the approximately 70,000 microbial natural products reported by Bérdy (2012), about 33,000 exhibit biological activity. Many of these have been developed into pharmaceuticals and agrochemicals. Bioherbicides are increasingly essential for both conventional and organic farming, as the development of new herbicides with novel modes of action and improved safety profiles becomes more challenging. Bioherbicides or bioinsecticides are generally categorized based on their chemical structure and mode of action, and their effectiveness depends heavily on both their mode of action and method of application. They typically combat weeds by disrupting cellular functions, inhibiting photosynthesis, or interfering with plant growth hormones (Won et al. 2016).

Streptomyces, a genus within actinobacteria, represents a fruitful source of structurally varied and biologically active secondary metabolites, making it a promising candidate for herbicide discovery. Despite limited commercial success in herbicide development from natural products compared to other pesticides, notable breakthroughs have occurred. Notably, *Streptomyces* has contributed significantly to the discovery of phytotoxic and herbicidal compounds (Table 8.4). About 60% of commercialized herbicides available till date are the bioactive products of *Streptomyces* (Pacios-Michelena et al. 2021). Key bioactive compounds contributing to these herbicidal effects include enzymes, amides, quinones, and antibiotics (Shi et al. 2019). These metabolites span a wide range of structural classes including amino acids, peptides, macrolides, nucleoside analogs, organic acids, amides, and polyethers and exhibit diverse modes of action (MOAs) (Radhakrishnan et al. 2018). Despite a decline in the proportion of new bioactive

Table 8.4 Effect of *Streptomyces* spp. on weed plants

Experimental weed	Experimental setup	*Streptomyces* used as mitigator	Effect of *Streptomyces*	References
Conyza canadensis	Petri-plate experiment	*Streptomyces* sp. Caat 7–52	Microbial metabolite inhibited seed germination with concentration at 6.25 μg/mL	Junior et al. (2025)
Amaranthus retroflexus, *Portulaca oleracea*, and *Echinochloa crusgalli*	1 week petri plate experiment with 30 seeds in each	*Streptomyces* sp. STD57	Spore suspension inhibited growth of roots and shoots of the weeds	He et al. (2024)
Setaria viridis and *A. retroflexus*	Pot experiment	*Streptomyces* sp. YJ-81	Inhibited seed germination at 100 mg/L concentration	Dai et al. (2023)
A. retroflexus	5 days Petri-plate experiment	*Streptomyces* strain NEAU-HV44	Post-emergence application of the 2 mg/mL extract significantly reduced weed height and fresh biomass by over 50%	Ling et al. (2023)
Digitaria ciliaris	10 days dose-dependent greenhouse experiment	*Streptomyces* sp. KR0005	Secondary metabolites produced exhibited significant phytotoxic effects that acted as cell membrane and photosynthesis inhibitor	Umurzokov et al. (2023)
D. ciliaris	10 days dose-dependent greenhouse experiment	*Streptomyces* sp. *KRA18–249*	At concentration of 500 μg/mL secondary metabolite showed 100% herbicidal potential	Umurzokov et al. (2022)
D. ciliaris	Soil experiment	*Streptomyces* sp. *W-200*	Reduced chlorophyll content by 65% and enhanced electrolyte leakage gradually within 24 h	Cho et al. (2022)
S. angulatus, *A. trifida*, and *A. artemisiifolia*	–	*S. formicae strain W-200*	Inhibited seed growth	Aung et al. (2021)
D. ciliaris	20–25 plants per cup in greenhouse experiment	*Streptomyces* sp. *KRA17-580*	Reduced growth, chlorophyll content and enhanced severe leaf bleaching in weed	Kim et al. (2020)
D. sanguinalis and *Sorghum bicolor*	Pot experiment	*Streptomyces* strain-329	Showed pre- and post-emergence herbicidal activity	Bo et al. (2019)

E. crusigalli, A. spinosus, and C. rotundus	4 days petri-dish experiment	*Streptomyces* sp. *DDBH019*	Metabolites produced completely inhibited the growth of the all tested species weed and also favored the growth of required plant rice	Priyadharsini et al. (2017)
A. theophrasti, A. retroflexus, D. ciliaris, and E. oryzoides	3 days in vitro experiment	*S. scopulidiris*	Exhibited herbicidal potential greater than 97.3% for all four assessed species which is attributed to cell membrane damage	Won et al. (2016)
A. theophrasti, A. retroflexus, D. ciliaris, and E. oryzoides	Petri-dish experiment	*S. scopulidiris*	Inhibited germination and promoted herbicidal activity	Won et al. (2015)
Raphanus sativus, Eruca sativa, Echinochloa crusgalli, and Lolium Multiflorum	Petri-dish experiment	*S. levis Strain LX-65*	Reduced weed germination from 100 to 0%	Sayed et al. (2014)
D. sanguinalis	Greenhouse and field experiment	*S. scopuliridis KR-001*	Caused wilting and burndown of leaves and stems, followed by discoloration and eventual plant death	Lee et al. (2013)
Chlorella vulgaris	Microplate experiment	*Streptomyces* sp. *AMC 23*	Bafilomycin compounds produced by strain showed phytotoxic activity	Crevelin et al. (2013)
C. occidentalis and C. rotundu	4 days petri-dish experiment	*Streptomyces* sp. *KA1-3*	Secondary metabolite N-phenylpropanamide from bacterial strain inhibited seed germination of both weeds by 80%	Priyadharsini et al. (2013)
Solanum nigrum, Sorghum halepense, Eleusine indica and Euphorbia hirta	Greenhouse experiments	*S. griseolus CGMCC 1370*	Inhibited the emergence of both root and shoot at 1 mg/L concentration of metabolite extract	Xu et al. (2009)
Digitaria sanguinalis and Echinochloa crus-galli	Pot experiment	*Streptomyces* sp. *8E-12*	Showed leaf bleaching, wilting, and eventual plant death	Lee et al. (2003)

compounds from actinobacteria, *Streptomyces* remains a crucial source of novel herbicidal agents. Continued exploration of secondary metabolites from these bacteria may uncover new MOAs aiding the development of environmentally safe herbicides.

Compounds like glufosinate and bialaphos, derived from *Streptomyces*, have been successfully commercialized. Recent advances in synthetic biology and high-throughput screening such as the bioassay developed to detect inhibitors of starch biosynthesis and the combinatorial biosynthesis of herbicidal thaxtomins by Jiang et al. (2018) highlight the untapped potential of *Streptomyces*. Another key compound, thaxtomin A, produced by *Streptomyces scabies*, inhibits cellulose biosynthesis by targeting the formation of cellulose synthase complexes. It shows strong pre-emergence herbicidal activity against *Lemna minor* and *Agrotis palustris* (Leep et al. 2016; King et al. 2001). Thaxtomin's herbicidal potential led to its patenting and further development. Also, nucleoside-based compounds from *Streptomyces* species have shown promising herbicidal activity. For instance, albucidin which was isolated from *Streptomyces albus subsp. chlorinus*, exhibited broad-spectrum post-emergence herbicidal activity and induced chlorosis and bleaching in weeds (Hahn et al. 2009). Another potent compound, hydantocidin, a spironucleoside from *Streptomyces hygroscopicus SANK 63584*, demonstrated non-selective herbicidal activity comparable to glyphosate. Its herbicidal effect is attributed to the inhibition of adenylosuccinate synthetase after 5′-phosphorylation (Ogawa 1973). Additionally, coformycin, isolated from an unidentified *Streptomyces* strain, showed strong activity against *Sorghum halepense*, *Digitaria sanguinalis,* and *Brassica juncea* at 6 kg/ha by disrupting ATP synthesis (Li et al. 2003; Isaac et al. 1991). These studies highlight the potential of *Streptomyces*-derived nucleosides as herbicides through unique biochemical mechanisms targeting plant metabolism.

Herbicidal potential of *Streptomyces*-derived peptides, also offers sustainable alternatives to synthetic herbicides. Several *Streptomyces* species produce potent herbicidal peptides. As exemplary, bialaphos (SF-1293) was independently isolated from *Streptomyces viridochromogenes* and *S. hygroscopicus* (Mallik 2024). Though initially its herbicidal properties were unrecognized, its hydrolysis product, phosphinothricin, was later developed into the commercial herbicide glufosinate (Basta) due to its inhibition of glutamine synthetase (Davies et al. 2025; Quinn 2018). Later, *S. hygroscopicus*-derived bialaphos itself was developed as a commercial herbicide via fermentation. *Streptomyces* species produce various non-proteinogenic amino acids with significant herbicidal activity. For instance, homoalanosine from *S. galilaeus* SC-1688 effectively controlled the growth of weed, *Xanthium strumarium,* and *Polygonum persicaria* (Fushimi et al. 1989). Pyridazocidin from an unclassified *Streptomyces* blocked $NADP^+$ electron transfer in photosystem I (Lamberth 2016). Similarly, acivicin, produced by *S. sviceus*, inhibited glutamine synthetase and was patented as a herbicide (Duke and Dayan 2011). Also, gabaculine from *S. toyacaenis* disrupted chlorophyll synthesis by inhibiting aminotransferase (Duke and Dayan 2015).

Several macrolide and ester-based compounds from *Streptomyces* species have also shown herbicidal potential. In a study, active compound cycloheximide

($C_{15}H_{23}NO_4$), isolated from a soil bacterium, *Streptomyces* sp. strain KR0005, was identified for its potent herbicidal activity. Bioassays demonstrated that the culture filtrate subdued the growth of *Digitaria ciliaris*, with a GR_{50} value of 1.37 g/L. Mode of action studies revealed that the filtrate disrupted photosynthesis and compromised cell membrane integrity in a dose-dependent manner holding a promising potential as bioherbicidal agent (Umurzokov et al. 2023). Another potent compound, diethyl 7-hydroxytrideca-2,5,8,11-tetraenedioate obtained from *Streptomyces* sp. DDBH019, showed stronger herbicidal effects than 2,4-D against *Echinochloa crus-galli*, *Amaranthus spinosus,* and *Cyperus rotundus*, without affecting rice (Priyadharsini et al. 2017). These findings highlight the selective and potent herbicidal action of *Streptomyces*-derived metabolites. In another study, fosmidomycin, from *S. lavendulae*, effectively inhibited the MEP pathway in chloroplasts, reducing photosynthesis and acted as a patented herbicide (Parkinson et al. 2019; Kuzuyama et al. 1998). Additionally, herboxidiene, produced by *S. chromofuscus* ATCC 49982, showed potent and selective herbicidal activity, particularly proved safe for biannual crops like wheat (Jha et al. 2014). Nigericin and geldanamycin, polyether antibiotics from *Streptomyces* sp. 11-1-2, inhibited radicle growth and caused visible phytotoxicity symptoms (Díaz-Cruz et al. 2022). Similarly, methoxhygromycin from *Streptomyces* sp. 8E-12 exhibited strong herbicidal effects on monocots like *Echinochloa crus-galli* (Lee et al. 2003). These verdicts underscore the herbicidal potential of *Streptomyces*-derived metabolites and these innovations are expected to accelerate the discovery of novel bioherbicidal agents from *Streptomyces*, contributing to sustainable weed management solutions.

8.7 Commercialization

Compared to synthetic agrochemicals, *Streptomyces*-derived products are more environmentally benign, biodegradable, and often more selective, reducing the risks to non-target organisms and ecosystems. Their resilience in diverse soil conditions and compatibility with plant microbiomes further enhance their value as candidates for integrated pest and weed management strategies. Commercialization of *Streptomyces*-derived products requires the development of a scalable fermentation technique that can generate significant amounts of these compounds. In order to ensure stability, usability, and effectiveness in field settings *Streptomyces*-based products might need to be made into certain formulations as granules, powders, liquids, or coatings. *Streptomyces* can greatly increase agricultural plant growth and enhance pest and disease management, making it a viable substitute for chemical fertilizers and pesticides (Sousa and Olivares 2016). Product formulated from *Streptomyces* may prove to be beneficial in sustainable agricultural production. However, product commercialization based on Actinobacteria has been limited due to difficulty in product formulations, market problems, lesser awareness, and technical problems including maintaining and then storing the required formulations (Reddy et al. 2016; Silva et al. 2022). For instance, a product marketed with the name of Actofit, Astur was the sole commercial treatment that showed efficiency against PPNs, web mites, colorado potato beetles, and some other plant pathogens (Gouli et al. 2015). To fully exploit the *Streptomyces*-based products for integrated pest management strategies in a sustainable way, an interdisciplinary approach that integrates scientific information, high-tech inventions, market development techniques, and regulatory agreements are essential (Alves Macedo et al. 2010; Mani et al. 2024; Wang et al. 2024). More research and development in this field is required in order to advance the efficiency of *Streptomyces* spp. and result in a long-term sustainable solution for the management of nematodes. Further, it is essential to perform detailed evaluation of behavioral properties, environmental fate, and hostile nature of possible microorganism to be used as biocontrol agent like *Streptomyces* (Vurukonda et al. 2021). It is necessary to understand that certain species though have the potential to act as nematicidal or herbicidal agent but may disrupt the biogeochemical cycles and eradication of essential native soil microbes that are beneficial for plant growth and development. Similarly, nodule formation by nitrogen-fixing bacteria was reported to be inhibited by *Streptomyces* spp. For instance, it was found that *S. kanamyceticus* inoculation reduced *Bradyrhizobium japonicum* nodule growth in soybeans, as a result of antagonistic chemicals (Gregor et al. 2003). Some of the *Streptomyces* spp. have the ability to promote nodulation and mycorrhizal development while inhibiting pathogenic growth, suggesting that not all of them have inhibitory effects. Negative side effects are therefore strain-specific, highlighting the necessity of carefully choosing and screening potential biocontrol species (Newitt et al. 2019). Figure 8.3 gives general illustration of various steps for *Streptomyces* commercialization.

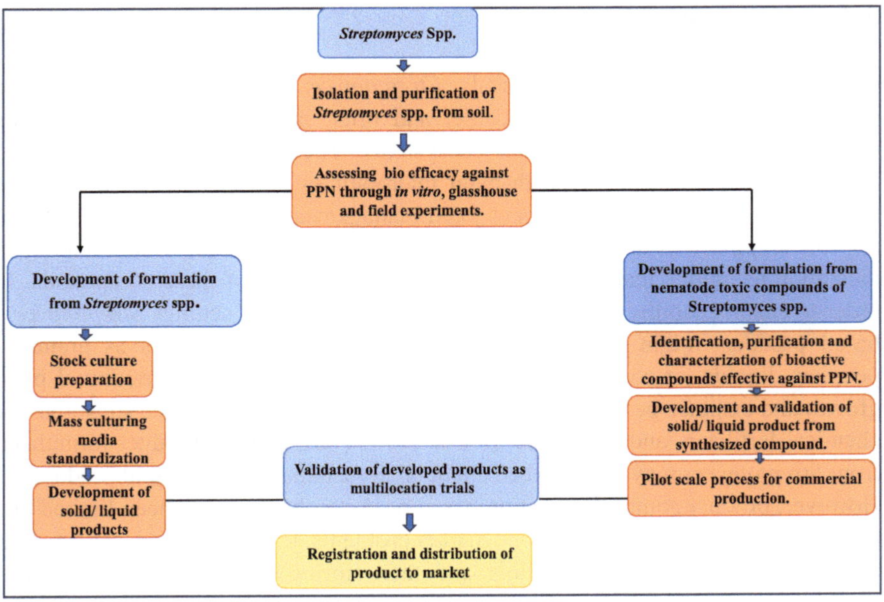

Fig. 8.3 Flow chart of industrial production process of *Streptomyces* spp. for commercial use

8.8 Conclusion and Future Aspects

The exploration of streptomyces as natural sources of biocontrol agents offers a promising avenue for the development of sustainable alternatives to conventional nematicides and herbicides. The metabolic flexibility of these soil-dwelling actinobacteria is well known, since they produce a broad range of bioactive chemicals with strong herbicidal and nematicidal effects. Through a variety of methods, such as the generation of phytohormones, siderophore, and chemotactic compounds, several studies have shown their capacity to reduce plant-parasitic nematodes and restrict weed development. The beneficial effects of *Streptomyces*-based compounds over synthetic agrochemicals include their eco-friendly, biodegradable, and often more selective nature. Additionally, these biological products also lower the ecosystem hazards. Various products from these microbes have been commercialized; however, their use is restricted to the laboratory conditions because of a number of hindrances on the way of effective use of these microbes in field conditions. Advanced formulations and delivery methods, better understanding of plant and microbial interaction, performance of microbe in lab and field conditions, and strict regulatory processes are some of these obstacles. Sealing these cracks using interdisciplinary research, advanced biotechnological practices, and supportive regulatory measures is required for effectively using the streptomyces for a sustainable eco-friendly agriculture.

References

Abbasi S, Kafi SA, Karimi E et al (2022) *Streptomyces* consortium improved quality attributes of bell pepper fruits, induced plant defense priming and changed microbial communities of rhizosphere under commercial greenhouse conditions. Rhizosphere 23:100570

Abo-Elyousr KAM, El-Metwally MM, El-Gamal NG (2023) Nematicidal potential of *Streptomyces* spp. against *Meloidogyne incognita* in eggplant. Egypt J Phytopathol 51(1):1–12

Abouziena HF, El-Saeid HM, Amin AA (2014) Water loss by weeds: a review. Int J ChemTech Res 7(01):323–336

Ádám B, Cocco P, Godderis L (2024) Hazardous effects of pesticides on human health. Toxics 12(3):186

Agale SV, Gopalakrishnan S, Gupta R et al (2018) Compatibility of *Streptomyces* sp., Metarhizium anisopliae and neem seed powder against pigeon pea pod borer complex. Int J Curr Microbiol Appl Sci 7(2):390–397

Aggarwal N, Thind SK, Sharma S (2016) Role of secondary metabolites of actinomycetes in crop protection. In: Plant growth promoting actinobacteria: a new avenue for enhancing the productivity and soil fertility of grain legumes. Springer Singapore, Singapore, pp 99–121

Ahamad L, Bhat AH, Kumar H et al (2023a) From soil to plant: strengthening carrot defenses against *Meloidogyne incognita* with vermicompost and *Arbuscular mycorrhizal* fungi biofertilizers. Front Microbiol 14:1206217

Ahamad L, Siddiqui ZA, Hashem A et al (2023b) Use of AM fungus *Rhizophagus irregularis* and silicon dioxide nanoparticles for the management of *Meloidogyne incognita*, *Alternaria dauci* and *Rhizoctonia solani* and growth of carrot. Arch Phytopathol Plant Protect 56:466–488

Alekhya G, Gopalakrishnan S (2016) Exploring plant growth-promotion actinomycetes from vermicompost and rhizosphere soil for yield enhancement in chickpea. Braz J Microbiol 47(1):85–95

Ali M, Ohri P (2023) Deciphering the synergistic effect of Jasmonic acid and Spermine in mitigating root-knot nematode stress in tomato plants through enhancing growth and activity of antioxidant enzymes. South Afr J Bot 161:21–35

Ali WM, Abdel-Mageed MA, Hegazy MGA et al (2023) Biocontrol agent of root-knot nematode *Meloidogyne javanica* and root-rot fungi, *Fusarium solani* in okra morphological, anatomical characteristics and productivity under greenhouse conditions. Sci Rep 13(1):11103

Aljuboori FK, Ibrahim BY, Mohamed AH (2022) Biological control of the complex disease of *Rhizoctonia solani* and root-knot nematode *Meloidogyne javanica* on chickpea by *Glomus* spp. and *Pseudomonas* sp. Iraq J Agric Sci 53:669–676

Alkaabi AKS (2022) Mangrove growth promotion by endophytic actinobacteria and seaweed extract. Theses 923

Alves Macedo J, Durães Sette L, Harumi Sato H (2010) A comparative biochemical characterization of microbial transglutaminases: commercial vs. a newly isolated enzyme from *Streptomyces* Sp. Food Bioprocess Technol 3(2):308–314

Amare T, Sharma JJ, Zewdie K (2014) Effect of weed control methods on weeds and wheat (*Triticum aestivum* L.) yield. World JAgric Res 2(3):124–128

Armin R, Zühlke S, Grunewaldt-Stöcker G et al (2021) Production of siderophores by an apple root-associated *Streptomyces* ciscaucasicus strain GS2 using chemical and biological OSMAC approaches. Molecules 26(12):3517

Atif AM, Elzamik FI, Mohamed GM et al (2023) Biological control of the root-knot nematode (*Meloidogyne incognita*) on eggplants with various chitinase-producing *Streptomyces* strains. Eur J Plant Pathol 167:371–394

Aung BB, Cho KM, Choi JS et al (2021) Herbicidal efficacy of *Streptomycesformicae* strain W-200 for the biocontrol of invasive weeds. Korean J Weed Sci 41(1):22

Aureliano M, De Sousa-Coelho AL, Dolan CC et al (2023) Biological consequences of vanadium effects on formation of reactive oxygen species and lipid peroxidation. Int J Mol Sci 24(6):5382

Azlay L, Oubassou EZ, Berr A et al (2024) In vitro assessment of the nematicidal potential of *Streptomyces violascens* strain AS2 against root-knot nematodes (*Meloidogyne* sp.). Arch Biol Sci 76:345–358

Bakr RA, Abd El-All AM, Salem MF et al (2022) Potential of different compost types in enhancement of physiological, biochemical parameters and control of *Meloidogyne javanica* in tomato plants. Egypt J Crop Prot 17:38–54

Balla A, Silini A, Cherif-Silini H et al (2021) The threat of pests and pathogens and the potential for biological control in forest ecosystems. Forests 12(11):1579

Bechtaoui N, Rabiu MK, Raklami A et al (2021) Phosphate-dependent regulation of growth and stresses management in plants. Front Plant Sci 12:679916

Behera P, Das N, Karan N (2024) Sustainable weed management strategies for crop production. In: Shweta, Navyashree R, Watane AA, Lakra B, Gill R (eds) Comprehensive insights into environmental science. Stella International Publication, Kurukshetra, pp 159–180

Behzadian S, Sahebani N, Karimi S et al (2022) Relationship between suppression of plant defence systems by *Meloidogyne javanica* with the distance of different parts of the plant from the nematode establishment site. Nematology 25:107–114

Bérdy J (2012) Thoughts and facts about antibiotics: where we are now and where we are heading. J Antibiot 65(8):385–395

Bo AB, Kim JD, Kim YS et al (2019) Isolation, identification and characterization of *Streptomyces* metabolites as a potential bioherbicide. PLoS One 14(9):e0222933

Butt UD, Khan S, Liu X et al (2024) Present status, limitations, and prospects of using *Streptomyces* bacteria as a potential probiotic agent in aquaculture. Probiotics Antimicrob Proteins 16(2): 426–442

Cerna-Chávez E, Rodríguez-Rodríguez JF, García-Conde KB et al (2024) Potential of Streptomyces avermitilis: a review on Avermectin production and its biocidal effect. Metabolites 14(7):374

Chandrawat BS, Siddiqui AU, Bhati SS et al (2020) Bio-agents: A source for initiation of defence enzymes in chilli infected with root-knot nematode, Meloidogyne incognita. J Entomol Zool Stud 8(6):1684–1688

Chater KF (2016) Recent advances in understanding *Streptomyces*. F1000Res 5:2795

Cho KM, Shin SC, Bo AB et al (2022) Isolation, screening and molecular identification of *Streptomyces* sp. W-200 and its bioherbicidal activity in weed control. Weed Turfgrass Sci 11(3):277–287

Chouyia FE, Ventorino V, Pepe O (2022) Diversity, mechanisms and beneficial features of phosphate-solubilizing *Streptomyces* in sustainable agriculture: a review. Front Plant Sci 13: 1035358

Crevelin EJ, Canova SP, Melo IS et al (2013) Isolation and characterization of phytotoxic compounds produced by *Streptomyces* sp. AMC 23 from red mangrove (Rhizophora mangle). Appl Biochem Biotechnol 171:1602–1616

Cui K, Xu T, Chen J et al (2022) Siderophores, a potential phosphate solubilizer from the endophyte *Streptomyces* sp. CoT10, improved phosphorus mobilization for host plant growth and rhizosphere modulation. J Clean Prod 367:133110

Dahal B, NandaKafle G, Perkins L et al (2017) Diversity of free-living nitrogen fixing *Streptomyces* in soils of the badlands of South Dakota. Microbiol Res 195:31–39

Dai Y, Chen K, Liu D et al (2023) Isolation and characterization of a herbicidal component produced by Streptomyces sp. YJ-81. Chin J Pesticide Sci 25(5):1104–1112

Das TK, Behera B, Nath CP et al (2024) Herbicides use in crop production: an analysis of cost-benefit, non-target toxicities and environmental risks. Crop Prot 181:106691

Davies AM, Trentham D, Sutton BJ et al (2025) Structure and activity of a phosphinothricin N-acetyltransferase (PSPTO_3321) from *Pseudomonas syringae* pv. tomato DC3000. Biochem Biophys Res Commun 755:151539

de Oliveira Costa SN, Ribeiro JM, da Cunha e Castro JM et al (2025) Tackling the plant parasitic nematode *Meloidogyne enterolobii*: challenges and strategies for control in relevant crops. Trop Plant Pathol 50(1):39

Demis E (2024) Mechanism of plant resistance to insects, weeds and pathogens. Middle East Res J Agri Food Sci 4(2):76–85

Díaz-Cruz GA, Liu J, Tahlan K et al (2022) Nigericin and geldanamycin are phytotoxic specialized metabolites produced by the plant pathogen *Streptomyces* sp. 11-1-2. Microbiol Spectr 10(2): e02314–e02321

Dimkpa CO, Svatoš A, Dabrowska P et al (2008) Involvement of siderophores in the reduction of metal-induced inhibition of auxin synthesis in *Streptomyces* spp. Chemosphere 74(1):19–25

Dimkpa CO, Merten D, Svatoš A et al (2009) Siderophores mediate reduced and increased uptake of cadmium by *Streptomyces* tendae F4 and sunflower (Helianthus annuus), respectively. J Appl Microbiol 107(5):1687–1696

Duke SO, Dayan FE (2011) Modes of action of microbially-produced phytotoxins. Toxins 3(8): 1038–1064

Duke SO, Dayan FE (2015) Discovery of new herbicide modes of action with natural phytotoxins. In: Maienfisch P, Stevenson TM (eds) Discovery and synthesis of crop protection products. American Chemical Society, pp 79–92

El-Serafy RS, El-Sheshtawy AA (2020) Effect of nitrogen fixing bacteria and moringa leaf extract on fruit yield, estragole content and total phenols of organic fennel. Sci Hortic 265:109209

Fitriani M, Arifiyanto A, Sumardi Lande ML et al (2022) *Streptomyces* hygroscopicus subspecies hygroscopicus strain I18: incubation time and tryptophan concentration effects on Indole-3-acetic acid (IAA) hormone production. In: AIP conference proceedings. AIP Publishing LLC, p 080001

Fu W, Pan Y, Shi Y et al (2022) Root morphogenesis of *Arabidopsis thaliana* tuned by plant growth-promoting *Streptomyces* isolated from root-associated soil of *Artemisia annua*. Front Plant Sci 12:802737

Fushimi S, Nishikawa S, Mito N et al (1989) Studies on a new herbicidal antibiotic, homoalanosine. J Antibiot 42(9):1370–1378

Ghareeb YR, Belal EB, El-Khateeb NM et al (2024) Utilizing bio-synthesis of nanomaterials as biological agents for controlling soil-borne diseases in pepper plants: root-knot nematodes and root rot fungus. BMC Plant Biol 24:110

Gouli VV, Gouli SY, Marcelino JA (2015) Concise illustrated dictionary of biocontrol terms. Academic

Green JM, Owen MD (2011) Herbicide-resistant crops: utilities and limitations for herbicide-resistant weed management. J Agric Food Chem 59(11):5819–5829

Gregor AK, Klubek B, Varsa EC (2003) Identification and use of actinomycetes for enhanced nodulation of soybean co-inoculated with *Bradyrhizobium japonicum*. Can J Microbiol 49(8): 483–491

Guo Q, Shi M, Chen L et al (2020) The biocontrol agent Streptomyces pactum increases Pseudomonas koreensis populations in the rhizosphere by enhancing chemotaxis and biofilm formation. Soil Biol Biochem 144:107755

Gupta RC, Gupta PK (2025) Toxicity of herbicides. In: Gupta RC (ed) Veterinary toxicology. Academic, pp 565–579

Gupta R, Singh A, Ajayakumar PV et al (2017) Microbial interference mitigates *Meloidogyne incognita* mediated oxidative stress and augments bacoside content in *Bacopa monnieri*. L. Microbiol Res 199:67–78

Hahn DR, Graupner PR, Chapin E et al (2009) Albucidin: a novel bleaching herbicide from *Streptomyces albus* subsp. chlorinus NRRL B-24108. J Antibiot 62(4):191–194

Hamdali H, Smirnov A, Esnault C et al (2010) Physiological studies and comparative analysis of rock phosphate solubilization abilities of Actinomycetales originating from Moroccan phosphate mines and of *Streptomyces* lividans. Appl Soil Ecol 44(1):24–31

He D, Gao C, Zhao S et al (2024) Antibacterial, herbicidal and plant growth-promoting properties of *Streptomyces* sp. STD57 from the rhizosphere of *Adenophora stricta*. Microorganisms 12(11): 2245

Hu Q, Yang M, Bo T et al (2022) Soluble macromolecules from two *Streptomyces* strains with potent nematicidal activity against *Meloidogyne incognita*. Rhizosphere 22:100529

Isaac BG, Ayer SW, Stonard RJ (1991) The isolation of α-methylene-β-alanine, a herbicidal microbial metabolite. J Antibiot 44(7):795–796

Jaizme-Vega MC, Rodriguez-Romero AS, Nunez LAB (2006) Effect of the combined inoculation of arbuscular mycorrhizal fungi and plant growth promoting rhizobacteria on papaya (*Carica papaya* L.) infected with the root-knot nematode *Meloidogyne incognita*. Fruits 61(3):151–162

Jang JY, Choi YH, Joo YJ et al (2015) Characterization of *Streptomyces* netropsis showing a nematicidal activity against *Meloidogyne incognita*. Res Plant Dis 21(2):50–57

Jha AK, Lamichhane J, Sohng JK (2014) Enhancement of herboxidiene production in *Streptomyces chromofuscus* ATCC 49982. J Microbiol Biotechnol 24(1):52–58

Ji X, Liu B, Fan M et al (2024) Biocontrol of *Meloidogyne incognita* by *Bacillus velezensis* TA-1 through induction of host resistance in tomato. J Pest Sci 97:2227–2236

Jiang G, Zhang Y, Powell M et al (2018) High-yield production of herbicidal thaxtomins and analogs in a nonpathogenic *Streptomyces* host. In: Natural product discovery and development in the genomic era

Jin N, Xue H, Li W et al (2017) Field evaluation of *Streptomyces rubrogriseus* HDZ-9-47 for biocontrol of *Meloidogyne incognita* on tomato. J Integr Agric 16:1347–1357

Jog R, Nareshkumar G, Rajkumar S (2012) Plant growth promoting potential and soil enzyme production of the most abundant *Streptomyces* spp. from wheat rhizosphere. J Appl Microbiol 113(5):1154–1164

Jog R, Pandya M, Nareshkumar G et al (2014) Mechanism of phosphate solubilization and antifungal activity of *Streptomyces* spp. isolated from wheat roots and rhizosphere and their application in improving plant growth. Microbiology 160(4):778–788

Junior OL, Oliveira RD, Ferreira HR et al (2025) Novel phytotoxic compounds from *Streptomyces* sp. Caat 7–52: a potential bioherbicide against *Conyza canadensis*. Pest Manag Sci. https://doi.org/10.1002/ps.8683

Kamaraju D, Chatterjee M, Papolu PK et al (2024) Host-induced RNA interference targeting the neuromotor gene FMRFamide-like peptide-14 (Mi-flp14) perturbs *Meloidogyne incognita* parasitic success in eggplant. Plant Cell Rep 43:178

Kang MK, Kim MH, Liu MJ et al (2021) Nematicidal activity of teleocidin B4 isolated from *Streptomyces* sp. against pine wood nematode, *Bursaphelenchus xylophilus*. Pest Manag Sci 77: 1607–1615

Kaur T, Manhas RK (2022) Evaluation of ACC deaminase and indole acetic acid production by *Streptomyces hydrogenans* DH16 and its effect on plant growth promotion. Biocatal Agric Biotechnol 42:102321

Kaushal K, Ali M, Ohri P (2023) Uncovering the synergistic interplay of melatonin and spermidine in the alleviation of nematode stress in *Solanum lycopersicum*. Pestic Biochem Physiol 195: 105574

Kavari M, Mahdikhani Moghadam E, Rouhani H (2022) Evaluation of the Trichoderma harzianum BI alone and in combination with thiamine, riboflavin and hexanoic acid effects on resistance induction against Meloidogyne javanica in tomato plants. J Iranian Plant Prot Res 36(2): 153–167

Khan A, Singh P, Srivastava A (2018) Synthesis, nature and utility of universal iron chelator–Siderophore: a review. Microbiol Res 212:103–111

Khan F, Siddique AB, Shabala S et al (2023a) Phosphorus plays key roles in regulating plants' physiological responses to abiotic stresses. Plants 12(15):2861

Khan S, Srivastava S, Karnwal A et al (2023b) *Streptomyces* as a promising biological control agents for plant pathogens. Front Microbiol 14:1285543

Khanal C, Land J (2023) Study on two nematode species suggests climate change will inflict greater crop damage. Sci Rep 13(1):14185

Kim HJ, Bo AB, Kim JD et al (2020) Herbicidal characteristics and structural identification of the potential active compounds from *Streptomyces* sp. KRA17-580. J Agric Food Chem 68(52): 15373–15380

King RR, Lawrence CH, Gray JA (2001) Herbicidal properties of the thaxtomin group of phytotoxins. J Agric Food Chem 49(5):2298–2301

Kochian LV (2012) Rooting for more phosphorus. Nature 488(7412):466–467

Kostina-Bednarz M, Płonka J, Barchanska H (2023) Allelopathy as a source of bioherbicides: challenges and prospects for sustainable agriculture. Rev Environ Sci Biotechnol 22(2): 471–504

Kumar S, Bhowmick MK, Ray P (2021) Weeds as alternate and alternative hosts of crop pests. Indian J Weed Sci 53(1):14–29

Kumar M, Kumar P, Das P et al (2022) Proactive role of *Streptomyces* spp. in plant growth stimulation and management of chemical pesticides and fertilizers. Int J Environ Sci Technol 19(10):10457–10476

Kumar D, Manhas RK, Ohri P (2023) Deciphering the growth promoting and stress curtailing role of nitric oxide in *Meloidogyne incognita* infested *Solanum lycopersicum* seedlings. Sci Hortic 319:112147

Kumar D, Manhas RK, Ohri P (2025a) Scrutinising the stress subduing effect of *Serratia marcescens* in *Meloidogyne incognita*-stressed tomato seedlings. N Z J Crop Hortic Sci 53(5): 2875–2907

Kumar D, Manhas RK, Ohri P (2025b) Synergistic interaction of sodium nitroprusside and Serratia marcescens in mitigation of nematode stress in tomato. Nitric Oxide 157:1–18

Kuzuyama T, Shimizu T, Takahashi S et al (1998) Fosmidomycin, a specific inhibitor of 1-deoxy-D-xylulose 5-phosphate reductoisomerase in the nonmevalonate pathway for terpenoid biosynthesis. Tetrahedron Lett 39(43):7913–7916

Lamberth C (2016) Naturally occurring amino acid derivatives with herbicidal, fungicidal or insecticidal activity. Amino Acids 48(4):929–940

Lee HB, Kim CJ, Kim JS et al (2003) A bleaching herbicidal activity of methoxyhygromycin (MHM) produced by an actinomycete strain *Streptomyces* sp. 8E-12. Lett Appl Microbiol 36(6): 387–391

Lee B, Kim JD, Kim YS et al (2013) Identification of *Streptomyces scopuliridis* KR-001 and its herbicidal characteristics. Weed & Turfgrass Science 2(1):38–46

Lee H, Kim J, Kim J et al (2022) Streptomyces-derived bioactive pigments: ecofriendly source of bioactive compounds. Coatings 12(12):1858

Leep DC, Doricchi L, Baz MJ, Millan FR, Chimeno RI, Inventors; Marrone Bio Innovations Inc, assignee (2016) Use of thaxtomin for selective control of rice and aquatic based weeds. United States patent US 9,339,034

Li Y, Sun Z, Zhuang X, Xu L, Chen S, Li M (2003) Research progress on microbial herbicides. Crop Prot 22(2):247–252

Li Z, Alami MM, Tang H et al (2022) Applications of *Streptomyces* jingyangensis T. and *Bacillus mucilaginosus* A. improve soil health and mitigate the continuous cropping obstacles for *Pinellia ternata* (Thunb.) Breit. Ind Crops Prod 180:114691

Lima F, Silva J, Santos R (2017) Soluble macromolecules from two *Streptomyces* strains with potent nematicidal activity against *Meloidogyne incognita*. Biocatal Agric Biotechnol 10:1–7

Ling L, Wang H, Zhang S et al (2023) Herbicidal activity of secondary metabolites isolated from *Streptomyces* sp. NEAU-HV44 on selected weeds and crops. J Agric Food Chem 71(45): 17056–17066

Liu D, Lian B, Wang B (2016) Solubilization of potassium containing minerals by high temperature resistant *Streptomyces* sp. isolated from earthworm's gut. Acta Geochim 35(3):262–270

Lopes FC, Martinelli AH, John EB et al (2021) Microbial hydrolytic enzymes: powerful weapons against insect pests. In: Khan MA, Ahmad W (eds) Microbes for sustainable lnsect pest management: hydrolytic enzyme & secondary metabolite, pp 1–31

Mallik MA (2024) Selective isolation and screening of soil microorganisms for metabolites. Allelopathy Agroecosyst 20:219

Malviya D, Singh P, Singh UB et al (2023) *Arbuscular mycorrhizal* fungi-mediated activation of plant defense responses in direct seeded rice (*Oryza sativa* L.) against root-knot nematode *Meloidogyne graminicola*. Front Microbiol 14:1104490

Mani J, Kandasamy D, Vendan R et al (2024) Unlocking the potential of *Streptomyces* species as promising biological control agents against phytonematodes. Physiol Mol Plant Pathol 134: 102465

Mani J, Mannu J, Somasundaram P et al (2025a) Unveiling the nematotoxic effect of *Streptomyces* spp. against the root-knot nematode, *Meloidogyne enterolobii* in guava. J Pest Sci 98:1–17

Mani J, Nagachandrabose S, Mannu et al (2025b) Unleashing oxymatrine from *Streptomyces rochei* (GA) as a potent nematicidal agent against root-knot nematode, *Meloidogyne enterolobii*. Physiol Mol Plant Pathol 138:102679

Mansoor S, Ali Wani O, Lone JK et al (2022) Reactive oxygen species in plants: from source to sink. Antioxidants 11(2):225

Manullang W, Chuang HW (2020) *Streptomyces* sp. mitigates abiotic stress response and promotes plant growth. J Plant Prot Res 60:263–274

Marin-Bruzos M, Grayston SJ, Forge T et al (2021) Isolation and characterization of streptomycetes and pseudomonad strains with antagonistic activity against the plant parasitic nematode *Pratylenchus penetrans* and fungi associated with replant disease. Biol Control 158:104599

Meidani C, Savvidis A, Lampropoulou E et al (2020) The nematicidal potential of bioactive *Streptomyces* strains isolated from Greek rhizosphere soils tested on *Arabidopsis* plants of varying susceptibility to Meloidogyne spp. Plants 9:699

Mogollón-Ortiz ÁM, Monteiro TSA, de Freitas LG et al (2024) Potential of different species of actinobacteria in the management of *Meloidogyne javanica*. Arch Microbiol 206:160

Narayana KJ, Prabhakar P, Vijayalakshmi MUVVA et al (2007) Biological activity of phenylpropionic acid isolated from a terrestrial Streptomycetes. Pol J Microbiol 56(3):191

Nazari MT, Schommer VA, Braun JCA et al (2023) Using *Streptomyces* spp. as plant growth promoters and biocontrol agents. Rhizosphere 27:100741

Newitt JT, Prudence SM, Hutchings MI et al (2019) Biocontrol of cereal crop diseases using streptomycetes. Pathogens 8(2):78

Nofal MM, Attia MS (2023) Management of root-knot nematode (*Meloidogyne incognita*) and improve defensive ability of cucumber plant using *Serratia marcescens,* Salicylic acid and Fosthiazate (melotherin®) through adaptive processes and antioxidant capacity. Menoufia J Plant Prot 8:31–45

Nozari RM, Ortolan F, Astarita LV et al (2021) *Streptomyces* spp. enhance vegetative growth of maize plants under saline stress. Braz J Microbiol 52(3):1371–1383

Ogawa Y (1973) Studies on a new antibiotic SF-1293. II. Chemical structure of antibiotic SF-1293. Sci Rep Meiji Seika Kaisha 13:42–48

Pacios-Michelena S, Aguilar González CN, Alvarez-Perez OB (2021) Application of *Streptomyces* antimicrobial compounds for the control of phytopathogens. Front Sustain Food Syst 5:696518

Pang F, Solanki MK, Wang Z (2022) *Streptomyces* can be an excellent plant growth manager. World J Microbiol Biotechnol 38(11):193

Parkinson EI, Erb A, Eliot AC et al (2019) Fosmidomycin biosynthesis diverges from related phosphonate natural products. Nat Chem Biol 15(11):1049–1056

Parven A, Meftaul IM, Venkateswarlu K et al (2025) Herbicides in modern sustainable agriculture: environmental fate, ecological implications and human health concerns. Int J Environ Sci Technol 22(2):1181–1202

Passari AK, Chandra P, Mishra VK et al (2016) Detection of biosynthetic gene and phytohormone production by endophytic actinobacteria associated with *Solanum lycopersicum* and their plant-growth-promoting effect. Res Microbiol 167(8):692–705

Priya AK, Alagumalai A, Balaji D et al (2023) Bio-based agricultural products: a sustainable alternative to agrochemicals for promoting a circular economy. RSC Sustainability 1(4): 746–762

Priyadharsini P, Dhanasekaran D, Kanimozhi B (2013) Isolation, structural identification and herbicidal activity of N-phenylpropanamide from Streptomyces sp. KA1-3. Arch Phytopathol Plant Prot 46(3):364–373

Priyadharsini P, Dhanasekaran D, Gopinath PM et al (2017) Spectroscopic identification and molecular modeling of diethyl 7-hydroxytrideca-2, 5, 8, 11-tetraenedioate: a herbicidal compound from *Streptomyces* sp. Arab J Sci Eng 42:2217–2227

Quinn JP (2018) Interactions of the herbicides glyphosate and glufosinate (phosphinothricin) with the soil microflora. In: Pesticide Interact Crop Product, vol 8, pp 245–265

Radhakrishnan R, Alqarawi AA, AbdAllah EF (2018) Bioherbicides: current knowledge on weed control mechanism. Ecotoxicol Environ Saf 158:131–138

Radwan WH, Abdelhafez AA, Mahgoub AE et al (2024) *Streptomyces avermitilis* MICNEMA2022: a new biorational strain for producing abamectin as an integrated nematode management agent. BMC Microbiol 24:329

Rashad FM, Fathy HM, El-Zayat AS et al (2015) Isolation and characterization of multifunctional *Streptomyces* species with antimicrobial, nematicidal and phytohormone activities from marine environments in Egypt. Microbiol Res 175:34–47

Raza A, Salehi H, Rahman MA et al (2022) Plant hormones and neurotransmitter interactions mediate antioxidant defenses under induced oxidative stress in plants. Front Plant Sci 13: 961872.

Reddy KR, Jyothi G, Sowjanya C et al (2016) Plant growth-promoting actinomycetes: mass production, delivery systems, and commercialization. In: Plant growth promoting actinobacteria: a new avenue for enhancing the productivity and soil fertility of grain legumes. Springer Singapore, Singapore, pp 287–298

Rehan M, Barakat H, Almami IS et al (2022) Production and potential genetic pathways of three different siderophore types in *Streptomyces* tricolor strain HM10. Fermentation 8(8):46

Saha M, Sarkar S, Sarkar B et al (2016) Microbial siderophores and their potential applications: a review. Environ Sci Pollut Res 23:3984–3999

Sambangi P, Gopalakrishnan S (2023) *Streptomyces*-mediated synthesis of silver nanoparticles for enhanced growth, yield and grain nutrients in chickpea. Biocatal Agric Biotechnol 47:102567

Santos L, Silva RVD, Gondim JPE et al (2023) Organic residue controls *Meloidogyne javanica* and improves gas exchange and development in the gilo. Rev Cienc Agron 55:e20218054

Sarwar A, Latif Z, Zhang S et al (2019) A potential biocontrol agent *Streptomyces violaceusniger* AC12AB for managing potato common scab. Front Microbiol 10:202

Sayed MH, Aziz ZK, Abouzaid AM et al (2014) Efficacy of extracellular metabolite produced by *Streptomyces* levis strain LX-65 as a potential herbicidal agent. J Am Sci 10(11):169–180

Sharma M, Jasrotia S, Ohri P (2019) Nematicidal potential of *Streptomyces antibioticus* strain M7 against *Meloidogyne incognita*. AMB Express 9:1–8

Sharma N, Khanna K, Manhas RK et al (2020) Insights into the role of *Streptomyces hydrogenans* as the plant growth promoter, photosynthetic pigment enhancer and biocontrol agent against *Meloidogyne incognita* in *Solanum lycopersicum* seedlings. Plants 9:1109

Sharma N, Manhas RK, Bhardwaj R et al (2021) Bioefficacy of bio-metabolites produced by *Streptomyces* sp. strain MR-14 in ameliorating *Meloidogyne incognita* stress in *Solanum lycopersicum* seedlings. J Plant Growth Regul 41:1–13

Sharma N, Manhas RK, Ohri P et al (2022) *Streptomyces hydrogenans* strain DH-16 alleviates negative impact of *Meloidogyne incognita* stress by modifying physio-biochemical attributes in *Solanum lycopersicum* plants. Sci Rep 12:15214

Sharma N, Khanna K, Jasrotia S et al (2023) Metabolites and chemical agents in the plant roots: an overview of their use in plant-parasitic nematode management. Nematology 25(3):243–257

Shi L, Wu Z, Zhang Y et al (2019) Herbicidal secondary metabolites from actinomycetes: structure diversity, modes of action, and their roles in the development of herbicides. J Agric Food Chem 68(1):17–32

Sikandar A, Wu F, He H et al (2024) Growth, physiological and biochemical variations in tomatoes after infection with different density levels of *Meloidogyne enterolobii*. Plants 13:293

Silva GD, Kitano IT, Ribeiro IA et al (2022) The potential use of actinomycetes as microbial inoculants and biopesticides in agriculture. Front Soil Sci 2:833181

Singh SP, Gaur R (2017) Endophytic *Streptomyces* spp. underscore induction of defence regulatory genes and confers resistance against *Sclerotium rolfsii* in chickpea. Biol Control 104:44–56

Singh A, Gupta R, Srivastava M et al (2022) Augmentation of systemic resistance and secondary metabolites by chitinolytic microbes in *Withania somnifera* against *Meloidogyne incognita*. Biocontrol Sci Tech 32(5):510–526

Sousa JADJ, Olivares FL (2016) Plant growth promotion by streptomycetes: ecophysiology, mechanisms and applications. Chem Biol Technol Agric 3:1–12

Sun Y, Xie J, Tang L et al (2023) Isolation, identification and molecular mechanism analysis of the nematicidal compound spectinabilin from newly isolated *Streptomyces* sp. DT10. Molecules 28: 4365

Suriyachadkun C, Chunhachart O, Srithaworn M et al (2022) Zinc-solubilizing *Streptomyces* spp. as bioinoculants for promoting the growth of soybean (Glycine max (L.) Merrill). J Microbiol Biotechnol 32(11):1435

Tauseef A, Gupta J, Rehman A et al (2021) Differential response of cowpea towards the CuO nanoparticles under Meloidogyne incognita stress. S Afr J Bot 139:175-182

Thakur P, Singh I (2018) Biocontrol of soilborne root pathogens: an overview. In: Root biology. Springer International Publishing, pp 181–220

Tran A, Tang A, O'Loughlin et al (2017) C. elegans avoids toxin-producing *Streptomyces* using a seven transmembrane domain chemosensory receptor. elife 6:e23770

Tran TM, Ameye M, Devlieghere F et al (2021) *Streptomyces* strains promote plant growth and induce resistance against *Fusarium verticillioides* via transient regulation of auxin signaling and archetypal defense pathways in maize plants. Front Plant Sci 12:755733

Ujjwal AK, Tagi N, Nayak DK et al (2023) Effect of total phenolic content and proline content on resistant and susceptible chili varieties influenced by root knot nematode, *Meloidogyne incognita*. Pharma Innov 12:371–374

Umurzokov M, Lee YM, Kim HJ et al (2022) Herbicidal characteristics and structural identification of a potential active compound produced by *Streptomyces* sp. KRA18–249. Pestic Biochem Phys 187:105213

Umurzokov M, Kim YS, Kim HJ et al (2023) Potential bio-herbicide from *Streptomyces* sp. KR0005 to control weeds in a horticultural field. Biocontrol Sci Technol 33(4):297–313

Vurukonda SS, Giovanardi D, Stefani E (2018) Plant growth promoting and biocontrol activity of Streptomyces spp. as endophytes. Int J Mol Sci 19(4):952

Vurukonda SSKP, Giovanadri D, Stefani E (2021) Growth promotion and biocontrol activity of Endophytic *Streptomyces* spp. In: Prime archives in molecular sciences, vol 1, 2nd edn, pp 1–55

Wang P, Sun Y, Yang L et al (2019) Chemotactic responses of the root-knot nematode Meloidogyne incognita to *Streptomyces plicatus*. FEMS Microbiol Lett 366(19):fnz234

Wang J, Ding Z, Bian J et al (2021) Chemotaxis response of Meloidogyne incognita to volatiles and organic acids from root exudates. Rhizosphere 17:100320

Wang M, Li H, Li J et al (2024) *Streptomyces* strains and their metabolites for biocontrol of phytopathogens in agriculture. J Agric Food Chem 72(4):2077–2088

Won OJ, Kim YT, Kim JD et al (2015) Herbicidal activity of herbicidin from a strain of soil actinomycete *Streptomyces scopuliridis*. Weed Turfgrass Sci 4(3):219–224

Won OJ, Kim YT, Choi JS et al (2016) Herbicidal activity and mode of action of *Streptomyces scopuliridis* metabolites. J Fac Agric Kyushu Univ 61(1):47–51

Wu F, Shi Q, Wang XJ et al (2019) Streptomyces canus GLY-P2 degrades ferulic and p-hydroxybenzoic acids in soil and affects cucumber antioxidant enzyme activity and rhizosphere bacterial community. Plant Soil 436:71–89

Xu W, Tao L, Gu X et al (2009) Herbicidal activity of the metabolite SPRI-70014 from *Streptomyces griseolus*. Weed Sci 57(5):547–553

Xu X, Fan L, Wu C et al (2025) Nematicidal activity of *Streptomyces aquilus* JXGZ01 against the root-knot nematode Meloidogyne incognita. Ann Appl Biol 186:302–310

Yadav AN, Sharma D, Gulati S et al (2015) Haloarchaea endowed with phosphorus solubilization attribute implicated in phosphorus cycle. Sci Rep 5(1):12293

Yang T, Jin W, Zou J et al (2024) NBR1a mediates root-knot nematode resistance by modulating antioxidant system, jasmonic acid and selective autophagy in *Solanum lycopersicum*. Plant Stress 11:100390

Yoon GY, Lee YS, Lee SY et al (2012) Effects on Meloidogyne incognita of chitinase, glucanase and a secondary metabolite from *Streptomyces cacaoi* GY525. Nematology 14(2):175–184

Zeng Q, Huang H, Zhu J et al (2013) A new nematicidal compound produced by *Streptomycesalbogriseolus* HA10002. Antonie Van Leeuwenhoek 103:1107–1111

Endophytic *Streptomyces* spp.: Probiotics for Plant Health

9

Indu Sharma, Preeti K. Pathania, Yogesh Kumar, Deepika Sharma, Sonika Sharma, and Ashutosh Sharma

Abstract

As the worldwide population keeps rising dramatically, the demand for food production has increased, highlighting the need for sustainable strategies and bio-based alternatives to improve agricultural crop productivity. The endophytic microbial communities, which reside within plant tissues, have been suggested as a promising, eco-friendly alternative for improving plant health and productivity. Among these, *Streptomyces* spp., a crucial microbial genus known for producing a diverse range of bioactive secondary metabolites, has received considerable attention from the scientific community. This chapter is compiled to accommodate the recent advances in understanding the diverse roles of endophytic *Streptomyces* spp. in improving plant health. In this chapter, they have been described as the 'probiotics' for improving plant health, particularly under stressful environments (both biotic and abiotic). The possible mechanisms, through which they promote plant growth, such as the synthesis of PGRs (plant growth regulators), the enhancement of nutrient availability and their potential as

I. Sharma
Department of Life Sciences, University Institute of Sciences, Sant Baba Bhag Singh University, Jalandhar, Punjab, India

P. K. Pathania
Department of Microbiology, Amity University, Mohali, Punjab, India

Y. Kumar
Department of Botany, Central University of Jammu, Bagla (Raya-Suchani), Samba, Jammu and Kashmir, India

D. Sharma
Amity Institute of Organic Agriculture, Amity University, Noida, Uttar Pradesh, India

S. Sharma · A. Sharma (✉)
Faculty of Agricultural Sciences, DAV University, Jalandhar, Punjab, India

T. Kaur et al. (eds.), *Streptomyces: Biological Candidates for Sustainable Agriculture*,
https://doi.org/10.1007/978-981-95-5803-2_9

biocontrol agents against biotic stresses, have been described in detail. Furthermore, endophytic *Streptomyces* enhance plant tolerance to various abiotic stressors, including heavy metal pollution, drought and salt, by modulating host gene expression and improving physiological resilience. The potential of endophytic *Streptomyces* as a next-generation bio-inoculant, paving the way to achieve sustainability in agro-ecosystems, has also been discussed. Overall, the chapter provides the readers with a comprehensive overview of the potential role of important endophytic *Streptomyces* spp. as probiotics for plant health promotion as an alternative elixir for improving sustainability in farming systems.

Keywords

Bio-inoculants · Sustainable agriculture · Biotic stress · Abiotic stress · Plant disease management · Insect pest · Microbial inoculants

9.1 Introduction

Endophytes are essential category of the beneficial microorganisms found in plant tissues, which have recently gained attention in agricultural biotechnology as plant growth promoters and effective biocontrol agents (Akinsanya et al. 2015; Vurukonda et al. 2018, 2021). Among these endophytes, actinobacteria, particularly *Streptomyces* spp., play a vital role in promoting growth and improving disease resistance in plants (Barka et al. 2016; Gao et al. 2021; Sathya et al. 2017; Vurukonda et al. 2021). Since the endophytic bacteria flourish in the endospheric region of plants, particularly the endodermis xylem tissue, which is rich in nutrients (Kim and Kwak 2023); as a result, the endophytic rhizobacteria have evolved a symbiotic association with its plant hosts, in which the plant host acts as a source of nutrition and nourishment for the endophytic microbes. *Streptomyces* is a genus containing several beneficial endo-microorganisms, known for their metabolic adaptability (producing several bioactive compounds like antibiotics, siderophores and antifungals), that suppresses phytopathogens, modulates plant defence responses and facilitates nutrient exchange (Chen et al. 2025; Kim and Kwak 2023; Oyedoh et al. 2023). Endophytic *Streptomyces* produce several secondary metabolites, which in turn help improve plants' capabilities to tolerate stress and thus reduce reliance on some ecologically harmful agrochemicals. *Streptomyces* are thus crucial in sustainable agriculture and human health, as they can be a potential source of some innovative biologically active chemicals like pharmaceutically significant antibiotics (Sivalingam et al. 2024). They are also reported to possess probiotic, nutraceutical and therapeutic potential in humans due to the presence of bioactive substances, including polyketides, alkaloids, terpenes, macrolides and peptides, possessing antibiotic, antifungal, anticancer, antiviral and antimicrobial properties (Moussa 2024; Sivalingam et al. 2024; Sudheer et al. 2024). *S. griseorubens* strain CIBA-NS1 isolated from *Salicornia* sp., showed a considerable antagonistic efficacy against *Vibrio campbellii* (Sudheer et al. 2024).

Several studies revealed that *Streptomyces* also acts as a natural biofertilizer that can improve soil health, lessen dependence on chemical or synthetic fertilizers and enhance plant growth; making it a recommended addition to sustainable agriculture practices, promoting more resilient and productive farming environments (Buzón-Durán et al. 2020; Oyedoh et al. 2023; Villafañe et al. 2024; Vurukonda et al. 2021; Worsley et al. 2020). A microbial formulation developed using *Streptomyces cavourensis* strain PES4 showed remarkable antagonistic behaviour against *Phytophthora* blight and amended plant growth, yield and fruit quality in passion fruits (Chen et al. 2025). In soybean, the endophytic *Streptomyces* strain N2A was reported to possess in vitro PGPR (plant growth promoting rhizobacteria) traits and antagonistic activity against phytopathogenic fungi, which caused stimulation of growth and yield as well as seed quality of soybean (Villafañe et al. 2024). Another study by Devi et al. (2023a) reported that protein was extracted from *Streptomyces* sp. strain SP5 and when tested for the insecticidal property, this protein extract showed positive results against the *Zeugodacus cucurbitae*, melon fruit fly. Moreover, this protein extract was non-toxic and also had anti-mutagenic potential. Due to its non-toxic, insecticidal and anti-mutagenic properties, the protein extract may be recommended as an environmentally friendly biocontrol agent against melon fly.

Further, several studies (Devi et al. 2023b; Fouda et al. 2020; Li et al. 2008; Supong et al. 2016) highlighted the antimicrobial, larvicidal and antioxidant potential of endophytic *Streptomyces* spp. thereby, emphasizing their future application as microbial antagonists to combat plant pathogens. Moreover, *Streptomyces* spp. showed antifungal potential against *Magnaportheoryzae* B157, and *Rhizoctonia solani* in cereal crops while also enhancing the growth of above-ground plant parts (Patel et al. 2018). Burgeoning evidence revealed that endophytic *Streptomyces* spp. colonizes plant tissue and produces various metabolites, including antibiotics (Villafañe et al. 2024; Vurukonda et al. 2021; Worsley et al. 2020). Moreover, *Streptomyces* spp. have also been reported to release bioactive metabolites such as antibiotic compounds, enzymes and plant growth regulators (PGRs), which can be utilized in the organic agricultural practices (Olanrewaju and Babalola 2019). Thus, due to non-toxic behaviour and their immense potential for disease management and plant growth promotion, the endophytic *Streptomyces* spp. and their metabolites are promising probiotics, which can be employed as suitable chemical-free bio-stimulants and effective alternatives to chemical-based antimicrobial formulations for plant protection and disease management. Because of the phytopathogen suppression, mineral acquisition, plant stress protection and growth promotion potential of endophytic actinobacteria, particularly *Streptomyces* spp., they may become warranted bio-stimulants, which can mitigate the negative impacts of agro-chemicals, ensuring the crop resilience and environmentally sustainable agriculture (Bose et al. 2025). Therefore, the present chapter focuses on the probiotic potential of endophytic *Streptomyces* to stimulate plant health through acceleration of mineral nutrient uptake, management of plant abiotic and biotic stressors in plants for sustainable alternative of existing agricultural practices.

9.2 Endophytic Streptomyces

Endophytes are a diverse group of beneficial microbes that inhabit the internal plant tissues, primarily residing in rhizosphere and endospheric zones of soil. These microbes have drawn a significant attention in agricultural biotechnology due to their potential to act as biocontrol agents and plant growth promoters, therefore, emerging as a potential and reliable alternative to chemical pesticides (Pandey et al. 2022; Vurukonda et al. 2018). Actinobacteria represent a persistent and significant part of endophytic bacterial communities in various plant species, as evidenced by both traditional culture-dependent methods and modern culture-independent techniques such as *16S rRNA* gene sequencing. Their widespread presence in different plant species highlights their ecological adaptability and key functional role in promoting plant growth and resistance towards diseases (Akinsanya et al. 2015; Hardoim et al. 2015; Trujillo et al. 2015). The major uses of endophytic *Streptomyces* spp., which confirms their use as plant probiotic have been highlighted in Fig. 9.1.

Genus *Streptomyces* comprises Gram-positive filamentous actinobacteria with a high G + C content. It belongs to the family Streptomycetaceae, within the order actinomycetes with more than 500 identified species (Barka et al. 2016). Remarkably, *Streptomyces* spp. alone accounts for about half of the overall population of actinobacteria present in the soil, ubiquitously distributed across diverse habitats

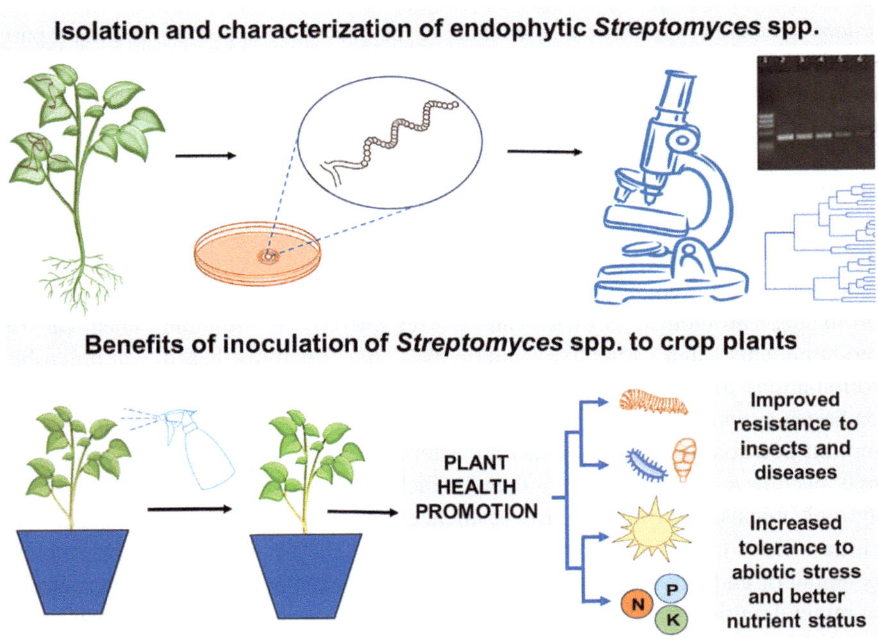

Fig. 9.1 An illustration describing the importance of the isolation and characterization of endophytic *Streptomyces* spp. for plant health promotion, as a plant probiotic

such as freshwater and marine ecosystems, rhizospheric soil and compost and vermicompost (Barka et al. 2016; Sathya et al. 2017; Seipke et al. 2012). The life cycle of a *Streptomyces* colony begins with the germination of a spore, which grows into the branched, thread-like filaments called hyphae, which form a complex network, known as mycelium. This filamentous structure provides a competitive edge to the *Streptomyces* spp. in soil colonization, out-competing other unicellular bacteria. In the nutrient-deprived conditions, these bacteria utilize a reproductive switch, which causes hyphae to rise above the surface. These aerial structures differentiate into chains of metabolically dormant spores, which help these microbes to survive under the harsh weather conditions (Zambri et al. 2022). This developmental switch is regulated by some internal signalling molecules like γ-butyrolactones, which trigger sporulation as well as the production of antibiotics and other secondary metabolites (Kato et al. 2007; Wang and Vining 2003). *Streptomyces* spp. have been traditionally described as soil-dwelling saprophytes, which aid in the decomposition of organic matter. However, in recent years they have been frequently isolated from the internal tissues of healthy plants also, such as roots, stems and leaves, which suggests that they have the important capability to adapt to an endophytic lifestyle or to colonize parallel ecological niches (Khamna et al. 2009; Qin et al. 2011; Viaene et al. 2016). They hold significant attention for their endophytic lifestyle in miscellaneous plant hosts, including cereals, legumes and medicinal plants via forming transient and long-term associations (Chaiharn et al. 2020; Goudjal et al. 2016; Htwe et al. 2018; Verma et al. 2011). *Streptomyces* interacts with both soil and plant root environments, resulting in the formation of mutualistic associations with host plants, which helps in soil stabilization, enhances uptake of mineral nutrients by plant roots and confers drought stress resistance (Liu et al. 2024a). Further, it has been found that *Streptomyces* spp. recognizes and colonizes the plant root surface, followed by the root penetration through natural root openings or micro-wounds (Viaene et al. 2016). Inside the root cells, *Streptomyces* spp. gets localized intercellularly or intracellularly, depending on the genotype of both plant and the bacterial species.

The genus *Streptomyces* is widely recognized for its remarkable metabolic adaptability, especially its capacity to release assorted biologically active compounds, including siderophores, antifungals, hydrolytic enzymes, antibiotics and small signalling molecules vital for plant and microbial communications. It accounts for more than 75% of naturally occurring antibiotics in nature (Barka et al. 2016; Bérdy 2012; Gopalakrishnan et al. 2020; Sathya et al. 2017; Van der Meij et al. 2017). The bioactive compounds synthesized by *Streptomyces* spp. suppress the growth of several phytopathogens, thereby modulating the plant defence responses and facilitating nutrient exchange. Several strains have shown possessing a considerable potential towards sustainable agriculture through plant growth promotion, biocontrol activity and alleviation of abiotic stress. They have been extensively studied and applied for their role in promoting plant growth, nutrient uptake, biological control of pests and plant pathogens and in improving mineral availability, contributing significantly to biofortification. Additionally, *Streptomyces* has shown potential in phytoremediation and bioremediation, aiding in the clean-up of contaminated soils and environments; thus, they are promising microbial candidates for sustainable agriculture (Gopalakrishnan et al. 2020).

Endophytic *Streptomyces* are prolific producers of many bioactive secondary metabolites like antimicrobial compounds (streptomycin, actinomycin, natamycin, etc.), which inhibit pathogenic bacteria and fungi. They also release siderophores that may chelate iron in the rhizosphere region, enhancing the iron uptake in plants (Bérdy 2012; Sathya et al. 2017). Besides, they are also involved in the production of phytohormones such as gibberellins (GAs), cytokinin (CK) and indole-3-acetic acid (IAA), which stimulate cell division, root elongation, influence seed germination and shoot growth in the host plants (Jog et al. 2014; Palaniyandi et al. 2013; Worsley et al. 2020). They have also been described for their ability to solubilize phosphate and zinc compounds and can act as nitrogen-fixing agents, thus can serve as effective biocontrol agents as well as plant growth promoters.

The root inhabiting *Streptomyces* spp. are known to release a wide array of important metabolites in plants that further play a crucial role in plant growth and development. These bacteria are reported to stimulate the signalling pathways of several vital PGRs such as jasmonic acid (JA), salicylic acid (SA) and ethylene (ET), which are also known to induce systemic acquired resistance (SAR) in plants (Nicolle 2023). Further studies revealed that *Streptomyces* spp. also induce the production of osmolytes, amino acids and enzymes, particularly antioxidant enzymes, which help in overcoming stress-induced oxidative damage and enhance abiotic stress tolerance in plants (Gopalakrishnan and Ganeshkumar 2013; Jog et al. 2014; Nazari et al. 2023). Taken together, the above reports reveal how endophytic *Streptomyces* has emerged as a potential candidate for plant growth and productivity enhancement, nutrient solubilization and acquisition, and plant defence stimulation.

9.3 Endophytic *Streptomyces* for Plant Health Promotion

The endophytic *Streptomyces* spp. live in close contact with the plants while living near its roots as well as also inhabit the roots and above-ground parts of the plants. *Streptomyces* spp. produces an array of biologically active compounds (Liu et al. 2019) having antifungal, antibacterial, larvicidal properties, which can kill the harmful microbes in the vicinity of plants and thus, protects plants even under stressed conditions. *Streptomyces* spp. may produce various metabolites like natural antibiotics and other compounds like PGRs that not only help protect plants but also help in the stimulation of growth and development of plants (Vurukonda et al. 2021). Because of the above benefits, *Streptomyces* are useful microbial candidates, particularly in organic agriculture.

9.3.1 Production of Plant Growth Regulators (PGRs)

Diverse *Streptomyces* spp. encourages growth and health of plants through the regulation of PGRs (or phytohormones), i.e. IAA, GA and CKs, which promote seed germination and root development. They also produce several enzymes, such as 1-aminocyclopropane-1-carboxylate (ACC) deaminase and nitrogenase, which help plants to withstand stress and increase nutrient availability. The endophytic *Streptomyces* may also possess strong biocontrol potential, as they can produce antibiotics and other secondary metabolites, which may kill pathogens and activate plant defence genes. In an investigation, endophytic *Streptomyces* strain DBT204 displayed considerable PGRs producing potential and biosynthetic genes for secondary metabolites, thus emphasizing its plant growth promoting (PGP) and biosynthetic potential (Passari et al. 2016). Further, in *S.* sp. strain DBT204, genes of 1-aminocyclopropane-1-carboxylate (ACC) deaminase (*acdS*), IAA (*iaaM*), two PGRs (i.e. IAA and kinetin) and six other antibiotics (namely, fluconazole, rifampicin, streptomycin, chloramphenicol, nalidixic acid and rifampicin) were discovered.

The crude PGP substances extracted from *Streptomyces* TM32 improved the shoot tip length of *Gymnema inodorum* tissue culture, reduced shoot growth and root development and resulted in higher total phenolic content and antioxidant activity (Sarin et al. 2018). Another investigation by Goudjal et al. (2013) reported that the endophytic PT2 strain of *Streptomyces* sp. produced the highest amount of IAA, with its cultivation in yeast extract-tryptone broth, promoting both seed germination and root elongation. Further, endophytic *Streptomyces rochei* strain ERY1 was found to produce antifungal metabolites against *Sclerotium rolfsii* and *Pythium aphanidermatum* and promote plant growth of Chinese cabbage seedlings (Suwitchayanon et al. 2018). Three PGRs, namely IAA, GA3 and CK were detected in *S. rochei* ERY1 culture filtrate, which further induced growth and increased the weight of Chinese cabbage seedlings. The study emphasized that *S. rochei* ERY1 strain can help in the management of fungal diseases, thus ensuring improved wellness and productivity of plants.

Endophytic *Streptomyces* spp. isolated from tea were found to produce ACC deaminase and IAA, which possess high metabolic potential for secondary metabolites with PGP, antifungal and antibacterial potential (Shan et al. 2018). Whereas, *Streptomyces tubercidicus* strain UAE1 isolated from a mangrove (*Avicennia marina*) was found to be an effective salt-tolerant isolate with the capability to produce different levels of PGRs, in vitro, viz., auxins (indole-3-pyruvic acid and IAA), GAs and CKs like isopentenyl adenoside, isopentenyl adenine and zeatin (Alkaabi et al. 2022). Besides this, *S. tubercidicus* strain UAE1 also produced enzymes such as ACC deaminase and nitrogenase with PGP potential. Moreover, Myo et al. (2019) focused on optimizing *Streptomyces fradiae* NKZ-259-mediated production of IAA and its employability as a PGP. The optimal conditions resulted in 82.363 μg/mL IAA yield, which was approximately four times more IAA secretion by NKZ-259 cells, thus recommending *S. fradiae* NKZ-259 as a possible candidate for IAA production and a considerable PGP potential. Further, *Streptomyces* sp. strain SSD49 suppressed eight pathogens, which included bacteria, pathogenic fungi affecting plants and human opportunistic microbes (Tian et al. 2019). It also stimulated the development of soybean, tomato, pepper and *Populus tomentosa* seedlings infested with SSD49. Two micro RNA libraries from *P. tomentosa* seedlings were found to control genes that encode transcription factors, such as auxin response factors (ARFs), phytohormone oxidase, short auxin up-regulated RNA (SAUR) genes, disease resistance proteins and GRAS proteins, as well as other response regulators.

The *Streptomyces* strain UPMRS4 has been identified as a promising PGP and biocontrol agent due to its strong antagonistic nature against *Pyricularia oryzae*, the causal organism of rice blast disease (Awla et al. 2017). Its treatment reduced disease severity by 67.9% and increased plant growth through the up-regulation of transcript levels of rice *pathogenesis-related* gene (*OsPR1a*), *glucanase* (*Gns1*), *chitinase* (*Cht-1*) and *salicylic acid-responsive* gene (*Oswrky45*). *OsWRKY45* is an important gene encoding a rice WRKY transcription factor that responds to SA, which has a critical role in plant protection against pathogens and abiotic stressors, and it ensures the balance between plant growth and defence mechanisms. Another endophytic *S. enissocaesilis* strain OM182843 was reported to possess antagonistic activity against *Meloidogyne incognita* (root-knot nematode) and had a considerable potential to produce PGRs such as IAA, SA and GA, as well as hydrolytic enzymes, with volatile organic compounds (VOCs) showing nematicidal activity (El-Akshar et al. 2022). Hence, this endophytic *S. enissocaesilis* strain can be employed both as a biocontrol agent and a potential PGP. Due to the immense potential of endophytic *Streptomyces* spp. in producing PGRs, ensuing the plant growth promotion, they may be employed as bio-agents to solve the double problem of securing global food security, while safeguarding ecological integrity in modern agriculture.

9.3.2 Improving Nutrient Uptake

Endophytic *Streptomyces* spp. are also known to improve nutrient absorption and uptake, mainly by the production of siderophores. The release of siderophores improves nutrient availability both to plant roots as well as the microbes nearby, while also stimulating the interactions between roots and microbes, ultimately promoting plant growth in turn. *Streptomyces albus* CINv1, an endophytic strain isolated from *Cinnamomum verum*, proved efficient against *Colletotrichum* sp., isolate CA0110, while also acting as a plant growth stimulator, safeguarding strawberry against anthracnose (Pupakdeepan et al. 2023). It is worth noting that it enhanced strawberry plant development by boosting nitrogen fixation, increasing phosphate solubilization and producing siderophores. The phosphate solubilization potential of endophytic *Streptomyces* sp. strain CoT10 isolated from *Camelliaoleifera* has also been reported by Cui et al. (2022). Further, the endophytic *Streptomyces* sp. strain CoT1 has been found to mobilize different phosphorus (P) sources, specifically in the acidic soils, through the production of siderophores. It also enhanced P contents, promoted plant growth by accumulation of higher levels of iron (Fe) and P through siderophore-mediated acquisition of P and Fe and enhanced microbial diversity. Besides this, the endophytic *Streptomyces* spp. are also known to enhance the microbial diversity and richness via increasing the number of root nodules in plants, which further affects the symbiotic association of these microbes, resulting in higher N (nitrogen) uptake by plants.

In another experiment, it was found that three different endophytic *Streptomyces* strains enhanced plant growth and microbial symbiosis in lucerne and its rhizobial partner mediated by greater number of root nodules, which resulted in improved N_2-fixation in lucerne plants (Le et al. 2016). The N_2 fixation rates were increased by the co-inoculation of endophytic *S. griseoflavus* P4 with *Bradyrhizobium japonicum* SAY3-7 in soybean (Htwe and Yamakawa 2016). Furthermore, the co-inoculation of *S. griseoflavus* P4 with *B. japonicum* SAY3-7 also showed a synergistic effect in enhancing the nodulation, N_2-fixation and nutrient uptake in soybean, which in turn resulted in increased soybean yield (Htwe et al. 2018). Moreover, similar to native *B. japonicum*, the *S. griseoflavus* P4 also enhanced the uptake and absorption of important phytonutrients such as N, P, potassium (K), calcium (Ca) and magnesium (Mg). It emphasized on the competitiveness of *S. griseoflavus* P4 strain even in the presence of native *Bradyrhizobia* spp., thus highlighting the application of *S. griseoflavus* P4 as a potential inoculant for biofertilizer production. Another study by Thenappan et al. (2024) reported the plant growth promoting potential of endophytic *Streptomyces* isolates through the increased production of IAA, total chlorophyll, carotenoids, siderophores and root N_2 content in rice.

Although N is highly crucial for plant development, however, plant growth is also significantly affected by the availability of other important nutrients like Fe. Due to the low solubility of Fe in aerobic soil conditions, the availability of Fe in the soils is often limited to plants, leading to severe deficiency symptoms. The endophytic *S. hygroscopicus* strain OsiSh-2 has been reported to confer plant protection against Fe-deficiency stress by modulating Fe-reduction and chelation, solubilization and translocation (Cao et al. 2021). Further, it enhanced Fe concentration, plant biomass, net leaf photosynthetic rate and chlorophyll content in rice plants under Fe-limiting conditions through the up-regulation of genes involved in Fe^{3+}-reduction and down-regulation of genes involved in Fe^{3+}-chelation. The endophytic *Streptomyces* strain JSA11 enhanced plant chemical composition and nutritional quality in *Chenopodium* sprouts (Almuhayawi et al. 2021). It was found that *Streptomyces* increased photosynthesis, mineral profile such as K, Ca, Fe, Mn, Mg, P, Na (sodium), Cu (copper) and Zn (zinc), antioxidant, vitamins, amino acid contents, fatty acid contents, leading to better nutrient use efficiency in sprouts. *Streptomycesalbus* and *S. griseus* strains improved pearl millet yield and nutrient content, including Fe, Zn and Ca levels, demonstrating their potential to complement conventional biofortification strategies (Srinivas et al. 2022). By combining endophytic *Streptomyces panayensis* inoculum with vermicompost, it enhanced available nutritional concentrations in soil and soil health, leading to improved farmland and maize productivity (Trang and Van Chuong 2025).

9.4 Endophytic *Streptomyces* for Abiotic Stress Tolerance

Several endophytic *Streptomyces* spp. have been reported to confer plant protection against a variety of abiotic stresses. A PGP endophytic *Streptomyces* sp. strain GKU 895 producing ACC (1-aminocyclopropane-1-carboxylate) deaminase has been demonstrated to increase salt stress tolerance in rice cultivars (Kruasuwan et al. 2023). Salinity increases both ACC synthase and ACC oxidase activities in plants, resulting in higher levels of ET. However, *Streptomyces* sp. strain GKU 895 can reduce stress-induced ET levels through ACC deaminase. This strain also stimulated the up-regulation of critical genes and transcription factors, increasing ion transport, photosynthesis, the formation of PGRs, reactive oxygen species (ROS) scavenging and maintaining cellular homeostasis. Further, *Streptomyces* sp. conferred stress resistance in *Arabidopsis thaliana* and *Brassica chinensis* by enhancing the production of IAA and improving phosphate solubility, which resulted in increased fresh biomass, levels of chlorophyll and root number under salt and heat-stressed conditions (Manullang and Chuang 2020). Another study on halotolerant *Streptomyces* sp. KLBMP5084 from *Limonium sinense* root revealed its potential as a plant growth stimulator (Gong et al. 2020). This strain, KLBMP5084, was found to inhibit various plant pathogenic fungi, including *Alternaria solani*. When applied to tomato seedlings, this *Streptomyces* sp. strain significantly enhanced their growth, leading to an increase in fresh biomass, length of shoots and the development of fibrous roots. Additionally, it also enhanced the activity of antioxidant enzymes, the level of soluble sugars and the proline content of tomato. It also accumulated K^+ and Ca^{2+} contents; whereas, it reduced Na^+ in tomato leaf. Further, *Streptomyces* sp. strain KLBMP5084 differentially expressed approximately 191 genes in tomato under salt stress, highlighting that strain KLBMP5084 is a budding biofertilizer that can promote tomato seedling growth during salt-stressed conditions. Table 9.1 presents the beneficial roles of endophytic *Streptomyces* spp. in conferring abiotic stress tolerance in plants.

9.5 Endophytic *Streptomyces* for Biotic Stress Tolerance

Endophytic *Streptomyces* are being explored for their potential role in conferring biotic stress tolerance to plants through the management of insect pests and pathogens of plant diseases, as a sustainable strategy to enhance plant resilience and improve agricultural productivity. The following sub-sections describe how various researchers are employing endophytic *Streptomyces* strains to fortify crops against biotic stresses, promoting healthier ecosystems and increased agricultural production. This not only enhances biotic stress tolerance but also reduces chemical pesticide use, improving soil health and biodiversity.

9.5.1 Endophytic *Streptomyces* for Insect Pest Management

Endophytic *Streptomyces* spp. has the tendency to colonize plant tissues and stimulate the synthesis of bioactive metabolites that have insecticidal and larvicidal potential. Due to this, these root-inhabiting bacteria has recently gained the attention of plant researchers to develop them as an effective and eco-friendly bio-alternative for insect pest management. Several strains of endophytic *Streptomyces* spp. are not

Table 9.1 Endophytic *Streptomyces* in abiotic stress management

Endophytic *Streptomyces* strain	Plant	Type of abiotic stress	Effect	References
Streptomyces sp. strain NRC21696	*Zea mays* L. (Maize)	Heavy metal stress (14 heavy metals in wastewater namely, Ni^{2+}, Zn^{2+}, Fe^{3+}, Cu^{2+}, Cd^{2+}, Ag^{+}, Al^{3+}, Hg^{2+}, Mn^{2+}, As^{3+}, As^{5+}, Cr^{6+}, Co^{2+} and Pb^{2+} and two heavy metals in Maize As^{3+} and Cr^{6+})	Removed 14 heavy metals from wastewater and improved maize productivity under As^{3+} and Cr^{6+} toxicity. It also increased maize yield by improving the germination % rate, and index, shoot and root tolerance indices, photosynthetic pigment levels, fresh and dry biomass, lengths of root, leaves and cobs	Al-Huqail and El-Bondkly (2022)
S. globisporus strain SP6C4, *S. parvulus* strain KPB2,*S. racemochromogenes* strain W1SF4, and *S. bacillaris* strain S8	Tomato	Salinity and drought stress	When *S. globisporus* SP6C4 colonized the rhizosphere and root endosphere, it gave tomatoes resistance to drought and salt stress	Kim and Kwak (2023)
Streptomyces sp. GKU 895	Rice cultivar IR29	Salt stress (150 mM NaCl)	Up-regulated gene expression and their corresponding transcription factors resulted in enhanced plant weight, growth and development through reduction of ROS by stimulation of ROS scavenging, ion transport and homeostasis in rice during salt stress	Kruasuwan et al. (2023)
S. albidoflavus strain OsiLf-2	Rice (*Oryza sativa*)	Drought and salt stress	Osmolyte-producing endophytic strain-induced drought and salt stress protection through enhanced accumulation of proline and soluble sugar content, stimulation of stress responses at physiological, biochemical and gene levels, thereby boosting rice yields	Niu et al. (2022)

(continued)

Table 9.1 (continued)

Endophytic *Streptomyces* strain	Plant	Type of abiotic stress	Effect	References
Streptomyces strain KLBMP 5084	*Limonium sinense*	Salt stress	Increased morphological parameters like weight, length and accumulation of pigments (total chlorophyll), osmolytes (proline) and antioxidant enzymes, lowered leaf malondialdehyde content and Na^+ accumulation in leaves and roots Also, up-regulated the gene expression of various PGP-related genes	Qin et al. (2017)
Streptomyces sp. GMKU 336	*Oryza sativa*	Salt stress	The strain significantly enhanced the contents of proline, K^+, Ca^+, chlorophyll and water, whereas it reduced ET, ROS, Na^+ and Na^+/K^+ ratio *ACO1* and *EREBP1* genes were down-regulated; whereas up-regulated *BADH1*, acdS encoding ACCD (1-aminocyclopropane-1-carboxylate deaminase) and stimulated Na^+ transporters, calmodulin and antioxidant enzymes	Jaemsaeng et al. (2018a)
Streptomyces sp. strain GMKU 336	Mung bean (*Vigna radiata*)	Flooding stress	Increased flooding tolerance in mung beans was conferred by the strain and its ACCD-deficient mutant, which also improved plant biomass, length and elongation, levels of chlorophyll, area and colour of leaf and formation of adventitious roots	Jaemsaeng et al. (2018b)

Streptomyces venezuelae strain ATCC 10712	Thai jasmine rice (*Oryza sativa* cv. KDML105)	Salt stress	Enhanced salt stress protection in rice by reducing ethylene, ROS and Na^+ contents, and by increasing the accumulation of chlorophyll, proline, relative water content and K^+	Yoolong et al. (2019)
Streptomyces spp. (*S. iranensis* strain HM 35, *S. rapamycinicus* strain NRRL 5491 and *S. violaceus niger* strain Tu 4113)	*Arabidopsis thaliana* ecotype Columbia (Col-0)	Drought and salinity stress	Alleviated the negative impact of both stresses by producing pteridic acid H (1) and its isomer F (2), which further promoted root growth in *Arabidopsis* under abiotic stress	Yang et al. (2023)

only found to play an immense role in plant growth promotion but also in the induction of SAR (Al-Quwaie 2024) through the release of crucial insecticidal compounds like macrolides, chitinases and polyketides that directly affect a wide range of plant insect pests (Arasu et al. 2013; El-Tarabily et al. 2021). It has been found that *Streptomyces* sp. strain AP123 and *S. hydrogenans* strain DH16 caused delayed insect development, induced deformities and reduced the rate of adult emergence of important insect pests, namely, *Helicoverpa armigera* and *Spodoptera litura* (Kaur et al. 2014; Arasu et al. 2013). These results emphasized the larvicidal and anti-feedant potential of *Streptomyces* strains against crucial plant insect pests.

Another independent investigation by Cerna-Chávez et al. (2024) and Balakrishnan et al. (2017) revealed the strong insecticidal activity of *S. avermitilis* and some marine-derived strains against *Plutella xylostella* and *Aedes aegypti*. The investigated *S. avermitilis* and marine-derived strains released avermectin-like compounds, which interfere with gamma-aminobutyric acid (GABA) neurotransmission of the feeding plant pests, namely, *P. xylostella* and *A. aegypti* and thus, result in feeding paralysis and rapid mortality of these pests. Further, Aggarwal et al. (2016) highlighted that *S. griseoviridis* produce antibiotics that suppress the insect gut flora and weaken the immune system and response in insect pests. *S. albidoflavus* was found to possess insecticidal properties against green peach aphids (Chen et al. 2018a), whereas *Streptomyces* sp. SA61 was reported to act against sap-feeding greenhouse whiteflies through the release of polyketides (Liu et al. 2024b). Thus, the *Streptomyces* spp. can either act directly as 'direct biopesticides' or indirectly act as 'indirect plant protectants' through the release of metabolites and their subsequent insecticidal action. Moreover, nanoparticles synthesized using endophytic microbes were also found to hinder the growth of stored grain pests (Goswami et al. 2010), and these microbe-based nanoparticles can be further utilized as effective nanoformulations in the form of nano-antimicrobial agents, nanopesticides and nanofungicides (Singh et al. 2023).

Reportedly, *Streptomyces* spp. like *S. griseoviridis* and *S. avermitilis*, produce secondary metabolites such as griseoviridin, actinobolin and avermectins, which interfere with the neuromuscular functions of the insect pest, or inhibit their feeding mechanisms by the disruption of their gut microbiome (Aggarwal et al. 2016; Cerna-Chávez et al. 2024). *Aedes aegypti* and other agricultural pests are found to be highly susceptible to polyketide-based metabolites (Arasu et al. 2013; Kaur et al. 2014). Some strains of *Streptomyces* spp. such as AP123 and *S. hydrogenans* DH16 are found to cause delayed development, adult deformities and larval mortality in *A. aegypti*, a mosquito vector and *Helicoverpa armigera* and *Spodoptera litura* through the production of polyketide-based metabolites. Some other species, like *Streptomyces rochei* and *S. lydicus*, also produce chitinases and other antifungal metabolites, enabling both direct insecticidal action through cuticle degradation and indirect protection via the suppression of plant pathogens (El-Tarabily et al. 2021; Yuan and Crawford 1995). Various reports pertaining to the insecticidal potential of important endophytic *Streptomyces* strains and their mechanism of action against plant insect pests are presented in Table 9.2.

Table 9.2 Endophytic *Streptomyces* for insect pest management

S. No.	*Streptomyces* species/ Strain	Mode of action	Target insect pest(s)	Effect	Reference
1.	*Streptomyces griseoviridis*	Production of griseoviridin and actinobolin	Insect pests	Antimicrobial agents May deter insects via antibiosis Suppress gut flora Weaken pest defence	Aggarwal et al. (2016)
2.	*Streptomyces* sp. strain K61	Indirect action via plant growth promotion	General soil-borne insects	Induced systemic resistance in plants Indirect pest suppression Reduces root herbivory by strengthening plant tissues	Al-Quwaie (2024)
3.	*Streptomyces* sp. strain KSF103	Ethyl-acetate extract	*Aedes*, *Culex* and *Anopheles* mosquito larvae	Larvicidal activity Potential biocontrol of mosquito vectors Reduces root herbivory by strengthening plant tissues	Amelia-Yap et al. (2023)
4.	*Streptomyces* sp. strain AP-123	Polyketide metabolite	*Helicoverpa armigera*, *Spodoptera litura*	Strong antifeedant (78%) Larvicidal properties Reduced survival and delayed adult emergence	Arasu et al. (2013)
5.	Marine *Streptomyces* sp.	Production of Avermectin-like metabolites	*Helicoverpa zea*, *Plutella xylostella*, *Spodoptera exigua*	Interfere with GABA neurotransmission Induce feeding paralysis Rapid larval mortality	Arasu et al. (2016)
6.	Marine *Streptomyces* sp.	Production of some key Actinomycete metabolites	*Aedes aegypti*	Rupture larval tissues Cause intestinal damage and dehydration Larvae fail to moult	Balakrishnan et al. (2017)

(continued)

Table 9.2 (continued)

S. No.	*Streptomyces* species/ Strain	Mode of action	Target insect pest(s)	Effect	Reference
7.	*S. hygroscopicus* and *S. griseochromogenes*	Production of milbemycins (macrolides)	Nematodes, ticks, fleas, some insects	Induced chloride ion channels Cause flaccid paralysis and death Affect neuromuscular junctions	Bhardwaj et al. (2020)
8.	*S. diastatochromogenes*	Production of unspecified bioactive compounds	*Mythimna separata*	Reduced feeding rate Lower pupation success High mortality in second and third instar larvae	Cao et al. (2019)
9.	*S. avermitilis*	Produces avermectins (neurotoxic to insects)	*Helicoverpa armigera*, mites, aphids	Irreversible neuromuscular paralysis Feeding inhibition, dehydration, eventual death Persistence on foliage reduces reinfestation	Cerna-Chávezet al. (2024)
10.	Endophytic *Streptomyces* sp. isolated from neem leaves	Unspecified actinomycete extracts	*Myzus persicae*	Insecticidal properties and disrupted feeding	Chen et al. (2018a)
11.	*Streptomyces* sp. strain CAI-155	Production of bioactive metabolites	*Helicoverpa armigera* larvae	Effective larval mortality	Gopalakrishnan et al. (2016)
12.	*S. hydrogenans* strain DH16	Ethyl-acetate and other secondary metabolites	*Spodoptera litura*	70% larval mortality Delayed growth Reduced pupation and adult emergence	Kaur et al. (2014)
13.	UV-mutant *Streptomyces* sp. strain KR0006	Enhanced production of insecticidal metabolites	Lab pests (test-stage)	Mutagenesis improved insecticidal activities	Kim et al. (2022a)

14.	*S. lactacystinicus*	Secretion of actinomycin and novel polyketide metabolite	*Plutella xylostella*	Increased larval mortality Feeding deterrence	Kim et al. (2022b)
15.	*Streptomyces* sp. SA61	Production of polyketides (strekingmycins, phenalinolactones)	*Trialeurodes vaporariorum*, *Myzus persicae* (Hemiptera: Aleyrodidae)	Increased mortality	Liu et al. (2024b)
16.	*Streptomyces phaeochromogenes* strain LD-37	Production of antibiotics/metabolites	*Ceratitis capitata* (Mediterranean fruit fly)	Mortality in larval, pupal and adult stages Induction of morphological deformities	Samri et al. (2017)
17.	*Streptomyces* sp. (bioformulation)	Part of microbial insecticide formulation	Stored grain pests (*Sitophilus oryzae*)	Grain protection through reduced infestation	Goswami et al. (2010)
18.	*Streptomyces* sp. from Asteraceae plants soil	Actinomycete metabolites	*Culex quinquefasciatus* larvae	Complete mortality in early instars Gut tissue lysis and impaired digestion	Tanvir et al. (2014)
19.	*Streptomyces bingchenggensis/nanchanggensis*	Milbemycin-type metabolites	Avermectin-resistant insects	Effective biological larvicide via chloride channel mechanism	Yan and Xia (2021)
20.	*Streptomyces microflavus* strain neau3	Seco-nemadectin (nematicide/pesticide)	*Caenorhabditis elegans*	Reduced pupal weight and adult emergence Deformed pupae/adults	Yang et al. (2013)
21.	*S. lydicus* strain WYEC108	Chitinase, natamycin, streptolydigin	Soft-bodied pests/fungal pathogens	Potential cuticle degradation Broad protection	Yuan and Crawford (1995)

In addition to these, certain isolates of *Streptomyces* also promote plant health and induced systemic resistance (ISR), as noted with *Streptomyces* sp. strain K61, indirectly reducing pest incidence by strengthening host defences (Al-Quwaie 2024). Notably, *Streptomyces* strains from marine environments and bio-formulations have shown promising efficacy against resistant insect populations and stored grain pests (Balakrishnan et al. 2017; Singh et al. 2023; Yan and Xia 2021). These multifaceted actions suggest that *Streptomyces* spp., hold immense potential as eco-friendly biopesticides, either alone or in integrated pest management (IPM) programmes.

9.5.2 Endophytic *Streptomyces* for Disease Management

Streptomyces spp. are the key players in plant disease prevention because they produce a variety of antibiotics and physiologically active chemicals that suppress pathogen development. Furthermore, these beneficial bacteria can improve plant health by stimulating growth and increasing nutrient availability in the soil. Table 9.3 highlights the role of *Streptomyces* spp. in managing diverse plant diseases. *Streptomyces* spp. have been reported to exhibit biological control potential against a wide array of phytopathogens, such as *Burkholderia glumae*, *Verticillium dahliae*, *Fusarium oxysporum*, *F. graminearum*, *Pythium ultimum*, *Phytophthora* sp. and *Scleritium rolfsii* (Colombo et al. 2019; Liu et al. 2019; Suárez-Moreno et al. 2019). Actinovate, Mycostop and Rhizovit are among the commercial bio-fungicides derived from *Streptomyces* spp. (Colombo et al. 2019; Liu et al. 2019). The selection of biological control agents (BCA) relies on the interactions between the BCA, the plant and the pathogen, which ensures the development of effective BCA for disease management. In an experiment, the endophytic *Streptomyces* strain DEF09, was found to be effective in inhibiting the *Fusarium* head blight in wheat by blocking the spread of *F. graminearum* in the infection site, thereby reducing disease severity up to 60% in field conditions (Colombo et al. 2019). Further, three *Streptomyces* isolates were found to possess antifungal and antibacterial potential and acted as the biocontrol agents for *Burkholderia glumae*, which causes bacterial ‘Panicle Blight’ disease in rice (Suárez-Moreno et al. 2019). These bacterial strains produced antimicrobial compounds, siderophores, enzymes and solubilizing phosphate, and also colonized rice plants, thereby stimulating rice plant growth. Moreover, *Streptomyces* sp. strain NEAU-S7GS2 was also reported to inhibit the mycelial growth and germination of *Sclerotinia sclerotiorum*, thus reducing sclerotinia stem rot in soybeans (Liu et al. 2019). This strain also improved plant growth via enhancing the solubilization of inorganic phosphates and ACC deaminase production.

Another endophytic *Streptomyces* spp. showed inhibitory effects on the phytopathogen *Magnaporthe oryzae* B157 and *Rhizoctonia solani*, in rice, sorghum and wheat against fungal diseases (Patel et al. 2018). In this case, the inhibition of fungal pathogen was caused due to the presence of bioactive compounds i.e. 2, 4-di-tert-butylphenol, 1-ethylthio-3-methyl-1, 3-butadiene and 2-(chloromethyl)-2-

Table 9.3 Role of *Streptomyces* spp. in plant disease management

S. No.	Plant	Disease	Pathogen	*Streptomyces* strain	Biocontrol mechanism	Active compound(s)	Reference
1.	*Agrostis stolonifera* (turfgrass)	Crown disease	*Rhizoctonia solani*	*Streptomyces violaceusniger* strain YCED9	Antibiosis (antifungal metabolites) + cell-wall degradation enzymes	Natamycin; 4-Phenyl-3-butenoic acid	Gao et al. (2021)
2.	*Allium cepa* (onion)	Bacterial rot	*Erwinia carotovora*	*S. lavendulae* strain HHFA1	Antibiosis through antifungal and possibly antibacterial action	Natamycin	Gopalakrishnan and Ganeshkumar (2013)
3.	*Arabidopsis thaliana*	Common scab-like symptoms (experimental model)	*Streptomyces scabies*	*Streptomyces* sp. strain IFB-A03	Activates salicylic acid-mediated defence responses	Thaxtomin A not produced; no known compound	Lin et al. (2012)
4.	*Arachis hypogaea* (groundnut)	Stem rot	*Sclerotium rolfsii*	*Streptomyces* sp. strain CBE	Antibiosis via 1H-pyrrole-2-carboxylic acid (PCA)	1H-Pyrrole-2-carboxylic acid (PCA)	Adhilakshmi et al. (2014)
5.	*Brassica oleracea* var. *capitata* (cabbage)	Alternaria leaf spot	*Alternaria brassicicola*	*Streptomyces* sp. strain MBCN152-1	Significant biocontrol activity; colonizes host tissues	Not specified	Hassan et al. (2017)
6.	*Capsicum annuum* (chilli pepper)	Root rot, blight	*Alternaria brassiceae* and *Colletotrichum gloeosporioides*	*Streptomyces* spp.	Antibiosis through volatile organic compound inhibiting pathogen growth	4-Phenyl-3-butenoic acid	Srividya et al. (2012)

(continued)

Table 9.3 (continued)

S. No.	Plant	Disease	Pathogen	*Streptomyces* strain	Biocontrol mechanism	Active compound(s)	Reference
		Blight	*Phytophthora capsica*	*Streptomyces* spp. strains 47 W08, 47 W10	Antibiosis via avermectins disrupting pathogen ion channels	Avermectins	Yandigeri (2021)
		Anthracnose	*C. gloeosporioides*	*S. cavurensis* strain NRRL 2740	Antibiosis via polyene macrolide affecting pathogen membranes	Lucensomycin	Intra et al. (2011)
		Anthracnose	*C. gloeosporioides*	*S. ambofaciens* strain S2	Antibiosis via ion channel disruption in fungus	Avermectins	Lyu et al. (2017)
7.	*Cicer arietinum* (chickpea)	Fusarium wilt	*Fusarium* spp.	*Streptomyces* spp. strains CAI-24, CAI-121, CAI-127, KAI-32 and KAI-90	Antifungal membrane disruptor	Natamycin	Gopalakrishnan and Ganeshkumar (2013)
		Basal rot	*Macrophomina phaseolina*	*Streptomyces* sp. strain RO3	Antibiosis via cellular respiration inhibition	Oligomycins A and C	Goudjal et al. (2014)
		Collar rot	*Sclerotium rolfsii*	Endophytic *Streptomyces* spp. namely, *S. diastaticus*, *S. Fradiae*, *S. olivochromogenes*, *S. collinus*, *S. ossamyceticus* and *S. griseus*	Stimulates defence-related enzymes like SOD, PO, PAL, APX, catalase, chitinase and β-glucanase, and accumulation of total phenolics and flavonoids under pathogenic stress	Not specified	Singh and Gaur (2017)

8.	*Citrus limon* (lemon)	Sour rot	*Geotrichum candidum*	*Streptomyces* sp. strain RO3	Antibiosis via oligomycins that inhibit pathogen respiration and mycelial growth	Oligomycins A	Goudjal et al. (2014)
9.	*Cucumis sativus* (cucumber)	Damping-off	*Pythium* spp.	*S. griseoviridis*	Antibiosis with VOC inhibition of hyphal growth	4-Phenyl-3-butenoic acid	Junaid et al. (2013)
		Anthracnose (leaf spot disease)	*Colletotrichum orbiculare*	*Streptomyces* sp. strain MBCu-56	Produces antifungal metabolites; colonizes leaf surfaces and penetrates beneath the cuticle	Not clearly identified	Shimizu et al. (2009)
		Fusarium wilt	*Fusarium oxysporum* f. sp. *cucumerinum*	*Streptomyces* isolate HAAG3-15 (sequence similarity with *S. sporoclivatus* strain NBRC 100767^{T})	Through broad-spectrum antifungal activity	Azalomycin B	Cao et al. (2020)
10.	*Fragaria* × *ananassa* (Strawberry)	–	*Botrytis cinerea*	*Streptomyces* sp. strain 3–10	Antibiosis by polyether compounds	Avermectins	Lyu et al. (2017)
11.	*Glycine max* (soybean)	Blight	*Xanthomonas campestris* pv. *Glycines*	*Streptomyces* sp.	Antibiosis via inhibition of bacterial respiration	Antimycin A17	Chaiharn et al. (2020)
12.	*Lactuca sativa* (lettuce)	Basal drop	*Sclerotinia minor*	*S. viridodiasticus*	Antibiosis through polyene and VOC production	Natamycin; 4-Phenyl-3-butenoic acid	Gao et al. (2021)

(continued)

Table 9.3 (continued)

S. No.	Plant	Disease	Pathogen	*Streptomyces* strain	Biocontrol mechanism	Active compound(s)	Reference
13.	*Lathyrus odoratus* (sweet pea)	Powdery mildew	*Oidium* sp.	*Streptomyces* sp.	Antifungal membrane interaction resulting in pathogen cell death	Natamycin	Sangmanee et al. (2009)
14.	*Malus domestica* (apple)	–	*Botryosphaeria dothedia*	*S. rochei* strain A1	Antibiosis via organic acid inhibiting pathogen growth	Propanoic acid	Zhang et al. (2016)
15.	*Musa* spp. (Banana)	Anthracnose	*C. musae*, *C. gloeosporioides*	*S. spectabilis* strain NBRC 13424	Antibiosis via membrane-active polyene	Lucensomycin	Chen et al. (2018b)
		Wilt	*Fusarium oxysporum*	*S. violaceusniger* strain G10	Antibiosis via acid-mediated membrane disruption	Propanoic acid	Chaiharn et al. (2020)
		Fusarium wilt of banana (Panama disease)	*F. oxysporum* f. sp. cubense TR4	*S. malaysiensis* strain 8ZJF-21	Produces antifungal metabolites; induces defence responses	Not specified	Zhang et al. (2022)
16.	*Oryza sativa* (Rice)	Blast	*Bipolaris oryzae*, *Curvularia oryzae*, *F. oxysporum*, *P. oryzae*, *R. solani* and *R. oryzae-sativae*	*S. hundungensis* strain MBRL 251; and *Streptomyces* sp. KH-strain 614	Antibiosis; mitochondrial disruption in pathogens	Antimycin A17	Nimaichand et al. (2013)
		Sheath blight	*R. solani*	*S. aurantiogriseus* strain VSMGT1014	Antibiosis via ATP synthase inhibition	Oligomycins A and C	Harikrishnan et al. (2014)
		Brown spot	*Helminthosporium oryzae* and *R. solani*	*S. misionensis* strain NBRC	Antibiosis targeting fungal development	Rapamycin	Poomthongdee et al. (2015)

		Brown spot	*Pyricularia* sp.	*Streptomyces* strains PC12, D4.1, D4.3 and W1	Antibiosis with membrane-active metabolites	Pyrroles [1,2-a] pyrazine-1,4-dione	Chaiharn et al. (2020)
17.	*Porphyra yezoensis*/ *Pyropia* spp.	Red rot	*Pythium porphyrae*	*Streptomyces* sp. strain AP77	Antibiosis via rapamycin disrupting fungal TOR signalling and growth	Rapamycin	Zhang et al. (2016)
18.	*Solanum lycopersicum* (Tomato)	Damping-off	*R. solani*	*Streptomyces* spp.	Antibiosis and induced systemic resistance through antibiotic Volatile Organic Compounds (VOCs)	Oligomycins A and C	Goudjal et al. (2014)
		Damping-off	*R. solani*	*Streptomyces* sp. strain S30	Antibiosis targeting fungal growth signalling pathways	Rapamycin	Gao et al. (2021)
		Fusarium wilt	*F. oxysporum* f. sp. *lycopersici*	*S. rochei* strain ACTA 1551	Antibiosis via pyrroles with membrane-disrupting properties	Pyrroles [1,2-a] pyrazine-1,4-dione	Kanini et al. (2013)
		Root rot	*R. solani*	*S. vinaceusdrappus* strain S5MW2	Antibiosis through 1H-pyrrole-2-carboxylic acid (PCA) disrupting fungal metabolism	1H-Pyrrole-2-carboxylic acid (PCA)	Yandigeri et al. (2015)
		Root rot	*R. solani*	*S. toxytricini* strain vh6	Antibiosis via fatty acid production, altering pathogen membrane integrity	Propanoic acid	Patil et al. (2011)

(continued)

Table 9.3 (continued)

S. No.	Plant	Disease	Pathogen	*Streptomyces* strain	Biocontrol mechanism	Active compound(s)	Reference
		Verticillium wilt	*Verticillium dahlia*	*Streptomyces* sp. strain DHV3-2	Exhibits strong antagonism; higher root colonization in infected plants	Possibly albocycline (unconfirmed)	Cao et al. (2016)
19.	*Triticum aestivum* (Wheat)	Root rot	*Fusarium cloumorum*	*S. aurantiogriseus* strain VSMGT1014	Antibiosis and induction of host resistance	Oligomycins A and C	Harikrishnan et al. (2014)
		Fusarium head blight (FHB) or scab	*Fusarium graminearum*	*Streptomyces* sp. strain DEF09	Produces chitinase; reduces mycelial growth; colonizes roots	Chitinase	Colombo et al. (2019)
20.	*Zingiber officinale* (Ginger)	Rhizome rot	*F. oxysporum* f. sp. *zingiberi*	*Streptomyces* spp.	Antibiosis via inhibition of fungal mitochondrial respiration	Antimycin A17	Manasa et al. (2013)

cyclopropyloxirane; and these BCA also up-regulated the gene expression of defensive responsive genes (*PR10a*, *NPR1*, *PAL* and *LOX2*) in *Streptomyces* colonized plants. Further, the seed treated with endophytic *Streptomyces* strain AzR-051 isolated from *Azadirachtaindica* significantly promoted plant growth and antagonized *Alternaria alternata*, a cause of early blight disease in tomato plants through production of IAA and siderophores (Verma et al. 2011).

9.6 Endophytic *Streptomyces* in Sustainable Agriculture

The recent traditional approaches of utilizing chemical-based inputs for sustainable and agricultural intensifications have failed to save both ecosystem and public health, which has shifted the research to trace out the beneficial microbes for ensuring the sustainable agriculture to suit the needs of both farmers and consumers (Oyedoh et al. 2023). Among various microbes, endophytic *Streptomyces* spp. have become a point of discussion because of their ability to live in rhizosphere, plant roots and aerial plant parts, secrete agroactive bio-metabolites and enhance plant defence against biotic and abiotic stressors. *Streptomyces* can be used as plant co-inoculants to enhance plant-microbe symbiosis, leading to sustainable agriculture production (Vurukonda et al. 2018). These eco-friendly microorganisms control pests and improve plant growth, potentially enhancing biofertilizers, biopesticides and crop protection through cooperative formulations. The endophytic *Streptomyces*, found in diverse plant species, enter plants through wounds, root hairs and epidermal conjunctions, and consequently result in enhanced plant growth, yield and health (Suman et al. 2016).

The genomes of five strains of endophytic *Streptomyces* that were isolated from the roots of *Arabidopsis thaliana*, were sequenced by Worsley et al. (2020) and their genomes were sequenced. The genome analysis revealed that all the *Streptomyces* strains secreted siderophores, and released ACC deaminase enzyme and IAA, an important auxin, involved in confirming plant growth. Out of these strains, one bacterial strain secreted filipins-resembling antifungal compounds, which conferred protection against *Gaeumannomyces graminis* var. *tritici*, which causes fungal disease in wheat seeds. The other three strains of endophytic *Streptomyces* spp. were found to improve the development of *A. thaliana* under both in vitro and in soil conditions. Buzón-Durán et al. (2020) reported that endophytic *Streptomyces* spp. release a wide range of commercially important antibiotics with antifungal, antibacterial, antimicrobial, insecticidal and larvicidal potential, which can fight various harmful plant pests and pathogens. During the composting process, *Streptomyces* spp. secreted enzymes, which hydrolytically degraded the lignocellulosic residues, and enhanced the composting process. Besides, application of these microbes as soil amendments in combination with organic composts augmented the growth and yield of the plants. Due to these properties and the capacity to produce bio-active antibiotic compounds, endophytic Streptomyces can be a bio-safe alternative to chemical-based fertilizers. By harnessing the natural capabilities of *Streptomyces* and utilizing organic compost, farmers can enhance soil health, promote plant growth and reduce reliance on synthetic fertilizers, ultimately leading to more resilient and productive farming environments.

9.7 Conclusion and Future Prospects

Recently, there has been a paradigm shift in the choice of consumers about the use of pesticide residue-free foodstuffs produced organically. Due to the increasing awareness about the toxicity and environmental concerns of indiscriminate use of agrochemicals, the current need has been felt to adopt ecologically safer, organic and sustainable technologies to increase crop production. In this regard, the use of endophytic *Streptomyces* spp. as plant probiotics is gaining attention from plant researchers. *Streptomyces* spp. constitue a root-inhabiting group of bacteria that helps in the uptake and absorption of minerals from soil by the plant roots. Also, these bacteria have been found to release several types of metabolites, enzymes and osmolytes which help in conferring disease resistance and plant stress tolerance, as well as insect pest management. Through the production of PGRs, the *Streptomyces* spp. promotes plant growth and enhances their productivity. Besides this, *Streptomyces*-based nano-formulations have also been confirmed positive against various plant pests and pathogens, thus emphasizing their possible implications in ensuring sustainable crop resilience. These endophytes, on successful inoculation, reside inside the plant tissue and strengthen the crops' resilience against abiotic and biotic stresses. Since these bacteria confer stress management and enhance plant productivity in an ecologically safer manner, therefore, *Streptomyces* spp. can be future plant probiotics that can be explored as a complete bio-package for healthier and sustainable agriculture.

Hitherto, studies are limited about the delivery, uptake, colonization and molecular aspects of exogenously applied commercial endophytic *Streptomyces* spp. to the target crops. With the advancements in next-generation sequencing and metabolomics, efforts are being made to study the intricate network of molecular communication between the endophytic *Streptomyces* spp. and their host species. However, there are certain challenges associated with their widespread adoption in the agricultural fields. One of the major challenges is how these microbes will interact with soil and the rhizosphere of the target plants, and whether the commercial formulations of endophytic *Streptomyces* spp. are effectively delivered to the target host or not. Further, the absorption of *Streptomyces* spp. followed by their colonization in the target host plant is also a matter of concern. Moreover, comprehensive field trials are needed to validate the efficacy of the bio-formulation developed from endophytic *Streptomyces* sp., to evaluate their impact on crop productivity, environmental sustainability and ecological safety.

The successful application of these beneficial endophytic *Streptomyces* spp. will ensure lower rate of crop loss because of lower impact of pathogen and pest infestation, resulting in higher yield and improved quality of agricultural produce, which is less reliant on chemical-based agro-chemicals, thus meeting the eco-safe and organic agro-products as per the requirements of the health-driven consumers.

Further research to develop novel genetically engineered and ecologically safe endophytic *Streptomyces* strains with enhanced capacity of antimicrobial production and synthesis of PGRs, siderophores and enzymes, might help enhance crop resilience and productivity, thus ensuring global food security.

References

Adhilakshmi M, Latha P, Paranidharan V, Balachandar D, Ganesamurthy K, Velazhahan R (2014) Biological control of stem rot of groundnut (*Arachis hypogaea* L.) caused by *Sclerotium rolfsii* Sacc. with actinomycetes. Arch Phytopathol Plant Protect 47(3):298–311

Aggarwal N, Thind SK, Sharma S (2016) Role of secondary metabolites of actinomycetes in crop protection. In: Plant growth promoting actinobacteria: A new avenue for enhancing the productivity and soil fertility of grain legumes, pp 99–121

Akinsanya MA, Goh JK, Lim SP, Ting ASY (2015) Metagenomics study of endophytic bacteria in *Aloe vera* using next-generation technology. Genomics Data 6:159–163

Al-Huqail AA, El-Bondkly AMA (2022) Improvement of *Zea mays* L. growth parameters under chromium and arsenic stress by the heavy metal-resistant *Streptomyces* sp. NRC21696. Int J Environ Sci Technol 19(6):5301–5322

Alkaabi AK, Ramadan GA, Elddin AMT, El-Tarabily KA, AbuQamar SF (2022) The multifarious endophytic actinobacterial isolate, *Streptomyces tubercidicus* UAE1, combined with the seaweed biostimulant further promotes growth of *Avicennia marina*. Front Mar Sci 9:896461

Almuhayawi MS, Abdel-Mawgoud M, Al Jaouni SK, Almuhayawi SM, Alruhaili MH, Selim S, Abdelgawad H (2021) Bacterial endophytes as a promising approach to enhance the growth and accumulation of bioactive metabolites of three species of *Chenopodium* Sprouts. Plants 10(12): 2745

Al-Quwaie DA (2024) The role of *Streptomyces* species in controlling plant diseases: a comprehensive review. Australas Plant Pathol 53(1):1–14

Amelia-Yap ZH, Low VL, Saeung A, Ng FL, Chen CD, Hassandarvish P, Tan GYA, AbuBakar S, Azman AS (2023) Insecticidal activities of *Streptomyces* sp. KSF103 ethyl acetate extract against medically important mosquitoes and non-target organisms. Sci Rep 13(1):4

Arasu MV, Al-Dhabi NA, Saritha V, Duraipandiyan V, Muthukumar C, Kim SJ (2013) Antifeedant, larvicidal and growth inhibitory bioactivities of novel polyketide metabolite isolated from *Streptomyces* sp. AP-123 against *Helicoverpa armigera* and *Spodoptera litura*. BMC Microbiol 13:1–6

Arasu MV, Esmail GA, Al-Dhabi NA, Ponmurugan K (2016) Managing pests and diseases of grain legumes with secondary metabolites from actinomycetes. In: Plant growth promoting actinobacteria: A new avenue for enhancing the productivity and soil fertility of grain legumes, pp 83–98

Awla HK, Kadir J, Othman R, Rashid TS, Hamid S, Wong MY (2017) Plant growth-promoting abilities and biocontrol efficacy of *Streptomyces* sp. UPMRS4 against *Pyricularia oryzae*. Biol Control 112:55–63

Balakrishnan S, Santhanam P, Srinivasan M (2017) Larvicidal potency of marine actinobacteria isolated from mangrove environment against *Aedes aegypti* and *Anopheles stephensi*. J Parasit Dis 41:387–394

Barka EA, Vatsa P, Sanchez L, Gaveau-Vaillant N, Jacquard C, Klenk HP, Clément C, Ouhdouch Y, van Wezel GP (2016) Taxonomy, physiology, and natural products of Actinobacteria. Microbiol Mol Biol Rev 80(1):1–43

Bérdy J (2012) Thoughts and facts about antibiotics: where we are now and where we are heading. J Antibiot 65(8):385–395

Bhardwaj K, Abraham J, Kaur S (2020) Natural product as avermectins and milbemycins for agriculture perspectives. In: Natural Bioactive Products in Sustainable Agriculture, pp 259–271

Bose R, Dubey H, Samadder A, Basak S, Dutta S, Dey RK, Dixit AK, Mathapati S, Kumar Y. R. S, Babu G, Srikanth N, Acharya R, Roy S (2025) Role of bacterial endophytes in bioremediation, crop improvement and stress management. J Crop Health 77(5):161

Buzón-Durán L, Pérez-Lebeña E, Martín-Gil J, Sánchez-Báscones M, Martín-Ramos P (2020) Applications of *Streptomyces* spp. enhanced compost in sustainable agriculture. In: Biology of composts, pp 257–291

Cao P, Liu C, Sun P, Fu X, Wang S, Wu F, Wang X (2016) An endophytic *Streptomyces* sp. strain DHV3-2 from diseased root as a potential biocontrol agent against *Verticillium dahliae* and growth elicitor in tomato (*Solanum lycopersicum*). Antonie Van Leeuwenhoek 109:1573–1582

Cao H, Liu W, Chang Y, Chen H, Zhang Y, Xu W, Tao L (2019) 5′-Epi-SPA-6952A, a new insecticidal 24-membered macrolide produced by *Streptomyces diastatochromogenes* SSPRC-11339. Nat Prod Res 33(5):659–664

Cao P, Li C, Wang H, Yu Z, Xu X, Wang X, Zhao J, Xiang W (2020) Community structures and antifungal activity of root-associated endophytic actinobacteria in healthy and diseased cucumber plants and *Streptomyces* sp. HAAG3-15 as a promising biocontrol agent. Microorganisms 8(2):236

Cao L, Gao Y, Yu J, Niu S, Zeng J, Yao Q, Wang X, Bu Z, Xu T, Liu X, Zhu Y (2021) *Streptomyces hygroscopicus* OsiSh-2-induced mitigation of Fe deficiency in rice plants. Plant Physiol Biochem 158:275–283

Cerna-Chávez E, Rodríguez-Rodríguez JF, García-Conde KB, Ochoa-Fuentes YM (2024) Potential of *Streptomycesavermitilis*: a review on Avermectin production and its biocidal effect. Meta 14(7):374

Chaiharn M, Theantana T, Pathom-Aree W (2020) Evaluation of biocontrol activities of *Streptomyces* spp. against rice blast disease fungi. Pathogens 9(2):126

Chen Y, Shafi J, Li M, Fu D, Ji M (2018a) Insecticidal activity of endophytic actinomycetes isolated from *Azadirachta indica* against *Myzus persicae*. Arch Biol Sci 70(2):349–357

Chen Y, Zhou D, Qi D, Gao Z, Xie J, Luo Y (2018b) Growth promotion and disease suppression ability of a *Streptomyces* sp. CB-75 from banana rhizosphere soil. *Front Microbiol* 8:2704

Chen YH, Sung KY, Tuan SJ, Huang JW, Lin YH, Huang TP (2025) A *Streptomyces* agent for biocontrol of *Phytophthora* blight and its modulation of rhizosphere microbiomes in passion fruit. Plant Dis. https://doi.org/10.1094/PDIS-01-25-0089-RE

Colombo EM, Kunova A, Pizzatti C, Saracchi M, Cortesi P, Pasquali M (2019) Selection of an endophytic *Streptomyces* sp. strain DEF09 from wheat roots as a biocontrol agent against *Fusarium graminearum*. Front Microbiol 10:2356

Cui K, Xu T, Chen J, Yang H, Liu X, Zhuo R, Peng Y, Tang W, Wang R, Chen L, Zhu Y (2022) Siderophores, a potential phosphate solubilizer from the endophyte *Streptomyces* sp. CoT10, improved phosphorus mobilization for host plant growth and rhizosphere modulation. J Clean Prod 367:133110

Devi S, Verma J, Sohal SK, Manhas RK (2023a) Insecticidal potential of endophytic *Streptomyces* sp. against *Zeugodacus cucurbitae* (Coquillett) (Diptera: Tephritidae) and biosafety evaluation. Toxicon 233:107246

Devi S, Sharma M, Manhas RK (2023b) Purification and biological analysis of antimicrobial compound produced by an endophytic *Streptomyces* sp. Sci Rep 13(1):15248

El-Akshar EA, Elmeihy RM, Tewfike TA, Abou-Aly HE (2022) Endophytic *Streptomyces enissocaesilis* as a nematicidal and biostimulant agent. Egypt Acad J Biol Sci G Microbiol 14(2):123–133

El-Tarabily KA, Ramadan GA, Elbadawi AA, Hassan AH, Tariq S, Ghazal EW, Abo Gamar MI, AbuQamar SF (2021) The marine endophytic polyamine-producing *Streptomycesmutabilis* UAE1 isolated from extreme niches in the Arabian Gulf promotes the performance of mangrove (*Avicennia marina*) seedlings under greenhouse conditions. Front Mar Sci 8:710200

Fouda A, Hassan SED, Abdo AM, El-Gamal MS (2020) Antimicrobial, antioxidant and larvicidal activities of spherical silver nanoparticles synthesized by endophytic *Streptomyces* spp. Biol Trace Elem Res 195(2):707–724

Gao Y, Ning Q, Yang Y, Liu Y, Niu S, Hu X, Pan H, Bu Z, Chen N, Guo J, Zhu Y (2021) Endophytic *Streptomyces hygroscopicus* OsiSh-2-mediated balancing between growth and disease resistance in host rice. MBio 12(4):10–1128

Gong Y, Chen LJ, Pan SY, Li XW, Xu MJ, Zhang CM, Xing K, Qin S (2020) Antifungal potential evaluation and alleviation of salt stress in tomato seedlings by a halotolerant plant growth-promoting actinomycete *Streptomyces* sp. KLBMP5084. Rhizosphere 16:100262

Gopalakrishnan S, Ganeshkumar P (2013) Systematic reviews and meta-analysis: understanding the best evidence in primary healthcare. J Family Med Prim Care 2(1):9–14

Gopalakrishnan S, Rajendran V, Arumugam S, Sharma HC, Vadlamudi S, Bhimineni RK, Gonzalez SV, Melø TM, Simic N (2016) Insecticidal activity of a novel fatty acid amide derivative from *Streptomyces* species against *Helicoverpa armigera*. Nat Prod Res 30(24): 2760–2769

Gopalakrishnan S, Srinivas V, Prasanna SL (2020) Streptomyces. In: Amaresan N, Annapurna K, Sankaranarayanan A, Senthil Kumar M, Kumar K (eds) Beneficial microbes in agro-ecology: bacteria and fungi, pp 55–71

Goswami A, Roy I, Sengupta S, Debnath N (2010) Novel applications of solid and liquid formulations of nanoparticles against insect pests and pathogens. Thin Solid Films 519(3): 1252–1257

Goudjal Y, Toumatia O, Sabaou N, Barakate M, Mathieu F, Zitouni A (2013) Endophytic actinomycetes from spontaneous plants of Algerian Sahara: indole-3-acetic acid production and tomato plants growth promoting activity. World J Microbiol Biotechnol 29(10):1821–1829

Goudjal Y, Toumatia O, Yekkour A, Sabaou N, Mathieu F, Zitouni A (2014) Biocontrol of *Rhizoctonia solani* damping-off and promotion of tomato plant growth by endophytic actinomycetes isolated from native plants of Algerian Sahara. Microbiol Res 169(1):59–65

Goudjal Y, Zamoum M, Sabaou N, Mathieu F, Zitouni A (2016) Potential of endophytic *Streptomyces* spp. for biocontrol of *Fusarium* root rot disease and growth promotion of tomato seedlings. Biocontrol Sci Tech 26(12):1691–1705

Hardoim PR, Van Overbeek LS, Berg G, Pirttilä AM, Compant S, Campisano A, Döring M, Sessitsch A (2015) The hidden world within plants: ecological and evolutionary considerations for defining functioning of microbial endophytes. Microbiol Mol Biol Rev 79(3):293–320

Harikrishnan H, Shanmugaiah V, Balasubramanian N (2014) Optimization for production of indole acetic acid (IAA) by plant growth promoting *Streptomyces* sp. VSMGT1014 isolated from rice rhizosphere. Int J Curr Microbiol Appl Sci 3(8):158–171

Hassan N, Nakasuji S, Elsharkawy MM, Naznin HA, Kubota M, Ketta H, Shimizu M (2017) Biocontrol potential of an endophytic *Streptomyces* sp. strain MBCN152-1 against *Alternaria brassicicola* on cabbage plug seedlings. Microbes Environ 32(2):133–141

Htwe AZ, Yamakawa T (2016) Low-density co-inoculation with *Bradyrhizobium japonicum* SAY3-7 and *Streptomyces griseoflavus* P4 promotes plant growth and nitrogen fixation in soybean cultivars. Am J Plant Sci 7(12):1652–1661

Htwe AZ, Moh SM, Moe K, Yamakawa T (2018) Effects of co-inoculation of *Bradyrhizobium japonicum* SAY3-7 and *Streptomyces griseoflavus* P4 on plant growth, nodulation, nitrogen fixation, nutrient uptake, and yield of soybean in a field condition. Soil Sci Plant Nutr 64(2): 222–229

Intra B, Mungsuntisuk I, Nihira T, Igarashi Y, Panbangred W (2011) Identification of actinomycetes from plant rhizospheric soils with inhibitory activity against *Colletotrichum* spp., the causative agent of anthracnose disease. BMC Res Notes 4:1–9

Jaemsaeng R, Jantasuriyarat C, Thamchaipenet A (2018a) Molecular interaction of 1-aminocyclopropane-1-carboxylate deaminase (ACCD)-producing endophytic *Streptomyces* sp. GMKU 336 towards salt-stress resistance of *Oryza sativa* L. cv. KDML105. *Sci Rep 8*(1): 1950

Jaemsaeng R, Jantasuriyarat C, Thamchaipenet A (2018b) Positive role of 1-aminocyclopropane-1-carboxylate deaminase-producing endophytic *Streptomyces* sp. GMKU 336 on flooding resistance of mung bean. Agric Nat Resour 52(4):330–334

Jog R, Pandya M, Nareshkumar G, Rajkumar S (2014) Mechanism of phosphate solubilization and antifungal activity of *Streptomyces* spp. isolated from wheat roots and rhizosphere and their application in improving plant growth. Microbiology 160(4):778–788

Junaid JM, Dar NA, Bhat TA, Bhat AH, Bhat MA (2013) Commercial biocontrol agents and their mechanism of action in the management of plant pathogens. Int J Modern Plant Anim Sci 1(2): 39–57

Kanini GS, Katsifas EA, Savvides AL, Karagouni AD (2013) *Streptomyces rochei* ACTA1551, an indigenous Greek isolate studied as a potential biocontrol agent against *Fusarium oxysporum* f. sp *lycopersici*. BioMed Res Int 2013(1):387230

Kato JY, Funa N, Watanabe H, Ohnishi Y, Horinouchi S (2007) Biosynthesis of γ-butyrolactone autoregulators that switch on secondary metabolism and morphological development in *Streptomyces*. Proc Natl Acad Sci 104(7):2378–2383

Kaur T, Vasudev A, Sohal SK, Manhas RK (2014) Insecticidal and growth inhibitory potential of *Streptomyceshydrogenans* DH16 on major pest of India, *Spodoptera litura* (Fab.) (Lepidoptera: Noctuidae). BMC Microbiol 14:1–9

Khamna S, Yokota A, Lumyong S (2009) Actinomycetes isolated from medicinal plant rhizosphere soils: diversity and screening of antifungal compounds, indole-3-acetic acid and siderophore production. World J Microbiol Biotechnol 25:649–655

Kim DR, Kwak YS (2023) Endophytic *Streptomyces* population induced by L-glutamic acid enhances plant resilience to abiotic stresses in tomato. Front Microbiol 14:1180538

Kim YS, Umurzokov M, Cho KM, Choi JS, Park KW (2022a) Insecticidal characteristics and structural identification of the potential active compounds from *Streptomyces* sp. KR0006: Strain improvement through mutagenesis. PLoS One 17(9):e0274766

Kim JY, Choi JY, Park DH, Park MG, Wang M, Kim HJ, Kim SH, Lee HY, Je YH (2022b) Juvenile hormone antagonistic activity of secondary metabolites from *Streptomyceslactacystinicus* and their insecticidal activity against *Plutella xylostella*. J Asia-Pac Entomol 25(2):101870

Kruasuwan W, Lohmaneeratana K, Munnoch JT, Vongsangnak W, Jantrasuriyarat C, Hoskisson PA, Thamchaipenet A (2023) Transcriptome landscapes of salt-susceptible rice cultivar IR29 associated with a plant growth promoting endophytic *Streptomyces*. Rice 16(1):6

Le XH, Ballard RA, Franco CM (2016) Effects of endophytic *Streptomyces* and mineral nitrogen on Lucerne (*Medicago sativa* L.) growth and its symbiosis with rhizobia. Plant Soil 405(1):25–34

Li J, Zhao GZ, Chen HH, Wang HB, Qin S, Zhu WY, Xu LH, Jiang CL, Li WJ (2008) Antitumour and antimicrobial activities of endophytic *Streptomycetes* from pharmaceutical plants in rainforest. Lett Appl Microbiol 47(6):574–580

Lin L, Ge HM, Yan T, Qin YH, Tan RX (2012) Thaxtomin A-deficient endophytic *Streptomyces* sp. enhances plant disease resistance to pathogenic *Streptomyces scabies*. Planta 236:1849–1861

Liu D, Yan R, Fu Y, Wang X, Zhang J, Xiang W (2019) Antifungal, plant growth-promoting, and genomic properties of an endophytic actinobacterium *Streptomyces* sp. NEAU-S7GS2. Front Microbiol 10:2077

Liu H, Li J, Singh BK (2024a) Harnessing co-evolutionary interactions between plants and *Streptomyces* to combat drought stress. Nat Plants 10(8):1159–1171

Liu F, Wang N, Wang Y, Yu Z (2024b) The insecticidal activity of secondary metabolites produced by *Streptomyces* sp. SA61 against *Trialeurodes vaporariorum* (Hemiptera: Aleyrodidae). *Microorganisms 12*(10):2031

Lyu A, Liu H, Che H, Yang L, Zhang J, Wu M, Chen W, Li G (2017) Reveromycins A and B from *Streptomyces* sp. 3–10: antifungal activity against plant pathogenic fungi *in vitro* and in a strawberry food model system. Front Microbiol 8:550

Manasa M, Kambar Y, Pallavi S, Vivek MN, Onkarappa R, Kekuda PTR (2013) Biocontrol potential of *Streptomyces* species against *Fusarium oxysporum* f. sp. *zingiberi* (causal agent of rhizome rot of ginger). J Adv Sci Res 4(04):1–3

Manullang W, Chuang HW (2020) *Streptomyces* sp. mitigates abiotic stress response and promotes plant growth. J Plant Protect Res 60(3):263–274

Moussa AY (2024) *Streptomyces* endophytes in edible plants: new insights into their chemistry and health benefits. Chem Biodivers 21(10):e202400888

Myo EM, Ge B, Ma J, Cui H, Liu B, Shi L, Jiang M, Zhang K (2019) Indole-3-acetic acid production by *Streptomyces fradiae* NKZ-259 and its formulation to enhance plant growth. BMC Microbiol 19(1):155

Nazari MT, Schommer VA, Braun JCA, dos Santos LF, Lopes ST, Simon V, Machado BS, Ferrari V, Colla LM, Piccin JS (2023) Using *Streptomyces* spp. as plant growth promoters and biocontrol agents. Rhizosphere 27:100741

Nicolle C (2023) Molecular mechanisms of the interaction of a rhizospheric Streptomyces sp. *strain with plant roots* (Doctoral dissertation, Université Paul Sabatier-Toulouse III)

Nimaichand S, Tamrihao K, Yang LL, Zhu WY, Zhang YG, Li L, Tang SK, Ningthoujam DS, Li WJ (2013) *Streptomyces hundungensis* sp. *nov*., a novel actinomycete with antifungal activity and plant growth promoting traits. J Antibiot 66(4):205–209

Niu S, Gao Y, Zi H, Liu Y, Liu X, Xiong X, Yao Q, Qin Z, Chen N, Guo L, Zhu Y (2022) The osmolyte-producing endophyte *Streptomyces albidoflavus* OsiLf-2 induces drought and salt tolerance in rice *via* a multi-level mechanism. Crop J 10(2):375–386

Olanrewaju OS, Babalola OO (2019) *Streptomyces*: implications and interactions in plant growth promotion. Appl Microbiol Biotechnol 103(3):1179–1188

Oyedoh OP, Yang W, Dhanasekaran D, Santoyo G, Glick BR, Babalola OO (2023) Sustainable agriculture: rare-actinomycetes to the rescue. Agronomy 13(3):666

Palaniyandi SA, Yang SH, Zhang L, Suh JW (2013) Effects of actinobacteria on plant disease suppression and growth promotion. Appl Microbiol Biotechnol 97:9621–9636

Pandey SS, Jain R, Bhardwaj P, Thakur A, Kumari M, Bhushan S, Kumar S (2022) Plant probiotics–endophytes pivotal to plant health. Microbiol Res 263:127148

Passari AK, Chandra P, Mishra VK, Leo VV, Gupta VK, Kumar B, Singh BP (2016) Detection of biosynthetic gene and phytohormone production by endophytic actinobacteria associated with *Solanum lycopersicum* and their plant-growth-promoting effect. Res Microbiol 167(8):692–705

Patel JK, Madaan S, Archana G (2018) Antibiotic producing endophytic *Streptomyces* spp. colonize above-ground plant parts and promote shoot growth in multiple healthy and pathogen-challenged cereal crops. Microbiol Res 215:36–45

Patil HJ, Srivastava AK, Singh DP, Chaudhari BL, Arora DK (2011) Actinomycetes mediated biochemical responses in tomato (*Solanum lycopersicum*) enhances bioprotection against *Rhizoctonia solani*. Crop Prot 30(10):1269–1273

Poomthongdee N, Duangmal K, Pathom-aree W (2015) Acidophilic actinomycetes from rhizosphere soil: diversity and properties beneficial to plants. J Antibiot 68(2):106–114

Pupakdeepan W, Termsung N, Ruangwong OU, Kunasakdakul K (2023) Plant growth promoting and colonization of endophytic *Streptomyces albus* CINv1 against strawberry anthracnose. Horticulturae 9(7):766

Qin S, Xing K, Jiang JH, Xu LH, Li WJ (2011) Biodiversity, bioactive natural products and biotechnological potential of plant-associated endophytic actinobacteria. Appl Microbiol Biotechnol 89:457–473

Qin S, Feng WW, Wang TT, Ding P, Xing K, Jiang JH (2017) Plant growth-promoting effect and genomic analysis of the beneficial endophyte *Streptomyces* sp. KLBMP 5084 isolated from halophyte *Limonium sinense*. Plant Soil 416:117–132

Samri SE, Baz M, Ghalbane I, El Messoussi S, Zitouni A, El Meziane A, Barakate M (2017) Insecticidal activity of a Moroccan strain of *Streptomyces phaeochromogenes* LD-37 on larvae, pupae and adults of the Mediterranean fruit fly, *Ceratitis capitata* (Diptera: Tephritidae). Bull Entomol Res 107(2):217–224

Sangmanee P, Bhromsiri A, Akarapisan A (2009) The potential of endophytic actinomycetes, (*Streptomyces* sp.) for the biocontrol of powdery mildew disease in sweet pea (*Pisum sativum*). As J Food Ag-Ind 93:e8

Sarin S, Nakaew N, Chidburee A (2018) Effects of growth regulators produced by *Methylobacterium radiotolerans* Ed5-9 and crude antimicrobial agents extracted from

Streptomyces TM32 on tissue culture of *Gymnema inodorum* (Lour.) Decne. Asian Health Sci Technol Rep 26(3):52–62

Sathya A, Vijayabharathi R, Gopalakrishnan S (2017) Plant growth-promoting actinobacteria: a new strategy for enhancing sustainable production and protection of grain legumes. 3 Biotech 7: 1–10

Seipke RF, Kaltenpoth M, Hutchings MI (2012) *Streptomyces* as symbionts: an emerging and widespread theme? FEMS Microbiol Rev 36(4):862–876

Shan W, Zhou Y, Liu H, Yu X (2018) Endophytic actinomycetes from tea plants (*Camellia sinensis*): isolation, abundance, antimicrobial, and plant-growth-promoting activities. Biomed Res Int 2018(1):1470305

Shimizu M, Yazawa S, Ushijima Y (2009) A promising strain of endophytic *Streptomyces* sp. for biological control of cucumber anthracnose. J Gen Plant Pathol 75:27–36

Singh SP, Gaur R (2017) Endophytic *Streptomyces* spp. underscore induction of defense regulatory genes and confers resistance against *Sclerotium rolfsii* in chickpea. Biol Control 104:44–56

Singh NA, Narang J, Garg D, Jain V, Payasi D, Suleman S, Swami RK (2023) Nanoparticles synthesis *via* microorganisms and their prospective applications in agriculture. Plant Nano Biol 5:100047

Sivalingam P, Easwaran M, Ganapathy D, Basha SF, Poté J (2024) Endophytic *Streptomyces*: an underexplored source with potential for novel natural drug discovery and development. Arch Microbiol 206(11):442

Srinivas V, Naresh N, Pratyusha S, Ankati S, Govindaraj M, Gopalakrishnan S (2022) Exploring plant growth-promoting Streptomyces spp. for yield and nutrition traits in pearl millet hybrids. Crop Pasture Sci 73(5):484–493

Srividya S, Thapa A, Bhat DV, Golmei K, Dey N (2012) *Streptomyces* sp. 9p as effective biocontrol against chilli soilborne fungal phytopathogens. Eur J Exp Biol 2(1):163–173

Suárez-Moreno ZR, Vinchira-Villarraga DM, Vergara-Morales DI, Castellanos L, Ramos FA, Guarnaccia C, Degrassi G, Venturi V, Moreno-Sarmiento N (2019) Plant-growth promotion and biocontrol properties of three *Streptomyces* spp. isolates to control bacterial rice pathogens. Front Microbiol 10:290

Sudheer NS, Biju IF, Balasubramanian CP, Panigrahi A, Kumar TS, Kumar S, Mandal B, Das S, De D (2024) Probiotic potential of a novel endophytic *Streptomyces griseorubens* CIBA-NS1 isolated from *Salicornia* sp. against *Vibrio campbellii* infection in shrimp. Microb Pathog 191:106677

Suman A, Yadav AN, Verma P (2016) Endophytic microbes in crops: diversity and beneficial impact for sustainable agriculture. In: Microbial inoculants in sustainable agricultural productivity: Vol. 1: research perspectives, pp 117–143

Supong K, Thawai C, Choowong W, Kittiwongwattana C, Thanaboripat D, Laosinwattana C, Koohakan P, Parinthawong N, Pittayakhajonwut P (2016) Antimicrobial compounds from endophytic *Streptomyces* sp. BCC72023 isolated from rice (*Oryza sativa* L.). Res Microbiol 167(4):290–298

Suwitchayanon P, Chaipon S, Chaichom S, Kunasakdakul K (2018) Potentials of *Streptomyces rochei* ERY1 as an endophytic actinobacterium inhibiting damping-off pathogenic fungi and growth promoting of cabbage seedling. Chiang Mai J Sci 45:692–700

Tanvir R, Sajid I, Hasnain S (2014) Larvicidal potential of Asteraceae family endophytic actinomycetes against *Culex quinquefasciatus* mosquito larvae. Nat Prod Res 28(22): 2048–2052

Thenappan DP, Pandey R, Hada A, Jaiswal DK, Chinnusamy V, Bhattacharya R, Annapurna K (2024) Physiological basis of plant growth promotion in rice by rhizosphere and Endosphere associated *Streptomyces* isolates from India. Rice 17(1):60

Tian W, Ge Y, Liu X, Dou G, Ma Y (2019) Identification and characterization of *Populus* microRNAs in response to plant growth-promoting endophytic *Streptomyces* sp. SSD49. World J Microbiol Biotechnol 35(7):97

Trang NNP, Van Chuong N (2025) The enhancement of soil fertility and baby maize output by *Streptomyces panayensis* and vermicompost. Eurasian J Soil Sci 14(2):140–148

Trujillo ME, Riesco R, Benito P, Carro L (2015) Endophytic actinobacteria and the interaction of *Micromonospora* and nitrogen fixing plants. Front Microbiol 6:1341

Van der Meij A, Worsley SF, Hutchings MI, van Wezel GP (2017) Chemical ecology of antibiotic production by actinomycetes. FEMS Microbiol Rev 41(3):392–416

Verma VC, Singh SK, Prakash S (2011) Bio-control and plant growth promotion potential of siderophore producing endophytic *Streptomyces* from *Azadirachta indica* A. Juss. Basic Microbiol 51(5):550–556

Viaene T, Langendries S, Beirinckx S, Maes M, Goormachtig S (2016) *Streptomyces* as a plant's best friend? FEMS Microbiol Ecol 92(8):fiw119

Villafañe DL, Maldonado RA, Bianchi JS, Kurth D, Gramajo H, Chiesa MA, Rodríguez E (2024) *Streptomyces* N2A, an endophytic actinobacteria that promotes soybean growth and increases yield and seed quality under field conditions. Plant Sci 343:112073

Vurukonda SSKP, Giovanardi D, Stefani E (2018) Plant growth promoting and biocontrol activity of *Streptomyces* spp. as endophytes. Int J Mol Sci 19(4):952

Vurukonda SSKP, Giovanardi D, Stefani E (2021) Growth promotion and biocontrol activity of Endophytic *Streptomyces* spp. In: Prime archives in molecular sciences, vol 1, 2nd edn, pp 1–55

Wang L, Vining LC (2003) Control of growth, secondary metabolism and sporulation in *Streptomyces venezuelae* ISP5230 by jadW 1, a member of the afsA family of γ-butyrolactone regulatory genes. Microbiology 149(8):1991–2004

Worsley SF, Newitt J, Rassbach J, Batey SF, Holmes NA, Murrell JC, Wilkinson B, Hutchings MI (2020) *Streptomyces* endophytes promote host health and enhance growth across plant species. Appl Environ Microbiol 86(16):e01053–e01020

Yan YS, Xia HY (2021) Recent advances in the research of milbemycin biosynthesis and regulation as well as strategies for strain improvement. Arch Microbiol 203(10):5849–5857

Yandigeri MS (2021) Role of actinomycetes in insect pest and plant disease management. In: Biopesticides in horticultural crops. CRC Press, pp 152–172

Yandigeri MS, Malviya N, Solanki MK, Shrivastava P, Sivakumar G (2015) Chitinolytic *Streptomyces vinaceusdrappus* S5MW2 isolated from Chilika lake, India enhances plant growth and biocontrol efficacy through chitin supplementation against *Rhizoctonia solani*. World J Microbiol Biotechnol 31:1217–1225

Yang LY, Wang JD, Zhang J, Xue CY, Zhang H, Wang XJ, Xiang WS (2013) New nemadectin congeners with acaricidal and nematocidal activity from *Streptomyces microflavus* neau3 Y-3. Bioorg Med Chem Lett 23(20):5710–5713

Yang Z, Qiao Y, Konakalla NC, Strøbech E, Harris P, Peschel G, Agler-Rosenbaum M, Weber T, Andreasson E, Ding L (2023) *Streptomyces* alleviate abiotic stress in plant by producing pteridic acids. Nat Commun 14(1):7398

Yoolong S, Kruasuwan W, Thanh Phạm HT, Jaemsaeng R, Jantasuriyarat C, Thamchaipenet A (2019) Modulation of salt tolerance in Thai jasmine rice (*Oryza sativa* L. cv. KDML105) by *Streptomyces venezuelae* ATCC 10712 expressing ACC deaminase. Sci Rep 9(1):1275

Yuan WM, Crawford DL (1995) Characterization of *Streptomyces lydicus* WYEC108 as a potential biocontrol agent against fungal root and seed rots. Appl Environ Microbiol 61(8):3119–3128

Zambri MP, Williams MA, Elliot MA (2022) How *Streptomyces* thrive: advancing our understanding of classical development and uncovering new behaviors. In: Advances in microbial physiology, vol 80. Academic, pp 203–236

Zhang Q, Yong D, Zhang Y, Shi X, Li B, Li G, Liang W, Wang C (2016) *Streptomyces rochei* A-1 induces resistance and defense-related responses against *Botryosphaeria dothidea* in apple fruit during storage. Postharvest Biol Technol 115:30–37

Zhang L, Liu Z, Wang Y, Zhang J, Wan S, Huang Y, Yun T, Xie J, Wang W (2022) Biocontrol potential of endophytic *Streptomyces malaysiensis* 8ZJF-21 from medicinal plant against banana *Fusarium* wilt caused by *Fusarium oxysporum* f. sp. *cubense* tropical race 4. Front Plant Sci 13:874819

Plant Growth-Promoting Streptomycetes: Mechanisms and Applications

10

Satish Koundal, Avinash Sharma, Jaya Verma, and Sapna Devi

Abstract

Streptomycetes are a diverse group of filamentous, Gram-positive actinobacteria, that have been emerged as promising plant growth-promoting microorganisms either by directly enhancing plant growth or by protecting crops from pathogens. In the rhizosphere, *Streptomyces* species promote beneficial symbiotic relationships that improve soil health and enhance nutrient uptake. Further, these bacteria are known for their ability to synthesize a wide variety of bioactive compounds through secondary metabolism. By fixing nitrogen, solubilizing phosphates, producing siderophores and synthesizing phytohormones, these actinobacteria contribute to plant health. Approximately one third of all antimicrobial compounds have been reported from *Streptomyces* species. This capability of *Streptomyces* spp. has attracted considerable attention from researchers seeking solutions for sustainable agriculture. *Streptomyces* have been reported to increase crop resistance to biotic and abiotic stressors, thereby lowering the need for chemical inputs. When *Streptomyces* are used as biofertilizers and biopesticides, they promote environment friendly farming practices while improving crop production and quality. However, the challenges such as the selection of highly efficient strains, optimization for large-scale production and their stability and efficacy under diverse field conditions continue to limit their widespread application. This chapter aims to provide an overview of

S. Koundal · A. Sharma
Department of Microbiology, DAV University, Jalandhar, Punjab, India

J. Verma
Department of Microbiology, Guru Nanak Dev University, Amritsar, Punjab, India

S. Devi (✉)
Department of Microbiology, School of Bioengineering and Biosciences, Lovely Professional University, Phagwara, Punjab, India

T. Kaur et al. (eds.), *Streptomyces: Biological Candidates for Sustainable Agriculture*,
https://doi.org/10.1007/978-981-95-5803-2_10

Streptomyces in light of their potential as plant growth-promoting bacteria (PGPBs), their secondary metabolism, plant interactions and underlying molecular mechanisms.

Keywords

Streptomyces · Biofertilizers · Biotic stress · Sustainability · Plant growth promotion

10.1 Introduction

Since the start of the green revolution in the early 1970s due to the use of chemical fertilizers and pesticides, agricultural productivity increased significantly to feed the world's struggling economy (Hamdan et al. 2022). This time period marked as a crucial turning point in global food production, ensuring food security for a rapidly increasing population (Oluwole et al. 2023). The agricultural revolutions led to significant changes in agricultural methods, emphasizing use of high-yield seed varieties, advanced irrigation techniques, mechanization and, more specifically, use of synthetic fertilizers and pesticides (Prabhakar and Brar 2024). This led to the widespread use of agrochemicals, as the majority of agricultural activities today are conducted using synthetic chemicals. However, their continuous and indiscriminate use has resulted in ecological imbalance, soil and water pollution, contamination of the food chain along with emergence of resistant plant pests (Zhou et al. 2024). Furthermore, long-term exposure to pesticide residues has resulted in ill-effects on human health, including neurological disorders, cancer and endocrine disruption (Kim et al. 2017).

Excessive use of chemical fertilizers has also led to reduced soil fertility due to depletion of organic matter, and destruction of beneficial soil microorganisms (Baweja et al. 2020; Pahalvi et al. 2021). All these concerns have prompted scientists to explore safe, environment-friendly natural farming practices. These natural farming methods place a strong emphasis on using renewable resources, controlling plant pests biologically and comprehending plant-microbe interactions in order to increase crop productivity and plant health (Akanmu et al. 2023; Badiyal et al. 2024). Thus, to reduce the negative impacts of the chemical fertilizers and to promote sustainable agriculture practices, plant growth-promoting microorganisms (PGPMs) have been emerged as a promising alternative to agrochemicals.

There are many beneficial microorganisms known as plant PGPMs, which colonize plant roots and promote plant growth through a multitude of mechanisms. They fix nitrogen, solubilize phosphate, produce plant hormones and suppress pathogens through the secretion of antimicrobial compounds, volatile organic compounds and lytic enzymes (Ansabayeva et al. 2025). PGPMs can significantly reduce the dependence on chemical inputs by enhancing nutrient uptake and improving plants' tolerance to stress, resulting in lower production cost and reduced environmental pollution. Among PGPMs, *Streptomyces*, a genus belonging to the phylum actinobacteria holds significant potential as plant growth promoters. *Streptomyces* species are Gram-positive filamentous, non-motile, aerobic, catalase-positive, non-acid-fast bacteria and are similar to fungi in their mycelia appearance and behaviour (Law et al. 2019). *Streptomyces* genus is a well-known prolific antibiotic-producer, discovered so far. They have a high GC content of 69–78% in their genomes (Mitousis et al. 2020). In natural ecosystems, they are widely distributed, and inhabit and thrive in a wide range of environments, including soil, compost, decaying vegetation, freshwater and marine ecosystems; and even in association with plants, *Streptomyces* can colonize the rhizosphere of a plant, and the internal plant tissues as endophytes, particularly in the root system and thus establish a symbiotic relationship without negatively affecting the host (Viaene et al. 2016). During colonization by streptomycetes, the host plant is not altered morphologically or physiologically, thereby distinguishing them from harmful pathogenic microorganisms. Plants benefit from these endophytic interactions in many ways, including improved nutrient uptake, increased resistance to biotic and abiotic stresses and protection against soil-borne pathogens by secreting bioactive compounds (Rey and Dumas 2017).

Ecologically, *Streptomyces* play a crucial role in nutrient cycling by decomposing complex organic substances like cellulose, lignin and chitin into simpler forms that plants can easily absorb (Javed et al. 2021). The decomposition process enriches the soil with nutrients, as well as fosters the growth of beneficial microorganisms that improve overall health of soil (Beroigui and Errachidi 2023). *Streptomyces* species are also capable of influencing the structures of rhizospheric and endospheric microbial communities by producing a variety of secondary metabolites, which possess antimicrobial properties (Vurukonda et al. 2018). These compounds can reduce the disease incidence by suppressing the growth of phytopathogenic bacteria and fungi. Even under different stress conditions, viz., drought, salinity and heavy metal contamination, *Streptomyces* species have shown enhanced rate of seed germination, increased crop yields and improved plant health. As a result of antagonistic and plant growth-promoting activities, streptomycetes are great candidates for the development of safe, effective and eco-friendly biopesticides and biofertilizers, respectively. Moreover, the growing demand for sustainable agriculture has resulted in increased research on development of *Streptomyces*-based bioinoculants (Samantaray et al. 2024). These findings highlight the potential of *Streptomyces* as a natural alternative to chemical inputs in modern agriculture.

10.2 Mechanisms of Plant Growth Promotion by *Streptomyces*

The *Streptomyces* genus is widely recognized for its ability to synthesize diverse bioactive metabolites. Besides its antibiotic-producing ability, certain species within this genus function as potent plant growth-promoting rhizobacteria (Fig. 10.1). These beneficial microbes contribute to plant development through both direct and indirect mechanisms. They directly enhance growth by synthesizing phytohormones like indole-3-acetic acid (IAA), gibberellins and cytokinins which stimulate root

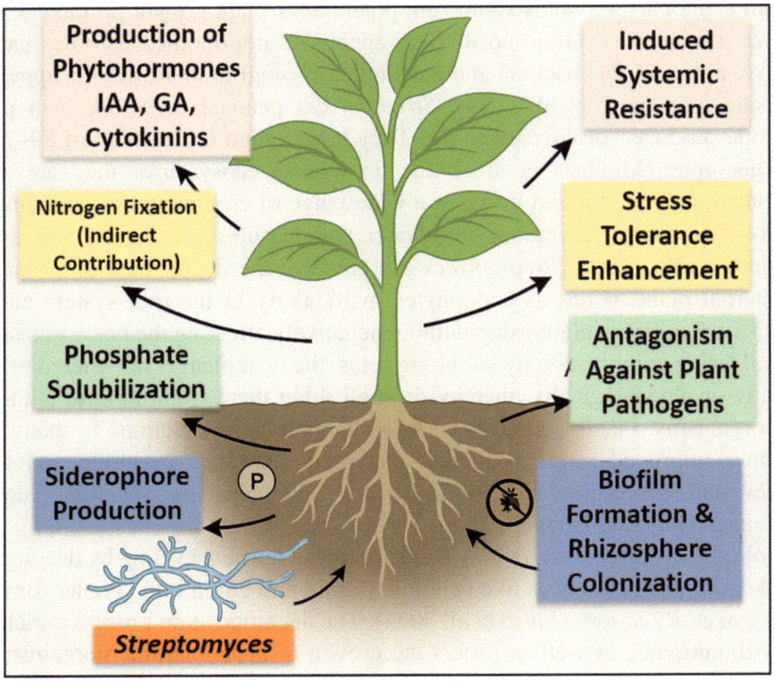

Fig. 10.1 Mechanism of plant-growth promotion by *Streptomyces*

elongation, cell division and nutrient absorption (Table 10.1). Streptomycetes are not classical nitrogen fixers, but they support nitrogen availability by mineralizing organic nitrogen and facilitating the activity of associative nitrogen-fixing microbes. They use phosphatases and organic acids to solubilize phosphate, enhancing the bioavailability of phosphorus. Additionally, they produce siderophores (iron sequestering molecules) that chelate iron from the soil, enhancing plant uptake of iron while limiting pathogen access. *Streptomyces* function as biocontrol agents by producing volatile chemical compounds, cell wall degrading hydrolytic enzymes and antibiotics that efficiently inhibit phytopathogens. Furthermore, by activating defence mechanisms and enhancing resistance to biotic and abiotic stressors, they induce systemic resistance (ISR) in plants. They mitigate the stresses by the production of antioxidant enzymes and reduction of ethylene levels via 1-Aminocyclopropane-1-carboxylate (ACC) deaminase. Furthermore, due to root colonization and filamentous structure, streptomycetes improve nutrient exchange in the rhizosphere and create biofilms. Thus, *Streptomyces* are useful as biofertilizers and bioprotective agents for sustainable agriculture because of their many diverse characteristics (Chouyia et al. 2022).

Table 10.1 Plant growth-promoting activities of *Streptomyces* species

Streptomyces species	PGP activity	Treated plant	Reference
Streptomyces sp. *GMKU 3100*	Production of siderophore	*Oryza sativa* and *Vigna radiate*	Rungin et al. (2012)
Streptomyces mhcr0816	Phytase, siderophore and IAA production, phosphate solubilization	*Triticum aestivum*	Jog et al. (2014)
Streptomyces albus	IAA production	*Piper nigrum*	Bhai et al. (2016)
Streptomyces manipurensis	Siderophore production, ACC deaminase activity, production of antifungal metabolites and IAA production	*Oryza sativa*	Tamreihao et al. (2019)
Streptomyces cameroonensis sp. *nov.*	Phosphate solubilization, ACC deaminase, production of siderophores and IAA	*Theobroma cacao*	Boudjeko et al. (2017)
Streptomyces rochei	IAA, siderophore, ammonia and AAC deaminase production and phosphate solubilization	*Triticum aestivum*	Singh et al. (2019)
Streptomyces palmae	Antifungal activity	*Oryza sativa*	Chaiharn et al. (2020)
Streptomyces sp.	Phosphate solubilization and IAA production	*Arabidopsis thaliana*	Manullang and Chuang (2020)
Streptomyces strain N2	Antifungal activity against *Gaeumannomyces graminis* var. *tritici*	*Triticum aestivum*	Worsley et al. (2020)
Streptomyces iakyrus, *Streptomyces thinghirensis*	Phosphate solubilization and IAA production	*Triticum durum*	Djebaili et al. (2021)
Streptomyces sp. PERM2	Solubilization of phosphate, siderophore and AAC deaminase production	*Xanthosoma sagittifolium*	Djuidje et al. (2022)
Streptomyces sp. SP5	IAA biosynthesis, production of antifungal metabolites	*Solanum lycopersicum*	Devi et al. (2022)
Streptomyces thinghirensis, *Streptomyces* sp. strain HM8	IAA and siderophore production, and phosphate solubilization	*Cucumis sativus* (cucumber)	Omar et al. (2022)
Streptomyces albidoflavus St-220	Phosphate solubilization, siderophore production, IAA production, antifungal activity and nitrogen fixation activity	*Salvia miltiorrhiza*	Du et al. (2022)
Streptomyces sp. STD57	IAA, antibacterial and herbicidal	Wheat, rice and tomato	He et al. (2024)
Streptomyces pratensis	Production of IAA, ACC and siderophores, phosphate solubilization and nitrogen fixation	Wheat	Tian et al. (2024)

(continued)

Table 10.1 (continued)

Streptomyces species	PGP activity	Treated plant	Reference
Streptomyces sp.	Nitrogen fixation and IAA production	Capsicum	Li et al. (2024)
Streptomyces sp. N2A	Phosphate solubilization and siderophores production	Soybean	Villafañe et al. (2024)
Streptomyces sp. PR69	Nitrogen fixation and siderophores production	Bell pepper	López-Reyes et al. (2024)
Streptomyces sp. GD-4	Ammonia production	*Lolium perenne* L., *Elymus dahuricus* Turcz. and *Elymus sibiricus* L.	Xu et al. (2025)
Streptomyces sp. Jrh8-9	Phosphate solubilization, nitrogen fixation and IAA production	Alfalfa	Guo et al. (2025)
Streptomyces sp. RT34, *Streptomyces* sp. AR39	Phosphate solubilization, IAA production and nitrogenase activity	*Glycine max*	Erdiandini et al. (2025)
Streptomyces sp. CLV179	Siderophores and ammonia production	Maize	Pedó et al. (2025)
Streptomyces misionensis, Streptomyces rhizoryzae	Nematicidal activity, phosphate solubilization, IAA production and ammonia production	*Oryza sativa* L.	Ruanpanun et al. (2025)

10.2.1 Phytohormones Production and Nutrient Mobilization by Streptomyces

Phytohormones are chemical messengers which significantly contribute in the growth and development of the plant. Even a very small concentration of these organic substances can trigger complex physiological process which regulates everything from the germination of seeds to flowering and senescence. Major classes of phytohormones include auxins, gibberellins, cytokinins, abscisic acid and ethylene. Some other molecules, such as brassinosteroids, jasmonic acid, salicylic acid and strigolactones with similar bioactivities to phytohormones have also been identified (Agudelo-Morales et al. 2021).

Indole-3-acetic acid (IAA) is the main representative of auxins. Auxin has been reported to contribute significantly to different processes, including cell division, root system differentiation, cell growth by expansion and gravitropic reaction of root (Kulaeva and Prokoptseva 2004). Cytokinins regulate different plant growth processes such as cell division, chloroplast differentiation, photosynthesis, leaf senescence and nutritional metabolism. Moreover, cytokinins also play an important role in the root and shoot growth, and they also function as positive and negative regulator of the meristem activity of stem and root, respectively (Li et al. 2020a, b). Gibberellins are another important class of phytohormones that significantly regulate different developmental processes, including shoot elongation, root development, flowering and seed germination (Castro-Camba et al. 2022). Abscisic acid play a vital role in plant growth, mainly by regulating various important processes including maturation of seed, bud dormancy, germination, seedling growth, stomatal closure, cuticular wax accumulation and leaf senescence (Chen et al. 2020; Mo et al. 2024). From germination to senescence various physiological processes have also been influenced by multifunctional phytohormone ethylene. These processes include cell division and elongation, root hair development, leaf and flower development, fruit ripening, photosynthesis and senescence (Khan et al. 2024). Thus, these phytohormones play an integral role in regulating various physiological processes, which ultimately promote plant growth and productivity.

The rapidly growing global population demands a substantial increase in agricultural productivity. Additionally, global warming and its associated climate abnormalities have led to an increase in the incidence of biotic and abiotic stress combinations, which adversely affect plant growth and productivity (Pandey et al. 2017; Das et al. 2025). Abiotic stress mainly includes drought, heavy metal toxicity, high temperature and salinity conditions, whereas biotic stresses can be defined as the negative effects of the living organisms including nematodes, arachnids, weeds viruses, fungi and bacteria on the overall growth of the plant (Talaat and Shawky 2017; Hashem et al. 2019). To alleviate negative effects of the stress, several strategies, including development of resistant varieties, use of chemicals and shifting of crop calendars, have been developed worldwide. Some of these strategies are time consuming, costly and harmful to the environment (use of chemicals), which necessitates the development of effective, low-cost, environmental friendly approaches to increase the stress tolerance in plants. Use of microorganisms to increase stress tolerance in plants is one of the promising approaches as these are economical, effective and environmental friendly (Grover et al. 2011; Talaat and Shawky 2017).

In different studies microbes have been found to help the plant to resist against biotic and abiotic stresses by using various mechanisms. Microorganisms present in the rhizosphere form complex relationships with plants and help in nutrients assimilation, and activation and development of the defence system (Meena et al. 2017). *Streptomyces* is genus of Gram-positive filamentous bacteria which are commonly present in zone of rhizosphere and found to promote plant growth through different mechanisms (Pang et al. 2022). An endophytic strain of *Streptomyces* has been documented to possess the ability to synthesize IAA (Verma et al. 2011). In another study, production of gibberellic acid, indole acetic acid, abscisic acid and kinetin from *Streptomyces* species has also been reported (Rashad et al. 2015). *Streptomyces olivaceoviridis*, *Streptomyces rimosus* and *Streptomyces rochei* have been documented for synthesis of auxins, gibberellins and cytokinin (Aldesuquy et al. 1998). Expression of the stress-responsive genes in plants has also been regulated by the phytohormones which help the plants to resist against biotic and abiotic stresses

(Khan 2025). *Streptomyces* spp. showed the ability to produce and to induce phytohormone production under stress conditions in different studies (Sadeghi et al. 2012; Yandigeri et al. 2012). Tryptophan exudates released by the plants under abiotic stress serves as precursor molecule for the synthesis of IAA by microbes. This microbial IAA is taken up by plants which boosts the plant growth (Jacob and Sudini 2016).

In different studies *Streptomyces* spp. have also been reported to possess various plant growth-promoting traits, including 1-Aminocyclopropane-1-carboxylate (ACC) deaminase activity, phosphate solubilization, ammonia production and siderophore production, along with the production of phytohormones (Tamreihao et al. 2016). Phytohormone ethylene is also known as the stress hormone, and it has been reported that under stress conditions, its production increases manifold in plants. This increased level of stress hormone exerts negative effect on plants, particularly by increasing the rate of cell senescence. ACC is the precursor of ethylene. ACC deaminase enzyme metabolizes this precursor into ammonia and α-ketobutyrate, thus regulating the biosynthesis of ethylene (Jacob and Sudini 2016; Jaemsaeng et al. 2018). *Streptomyces* spp. have been reported to exhibit plant growth-promoting potential via synthesizing ACC deaminase under various stress conditions such as flooding and salt conditions (Jaemsaeng et al. 2018; Yoolong et al. 2019). ACC deaminase activity of *Streptomyces filipinensis* and *Streptomyces atrovirens* isolated from tomato rhizosphere has been reported (El-Tarabily 2008). In another study, *Streptomyces hydrogenans* has been reported to exhibit ACC deaminase activity and IAA production potential (Kaur and Manhas 2022).

In plants, phosphorus is the second most important macronutrient after nitrogen which is required for various metabolic processes, including cell division, macromolecules synthesis, membrane integrity, energy generation, photosynthesis and signal transduction (Rawat et al. 2020). In agricultural soil, phosphorus deficiency is common throughout the world; thus, farmers use phosphorus fertilizers regularly to fulfil the phosphorus requirement of the plants. A large portion of the applied phosphorus precipitates into non-available complexes. Phosphate solubilizing bacteria transform this non-available form of phosphorus into usable form for the plants and ultimately limit the requirement of chemical fertilizers (Chouyia et al. 2020a, b; Rawat et al. 2020; Chouyia et al. 2022). Various phosphate solubilizing species of *Streptomyces*, including *Streptomyces roseocinereus*, *Streptomyces asenjonii*, *Streptomyces violascens*, *Streptomyces carpinensis*, *Streptomyces alboviridis*, *Streptomyces fulvissimus*, *Streptomyces cellulosae* and *Streptomyces olivoverticillatum* have been documented in different studies (Balakrishna et al. 2012; Chouyia et al. 2020a, b; Chouyia et al. 2022).

In iron deficient conditions, microorganisms have been documented to synthesize low molecular weight molecules known as siderophores. Production of siderophores increases iron uptake and thus promotes plant growth (Singh et al. 2022). Several species of *Streptomyces* have been found to possess the ability to produce siderophores. Dimkpa et al. (2008) have documented the hydroxamate siderophores production and plant growth promotion potential of *Streptomyces acidiscabies* under nickel stress. *Streptomyces pactum* isolated from pesticide-contaminated agricultural soil has been found to synthesize siderophores and to promote the growth of *Arachis hypogaea* (groundnut) under cadmium stress (Musani and Das 2024). In another study, desferrioxamine E type of siderophore production has been reported from *Streptomyces parvulus* (Gáll et al. 2016). Alinejad et al. (2024) have documented the catecholate siderophore production from two *Streptomyces* spp. isolated from the rhizosphere of medicinal plants.

Potassium and zinc are other important nutrients required for the growth of the plants. Potassium plays an important role in solute transport, protein synthesis, energy metabolism, photosynthesis, etc. Zinc acts as a regulatory cofactor for different important enzymes, and is also involved in carbohydrate metabolism, protein synthesis, chlorophyll synthesis, maintenance of membrane integrity and phytohormone production. Potassium and zinc solubilizing bacteria can convert unavailable forms of potassium and zinc into plant usable forms (Suriyachadkun et al. 2022; Yadav 2022). *Streptomyces* possessing zinc and potassium solubilizing ability have also been documented to promote plant growth (Boubekri et al. 2021; Suriyachadkun et al. 2022).

Streptomyces also possess some other mechanisms through which it promotes plant growth under normal and stress conditions. *Streptomyces hygroscopicus* have been documented to synthesize compounds pteridic acids A and B which exhibit auxin-like plant growth-promoting activity (Igarashi et al. 2002). In another study, pteridic acids H and F have been isolated from *S. iranensis* fermentation broth and found to promote plant growth under abiotic stress (Yang et al. 2023). Abiotic stress is also known to induce oxidative stress in plants, which compromises plant growth and productivity (Sachdev et al. 2021).

Production of ammonia is another important plant growth-promoting trait of *Streptomyces*. Accumulation of nitrogen by the production of ammonia promotes plant root and shoot growth (Dutta and Thakur 2017). Nonthakaew et al. (2022) isolated four species of *Streptomyces* from rhizospheric soil and reported ammonia production along with IAA, siderophore production and phosphate solubilization potential. In different studies, various species of *Streptomyces*, including *S. nobilis*, *S. Kunmingenesis*, *S. enissocaesilis*, *S. djakartensis*, *S.alboviridis*, *S. griseorubens* and *S. pseudovenezuelae* have been documented to produce ammonia and improve the growth of different plants under normal and stress conditions (Anwar et al. 2016; Chukwuneme et al. 2020; Boubekri et al. 2021). *Streptomyces* spp. have been found to exhibit antioxidant potential, and plants inoculated with them displayed significant reduction in the reactive oxygen species synthesis and increased levels of antioxidant enzymes under stress conditions (Manullang and Chuang 2020; Nivedita et al. 2024).

10.2.2 Indirect Contribution of Streptomyces to Nitrogen Fixation

Nitrogen is a vital macronutrient for plant growth, constituting a major element of amino acids, proteins and nucleic acids. Certain bacteria, like *Rhizobium*, *Azotobacter* and *Azospirillum*, can fix atmospheric nitrogen. *Streptomyces* species, although non-diazotrophic, play a crucial indirect role in nitrogen fixation and availability in the rhizosphere (Jog et al. 2012; Palaniyandi et al. 2013). These filamentous actinobacteria enhance nitrogen fixation by forming synergistic interactions with nitrogen-fixing microbes and by mineralizing organic nitrogen sources. *Streptomyces* increases nitrogen availability primarily by inducing rhizosphere-associated nitrogen-fixing microorganisms. *Streptomyces* suppresses pathogenic microorganisms by generating antibiotics, siderophores and signalling chemicals, and thus forms a microecological niche that promotes the growth of nitrogen fixers, *Azospirillum* and *Herbaspirillum* (Gopalakrishnan et al. 2011; Chouyia et al. 2020a, b). The biological nitrogen fixation (BNF) activity close to plant roots is increased by this mutualistic relationship, which helps the host plant absorb nitrogen more effectively. Additionally, *Streptomyces* is essential for the mineralization of organic nitrogen. They have a variety of extracellular enzymes, including amidases, ureases, deaminases and proteases, which decompose complex nitrogenous organic matter into simpler forms such as nitrate (NO_3^-) and ammonium (NH_4^+) that plants can utilize, (El-Tarabily and Sivasithamparam 2006). In the rhizosphere and bulk soil, this breakdown releases nitrogen from organic wastes, allowing plants to absorb it. Furthermore, their enzymatic activity adds nitrogen to the soil-plant system, improving overall soil fertility (Schrey and Tarkka 2008).

Streptomyces also exert indirect impact on nitrogen fixation by influencing root exudation and recruitment of rhizosphere microbiome. Their rhizospheric invasion frequently causes the exudation of root materials rich in sugars, flavonoids and amino acids, which act as chemical markers that attract nitrogen-fixing bacteria (Viaene et al. 2016). In this way, *Streptomyces* function as facilitators for nitrogen-fixers in the root zone of plants, to improve their functional effectiveness. Additionally, *Streptomyces* can physically attach to root surfaces due to their filamentous and mycelial growth pattern, which enhances soil aggregation, aeration and nutrient availability (Gopalakrishnan et al. 2013). This structural adaptation enhances nutrient cycling efficiency by enabling closer contact between roots and microbes that transform nitrogen. The synergistic effects of these mechanisms include better nitrogen nutrition, increased biomass accumulation and increased yield in a variety of crops, especially in soils with low or inadequate nutrient inputs. It has been reported that, even in the absence of direct nitrogen fixation, *Streptomyces* inoculation of crops can raise the total nitrogen content in plant tissues (Jog et al. 2012). *Streptomyces lydicus* WYEC108, which colonized and sporulated within the nodule's surface cell layers of pea root, influenced nodulation by increasing the average size of the nodules, improving the vigour of bacteroids within the nodules by enhancing nodular iron and possibly other soil nutrient assimilation (Tokala et al. 2002). The impact of *Streptomyces* spp. isolated from the rhizosphere on five legumes (soybean, kidney bean, chickpea, lentil and pea) demonstrated that soil microbial populations were boosted, while soil nutrients and organic matter content were also increased (Abdelgawad et al. 2020). Soil enrichment with *Streptomyces* sp. boosted photosynthesis, which subsequently increased legume production. *Streptomyces* sp. also boosted nitrogen availability in soil, legume tissue and seeds, which activated important nitrogen-metabolizing enzymes such as glutamine synthetase, glutamate synthetase and nitrate reductase (Oyedoh et al. 2023). In addition to higher amounts of nitrogen-containing amino acids in actinobacterial-treated legume seeds, significant quantities of sugar, organic acids and fatty acids, as well as antioxidant phenolics, minerals and vitamins were also observed.

10.2.3 Induced Systemic Resistance and Stress Tolerance Enhancement

Induction of systemic resistance (ISR) and improvement of stress tolerance are two important plant-beneficial features associated with *Streptomyces* that support plant health and production during a range of biotic and abiotic stress situations (Viaene et al. 2016; Palaniyandi et al. 2013). ISR is a type of increased defence that plants generate when *Streptomyces* and other helpful bacteria colonize them. ISR includes the activation of the ethylene (ET) and jasmonic acid (JA) signalling pathways and is triggered without producing diseases, in contrast to systemic acquired resistance, which is induced by pathogens (Conn et al. 2008). ISR is stimulated by *Streptomyces* species in a number of mechanisms. They activate the plant's innate immune system by producing elicitor molecules such as lipopolysaccharides, volatile organic compounds (VOCs), siderophores and secondary metabolites (Viaene et al. 2016). After a pathogen attack, priming quickly activates defence-related genes, minimizing the impact of the disease. Several studies documented that plants treated with *Streptomyces* show enhanced resistance to fungal, bacterial and viral pathogens even in the absence of a direct microbial infection (Gopalakrishnan et al. 2013). Furthermore, *Streptomyces* induce systemic resistance, which means that plant components exhibit increased resistance even in the absence of their close contact with the beneficial bacteria. Palaniyandi et al. (2013) described that broad-spectrum defence approaches exhibited by plants (treated with *Streptomyces* spp.) lessens the need for chemical pesticides and also enhances plant resistance to a variety of diseases. Additionally, *Streptomyces* spp. enhances plant tolerance to abiotic stresses (drought, salinity, heavy metal toxicity) and oxidative stresses. One main mechanism to combat these stresses involves production of osmolytes like proline, trehalose and glycine betaine, which help plants maintain cellular homeostasis under osmotic stress conditions (Jog et al. 2012). Drought and salinity are significant stressors that limit global crop productivity, necessitating sustainable solutions. Niu et al. (2022) isolated *Streptomyces albidoflavus* OsiLf-2, a salt-tolerant endophytic actinomycete, in enhancing rice stress tolerance by boosting osmolyte production (proline, polysaccharides and ectoine) and stress-related gene expression. Its application improved rice growth and yield under drought and saline conditions, presenting a promising biotechnological strategy for resilient agriculture.

Another key mechanism involves activating antioxidant defence systems. *Streptomyces* can upregulate enzymes like superoxide dismutase, catalase and peroxidase in plants, which neutralize reactive oxygen species produced under stress (El-Tarabily and Sivasithamparam 2006). Under unfavourable conditions, these enzymes reduce cellular damage and support sustained plant metabolism. Moreover,

Streptomyces produce 1-aminocyclopropane-1-carboxylate deaminase, an enzyme that degrades the ethylene precursor ACC. Under stress, ethylene levels frequently rise and may retard plant growth. *Streptomyces* promote root elongation and biomass accumulation while reducing ethylene concentrations, which minimizes stress-induced growth retardation (Glick 2014). In contrast to non-inoculated controls, plants inoculated with *Streptomyces* strains exhibit enhanced water-use efficiency, chlorophyll content and yield performance under stress, according to field and greenhouse experiments (Gopalakrishnan et al. 2011). *Streptomyces*, therefore, serves as a useful way to establish agricultural systems that are climate resilient.

10.2.4 Root Colonization by Streptomyces

Soil is a habitat for billions of different microorganisms, many of which develop dynamic interactions with plants that range from beneficial to pathogenic (Bardgett and Van Der Putten 2014). Microbe penetration or colonization of host tissue is a complicated process mediated by signalling molecules, secretory products of both plants and microbes. The first phase of colonization is usually defined as the adherence of a specific microbial strain to the host surface. Following that, the bacteria migrate towards the host surface via chemotactic migration, which occurs before attachment, in response to root exudates (Kumar et al. 2020). Root colonization by *Streptomyces* bacteria is affected by plant genotype. There have also been findings that root colonization by *Streptomyces* species has a function in biocontrol and plant growth promotion. Tokala et al. (2002) hypothesized that nodule and root colonization is one of the mechanisms by which *Streptomyces* promotes pea plant growth and possibly other leguminous plants.

Getha et al. (2005) reported that *Streptomyces* strain g10 stimulated ISR in potted banana plantlets and significantly increased its rhizospheric population by more than two times by the end of the third week following inoculation. Furthermore, it also successfully managed Fusarium wilt disease in banana plantlets. In the year 2008, El-Tarabily found that rhizosphere-competent streptomycete isolates, *S. filipinensis* and *S. atrovirens*, promoted tomato plant growth more than non-rhizosphere-competent isolates. The efficacy of six *Streptomyces* strains on chickpea growth was demonstrated by Gopalakrishnan et al. (2015). Scanning electron microscopic (SEM) analysis of chickpea roots revealed a tremendous extent of root colonization by all six *Streptomyces* strains, which further resulted in increased chickpea plant growth, total nitrogen, dehydrogenase activity, available phosphorus and organic carbon, when compared to the uninoculated control in the rhizosphere of mature crop. Patel et al. (2018) characterized *Streptomyces* spp. for their colonization ability, plant growth promotion and protection against fungal disease in three cereal crops. These *Streptomyces* strains showed profuse colonization in roots and aerial parts of cereal plants and exhibited plant growth-promoting and biocontrol potential. Later in the year 2021, Saraylou et al. reported plant growth-promoting traits of four *Streptomyces* isolates (*Streptomyces griseus* strain 52-1, *Streptomyces albogriseolus* strain DSM 4, *Streptomyces aurantiacus* strain 4683 and *Streptomyces kanamyceticus* strain 22-5) isolated from rhizospheres of tomatoes and sunflowers. These isolates showed high root colonization ability in spinach (*Spinacia oleracea l.*). Zhou et al. (2024) isolated *Streptomyces rochei* JK1 from the soybean rhizosphere, which colonized both soybean and ginger roots, indicating adaptability to diverse plant hosts. This strain produces borrelidin, a macrolide compound with strong inhibitory effects on *Pythium myriotylum* and *Phytophthora sojae* (oomycetes). This compound is stable under different environmental conditions, highlighting its potential as a robust biopesticide. Recently, Tigani et al. (2025) reported the root colonizing ability of *Streptomyces* sp. PG2 in *Arabidopsis thaliana*. This interaction triggers the production of an antifungal compound DHP (2-5 dihydro-phenylalanine) in *A. thaliana*, thereby enhancing its resistance against *Rhizoctonia solani*. Similar plant-induced biosynthesis was also observed in sugar beet and in heterologous hosts. These findings reveal that plant colonization activates microbial biosynthetic pathways, strengthening rhizosphere resilience.

10.2.5 Biocontrol of Plant Pathogens

Streptomyces are well known for producing a variety of substances that can directly stimulate plant growth by producing phytohormones like gibberellins, cytokinins and indole acetic acid (IAA) and by increasing the uptake of potassium, phosphorus and nitrogen as well as other vital minerals. In addition to these direct mechanisms, *Streptomyces* species also promote plant growth by indirect mechanisms. These bacteria as producers of diverse secondary metabolites (antibiotics, antifungals, insecticidals, etc.) inhibit the growth of plant pests and pathogens (Kinkel et al. 2012; Bekiesch et al. 2021) (Table 10.2).

10.2.5.1 Production of Antifungal Compounds

The antagonistic effect of *Streptomyces* against the different fungal diseases is related to the synthesis of antifungal metabolites (Ghanem et al. 2022). Streptomycin and cycloheximide were the first antibiotics produced by *Streptomyces griseus* and used for the treatment of different bacterial and fungal infections of plants (Leben and Keitt 1954). Several antifungal metabolites produced by *Streptomyces* species include streptomycin, natamycin, nystatin, rimocidin and candicidin. Three antifungal chemicals (nigrecine, geltanamycine and guanidylfingine) produced by *Streptomyces violaceusniger* YCED9 and their effects on several plant illnesses were examined by Trejo-Estrada et al. (1998). Taechowisan et al. (2005) recovered two antifungal compounds 5, 7-dimethoxy-4-pmethoxylphenylcoumarin and 5, 7-dimethoxy-4-phenylcoumarin, from the culture filtrate of *Streptomyces aureofaciens* CMUAc130 which were found to be effective against banana anthracnose and wheat wilt caused by *Colletotrichum musae* and *Fusarium oxysporum*, respectively. Nguyen et al. (2015) isolated an antifungal metabolite from *Streptomyces griseus* H7602 cell-free culture filtrate, which was identified as 1H-pyrrole-2-carboxylic acid (PCA) by spectral studies. PCA showed antifungal potential against *Phytophthora capsici*. Zheng et al. (2019) also reported significant growth-

Table 10.2 Biocontrol activities of *Streptomyces* spp. and their effects on pathogens and host plants

Activity	Representative strain	Active compound/ enzyme/VOC	Effect on pathogen	Effect on host	References
Production of antifungal compounds	*Streptomyces lydicus* JCK-6019	Natamycin and other diffusible metabolites	Inhibition of broad-spectrum soil-borne fungi (*Fusarium*, *Rhizoctonia*, *Pythium*) and oomycetes	Promote growth in cucumber and reduced disease incidence; safe for host tissue	Luo et al. (2025)
	Streptomyces sp. (oligomycin producers)	Oligomycin-type polyether antibiotics	Inhibited spore germination and mycelial growth of *Botrytis cinerea* and related pathogens	Decreased grey mould severity in tomato; improved plant vigour	Louviot et al. (2024)
	Streptomyces graminearus STR-1	Secondary metabolites with antifungal metabolites	Antagonistic to several phytopathogenic fungi	Enhanced seedling protection in greenhouse trials	Shen et al. (2024)
Lytic enzyme production	*Streptomyces roseolus*	Secreted chitinases (GH18 family)	Hydrolyzed fungal cell wall chitin; inhibited spore germination	Enhanced disease resistance and root colonization	Xiayun et al. (2012)
	Rhizospheric *Streptomyces* spp.	Chitinase, β-1,3-glucanase, protease	Degraded hyphal walls of *Fusarium*, *Rhizoctonia*, *Pythium*	Reduced damping-off and promoted seedling vigour	Vurukonda et al. (2021)
Production of volatile organic compounds (VOCs)	*Streptomyces* sp. TOR3209	Short-chain alcohols, ketones, terpenes (complex volatilome)	Inhibited pathogen growth and spore germination in dual-plate assays	Induced systemic resistance and improved biomass in *Arabidopsis*	He et al. (2022)
	Streptomyces sporoverrucosus B-1662	Dibutoxybutane and other VOCs	> 90% reduction in anthracnose incidence in lab and field	Increased yield and no phytotoxicity in field-grown plants	Kim et al. (2024)
	Streptomyces scabiei	Complex VOC mixture (terpenoids, sulfur compounds)	Some VOCs suppressed growth of other microbes	Demonstrated dual ecological roles (signalling and inhibition)	Stulanovic et al. (2025)

promoting potential of *Steptomycetes* sp. strain FJAT-31547 which exhibited broad-spectrum antibacterial as well as antifungal activities with high biocontrol efficacy against bacterial wilt and *Fusarium* wilt of tomato. This *Streptomyces* sp. played dual function both as a biocontrol agent against *Fusarium* infection and as a plant growth-promoting rhizobacterium (PGPR), thereby contributing to improved plant health and vigour. It was found that ethyl acetate extracts of *Streptomyces* spp. showed greater control against *F. oxysporum*, the cause of tomato wilt disease. The extracts also significantly increased tomato seedling biomass in terms of leaf count, plant height, and dry and fresh weight (Puppala et al. 2019). It was found that in addition to maintaining the soil bacterial community, *Streptomyces jietaisiensis* strain A034 promotes plant growth and prevents root knot disease (Ruanpanun and Nimnoi 2020) and *Streptomyces* strains can increase the growth of wheat in saline conditions (Cao et al. 2021). Secondary metabolites produced by *Streptomyces amritsarensis* V31 and *Kribella karoonensis* MSCA185 were tested for antifungal activity against different fungal phytopathogens (Shahid et al. 2021). In response to *S. amritsarensis* V31 metabolites, *Rhizoctonia solani*, *Alternaria alternata*, *Aspergillus flavus*, *Fusarium oxysporum f.* sp. *lycopersici*, *Sarocladium oryzae* and *Sclerotinia sclerotiorum* displayed suppressed growth. In another study, *Streptomyces angustmyceticus* strain TH23-7 demonstrated antifungal activity against *Lasiodiplodia theobromae* which suggested its potential to manage the spadix rot disease of flamingo flowers (Ruangwong et al. 2022). A year later, Kadaikunnan et al. (2023) documented that the addition of *Streptomyces parvulus* VRR3 resulted in decreased incidence of *Fusarium* wilt disease signs on green gram plant (*Vigna radiata* L.), and induced growth of green gram seedlings by enhancing root and shoot length.

Apart from direct antagonistic effects, *Streptomyces* species can modulate plant immunity by enhancing the innate defences of plants through systemic resistance (Mun et al. 2020). Furthermore, their rhizosphere competence enables them to colonize plant roots efficiently and establish a protective barrier that prevents pathogen invasion. In addition to production of antifungal metabolites, improving plant microenvironment and bioavailability of nutrients, root colonization also enhances effectiveness of pathogen suppression. Ngalimat et al. (2023) reported that *Streptomyces* spp. attenuated the unfavourable effects of bacterial panicle blight, enhanced rice yield qualities, stimulated defence-related gene transcript levels and showed no phytotoxic influence on rice plants. According to some reports *Streptomyces* sp. can produce thermostable and acidic phytase, and can promote plant growth-promoting characteristics of tomato plants (Puppala et al. 2019).

10.2.5.2 Lytic Enzyme Production by Streptomyces

The cell walls of species in the Chromista kingdom, like oomycetes, are composed of cellulose and 1, 3-glucan, while the cell walls of higher fungi are composed of layers of chitin and 1, 3-glucan (Singh 2023). When the cell wall breaks down, the cell contents flow out and the cell collapses. Numerous bacteria produce lytic enzymes that are capable of dissolving the cell walls of other organisms (Cunha et al. 2010). The production of lytic enzymes by bacteria has been demonstrated as one of the biocontrol methods for plant diseases (Kaur et al. 2011). The direct action of chitinase, either by itself or in combination with other antifungal compounds produced by an antagonist, makes chitinolytic bacteria more effective fungal pathogen antagonists (Narayanasamy and Narayanasamy 2013).

Several reports provide a significant role of lytic enzymes, particularly chitinases and β-1, 3-glucanases produced by *Streptomyces* species, in exhibiting broad-spectrum antifungal activity against various phytopathogens (Mishra et al. 2020). Under in vitro conditions, all the *Streptomyces* species produce chitinases that inhibit the germination of *Moniliophthora perniciosa*. Chitinase produced by *Streptomyces viridlfcans* was reported with the ability to lyse the cell wall of *Rhizoctonia solani, Aspergillus* sp., *F. oxysporum*, *Curvularia lunata* and *Sclerotinia oryzae* (Gupta et al. 1995). Chitinase produced by *Streptomyces* significantly degrades fungal cell walls, highlighting its potential role in biocontrol against multiple phytopathogenic fungi. The chitinase synthesis in *Streptomyces lydicus* WYEC108 culture filtrates and its antagonistic potential against the cell walls of *R. solani*, *Aphanomyces* sp., *F. solani* and *Pythium ultimum* were investigated by Mahadevan and Crawford (1997).

A thermostable endochitinase from *Streptomyces* RC1071 active against different phytopathogenic fungi such as *Aspergillus parasiticus*, *F. solani*, *F. graminearum*, *F. oxysporium*, *Magnaporthe grisea* and *C. gloeosporioides* was purified by Gomes et al. (2001). Joo (2005) purified and characterized an extracellular chitinase from *Streptomyces halstedii* AJ-7 and the enzyme showed biocontrol potential against various fungal phytopathogens. The highest antifungal activity was observed against *F. oxysporum*. Endophytic *Streptomyces* spp. reported by Quecine et al. (2008) showed chitinolytic activity against *Phytophthora parasitica*, *Guignardia citricarpa*, *R. solani*, *C. sublineolum*, *Pythium* sp. and *F. oxysporum*. The potential of different *Streptomyces* spp. (*S. albovinaceus*, *S. griseus*, *S. setonii*, *S. caviscabies* and *S. virginiae*) as antagonists of *Moniliophthora* (ex Crinipellis) *perniciosa* was examined by Macagnan et al. (2008).

According to Lee et al. (2012), *Streptomyces cavourensis* subsp. *cavourensis* (coded as SY224) produces lytic enzymes chitinase, β-1, 3-glucanase, lipase and protease, which decreased the anthracnose in pepper. Awad et al. (2014) discovered *Streptomyces glauciniger* WICC-A03, another strong chitinase-producing strain, which showed antifungal efficacy against phytopathogenic fungi. Both brown rot and white rot fungi were effectively inhibited by the broad-spectrum antifungal activity of *Streptomyces violaceusniger* MTCC 3959 cell-free culture filtrate (Nagpure et al. 2014). According to Tiwari et al. (2021), *Streptomyces griseus* H7602 produces the lytic enzymes chitinase, 1, 3-glucanase, lipase and protease, which have antifungal properties against *P. capsici*, which causes pepper root rot. *Streptomyces* strains are used in controlling rot-causing phytopathogens, making them important for managing both brown and white rot diseases in plants. The presence of multiple lytic enzymes underscores the strong biocontrol potential of the *Streptomyces* strain, enabling effective degradation of fungal cell walls and inhibiting the mycelial growth of a wide range of fruit-rotting fungi. One 1, 3-glucanase and four protease isoenzymes were observed in the zymogram analyses of cell-free culture filtrate. *S. exfoliatus* strain MT9 exhibits significant and wide-ranging antagonism by producing several cell wall lytic enzymes and inhibiting the mycelial growth of numerous fruit-rotting fungi. Cell culture filtrate of *Streptomyces angustmyceticus* NR8-2 produces cell wall-degrading enzyme β-1, 3-glucanase, involved in the antifungal activity against *Colletotrichum* sp. and *C. lunata* (Wonglom et al. 2019).

10.2.5.3 Production of Volatile Organic Compounds (VOCs)

Volatile organic compounds are low-molecular-weight (100–500 Da) carbon-containing lipophilic chemicals that easily evaporate at normal temperature and pressure (Geranios 2018). VOCs can permeate into the atmosphere and soil, and approximately 1300 VOCs from diverse microorganisms have been identified so far (Chatterjee et al. 2024). Numerous literature reports have shown that VOCs produced by *Streptomyces* spp. have significant antifungal activity against fungal pathogens (Li et al. 2020a, b). Some previous studies have reported that the conidiophores and hyphae of five fungi, viz., *Aspergillus giganteus, Fusarium oxysporum, Penicillium viridicatum, Trichoderma viride* and *Zygorhynchus vuilleminii*, show abnormal morphology when treated with the VOCs of nine bacteria and an actinobacteria (Wang et al. 2013).

It has been reported that *Streptomyces*-derived volatile organic compounds are effective biocontrol agents against a broad spectrum of post-harvest fungal pathogens in fruits and vegetables. Volatile compounds derived from *Streptomyces globisporus*, including dimethyl sulphide, acetophenone and dimethyltrisulfide, showed antifungal activity by suppressing *Penicillium italicum* infection of *Citrus microcarpa* (Li et al. 2010). According to Wang et al. (2013), *Streptomyces alboflavus* TD-1 produced 27 volatile chemicals that inhibited *Penicillium citrinum, Aspergillus ochraceus, A. flavus, A. niger* and *Fusarium moniliforme*. Volatile organic compounds derived from *Streptomyces* have been demonstrated to be effective biocontrol agents against major phytopathogens, including *R. solani, F. oxysporum* and *S. sclerotiorum* in several studies. Cordovez et al. (2015) recovered five different volatile compounds from two *Streptomyces strains* (W47 and W214) which showed inhibitory effects on *R. solani*. These compounds were identified as methyl butanoate, methyl 2-methylpentanoate, methyl 3-methylpentanoate, 1, 3, 5-trichloro-2-methoxy benzene and 3-octanone. Wu et al. (2015) isolated *Streptomyces albulus* strain NJZJSA2 from a forest soil sample and evaluated its ability to produce different volatile organic compounds (VOCs), antagonistic against *F. oxysporium* and *S. sclerotiorum* and was found to produce 13 VOCs as evident from the GC/MS analyses.

In vitro and in vivo studies demonstrated that the volatile organic compounds produced by *S. globisporus* JK-1 inhibited *B. cinerea*, a post-harvest grey mould of tomato fruit (Zhao et al. 2022). Lyu et al. (2020) reported that *Streptomyces*-derived volatile organic compounds displayed significant biocontrol potential by targeting key developmental and toxin biosynthesis pathways of *Aspergillus* species. For instance, *Streptomyces yanglinensis* 3–10 produced a variety of volatile chemicals that inhibit the expression of genes involved in aflatoxin production, conidial germination and sporulation in *A. parasiticus* and *A. flavus*. *S. philanthi* RL-1-178 strain was reported by Boukaew and Prasertsan (2020) with the ability to produce 39 volatile organic compounds, suppressing two aflatoxin-producing fungi, *A. parasiticus* and *A. flavus*. The biocontrol of spadix rot (induced by *L. theobromae*) by 2, 2-dimethyl-4-(3-methylbut-2-enyl)-6-methylidenecyclohexyl, a volatile chemical generated by *S. angustmyceticus* TH23–7, was studied by Ruangwong et al. (2022). *Streptomyces* sp. strain S97 produced volatile chemicals that decreased the mycelial growth of *R. solani* and *Phoma* by 95%, *F. solani* by 69% and *F. oxysporum* and *S. rolfsii* by 20% (Ayed et al. 2021). Thus, these diverse studies documented that biological control of fungal pathogens in agriculture could be achieved by using microbial volatile organic compounds (VOCs).

10.3 Commercial Plant Growth Promoting Agents

Commercial plant growth-stimulating agents are bioformulations that use streptomycetes and other helpful microbes to improve the productivity and health of plants (Aamir et al. 2020). For sustainable agriculture, these agents are being developed as biostimulants, biofertilizers and biocontrol products (Elnahal et al. 2022) (Table 10.3). They enhance nutrient intake, promote root development and shield plants from diseases through natural processes. The stability, effectiveness and environmental friendliness of *Streptomyces*-based products are driving their growing appeal. The commercialization of these microbial inoculants is a significant step in promoting sustainable agricultural production and lowering dependence on chemical fertilizers.

10.4 Conclusion and Future Prospects

Streptomyces are group of beneficial soil bacteria that helps plants to grow better, especially under challenging conditions such as drought, poor soil or attacks by harmful organisms. One of the most important mechanisms to support plant growth is by producing natural plant hormones like auxins, gibberellins, cytokinins and abscisic acid. These hormones control many aspects of plant development from seed germination to flowering and stress responses. In addition to producing hormones, *Streptomyces* also transforms vital elements like zinc, potassium and phosphorus into forms that plants can readily absorb. Additionally, they also produce siderophores, molecules that bind to iron and make it available to plants, particularly in soils with limited iron availability. The production of the enzyme ACC deaminase is another important characteristic of *Streptomyces*. ACC deaminase aids in reducing ethylene levels, a stress hormone that accumulates in plants during difficult stress periods and can stunt their growth. *Streptomyces* helps plants remain healthier and more productive by lowering ethylene. *Streptomyces* play an important supportive role, even if they don't directly fix nitrogen like some other bacteria do. In addition to providing a favourable environment for nitrogen-fixing bacteria to grow rapidly,

Table 10.3 Commercial plant growth-promoting agents

Name	Strain	Target pathogens/Pests	Mechanism of action	Manufacturer/ Country	References
Micro108®	*Streptomyces lydicus* WYEC108	Soil-borne fungal pathogens	Competitive root colonization, antifungal secondary metabolites, improved nutrient uptake	Novozymes biologicals (USA)	Yuan and Crawford (1995)
Arzent™	*S. hygroscopicus*	Soil-borne fungal pathogens	Plant growth promotion and biocontrol	Innovative biosystems, USA	Hamby and Crawford (2000)
Spinosad	*S. spinosa*	Insect pests	Acts on nicotinic acetylcholine receptors (nAChRs) and GABA receptors in insects	ChemWerth, USA	Thompson et al. (1999)
Mycostop®/ MycoStop® PRO	*Streptomyces griseoviridis* K61	*Fusarium*, *Pythium*, *Alternaria*, *Rhizoctonia*	Rhizosphere colonization, antibiosis, competition for nutrients, secretion of antifungal compounds	Lallemand Plant Care (Finland)	Minuto et al. (2006)
Rhizovit®	*Streptomyces rimosus*	Soil-borne fungal pathogens	Production of antifungal metabolites, enzyme secretion and competition in rhizosphere	Biovet AG, Switzerland	Yu et al. (2008)
Actinovate®/ Actinovate SP	*Streptomyces lydicus* WYEC108	*Pythium*, *Rhizoctonia*, *Fusarium*, *Phytophthora*, *Botrytis*	Root and leaf colonization, production of antifungal metabolites and enzymes, competition and induced resistance	Novozymes Biologicals (USA)	Sekar et al. (2016)
AFFIRMWDG	*S. cacaoi* var. asoensis)	Turfgrass, ornamentals and certain fruit/vegetable crops	Fungicide for turfgrass fungi	Nufarm Limited, USA	Shrivastava and Kumar (2018)
Agri-Mycin® 17/Keystrepto™	Streptomycin (from *S. griseus*)	*Xanthomonas*, *Erwinia*, *Pseudomonas*	Inhibits bacterial protein synthesis (30S ribosomal subunit binding)	Nufarm/ Loveland Products	Pacios-Michelena et al. (2021)
YAN TEN Streptomyces®	*Streptomyces saraceticus* KH400	*Phytophthora*, *Rhizoctonia*, *Fusarium*	Secretion of antibiotics and hydrolytic enzymes; suppressive rhizosphere	YAN Agricultural BioTech, Taiwan	Alam et al. (2022)
Avermectin®/ Abamectin formulations	Avermectins (from *S. avermitilis*)	Insects, mites, nematodes	Binds glutamate-gated Cl^- channels → paralysis of invertebrates	Syngenta, BASF, etc.	Radwan et al. (2024)

they also help in the breakdown of organic matter into simpler forms of nitrogen that plants can use. In the soil, *Streptomyces* collectively function as silent companions. They increase nutrient availability, promote plant health and aid in the environmentally responsible survival of crops under stress. *Streptomyces* provide a natural, practical and cost-effective way to boost soil health and agricultural productivity as we search for more sustainable farming practices to feed the growing world population. They are useful companions in both conventional and modern farming because of their function in promoting plant growth. Future studies should focus on exploring diverse mechanisms *Streptomycetes* use to enhance the plant growth and stress tolerance. Integration of *Streptomycetes*-based bioinoculants with precision and sustainable technologies and methods could transform eco-friendly agriculture. Novel *Streptomyces* strains isolated from extreme environment may be explored for different metabolites beneficial for the plant. Developing climate resilient crops by using *Streptomyces* also present a promising approach for challenges including global food security.

References

Aamir M, Rai KK, Zehra A, Dubey MK et al (2020) Microbial bioformulation-based plant biostimulants: a plausible approach toward next generation of sustainable agriculture. In: Microbial endophytes. Woodhead Publishing, pp 195–225

AbdElgawad H, Abuelsoud W, Madany MM et al (2020) Actinomycetes enrich soil rhizosphere and improve seed quality as well as productivity of legumes by boosting nitrogen availability and metabolism. Biomolecules 10(12):1675

Agudelo-Morales CE, Lerma TA, Martínez JM et al (2021) Phytohormones and plant growth regulators – a review. J Sci Technol Appl 10:27–65

Akanmu AO, Olowe OM, Phiri AT et al (2023) Bioresources in organic farming: implications for sustainable agricultural systems. Horticulturae 9(6):659

Alam K, Mazumder A, Sikdar S, Zhao YM, Hao J, Song C, Li A (2022) Streptomyces: the biofactory of secondary metabolites. Front Microbiol 13:968053

Aldesuquy HS, Mansour FA, Abo-Hamed SA (1998) Effect of the culture filtrates of *Streptomyces* on growth and productivity of wheat plants. Folia Microbiol 43:465–470

Alinejad F, Khodakaramian G, Chandramowli D (2024) Catechol-type siderophore producer *Streptomyces* spp. with quorum-quenching activity isolated from wild medicinal plants rhizosphere. Microbe 2:100033

Ansabayeva A, Makhambetov M, Rebouh NY et al (2025) Plant growth-promoting microbes for resilient farming systems: mitigating environmental stressors and boosting crops productivity—a review. Horticulturae 11(3):260

Anwar S, Ali B, Sajid I (2016) Screening of rhizospheric actinomycetes for various in-vitro and in-vivo plant growth promoting (PGP) traits and for agroactive compounds. Front Microbiol 7:1334

Awad HM, El-Enshasy HA, Hanapi SZ et al (2014) A new chitinase-producer strain *Streptomyces glauciniger* WICC-A03: isolation and identification as a biocontrol agent for plant phytopathogenic fungi. Nat Prod Res 28(24):2273–2277

Ayed A, Kalai-Grami L, Ben Slimene I et al (2021) Antifungal activity of volatile organic compounds from *Streptomyces* sp. strain S97 against *Botrytis cinerea*. Biocontrol. Sci Technol 31(12):1330–1348

Badiyal A, Mahajan R, Rana RS et al (2024) Synergizing biotechnology and natural farming: pioneering agricultural sustainability through innovative interventions. Front Plant Sci 15: 1280846

Balakrishna G, Shiva Shanker A, Pindi PK (2012) Isolation of phosphate solibulizing actinomycetes from forest soils of Mahabubnagar district. IOSR J Pharm 2:271–275

Baweja P, Kumar S, Kumar G (2020) Fertilizers and pesticides: their impact on soil health and environment. In: Soil health, pp 265–285. https://doi.org/10.1007/978-3-030-44364-1_15

Bekiesch P, Oberhofer M, Sykora C et al (2021) Piperazic acid containing peptides produced by an endophytic *Streptomyces* sp. isolated from the medicinal plant Atropa belladonna. Nat Prod Res 35(7):1090–1096

Beroigui O, Errachidi F (2023) *Streptomyces* at the heart of several sectors to support practical and sustainable applications: a review. Prog Microbes Mol Biol 6(1). https://doi.org/10.36877/pmmb.a0000345

Bhai RS, Lijina A, Prameela TP, Krishna PB, Thampi AN (2016) Biocontrol and growth promotive potential of Streptomyces spp. in black pepper (Piper nigrum L.). J Biol Contr 30:177–189

Boubekri K, Soumare A, Mardad I (2021) The screening of potassium-and phosphate-solubilizing actinobacteria and the assessment of their ability to promote wheat growth parameters. Microorganisms 9:470

Boudjeko T, Tchinda RA, Zitouni M, Nana JA, Lerat S, Beaulieu C (2017) Streptomyces cameroonensis sp. nov., a geldanamycin producer that promotes Theobroma cacao growth. Microbes Environ 32:24–31

Boukaew S, Prasertsan P (2020) Efficacy of volatile compounds from *Streptomyces philanthi* RL-1-178 as a biofumigant for controlling growth and aflatoxin production of the two aflatoxin-producing fungi on stored soybean seeds. J Appl Microbiol 129(3):652–664

Cao L, Gao Y, Yu J et al (2021) Streptomyces hygroscopicus OsiSh-2-induced mitigation of Fe deficiency in rice plants. Plant Physiol Biochem 158:275–283

Castro-Camba R, Sánchez C, Vidal N et al (2022) Plant development and crop yield: the role of gibberellins. Plants 11:2650

Chaiharn M, Theantana T, Pathom-Aree W (2020) Evaluation of biocontrol activities of Streptomyces spp. against rice blast disease fungi. Pathogens 9:126

Chatterjee A, Saikia R, Saikia D, Kaman PK (2024) Molecular interactions between opportunistic fungi, nematodes, and plants. In: Opportunistic fungi, nematode and plant interactions: interplay and mechanisms. Springer Nature Singapore, Singapore, pp 165–186

Chen K, Li GJ, Bressan RA et al (2020) Abscisic acid dynamics, signaling, and functions in plants. J Integr Plant Biol 62:25–54

Chouyia FE, Ouhdouch Y, Barakate M (2020a) *Streptomyces* spp. as plant growth-promoting rhizobacteria: potential and mechanisms. Arch Microbiol 202(4):665–672

Chouyia FE, Romano I, Fechtali T et al (2020b) P-solubilizing *Streptomyces* roseocinereus MS1B15 with multiple plant growth-promoting traits enhance barley development and regulate rhizosphere microbial population. Front Plant Sci 7(11):1137

Chouyia FE, Ventorino V, Pepe O (2022) Diversity, mechanisms and beneficial features of phosphate-solubilizing *Streptomyces* in sustainable agriculture: a review. Front Plant Sci 13: 1035358

Chukwuneme CF, Babalola OO, Kutu FR et al (2020) Characterization of actinomycetes isolates for plant growth promoting traits and their effects on drought tolerance in maize. J Plant Interact 15:93–105

Conn VM, Walker AR, Franco CM (2008) Induced systemic resistance in plants. Plant Signal Behav 3(8):598–600

Cordovez V, Carrion VJ, Etalo DW et al (2015) Diversity and functions of volatile organic compounds produced by *Streptomyces* from a disease-suppressive soil. Front Microbiol 6:1081

Cunha A, Almeida A, Coelho FJRC et al (2010) Bacterial extracellular enzymatic activity in globally changing aquatic ecosystems. Curr Res Technol Educ Top Appl Microbiol Microb Biotechnol 1:124–135

Das S, Shil S, Rime J et al (2025) Phytohormonal signaling in plant resilience: advances and strategies for enhancing abiotic stress tolerance. Plant Growth Regul 105:329–360

Devi S, Sharma M, Manhas RK (2022) Investigating the plant growth promoting and biocontrol potentiality of endophytic Streptomyces SP. SP5 against early blight in Solanum lycopersicum seedlings. BMC Microbiol 22:285

Dimkpa C, Svatoš A, Merten D et al (2008) Hydroxamate siderophores produced by *Streptomyces acidiscabies* E13 bind nickel and promote growth in cowpea (*Vigna unguiculata* L.) under nickel stress. Can J Microbiol 54:163–172

Djebaili R, Pellegrini M, Rossi M, Forni C, Smati M, Del Gallo M, Kitouni M (2021) Characterization of plant growth-promoting traits and inoculation effects on Triticum durum of actinomycetes isolates under salt stress conditions. Soil Syst 5:26

Djuidje PK, Asultan W, Beaulieu C, Wong MY, Boudjeko T (2022) Characterization of endophytic Streptomyces strains from roots of cocoyam (Xanthosoma sagittifolium L. Schott) in the South West region of Cameroon, their in vitro plant growth promoting abilities and biocontrol efficacy against Pythium myriotylum. S Afr J Bot 144:145–155

Dutta J, Thakur D (2017) Evaluation of multifarious plant growth promoting traits, antagonistic potential and phylogenetic affiliation of rhizobacteria associated with commercial tea plants grown in Darjeeling, India. PloS One 12:e0182302

Du Y, Wang T, Jiang J, Wang Y, Lv C, Sun K, Sun J, Yan B, Kang C, Guo L, Huang L (2022) Biological control and plant growth promotion properties of Streptomyces albidoflavus St-220 isolated from Salvia miltiorrhiza rhizosphere. Front Plant Sci 13:976813

Erdiandini I, Tjahjoleksono A, Astuti RI, Husen E, Wahyudi AT (2025) Peat-derived Streptomyces spp. isolated from edamame rhizosphere with plant growth-promoting properties. Biodiversitas 26:326–334

Elnahal AS, El-Saadony MT, Saad AM et al (2022) The use of microbial inoculants for biological control, plant growth promotion, and sustainable agriculture: a review. Eur J Plant Pathol 162(4):759–792

El-Tarabily KA (2008) Promotion of tomato (*Lycopersicon esculentum* Mill.) plant growth by rhizosphere competent 1-aminocyclopropane-1-carboxylic acid deaminase-producing streptomycete actinomycetes. Plant Soil 308:161–174

El-Tarabily KA, Sivasithamparam K (2006) Non-streptomycete actinomycetes as biocontrol agents of soil-borne fungal plant pathogens and as plant growth promoters. Soil Biol Biochem 38(7): 1505–1520

Gáll T, Lehoczki G, Gyémánt G (2016) Optimization of desferrioxamine E production by *Streptomyces parvulus*. Acta Microbiol Immunol Hung 63:475–489

Geranios P (2018) Negative cell autotropisms in Aspergillus fumigatus. The University of Manchester

Getha K, Vikineswary S, Wong WH, Seki T, Ward A, Goodfellow M (2005) Evaluation of Streptomyces sp. strain g10 for suppression of Fusarium wilt and rhizosphere colonization in pot-grown banana plantlets. J Ind Microbiol Biotechnol 32(1):24–32

Ghanem GA, Gebily DA, Ragab MM et al (2022) Efficacy of antifungal substances of three *Streptomyces* spp. against different plant pathogenic fungi. Egypt J Biol Pest Control 32(1):112

Glick BR (2014) Bacteria with ACC deaminase can promote plant growth and help to feed the world. Microbiol Res 169(1):30–39

Gomes RC, Semedo LTAS, Soares RMA et al (2001) Purification of a thermostable endochitinase from *Streptomyces* RC1071 isolated from a cerrado soil and its antagonism against phytopathogenic fungi. J Appl Microbiol 90(4):653–661

Gopalakrishnan S, Srinivas V, Vidya MS et al (2011) Evaluation of actinomycetes for biological control of Fusarium wilt and plant growth-promotion in chickpea. Ann Microbiol 61:623–633

Gopalakrishnan S, Sathya A, Vijayabharathi R et al (2013) Effect of plant growth-promoting actinomycetes on plant growth, rhizosphere microbial populations, and soil health of rice. Indian J Microbiol 53(4):447–453

Gopalakrishnan S, Vadlamudi S, Samineni S, Sameer Kumar C (2015) Plant growth-promotion and biofortification of chickpea and pigeonpea through inoculation of biocontrol potential bacteria, isolated from organic soils. Springerplus 5(1):1882

Grover M, Ali SZ, Sandhya V et al (2011) Role of microorganisms in adaptation of agriculture crops to abiotic stresses. World J Microbiol Biotechnol 27:1231–1240

Guo L, Zhang X, Liu Y et al (2025) Salt-alkali-tolerant growth-promoting *Streptomyces* sp. Jrh8-9 enhances alfalfa growth and resilience under saline-alkali stress through integrated modulation of photosynthesis, antioxidant defense, and hormone signaling. Microbiol Res 296:128158

Gupta R, Saxena RK, Chaturvedi P, Virdi JS (1995) Chitinase production by *Streptomyces viridificans*: its potential in fungal cell wall lysis. J Appl Bacteriol 78(4):378–383

Hamby MK, Crawford DL (2000) The enhancement of plant growth by selected Streptomyces species. In: American Society for Microbiology, 100th general meeting, Los Angeles, CA. Abstract (No. 567)

Hamdan MF, Mohd Noor SN, Abd-Aziz N, Pua TL, Tan BC (2022) Green revolution to gene revolution: technological advances in agriculture to feed the world. Plants 11(10):1297

Hashem A, Tabassum B, Abd_Allah EF (2019) *Bacillus subtilis*: a plant-growth promoting rhizobacterium that also impacts biotic stress. Saudi J Biol Sci 26:1291–1297

He Y, Guo W, Peng J, Guo J, Ma J, Wang X, Wang Z (2022) Volatile organic compounds of *Streptomyces* sp. TOR3209 stimulated tobacco growth by up-regulating the expression of genes related to plant growth and development. Front Microbiol 13:891245

He D, Gao C, Zhao S et al (2024) Antibacterial, herbicidal, and plant growth-promoting properties of *Streptomyces* sp. STD57 from the rhizosphere of *Adenophora stricta*. Microorganisms 12: 2245

Igarashi Y, Iida T, Yoshida R et al (2002) Pteridic acids A and B, novel plant growth promoters with auxin-like activity from *Streptomyces hygroscopicus* TP-A0451. J Antibiot 55:764–767

Jacob S, Sudini HK (2016) Indirect plant growth promotion in grain legumes: role of Actinobacteria. In: Subramaniam G, Arumugam S, Rajendran V (eds) Plant growth promoting Actinobacteria. Springer, Singapore, pp 17–32

Jaemsaeng R, Jantasuriyarat C, Thamchaipenet A (2018) Positive role of 1-aminocyclopropane-1-carboxylate deaminase-producing endophytic *Streptomyces* sp. GMKU 336 on flooding resistance of mung bean. Agric Nat Resour 52:330–334

Javed Z, Tripathi GD, Mishra M, Dashora K (2021) Actinomycetes–the microbial machinery for the organic-cycling, plant growth, and sustainable soil health. Biocatal Agric Biotechnol 31:101893

Jog R, Pandya M, Nareshkumar G, Rajkumar S (2012) *Streptomyces* spp.: a novel plant growth-promoting rhizobacteria. Indian J Microbiol 52(1):124–131

Jog R, Pandya M, Nareshkumar G, Shalini G (2014) Mechanism of phosphate solubilization and antifungal activity of Streptomyces spp. isolated from wheat roots and rhizosphere and their application in improving plant growth. Microbiology 160:778–788

Joo GJ (2005) Purification and characterization of an extracellular chitinase from the antifungal biocontrol agent *Streptomyces halstedii*. Biotechnol Lett 27:1483–1486

Kadaikunnan S, Alharbi NS, Khaled JM, Alobaidi AS (2023) Biocontrol property of *Streptomyces parvulus* VRR3 in green gram plant (*Vigna radiata* L.) against *Fusarium solani* in greenhouse. Physiol Mol Plant 128:102128

Kaur T, Manhas RK (2022) Evaluation of ACC deaminase and indole acetic acid production by *Streptomyces hydrogenans* DH16 and its effect on plant growth promotion. Biocatal Agric Biotechnol 42:102321

Kaur R, Kaur J, Singh RS (2011) Non-pathogenic Fusarium as a biological control agent. Plant Pathol J 9(3):79–91

Khan N (2025) Decoding Phytohormone signaling in plant stress physiology: insights, challenges and future directions. Environ Exp Bot 5:106099

Khan S, Alvi AF, Khan NA (2024) Role of ethylene in the regulation of plant developmental processes. Stresses 4:28–53

Kim KH, Kabir E, Jahan SA (2017) Exposure to pesticides and the associated human health effects. Sci Total Environ 575:525–535

Kim D, Kim J, Lee Y, Balaraju K, Hwang YJ, Lee MH, Jeon Y (2024) Evaluation of *Streptomyces sporoverrucosus* B-1662 for biological control of red pepper anthracnose and apple bitter rot diseases in Korea. Front Microbiol 15:1429646

Kinkel LL, Schlatter DC, Bakker MG, Arenz BE (2012) *Streptomyces* competition and co-evolution in relation to plant disease suppression. Res Microbiol 163(8):490–499

Kumar A, Kumar R, Kumari M, Goldar S (2020) Enhancement of plant growth by using PGPR for a sustainable agriculture: a review. Int J Curr Microbiol App Sci 9(2):152–165

Kulaeva ON, Prokoptseva OS (2004) Recent advances in the study of mechanisms of action of phytohormones. Biochem Mosc 69:233–247

Law JWF, Pusparajah P, Ab Mutalib NS et al (2019) A review on mangrove actinobacterial diversity: the roles of *Streptomyces* and novel species discovery. Prog Microbes Mol Biol 2(1). https://doi.org/10.36877/pmmb.a0000024

Leben C, Keitt GW (1954) Antibiotics and plant disease, effects of antibiotics in control of plant diseases. J Agric Food Chem 2(5):234–239

Lee SY, Tindwa H, Lee YS et al (2012) Biocontrol of anthracnose in pepper using Chitinase, β-1, 3 Glucanase, and 2-Furancarboxaldehyde produced by *Streptomyces cavourensis* SY224. J Microbiol Biotechnol 22(10):1359–1366

Li Q, Ning P, Zheng L, Huang J, Li G, Hsiang T (2010) Fumigant activity of volatiles of *Streptomyces globisporus* JK-1 against *Penicillium italicum* on *Citrus microcarpa*. Postharvest Biol Technol 58(2):157–165

Li SM, Zheng HX, Zhang XS et al (2020a) Cytokinins as central regulators during plant growth and stress response. Plant Cell Rep 40:271–282

Li X, Li B, Cai S et al (2020b) Identification of rhizospheric actinomycete *Streptomyces lavendulae* sps-33 and the inhibitory effect of its volatile organic compounds against *Ceratocystis fimbriata* in postharvest sweet potato (Ipomoea batatas (L.) Lam.). Microorganisms 8(3):319

Li S, Liang W, Huang H et al (2024) Antagonistic and plant growth-promoting properties of *Streptomyces* F2 isolated from vineyard soil. Agronomy 14:1489

López-Reyes PK, De la Torre-Zavala S, Cortés-González MM et al (2024) Biological control of *Streptomyces* sp. PR69 against *Phytophthora capsici* and its growth-promoting effects on plants. Horticulturae 10:1365

Louviot F, Abdelrahman O, Abou-Mansour E et al (2024) Oligomycin-producing Streptomyces sp. newly isolated from Swiss soils efficiently protect *Arabidopsis thaliana* against *Botrytis cinerea*. Msphere 9(7):e00667–e00623

Luo F, Gwak H, Park AR, Nguyen VT, Kim JC (2025) Biocontrol potential of natamycin-producing *Streptomyces lydicus* JCK-6019 against soil-borne fungal diseases of cucumber and characterization of its biocontrol mechanism. Pest Manag Sci 81(4):1971–1987

Lyu A, Yang L, Wu M, Zhang J, Li G (2020) High efficacy of the volatile organic compounds of *Streptomyces yanglinensis* 3-10 in suppression of *Aspergillus* contamination on peanut kernels. Front Microbiol 11:142

Macagnan D, Romeiro RDS, Pomella AW, de Souza JT (2008) Production of lytic enzymes and siderophores, and inhibition of germination of basidiospores of *Moniliophthora* (ex Crinipellis) *perniciosa* by phylloplane actinomycetes. Biol Control 47(3):309–314

Mahadevan B, Crawford DL (1997) Properties of the chitinase of the antifungal biocontrol agent *Streptomyces lydicus* WYEC108. Enzym Microb Technol 20(7):489–493

Manullang W, Chuang HW (2020) *Streptomyces* sp. mitigates abiotic stress response and promotes plant growth. J Plant Prot Res 60:263–274

Meena M, Swapnil P, Zehra A (2017) Beneficial microbes for disease suppression and plant growth promotion. In: Singh D, Singh H, Prabha R (eds) Plant-microbe interactions in agro-ecological perspectives. Springer, Singapore, pp 395–432

Minuto A, Spadaro D, Garibaldi A, Gullino ML (2006) Control of soilborne pathogens of tomato using a commercial formulation of *Streptomyces griseoviridis* and solarization. Crop Prot 25(5): 468–475

Mishra P, Mishra J, Dwivedi SK, Arora NK (2020) Microbial enzymes in biocontrol of phytopathogens. In: Microbial enzymes: roles and applications in industries, pp 259–285

Mitousis L, Thoma Y, Musiol-Kroll EM (2020) An update on molecular tools for genetic engineering of actinomycetes—the source of important antibiotics and other valuable compounds. Antibiotics 9(8):494

Mo W, Zheng X, Shi Q et al (2024) Unveiling the crucial roles of abscisic acid in plant physiology: implications for enhancing stress tolerance and productivity. Front Plant Sci 15:1437184

Mun BG, Lee WH, Kang SM, Lee SU, Lee SM, Lee DY, Shahid M, Yun BW, Lee IJ (2020) *Streptomyces* sp. LH 4 promotes plant growth and resistance against *Sclerotinia sclerotiorum* in cucumber via modulation of enzymatic and defense pathways. Plant Soil 448:87–103

Musani T, Das M (2024) Effect of Siderophore producing cadmium tolerant *Streptomyces pactum* OR958669 on plant growth of groundnut (*Arachis hypogaea* L.). Curr Agric Res J 12:485–495

Nagpure A, Choudhary B, Kumar S, Gupta RK (2014) Isolation and characterization of chitinolytic *Streptomyces* sp. MT7 and its antagonism towards wood-rotting fungi. Ann Microbiol 64:531–541

Narayanasamy P, Narayanasamy P (2013) Mechanisms of action of fungal biological control agents. In: Biological management of diseases of crops: volume 1: characteristics of biological control agents, pp 99–200

Ngalimat MS, Mohd Hata E, Zulperi D et al (2023) A laudable strategy to manage bacterial panicle blight disease of rice using biocontrol agents. J Basic Microbiol 63(11):1180–1195

Nguyen XH, Naing KW, Lee YS, Kim YH, Moon JH, Kim KY (2015) Antagonism of antifungal metabolites from *Streptomyces griseus* H7602 against *Phytophthora capsici*. J Basic Microbial 55(1):45–53

Niu S, Gao Y, Zi H, Liu Y, Liu X, Xiong X, Yang Y, Zhu Y (2022) The osmolyte-producing endophyte *Streptomyces albidoflavus* OsiLf-2 induces drought and salt tolerance in rice via a multi-level mechanism. Crop J 10(2):375–386

Nivedita S, Behera SS, Behera PK (2024) Salt-resistant *Streptomyces* consortia promote growth of rice (*Oryza sativa* var. Swarna) alleviating salinity and drought stress tolerance by enhancing photosynthesis, antioxidant function, and proline content. Microbe 4:100124

Nonthakaew N, Panbangred W, Songnuan W et al (2022) Plant growth-promoting properties of *Streptomyces* spp. isolates and their impact on mung bean plantlets' rhizosphere microbiome. Front Microbiol 13:967415

Oluwole O, Ibidapo O, Arowosola T, Raji F, Zandonadi RP, Alasqah I et al (2023) Sustainable transformation agenda for enhanced global food and nutrition security: a narrative review. Front Nutr 10:1226538

Omar AF, Abdelmageed AH, Al-Turki A, Abdelhameid NM, Sayyed RZ, Rehan M (2022) Exploring the plant growth-promotion of four Streptomyces strains from rhizosphere soil to enhance cucumber growth and yield. Plants 11:3316

Oyedoh OP, Yang W, Dhanasekaran D, Santoyo G, Glick BR, Babalola OO (2023) Rare rhizo-Actinomycetes: a new source of agroactive metabolites. Biotechnol Adv 67:108205

Pacios-Michelena S, Aguilar González CN, Alvarez-Perez OB, Rodriguez-Herrera R, Chávez-González M, Arredondo Valdés R, Ilyina A (2021) Application of Streptomyces antimicrobial compounds for the control of phytopathogens. Front Sustainable Food Syst 5:696518

Pahalvi HN, Rafiya L, Rashid S, Nisar B, Kamili AN (2021) Chemical fertilizers and their impact on soil health. In: Microbiota and biofertilizers, vol 2: ecofriendly tools for reclamation of degraded soil environs, pp 1–20

Palaniyandi SA, Damodharan K, Yang SH, Suh JW (2013) *Streptomyces* sp. as efficient plant growth-promoting rhizobacteria: mechanisms and future prospects. World J Microbiol Biotechnol 29(3):427–435

Pandey P, Irulappan V, Bagavathiannan MV et al (2017) Impact of combined abiotic and biotic stresses on plant growth and avenues for crop improvement by exploiting physio-morphological traits. Front Plant Sci 8:537

Pang F, Solanki MK, Wang Z (2022) *Streptomyces* can be an excellent plant growth manager. World J Microbiol Biotechnol 38(11):193

Patel P, Trivedi G, Saraf M (2018) Iron biofortification in mungbean using siderophore producing plant growth promoting bacteria. Environ Sustain 1(4):357–365

Pedó CJ, Ramos LM, Fick GP et al (2025) Evaluation of the tolerance of plant growth-promoting *Streptomyces* spp. to glyphosate-based herbicide. S Afr J Bot 186:293–302

Prabhakar AC, Brar GP (2024) Green revolution, agricultural performance with sustainability and bio-diversity: special reference to India. Int J Econ Perform 7(1):281–318

Puppala KR, Bhavsar K, Sonalkar V, Khire JM, Dharne MS (2019) Characterization of novel acidic and thermostable phytase secreting *Streptomyces* sp. (NCIM 5533) for plant growth promoting characteristics. Biocatal Agric Biotechnol 18:101020

Quecine MC, Araujo WL, Marcon J, Gai CS, Azevedo JLD, Pizzirani-Kleiner AA (2008) Chitinolytic activity of endophytic *Streptomyces* and potential for biocontrol. Lett Appl Microbiol 47(6):486–491

Radwan WH, Abdelhafez AA, Mahgoub AE, Zayed MS (2024) Streptomyces avermitilis MICNEMA2022: a new biorational strain for producing abamectin as an integrated nematode management agent. BMC Microbiol 24(1):329

Rungin S, Indananda C, Suttiviriya P, Kruasuwan W, Jaemsaeng R, Thamchaipenet (2012) Plant growth enhancing effects by a siderophore-producing endophytic streptomycete isolated from a Thai jasmine rice plant (Oryza sativa L. cv. KDML105). Antonie Van Leeuwenhoek 102(3): 463–472

Rashad FM, Fathy HM, El-Zayat AS et al (2015) Isolation and characterization of multifunctional *Streptomyces* species with antimicrobial, nematicidal and phytohormone activities from marine environments in Egypt. Microbiol Res 175:34–47

Rawat P, Das S, Shankhdhar D et al (2020) Phosphate-solubilizing microorganisms: mechanism and their role in phosphate solubilization and uptake. J Soil Sci Plant Nutr 21:49–68

Rey T, Dumas B (2017) Plenty is no plague: *Streptomyces* symbiosis with crops. Trends Plant Sci 22(1):30–37

Ruangwong OU, Kunasakdakul K, Daengsuwan W, Wonglom P, Pitija K, Sunpapao A (2022) A *Streptomyces* rhizobacterium with antifungal properties against spadix rot in flamingo flowers. Physiol Mol Plant 117:101784

Ruanpanun P, Nimnoi P (2020) Evaluation on the efficiency and persistence of *Streptomyces jietaisiensis* strain A034 in controlling root knot disease and promoting plant growth in the plant-parasitic nematode infested soils. Biol Control 144:104221

Ruanpanun P, Pirankham P, Dilokpimol A et al (2025) Rice rhizosphere bacteria displayed nematicidal activities against the rice root-knot nematode *Meloidogyne graminicola*. Rhizosphere 36:101183

Sachdev S, Ansari SA, Ansari MI et al (2021) Abiotic stress and reactive oxygen species: generation, signaling, and defense mechanisms. Antioxidants 10:277

Sadeghi A, Karimi E, Dahaji PA (2012) Plant growth promoting activity of an auxin and siderophore producing isolate of *Streptomyces* under saline soil conditions. World J Microbiol Biotechnol 28:1503–1509

Samantaray A, Chattaraj S, Mitra D et al (2024) Advances in microbial based bio-inoculum for amelioration of soil health and sustainable crop production. Curr Res Microb Sci 7:100251

Schrey SD, Tarkka MT (2008) Friends and foes: Streptomycetes as modulators of plant disease and symbiosis. Antonie Van Leeuwenhoek 94:11–19

Sekar J, Raj R, Prabavathy VR (2016) Microbial consortial products for sustainable agriculture: commercialization and regulatory issues in India. In: Agriculturally important microorganisms: commercialization and regulatory requirements in Asia. Springer Singapore, Singapore, pp 107–132

Shahid M, Singh BN, Verma S, Choudhary P, Das S, Chakdar H, Murugan K, Goswami SK, Saxena AK (2021) Bioactive antifungal metabolites produced by *Streptomyces* amritsarensis V31 help to control diverse phytopathogenic fungi. Braz J Microbiol 52:1687–1699

Shen W, Liu R, Wang J, Yang M, Qi T, Shu G, Chen X (2024) Characterization of a broad-spectrum antifungal strain, *Streptomyces graminearus* STR-1, against *Magnaporthe oryzae.* Front Microbiol 15:1298781

Shrivastava P, Kumar R (2018) Actinobacteria: eco-friendly candidates for control of plant diseases in a sustainable manner. In: New and future developments in microbial biotechnology and bioengineering. Elsevier, pp 79–91

Singh Y (2023) A textbook of fungi, bacteria and viruses. Academic Guru Publishing House

Singh P, Chauhan PK, Upadhyay SK et al (2022) Mechanistic insights and potential use of siderophores producing microbes in rhizosphere for mitigation of stress in plants grown in degraded land. Front Microbiol 13:898979

Singh RP, Manchanda G, Maurya IK, Maheshwari NK, Tiwari PK, Rai AR (2019) Streptomyces from rotten wheat straw endowed the high plant growth potential traits and agro-active compounds. Biocatal Agric Biotechnol 17:507–513

Stulanovic N, Bensaada D, Belde L, Adam D, Hanikenne M, Focant JF, Rigali S (2025) Antifungal and other bioactive properties of the volatilome of *Streptomyces scabiei.* Appl Environ Microbiol 91:e00863-25

Suriyachadkun C, Chunhachart O, Srithaworn M (2022) Zinc-solubilizing *Streptomyces* spp. as bioinoculants for promoting the growth of soybean (Glycine max (L.) Merrill). J Microbiol Biotechnol 32:1435

Taechowisan T, Lu C, Shen Y, Lumyong S (2005) Secondary metabolites from endophytic *Streptomyces aureofaciens* CMUAc130 and their antifungal activity. Microbiology (Reading) 151(5):1691–1695

Talaat NB, Shawky BT (2017) Microbe-mediated induced abiotic stress tolerance responses in plants. In: Singh D, Singh H, Prabha R (eds) Plant-microbe interactions in agro-ecological perspectives. Springer, Singapore, pp 101–133

Tamreihao K, Ningthoujam DS, Nimaichand S (2016) Biocontrol and plant growth promoting activities of a *Streptomyces corchorusii* strain UCR3-16 and preparation of powder formulation for application as biofertilizer agents for rice plant. Microbiol Res 192:260–270

Tamreihao K, Nimaichand S, Chanu SB, Devi KA, Lynda R, Jeeniita N, Ningthoujam DS, Roy SS (2019) Streptomyces manipurensis MBRL 201T as potential candidate for biocontrol and plant growth promoting agent for rice. Indian J Exp Biol 57:741–749

Thompson GD, Hutchins SH, Sparks TC (1999) Development of spinosad and attributes of a new class of insect control products. In: Radcliffe EB, Hutchinson WD (eds) Radcliffe's IPM World textbook. University of Minnesota, St. Paul. http:llipmworldMmn.edu

Tian X, Hu L, Jia R et al (2024) *Streptomyces pratensis* S10 promotes wheat plant growth and induces resistance in wheat seedlings against *Fusarium graminearum.* J Fungi 10:578

Tigani W, Du C, Elsayed SS, Innocenti P et al (2025) Antifungal biosynthesis by root-associated *Streptomyces* and *Pseudomonas* is elicited upon plant colonization. bioRxiv. 2025:2025-04

Tiwari PK, Mandal S, Anchal A. (2021) The screening, characterization and identification of chitinolytic actinomycetes *Streptomyces* sp. ANU5a. https://doi.org/10.21203/rs.3.rs-473198/v1

Tokala RK, Strap JL, Jung CM, Crawford DL, Salove MH, Deobald LA, Bailey JF, Morra MJ (2002) Novel plant-microbe rhizosphere interaction involving *Streptomyces lydicu*s WYEC108 and the pea plant (*Pisum sativum*). Appl Environ Microbiol 68:2161–2171

Trejo-Estrada SR, Paszczynski A, Crawford DL (1998) Antibiotics and enzymes produced by the biocontrol agent *Streptomyces violaceusniger* YCED-9. J JIMB 21(1–2):81–90

Verma VC, Singh SK, Prakash S (2011) Bio-control and plant growth promotion potential of siderophore producing endophytic *Streptomyces* from *Azadirachta indica* A. Juss. J Basic Microbiol 51:550–556

Viaene T, Langendries S, Beirinckx S, Maes M, Goormachtig S (2016) *Streptomyces* as a plant's best friend? FEMS Microbiol Ecol 92(8):fiw119

Villafañe DL, Maldonado RA, Bianchi JS et al (2024) *Streptomyces* N2A, an endophytic actinobacteria that promotes soybean growth and increases yield and seed quality under field conditions. Plant Sci 343:112073

Vurukonda SSKP, Giovanardi D, Stefani E (2018) Plant growth promoting and biocontrol activity of *Streptomyces* spp. as endophytes. Int J Mol Sci 19(4):952

Vurukonda SSKP, Giovanadri D, Stefani E (2021) Growth promotion and biocontrol activity of Endophytic Streptomyces spp. In: Prime archives in molecular sciences, vol 1, 2nd edn, pp 1–55

Wang C, Wang Z, Qiao X, Li Z, Li F, Chen M, Wang Y, Huang Y, Cui H (2013) Antifungal activity of volatile organic compounds from *Streptomyces alboflavus* TD-1. FEMS Microbiol Lett 341(1):45–51

Wonglom P, Suwannarach N, Lumyong S, Ito SI, Matsui K, Sunpapao A (2019) *Streptomyces angustmyceticus* NR8-2 as a potential microorganism for the biological control of leaf spots of *Brassica rapa* subsp. pekinensis caused by *Colletotrichum* sp. and *Curvularia lunata*. Biol Control 138:104046

Worsley SF, Newitt J, Rassbach J, Batey SF, Holmes NA, Murrell JC, Wilkinson B, Hutchings MI (2020) Streptomyces endophytes promote host health and enhance growth across plant species. Appl Environ Microbiol 86:e01053–20

Wu Y, Yuan JEY, Raza W, Shen Q, Huang Q (2015) Effects of volatile organic compounds from *Streptomyces albulus* NJZJSA2 on growth of two fungal pathogens. J Basic Microbiol 55(9): 1104–1117

Xiayun J, Chen D, Shenle H, Wang W, Chen S, Zou S (2012) Identification, characterization and functional analysis of a GH-18 chitinase from *Streptomyces roseolus*. Carbohydr Polym 87(4): 2409–2415

Xu W, Liu Y, Cheng Y et al (2025) Plant growth-promoting effect and complete genomic sequence analysis of the beneficial rhizosphere *Streptomyces* sp. GD-4 isolated from *Leymus secalinus*. Microorganisms 13:286

Yadav AN (2022) Potassium-solubilizing microorganisms for agricultural sustainability. J Appl Biol Biotechnol 10:1–4

Yandigeri MS, Meena KK, Singh D (2012) Drought-tolerant endophytic actinobacteria promote growth of wheat (*Triticum aestivum*) under water stress conditions. Plant Growth Regul 68:411–420

Yang Z, Qiao Y, Konakalla NC et al (2023) *Streptomyces alleviate* abiotic stress in plant by producing pteridic acids. Nat Commun 14:7398

Yoolong S, Kruasuwan W, Thanh Phạm HT (2019) Modulation of salt tolerance in Thai jasmine rice (*Oryza sativa* L. cv. KDML105) by *Streptomyces venezuelae* ATCC 10712 expressing ACC deaminase. Sci Rep 9:1275

Yu J, Liu Q, Liu Q, Liu X, Sun Q, Yan J, Fan S (2008) Effect of liquid culture requirements on antifungal antibiotic production by *Streptomyces rimosus* MY02. Bioresour Technol 99(6): 2087–2091

Yuan WM, Crawford DL (1995) Characterization of *Streptomyces lydicus* WYEC108 as a potential biocontrol agent against fungal root and seed rots. Appl Environ Microbiol 61(8):3119–3128

Zhao X, Zhou J, Tian R, Liu Y (2022) Microbial volatile organic compounds: antifungal mechanisms, applications, and challenges. Front Microbiol 13:922450

Zheng X, Wang J, Chen Z, Zhang H, Wang Z, Zhu Y, Liu B (2019) A *Streptomyces* sp. strain: isolation, identification, and potential as a biocontrol agent against soilborne diseases of tomato plants. Biol Control 136:104004

Zhou W, Li M, Achal V (2024) A comprehensive review on environmental and human health impacts of chemical pesticide usage. Emerg Contam 11:100410

Plant Growth-Promoting Streptomyces: Production, Delivery System, and Commercialization

11

Anand Dave

Abstract

Streptomyces species, a prominent group within the Actinobacteria, thrive in the plant rhizosphere where they exert diverse influences on plant development and physiological processes. These organisms play a significant role in protecting plants against various diseases, primarily through their capacity to act as plant growth-promoting Streptomyces (PGPS). Widely applied in sustainable agriculture, PGPS function as both biofertilizers and biocontrol agents. They are typically produced via fermentation and can be applied to crops through seed coating or soil inoculation methods. *Streptomyces* spp. and their secondary metabolites exhibit strong antagonistic activity against a wide range of fungal and bacterial phytopathogens, largely attributed to their production of antibiotics and volatile organic compounds (VOCs) in both soil and *in planta*. These bioactive compounds are instrumental in identifying potent antagonists of plant pathogens and have demonstrated efficacy across multiple cropping systems. Furthermore, the plant growth-promoting effects of *Streptomyces* spp. have been validated in several crops, thereby supporting their broad use as biofertilizers to enhance crop productivity. For practical field applications, formulations of *Streptomyces* strains using talcum and kaolin as carriers have been developed. Among these, talcum-based formulations exhibited superior cell viability after 6 months of storage, with 4 °C identified as the optimal storage temperature. Field trials utilizing the talcum-based powder formulation resulted in significant improvements in shoot length, shoot and root biomass, grain yield, and individual grain weight.

A. Dave (✉)
Department of Microbiology and Biotechnology Centre, The Maharaja Sayajirao University of Baroda, Vadoadara, Gujarat, India

T. Kaur et al. (eds.), *Streptomyces: Biological Candidates for Sustainable Agriculture*,
https://doi.org/10.1007/978-981-95-5803-2_11

Keywords

Plant growth-promoting Streptomyces (PGPS) · *Streptomyces* sp. · Biofertlizer · Talcum · Kaolin · Phytopathogens

11.1 Introduction

Streptomyces, a diverse genus within the phylum Actinobacteria, is widely recognized for its unparalleled capacity to biosynthesize a broad spectrum of bioactive secondary metabolites (Alam et al. 2022). The seminal discovery of streptomycin by Selman A. Waksman and colleagues in the 1940s marked a pivotal advancement in antimicrobial chemotherapy and firmly established *Streptomyces* as a cornerstone of natural product research (Woodruff 2014). Since then, species within this genus have been credited with the production of over two-thirds of clinically significant antibiotics, along with a myriad of antifungal, antiviral, immunosuppressive, and anticancer agents. Beyond their pharmaceutical relevance, *Streptomyces* spp. play critical ecological roles in terrestrial environments, particularly within the rhizosphere, where they facilitate organic matter decomposition, nutrient cycling, and the modulation of the microbial community structure, processes essential for soil health and plant productivity (Thepbandit and Athinuwat 2024).

The rhizosphere is a highly dynamic microenvironment at the soil–root interface, characterized by complex interactions between plant roots and soil microorganisms. Within this niche, *Streptomyces* species leverage their filamentous, hyphal-like morphology to effectively colonize root surfaces and explore the surrounding soil matrix (Olanrewaju et al. 2019; Thepbandit and Athinuwat 2024). This growth habit enhances spatial competitiveness and nutrient acquisition, while their ability to sporulate provides resilience against abiotic stress conditions, ensuring persistence in fluctuating soil environments. Moreover, *Streptomyces* are prolific producers of antimicrobial compounds that suppress phytopathogens, contributing to the maintenance of a balanced microbial community, and the promotion of plant health (Newitt et al. 2019; Dow et al. 2023; Khan et al. 2023).

Recent research has increasingly focused on plant growth-promoting *Streptomyces* (PGPS), which enhance plant development through multiple direct and indirect mechanisms. PGPS contribute to improved nutrient availability by solubilizing insoluble phosphate, and synthesizing siderophores that chelate iron, thereby increasing its bioavailability in iron-limited soils. Some strains are also involved in nitrogen metabolism, further supporting plant nutritional status (Bargaz et al. 2021; De Andrade et al. 2023). In addition, PGPS produce a suite of phytohormones, including indole-3-acetic acid (IAA), gibberellins, and cytokinins, which modulate root system architecture by stimulating lateral root development and root hair elongation. These changes collectively improve water and nutrient uptake efficiency, leading to enhanced plant growth and productivity (Sousa and Olivares 2016; Nonthakaew et al. 2022; Wang et al. 2023).

Beyond direct plant growth enhancement, plant growth-promoting *Streptomyces* (PGPS) exhibit robust biocontrol activity against a broad spectrum of phytopathogens. These effects are mediated through the production of antibiotics, lytic enzymes such as chitinases and β-1,3-glucanases, and volatile organic compounds (VOCs), which collectively suppress fungal and bacterial pathogens (El-Saadony et al. 2024; Villavicencio-Vásquez et al. 2025). In addition, PGPS can trigger induced systemic resistance (ISR) in plants, a defense mechanism governed primarily by jasmonic acid- and ethylene-dependent signaling pathways. ISR primes the plant immune system, enabling a more rapid and robust defense response upon subsequent pathogen exposure, even in the absence of a prior infection (Kour et al. 2024; Dave et al. 2024).

PGPS also contribute to the mitigation of abiotic stresses such as drought, salinity, and heavy metal toxicity. These microorganisms confer stress resilience by producing osmoprotectants, antioxidant enzymes, and other protective metabolites that maintain cellular homeostasis and integrity under adverse environmental conditions (Sousa and Olivares 2016; Nonthakaew et al. 2022). Moreover, the synthesis of extracellular polysaccharides and the formation of biofilms by PGPS enhance soil structure, water retention, and nutrient stabilization in the rhizosphere, thereby creating a more favorable microenvironment for root development and overall plant performance (Chen et al. 2024). The ecological versatility and multifunctionality of *Streptomyces* make them valuable candidates in sustainable agriculture. Their integrated roles in promoting plant growth, suppressing phytopathogens, and enhancing stress tolerance align with the growing demand for environmentally benign alternatives to synthetic agrochemicals. However, the successful translation of laboratory and greenhouse findings into consistent field performance remains a significant challenge, owing to strain-specific responses, environmental variability, and complex microbial interactions in the rhizosphere. Overcoming these constraints necessitates a deeper understanding of PGPS ecology, mechanisms of action, and formulation of strategies tailored to specific agroecosystems (Santoyo 2022; Copeland et al. 2025).

This chapter provides a comprehensive overview of current knowledge on plant growth-promoting *Streptomyces*, with particular emphasis on their functional roles in the rhizosphere and potential as bioinoculants for sustainable crop production. By integrating insights from microbial ecology, plant physiology, and biotechnological innovation, the chapter aims to inform future efforts to deploy these beneficial microorganisms effectively in field settings, contributing to resilient and high-performing agroecosystems.

11.2 Functional Mechanisms of Plant Growth-Promoting *Streptomyces* (PGPS)

Plant growth-promoting *Streptomyces* (PGPS) constitute a distinctive group of soil-dwelling actinobacteria whose multifaceted interactions with plants support growth, stress resilience, and productivity. Through diverse direct and indirect mechanisms,

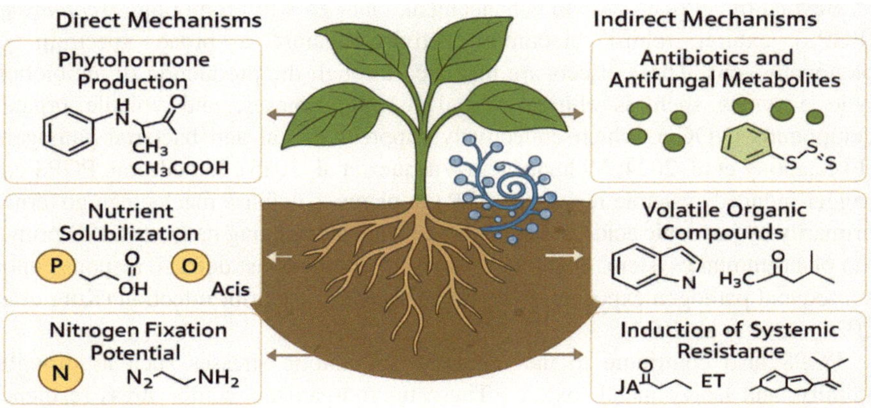

Fig. 11.1 Functional mechanisms (direct and indirect) of plant growth-promoting *Streptomyces* (PGPS)

PGPS enhance nutrient uptake, modulate phytohormone dynamics, and provide protection against a wide range of phytopathogens (Fig. 11.1 and Table 11.1). The intricate biochemical processes and ecological strategies deployed by PGPS highlight their significance as keystone taxa in rhizosphere ecology and their potential for sustainable crop production (Kumawat et al. 2022; Etesami 2025).

11.2.1 Direct Mechanisms

The direct contributions of PGPS to plant health are largely mediated by their capacity to synthesize phytohormones, solubilize essential nutrients, and, in some instances, facilitate nitrogen assimilation. These mechanisms are particularly beneficial under nutrient-deficient or stress-prone conditions.

11.2.1.1 Phytohormone Production and Regulation

One of the most well-documented traits of PGPS is their ability to synthesize indole-3-acetic acid (IAA), a key auxin that modulates root architecture and enhances root surface area for water and nutrient acquisition. IAA biosynthesis in *Streptomyces* predominantly follows tryptophan-dependent pathways, including the indole-3-pyruvate (IPA) and indole-3-acetamide (IAM) routes, regulated by genes such as *iaaM* and *iaaH* (Zhang et al. 2019; Tang et al. 2023). In addition to auxins, PGPS also produce gibberellins (GAs) and cytokinins, which collectively regulate cell division, elongation, and nutrient mobilization (Orozco-Mosqueda et al. 2023). Gibberellins promote seed germination and stem elongation, while cytokinins delay senescence and enhance sink strength in growing tissues (Janowska and Andrzejak 2022).

Table 11.1 Overview of plant growth promotion and biocontrol mechanisms mediated by *Streptomyces* spp

Mechanism category	Specific mechanism	Molecular/physiological basis	Representative compounds/processes	Ecological/functional outcomes	References
Direct mechanisms	Phytohormone biosynthesis	Tryptophan-dependent IAA production; GA and cytokinin biosynthesis; modulation of plant signaling	Indole-3-acetic acid, gibberellins, cytokinins	Enhanced root system architecture; shoot growth regulation	Etesami (2025)
	Phosphate solubilization	Organic acid secretion; phosphatase-mediated hydrolysis; rhizosphere acidification	Gluconic, oxalic acids; acid/alkaline phosphatases	Increased phosphorus bioavailability; improved nutrition	Li et al. (2023)
	Siderophore production	High-affinity Fe^{3+} chelation; siderophore receptor-mediated uptake	Desferrioxamines and hydroxamates	Improved iron uptake; pathogen iron competition	Timofeeva et al. (2022b)
	Nitrogen fixation and microbial synergies	nif-like gene expression; cooperative interaction with diazotrophs	Nitrogenase components; biofilm formation	Enhanced nitrogen availability; improved rhizosphere function	Nichio et al. (2025)
Indirect mechanisms	Antibiotic and antifungal metabolite production	Biosynthesis of secondary metabolites targeting pathogens' vital cellular processes	Streptomycin, actinorhodin, polyketides, lipopeptides	Suppression of fungal and bacterial phytopathogens	Butassi et al. (2021)
	Volatile organic compounds (VOCs)	Emission of antimicrobial volatiles; plant defense signaling modulation	Geosmin, methylisoborneol, 2-Nonanone	Pathogen inhibition; ISR priming	Chandrasekaran et al. (2022)
	Induced systemic resistance (ISR)	MAMP-triggered activation of JA and ET pathways; defense gene priming	Lipopolysaccharides, peptidoglycans, flagellin fragments	Broad-spectrum enhanced plant immunity	Yu et al. (2022)

11.2.1.2 Nutrient Solubilization and Siderophore Production

Phosphorus (P) and iron (Fe) are critical macronutrients that are often present in insoluble forms in soil. PGPS improve phosphorus availability through secretion of organic acids such as gluconic, citric, and oxalic acids, which chelate metal ions and liberate soluble orthophosphates (Li et al. 2023; Pang et al. 2024). Additionally, phosphatase enzymes hydrolyze organic phosphorus compounds, further expanding the pool of bioavailable phosphorus.

In iron-limiting conditions, particularly in aerobic soils where ferric iron (Fe^{3+}) is sparingly soluble, PGPS secrete high-affinity siderophores such as desferrioxamines and hydroxamates. These compounds chelate Fe^{3+}, enabling its uptake by both bacteria and host plants through specialized transporter systems (Timofeeva et al. 2022b). Moreover, by sequestering iron, PGPS deprive pathogenic microbes of this essential nutrient, thereby exerting a bioprotective effect.

11.2.1.3 Nitrogen Fixation Potential and Rhizosphere Synergies

Although *Streptomyces* are not classical diazotrophs, emerging metagenomic evidence indicates the presence of *nif*-like gene clusters in certain strains, suggesting a latent capacity for nitrogen fixation under microaerophilic conditions (Dai et al. 2014; Nichio et al. 2025). PGPS also engage in synergistic interactions with established nitrogen-fixing microbes such as *Rhizobium*, *Azospirillum*, and *Frankia*. These interactions facilitated through biofilm formation, quorum sensing, and modulation of rhizosphere chemistry, enhance the colonization and nitrogen-fixation efficiency of diazotrophs, indirectly contributing to plant nitrogen acquisition (Harjai and Sabharwal 2017; Thepbandit and Athinuwat 2024).

11.2.2 Indirect Mechanisms: Biocontrol and Immune Priming

Beyond nutrient facilitation, PGPS exert potent biocontrol effects through antimicrobial metabolite production and activation of plant defense pathways, reducing dependency on synthetic agrochemicals.

11.2.2.1 Production of Antibiotics and Antifungal Metabolites

Streptomyces are prolific producers of secondary metabolites with antimicrobial properties, including aminoglycosides (e.g., streptomycin), polyketides (e.g., actinorhodin, tetracycline), and nonribosomal peptides (e.g., bacitracin). These compounds interfere with vital processes such as cell wall biosynthesis, protein translation, and membrane integrity in pathogenic microbes (Alam et al. 2022). Antifungal activity is further reinforced by the production of chitinases, polyenes, and biosurfactants, which degrade fungal cell walls and inhibit biofilm formation (Human et al. 2016). These broad-spectrum metabolites effectively suppress pathogens such as *Fusarium oxysporum*, *Rhizoctonia solani*, *Pythium ultimum*, and *Phytophthora* spp (Butassi et al. 2021).

11.2.2.2 Volatile Organic Compounds as Biocontrol Agents

PGPS emit a diverse array of volatile organic compounds (VOCs), including geosmin, methylisoborneol, 2-nonanone, hydrogen cyanide (HCN), and *dimethyl* disulfide, which serve as both antimicrobial agents and inter-kingdom (Bacteria-plant) signaling molecules (Cuervo et al. 2023). These VOCs inhibit pathogen's spore germination, hyphal growth, and metabolic activity. Additionally, they modulate host plant responses by inducing the expression of defense-related genes and enhancing oxidative stress tolerance, effectively priming systemic resistance mechanisms.

11.2.2.3 Induction of Systemic Resistance (ISR)

PGPS can activate induced systemic resistance (ISR) in host plants through the release of microbe-associated molecular patterns (MAMPs) such as lipopolysaccharides, peptidoglycans, and flagellin derivatives. Recognition of these signals by plant pattern recognition receptors (PRRs) initiates a cascade of defense signaling mediated by jasmonic acid (JA) and ethylene (ET) pathways (El-Saadony et al. 2024; Villavicencio-Vásquez et al. 2025). ISR leads to a primed state characterized by rapid activation of defense responses, increased expression of pathogenesis-related (PR) genes, and accumulation of secondary metabolites upon pathogen encounter.

11.3 Isolation, Screening, and Production Technologies

The application of plant growth-promoting *Streptomyces* (PGPS) in sustainable agriculture depends upon systematic isolation, precise phenotypic and genotypic screening, and the development of scalable production systems. These processes not only enable the identification of elite strains with multifunctional traits but also ensure the reproducibility and quality assurance required for commercial bioinoculant formulations. This section delineates the contemporary methodologies employed in isolating PGPS strains, characterizing their functional capabilities, and advancing them toward formulation and deployment.

11.3.1 Isolation and Characterization of *Streptomyces* Strains

11.3.1.1 Source Selection and Pre-enrichment Strategies

Rhizospheric soils, particularly those associated with healthy and high-performing crops, serve as primary reservoirs for isolating PGPS strains, owing to the unique ecological niche and rich microbial diversity that characterize the root–soil interface (Meenakshi et al. 2024). To suppress fast-growing saprophytes and selectively enrich spore-forming actinobacteria, soil samples are often subjected to dry heat pretreatment (50–55 °C for 5–10 min), chemical treatments (e.g., 1.5% phenol), or air drying (Fig. 11.2). These methods exploit the intrinsic resilience of *Streptomyces*

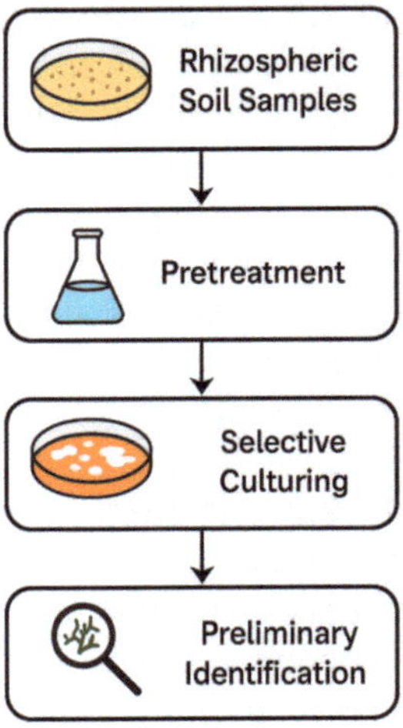

Fig. 11.2 Schematic workflow of PGPS isolation, screening, and identification from rhizospheric soil samples

spores, enhancing the selective isolation of target taxa (Al-Ansari et al. 2019; Fatima et al. 2019).

11.3.1.2 Selective Media and Culturing Conditions

Selective isolation is facilitated using actinomycete-specific media such as starch–casein agar (SCA), glycerol–asparagine agar, or humic acid–vitamin agar, often supplemented with antifungals (e.g., cycloheximide and nystatin) and antibacterial (e.g., nalidixic acid) (Mondal and Thomas 2022). These media influence the metabolic capabilities of *Streptomyces*, notably their ability to degrade complex polysaccharides and proteins to suppress microbial competition. Incubation is typically carried out at 28 ± 2 °C for 7–14 days to accommodate the slow-growing nature of actinobacteria.

11.3.1.3 Morphological and Biochemical Characterization

Isolates are initially screened based on colony morphology, pigmentation (aerial and substrate mycelia), and sporulation patterns. High-resolution imaging using scanning electron microscopy (SEM) is employed to distinguish spore chain types (e.g., rectiflexibiles, retinaculiaperti, straight, and coiled), aiding in species-level taxonomic inference under the International *Streptomyces* Project (ISP) guidelines (Khirennas et al. 2023). Biochemical characterization is performed via enzymatic assays to detect extracellular hydrolases such as amylase, cellulase, chitinase, and protease, which are key indicators of biocontrol and nutrient-mobilizing capacities. Additionally, carbon substrate utilization profiles, typically assessed using Biolog GEN III MicroPlates or customized basal media, provide a comprehensive metabolic fingerprint of each isolate (Balogová et al. 2025).

Table 11.2 Key functional genes and corresponding plant growth-promoting traits identified in PGPS

Functional gene(s)	Encoded function	Associated PGP trait	Example species	References
iaaM, *iaaH*	Indole-3-acetic acid biosynthesis	Root architecture modulation	*Streptomyces griseus*	Tang et al. (2023)
Phosphatases	Organic phosphorus mineralization	Phosphate solubilization	*Streptomyces coelicolor*	Timofeeva et al. (2022a)
PKS gene clusters	Polyketide antibiotic biosynthesis	Biocontrol via antimicrobial compounds	*Streptomyces avermitilis*	Risdian et al. (2019)
Siderophore biosynthesis genes	Iron chelation	Enhanced iron acquisition & pathogen suppression	*Streptomyces albidoflavus*	Miethke and Marahiel (2007)

11.3.1.4 Molecular and Genomic Characterization

Taxonomic identification is conventionally achieved through amplification and sequencing of the 16S rRNA gene using universal primers (e.g., 27F/1492R), with subsequent sequence alignment against curated databases such as or NCBI (The National Center for Biotechnology Information) EzBioCloud or SILVA (Johnson et al. 2019). However, given the limitations of 16S rRNA in resolving closely related taxa, multilocus sequence analysis (MLSA) or whole-genome sequencing (WGS) is increasingly utilized for precise phylogenetic placement and functional annotation (Caudill and Brayton 2022).

Genome mining using platforms like antibiotics and Secondary Metabolites Analysis Shell (antiSMASH) and BActeriocin GEnome mining tooL (BAGEL) reveals biosynthetic gene clusters (BGCs) associated with secondary metabolite production, including non-ribosomal peptide synthetases (NRPS), polyketide synthases (PKS), siderophores, and terpenoids. Functional genes implicated in plant growth-promotion such as *iaaM* and *iaaH* (auxin biosynthesis), *phoD* (phosphate solubilization), and *acdS* (1-aminocyclopropane-1-carboxylate deaminase) are critical markers for prioritizing strains with biofertilization and bioprotection potential (Duan et al. 2021; Kumar et al. 2022) (Table 11.2).

11.3.2 Functional Screening of Plant Growth-Promoting Traits

Screening for plant growth-promoting traits employs a battery of functional assays: IAA production is quantified via Salkowski's reagent colorimetry or high-performance liquid chromatography (HPLC) (Guardado-Fierros et al. 2024); phosphate solubilization is assessed by halo formation on Pikovskaya's agar and spectrophotometric quantification of soluble phosphorus; siderophore production is detected using chrome azurol S (CAS) assays (Ghosh et al. 2016); and antagonistic activity is evaluated through dual-culture inhibition assays against key phytopathogens. Integration of these screening results with omics data facilitates the identification of strains with multifunctional PGP potential (Nysanth et al. 2022; Zhao et al. 2022).

11.3.3 Fermentation Approaches

The scalable production of plant growth-promoting *Streptomyces* (PGPS) bioinoculants necessitates robust fermentation methodologies that optimize biomass accumulation, metabolite biosynthesis, cost efficiency, and environmental sustainability. Two principal fermentation modalities, submerged fermentation (SmF) and solid-state fermentation (SSF), are widely employed, each with distinct operational advantages aligned to specific product formulations and end-use applications.

11.3.3.1 Submerged Fermentation (SmF)

SmF involves cultivation of *Streptomyces* strains in liquid nutrient media under rigorously controlled physicochemical conditions, including pH, temperature, dissolved oxygen concentration, and agitation speed. These parameters are dynamically monitored and adjusted via advanced bioreactor control systems integrating real-time sensors and automated feedback loops (Yang et al. 2024). The aqueous environment promotes rapid cellular proliferation and facilitates the production of extracellular bioactive metabolites such as antibiotics, hydrolytic enzymes, and phytohormones. Bioreactor configurations supporting SmF are designed to optimize oxygen transfer rates and substrate uptake kinetics, thereby maximizing biomass yield and metabolite productivity (Fig. 11.3a). The resulting biomass is amenable to downstream processing steps, including centrifugation, filtration, and formulation into liquid bioinoculants or sterile spore suspensions.

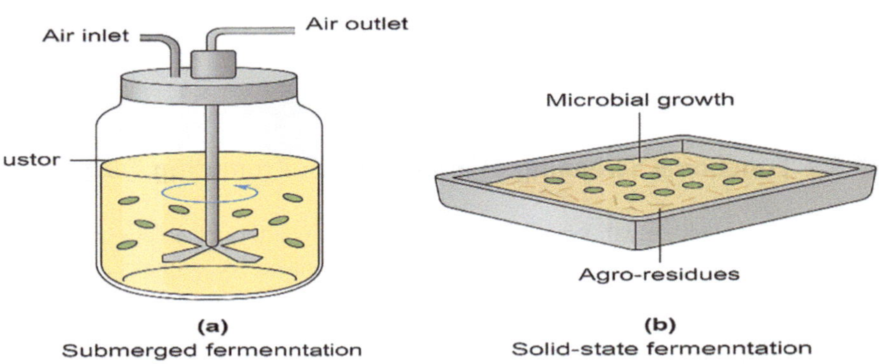

Fig. 11.3 (**a**) Submerged fermentation setup illustrating controlled aeration and agitation in a bioreactor; (**b**) Solid-state fermentation schematic depicting microbial growth on agro-residues

SmF-based production thus ensures batch-to-batch consistency and high purity, attributes critical for regulatory approval and commercial deployment.

11.3.3.2 Solid-State Fermentation (SSF)

SSF entails the growth of *Streptomyces* on solid agro-industrial substrates such as wheat bran, rice husk, or sugarcane bagasse, which emulate the natural ecological niche of actinobacteria (Hölker and Lenz 2005; Ren et al. 2025). This fermentation mode offers significant advantages, including reduced water and energy consumption, minimal effluent generation, and simplified downstream recovery, aligning with principles of green and sustainable bioprocessing (Fig. 11.3b). The heterogeneous and particulate nature of SSF substrates stimulates sporulation and enhances the biosynthesis of diverse secondary metabolites, often resulting in an expanded antimicrobial spectrum and improved biocontrol efficacy. SSF-derived spores exhibit superior shelf stability and resilience to abiotic stresses such as desiccation and temperature fluctuations, rendering them particularly suitable for seed-coating applications or soil amendments. Despite its environmental and economic benefits, SSF faces challenges related to scale-up, uniform aeration, and process standardization, necessitating continued technological innovation to fully harness its potential for PGPS production (Mattedi et al. 2023).

Optimization efforts include manipulating parameters such as moisture content, particle size, inoculum density, pH, temperature, and aeration in SSF, alongside nutrient composition and bioreactor design in SmF. Both types of fermentation are compared in Table 11.3. Recent innovations integrate co-cultivation techniques, immobilized cell reactors, and continuous fermentation processes to further enhance yield, strain stability, and metabolite profile. Quality control protocols encompass spore viability assays, metabolite quantification, and contaminant screening, ensuring reproducibility and field efficacy of commercial bioinoculants (Clausen and Junne 2025; Clara Ivette et al. 2025).

Table 11.3 Comparison of submerged and solid-state fermentation for PGPS production

Parameter	Submerged (SmF)	Solid-state (SSF)
Growth medium	Liquid nutrient broth	Agro-industrial solid residues
Water requirement	High	Low
Aeration and mixing	Controlled bioreactors with agitation	Passive aeration or forced aeration
Biomass yield	High, rapid growth	Moderate to high, slower growth
Metabolite profile	Consistent, optimized for target metabolites	Variable, often broader secondary metabolite spectrum
Cost and sustainability	Higher energy and water costs	Lower cost, environmentally sustainable
Formulation type	Liquid or spore suspensions	Spore powders, granules
Application	Seed treatment, foliar sprays, soil drench	Seed coating, soil amendment

11.4 Formulation and Delivery Strategies

The formulation and delivery of microbial inoculants are critical determinants of their efficacy and consistency across diverse agricultural ecosystems. These strategies must not only ensure microbial viability and stability but also facilitate controlled release, targeted colonization, and functional integration within complex plant–soil–microbe interfaces (Díaz-Rodríguez et al. 2025). Recent advances have leveraged interdisciplinary insights from material science, microbial physiology, and precision agriculture to develop sophisticated delivery platforms. This section highlights current innovations, dynamic interplay between carrier design, application modalities, and agroecological compatibility.

11.4.1 Carrier-Based Formulations

Carriers transcend their traditional roles as inert supports, functioning as engineered microenvironments that modulate microbial physiology, protect against abiotic stresses, and fine-tune interaction kinetics with host plants. Advances in biopolymer engineering, colloidal chemistry, and surface functionalization have propelled carrier technology toward programmable release systems and environmental responsiveness (Trivedi et al. 2005).

11.4.1.1 Mineral Carriers: Talcum and Kaolin

Mineral carriers such as talcum (magnesium silicate) and kaolin (aluminum silicate) remain industry benchmarks due to their physicochemical inertness, large specific surface areas, and minimal interference with bioactive agents. Talcum's hydrophobic lamellar architecture confers superior moisture exclusion, thereby preserving microbial viability ($>10^8$ CFU g^{-1}) and preventing oxidative degradation during storage, with stability maintained for up to 12 months under both ambient and refrigerated conditions (Saharan et al. 2010; Kumar et al. 2023). Conversely, kaolin's hydrophilic and mesoporous structure facilitates moisture retention and gradual hydration, which enhances spore viability but requires stringent moisture control to avert premature germination (Khalil et al. 2024). Surface modification techniques such as silane coupling and electrostatic charge tailoring have further enhanced microbial adhesion, dispersion homogeneity, and compatibility with polymer additives (Jiang et al. 2023; Niranjana et al. 2025).

11.4.1.2 Physicochemical and Microstructural Optimization

Optimization of carrier moisture content (ideally 8–10% w/w), particle size distribution (50–200 μm), and surface charge density critically influences microbial stabilization and delivery efficacy. Particle size optimization balances maximal surface area for microbial adhesion with aerodynamic and hydrodynamic properties suited for seed coating, soil incorporation, or foliar application. Cutting-edge nanocarriers, including mesoporous silica nanoparticles and graphene oxide sheets, alongside biopolymeric encapsulants such as chitosan and alginate, provide enhanced control over microbial release kinetics, UV protection, oxidative stress mitigation, and targeted delivery to plant tissues (Di Rago et al. 2025).

11.4.1.3 Functional Additives and Encapsulation Technologies

The incorporation of functional additives such as osmoprotectants (trehalose and sorbitol), antioxidants (ascorbate and glutathione), and quorum sensing modulators synergistically enhances microbial resilience and bioactivity. Advanced encapsulation methodologies, including layer-by-layer (LbL) self-assembly, microfluidic droplet generation, and spray drying with matrix formers, enable precise tuning of encapsulant composition, mechanical robustness, and release profiles. Biodegradable polyester matrices (e.g., PLGA and PCL) and stimuli-responsive hydrogels (e.g., pH- or redox-sensitive systems) have demonstrated superior field-level protection and controlled release capabilities (Balla et al. 2022).

11.4.2 Application Methods

The delivery modality profoundly influences microbial colonization efficiency, modulation of plant signaling pathways (e.g., systemic acquired resistance [SAR], induced systemic resistance [ISR]), and rhizosphere microbiome structuring. The integration of formulation characteristics with crop physiology is essential to maximize functional outcomes under field conditions.

11.4.2.1 Seed Coating and Pelleting

Seed-applied bioinoculants localize microbial populations to the spermosphere and rhizosphere, leveraging early root exudates for successful colonization. Innovations in bioadhesive polymers (polyvinyl alcohol, carboxymethyl cellulose) and nanocomposite coatings have markedly improved adherence, hydration control, and microbial viability during storage and upon sowing. Seed pelleting incorporates multi-layered microcapsules embedding nutrients and growth stimulators to synchronize microbial release with root development. Co-inoculation with microbial consortia promotes synergistic interactions, enhancing plant stress tolerance and growth (Ma 2019).

11.4.2.2 Soil Application: Granules, Powders, and Liquid Suspensions

Soil-based applications including broadcasting, banding, and side-dressing introduce inoculants directly into the rhizosphere. Granules composed of bentonite, vermiculite, or alginate matrices facilitate controlled hydration and sustained microbial release, whereas powders enable rapid dispersal but are vulnerable to desiccation and leaching. Liquid formulations compatible with fertigation systems offer uniform distribution but necessitate advanced stabilization techniques (e.g., surfactants, dispersants, and rheology modifiers) to prevent sedimentation and viability loss. Emerging biochar-based carriers and soil-conditioning delivery systems harness porous sorptive matrices to enhance microbial persistence and rhizosphere microhabitat optimization (Kabir et al. 2023).

11.4.2.3 Foliar Sprays and Endophytic Colonization

Foliar delivery exploits plant systemic defense pathways (SAR, ISR), enabling broad-spectrum protection and growth stimulation. Formulations enriched with surfactants (e.g., non-ionic ethoxylates) and wetting agents improve spray coverage and stomatal penetration. Nanoemulsions and electrostatic spraying enhance adhesion and uptake efficiency. Notably, certain *Streptomyces* and *Bacillus* strains demonstrate the capacity for endophytic colonization following foliar application, establishing systemic populations that modulate phytohormonal signaling and defense mechanisms (Chaudhary et al. 2022; Niranjana et al. 2025). Integration of RNA-based biocontrols and phage therapies via foliar routes heralds next-generation precision plant protection.

11.4.2.4 Precision Agriculture and Next-Generation Delivery Systems

Emerging delivery platforms including UAV-assisted application, electrostatic field-driven deposition, and hydrogel microbeads with programmable release profiles offer unparalleled precision and scalability. Coupling these technologies with Internet of Things (IoT)-enabled environmental sensors enables real-time parameter monitoring and dynamic delivery optimization, enhancing efficacy while minimizing ecological footprints (Rajak et al. 2023). Multi-modal delivery systems that combine soil, seed, and foliar treatments have demonstrated additive and synergistic benefits for crop yield, stress resilience, and soil health.

11.4.3 Integration with Agroecological Systems

Holistic formulation and delivery strategies must accommodate the intrinsic complexity of agroecosystems, encompassing soil heterogeneity, microbial community dynamics, climatic variability, and crop-specific physiological traits. Tailoring formulations to agroecological zones, supported by computational modeling techniques such as computational fluid dynamics (CFD) and agent-based modeling (ABM), can improve predictions of microbial dispersal and colonization success (Stavrakakis et al. 2023). Regulatory frameworks should emphasize multi-scale validation, biosafety evaluation, and stringent quality control to ensure environmental sustainability and promote farmer adoption.

11.5 Field Evaluations and Agronomic Impact of PGPR

11.5.1 Crop-Specific Trials

Extensive field evaluations have demonstrated that inoculation with *Streptomyces* species as plant growth-promoting rhizobacteria (PGPR) significantly enhances key agronomic traits across diverse crops, underscoring their potential as sustainable bioinoculants. Application of *Streptomyces* strains consistently results in shoot length increases ranging from 25% to 30% relative to uninoculated controls

(Nonthakaew et al. 2022; Suriyachadkun et al. 2022; Aallam et al. 2023). These growth enhancements are primarily driven by *Streptomyces*-mediated modulation of phytohormones including indole-3-acetic acid (IAA), gibberellins, and cytokinins that stimulate cell division and elongation in shoot tissues. Field trials involving cereals such as maize and wheat report not only increased shoot biomass but also improved overall plant vigor and health (Nozari et al. 2021; Boubekri et al. 2022).

Root system architecture is similarly improved, with inoculated plants exhibiting 20–40% increases in root dry weight (Gopalakrishnan et al. 2014). *Streptomyces* enhances root development by promoting lateral root proliferation and root hair density, facilitating improved soil exploration and nutrient uptake. Their ability to solubilize phosphate, produce siderophores for iron chelation, and engage in associative nitrogen fixation synergistically enhances nutrient bioavailability. Consequently, treated plants display increased resilience to abiotic stresses, including drought and nutrient-deficient soils.

Yield improvements of up to 40% have been documented in both cereals and legumes treated with *Streptomyces* inoculants. This yield gain reflects a multifaceted mechanism combining enhanced nutrient acquisition, suppression of soil-borne pathogens via antimicrobial metabolites (e.g., antibiotics and lytic enzymes), and induction of systemic resistance in host plants. Notably, *Streptomyces* strains confer protection against prevalent root diseases in crops such as wheat and soybean, while simultaneously promoting growth, highlighting their dual role as biofertilizers and biocontrol agents (Khan et al. 2023).

11.5.2 Environmental and Economic Benefits

The adoption of PGPR inoculants in agricultural systems offers substantial environmental benefits by reducing reliance on synthetic agrochemicals. Empirical data from field studies indicate that PGPR treatment enables reductions of up to 30% in nitrogen and phosphorus fertilizer inputs without sacrificing crop yields (Shah et al. 2021). This reduction mitigates adverse environmental consequences including eutrophication of water bodies, greenhouse gas emissions, and soil acidification linked to excessive fertilizer application.

PGPR inoculation also promotes soil health by enhancing microbial activity, accelerating organic matter decomposition, and improving nutrient cycling. The positive interactions between introduced PGPR and indigenous soil microbiota contribute to a resilient and functionally diverse rhizosphere ecosystem, supporting sustainable soil fertility over time. Improved microbial biodiversity further enhances soil structure and water retention, reducing erosion and nutrient leaching.

From an economic perspective, PGPR use confers significant cost savings to farmers by decreasing fertilizer and pesticide requirements while boosting crop productivity. Long-term cost–benefit analyses reveal net profitability increases of 20–35% for PGPR-treated crops, driven by reduced input costs and improved yield quality and quantity (Srivastava et al. 2024). These economic incentives foster broader adoption of PGPR-based biofertilizers across both smallholder and commercial farming systems, advancing sustainable intensification of agriculture.

11.6 Challenges and Regulatory Frameworks

11.6.1 Biosafety and Environmental Considerations

The deployment of *Streptomyces*-based plant growth-promoting formulations in agricultural ecosystems requires stringent biosafety assessments to safeguard environmental integrity and public health. Principal concerns include potential adverse interactions with non-target organisms, risks of horizontal gene transfer to native microbial populations, and the long-term persistence and ecological stability of introduced strains. Comprehensive risk assessment protocols must integrate both controlled laboratory studies and extensive field trials to characterize the environmental fate, dispersal dynamics, and ecological impacts of these bioinoculants. Such evaluations should consider cumulative and chronic effects across multiple cropping cycles and diverse agroecological zones to prevent unintended ecological disturbances. Robust biosafety monitoring frameworks are essential to ensure that the benefits of *Streptomyces* inoculants do not compromise ecosystem balance or biodiversity.

11.6.2 Regulatory Compliance and Harmonization

The regulatory environment governing microbial bioinoculants, including *Streptomyces*-based products, is complex and varies substantially across jurisdictions. Regulatory approval typically mandates comprehensive evidence demonstrating product efficacy in promoting plant growth and pathogen suppression, alongside stringent proof of environmental safety, non-pathogenicity, and the absence of deleterious effects on human and animal health (Khan et al. 2023). International agencies such as the Food and Agriculture Organization (FAO) and the Organisation for Economic Co-operation and Development (OECD) offer critical guidance frameworks to support the establishment of science-based, standardized regulatory protocols. However, heterogeneous national policies and discrepancies in biosafety criteria pose significant barriers to the global commercialization and adoption of PGPR technologies. Addressing these challenges requires coordinated international efforts to harmonize regulatory frameworks, streamline approval procedures, and promote transparency in risk assessment processes. Establishing such harmonization will facilitate responsible innovation, accelerate market access, and ensure the safe, effective integration of *Streptomyces*-based bioinoculants into sustainable agricultural practices worldwide.

11.7 Future Perspectives

11.7.1 Precision Agriculture and Digital Integration

The integration of *Streptomyces*-based bioinoculants with precision agriculture tools represents a promising frontier in sustainable farming. By combining these biological solutions with sensor networks, real-time monitoring systems, and advanced data analytics, it becomes possible to apply plant growth-promoting *Streptomyces* (PGPS) formulations in a site-specific, optimized manner. This synergy improves agronomic outcomes and resource-use efficiency, while reducing risks associated with over-application and environmental degradation. As digital agriculture continues to evolve, the fusion of biological and technological innovations is set to redefine sustainable crop production systems.

11.7.2 Omics-Driven Innovations and Synthetic Biology

Recent advances in omics technologies such as metagenomics, transcriptomics, proteomics, and metabolomics are unlocking new dimensions in our understanding of the interactions between *Streptomyces* strains, their plant hosts, and the surrounding soil microbiome. These insights provide a foundation for the rational development of next-generation PGPS strains with enhanced growth-promoting and biocontrol properties. Furthermore, genome-editing tools like CRISPR-Cas and synthetic biology platforms offer precise means to engineer *Streptomyces* functionalities tailored to specific crops, environmental conditions, and agronomic needs. Such innovations hold significant promise for expanding the applicability and effectiveness of PGPS technologies (Ansori et al. 2023).

11.7.3 Climate Resilience and Sustainable Intensification

Streptomyces-based PGPS show considerable potential in helping crops withstand climate-related stresses such as drought, salinity, and nutrient deficiencies. By enhancing root development, activating stress–response pathways, and supporting nutrient cycling, these bioinoculants contribute meaningfully to climate-resilient and resource-efficient agriculture. Their use aligns with global efforts toward sustainable intensification producing more food on existing farmland while reducing environmental impact. As a result, PGPS technologies not only support crop productivity but also promote soil health, reduce dependence on chemical inputs, and advance long-term ecological sustainability.

11.8 Conclusion

Streptomyces-based PGPS stand at the forefront of sustainable agricultural biotechnology, offering a wide array of benefits from boosting plant productivity and suppressing diseases to enhancing long-term soil health. Their integration into modern crop management systems addresses some of the most pressing global challenges, including food security, environmental degradation, and climate resilience. However, realizing their full potential requires sustained interdisciplinary research, supportive and harmonized regulatory frameworks, and active engagement from all stakeholders: researchers, policymakers, industry actors, and farmers alike. By embracing these innovative microbial tools, agriculture can take a decisive step toward a more resilient, productive, and ecologically balanced future.

References

Aallam Y, Dhiba D, Rasafi TE et al (2023) Assessment of two endemic rock phosphate solubilizing Streptomyces spp. on sugar beet (Beta vulgaris L.) growth under field conditions. Sci Hortic 316:112033. https://doi.org/10.1016/j.scienta.2023.112033

Alam K, Mazumder A, Sikdar S et al (2022) Streptomyces: the biofactory of secondary metabolites. Front Microbiol 13:968053. https://doi.org/10.3389/fmicb.2022.968053

Al-Ansari M, Alkubaisi N, Vijayaragavan P, Murugan K (2019) Antimicrobial potential of Streptomyces sp. to the Gram positive and Gram negative pathogens. J Infect Public Health 12:861–866. https://doi.org/10.1016/j.jiph.2019.05.016

Ansori ANM, Antonius Y, Susilo RJK et al (2023) Application of CRISPR-Cas9 genome editing technology in various fields: a review. Narra J 3:e184. https://doi.org/10.52225/narra.v3i2.184

Balla A, Silini A, Cherif-Silini H et al (2022) Recent advances in encapsulation techniques of plant growth-promoting microorganisms and their prospects in the sustainable agriculture. Appl Sci 12:9020. https://doi.org/10.3390/app12189020

Balogová M, Šimonovičová A, Vojtková H, Novák O (2025) The metabolic fingerprint of soil microbiota in an acidified environment using Biolog Microstationtm. Biologia (Bratisl) 80: 1589–1603. https://doi.org/10.1007/s11756-025-01934-3

Bargaz A, Elhaissoufi W, Khourchi S et al (2021) Benefits of phosphate solubilizing bacteria on belowground crop performance for improved crop acquisition of phosphorus. Microbiol Res 252:126842. https://doi.org/10.1016/j.micres.2021.126842

Boubekri K, Soumare A, Mardad I et al (2022) Multifunctional role of Actinobacteria in agricultural production sustainability: a review. Microbiol Res 261:127059. https://doi.org/10.1016/j.micres.2022.127059

Butassi E, Svetaz L, Carpinella MC et al (2021) Fungal biofilms as a valuable target for the discovery of natural products that cope with the resistance of medically important fungi—latest findings. Antibiotics 10:1053. https://doi.org/10.3390/antibiotics10091053

Caudill MT, Brayton KA (2022) The use and limitations of the 16S rRNA sequence for species classification of Anaplasma samples. Microorganisms 10:605. https://doi.org/10.3390/microorganisms10030605

Chandrasekaran M, Paramasivan M, Sahayarayan JJ (2022) Microbial volatile organic compounds: an alternative for chemical fertilizers in sustainable agriculture development. Microorganisms 11:42. https://doi.org/10.3390/microorganisms11010042

Chaudhary P, Singh S, Chaudhary A et al (2022) Overview of biofertilizers in crop production and stress management for sustainable agriculture. Front Plant Sci 13:930340. https://doi.org/10.3389/fpls.2022.930340

Chen Q, Song Y, An Y et al (2024) Soil microorganisms: their role in enhancing crop nutrition and health. Diversity 16:734. https://doi.org/10.3390/d16120734

Clara Ivette R-M, Luis Alberto M-G, Luis Galdino G-P et al (2025) Advances and challenges in the production and use of native bacteria as plant probiotics in agronomic applications: a Mexican review. J Agric Food Res 21:101917. https://doi.org/10.1016/j.jafr.2025.101917

Clausen AK, Junne S (2025) Recent advances in yeast and bacteria co-cultivation for bioprocess applications. World J Microbiol Biotechnol 41:170. https://doi.org/10.1007/s11274-025-04385-9

Copeland C, Schulze-Lefert P, Ma K-W (2025) Potential and challenges for application of microbiomes in agriculture. Plant Cell 37:koaf185. https://doi.org/10.1093/plcell/koaf185

Cuervo L, Álvarez-García S, Salas JA et al (2023) The volatile organic compounds of Streptomyces spp.: an in-depth analysis of their antifungal properties. Microorganisms 11:1820. https://doi.org/10.3390/microorganisms11071820

Dai Z, Guo X, Yin H et al (2014) Identification of nitrogen-fixing genes and gene clusters from metagenomic library of acid mine drainage. PLoS One 9:e87976. https://doi.org/10.1371/journal.pone.0087976

Dave A, Ingle S, Perveen K et al (2024) Harnessing plant growth–promoting and wilt-controlling biopotential of a consortium of actinomycetes and mycorrhizae in pigeon pea. J Phytopathol 172:e13399. https://doi.org/10.1111/jph.13399

De Andrade LA, Santos CHB, Frezarin ET et al (2023) Plant growth-promoting Rhizobacteria for sustainable agricultural production. Microorganisms 11:1088. https://doi.org/10.3390/microorganisms11041088

Di Rago S, Morandi F, Pizzetti F, Rossi F (2025) Advances in controlled release systems for sustainable crop production: a review of nano-, micro-, and macro-formulations. Advanced Agrochem:S2773237125000498. https://doi.org/10.1016/j.aac.2025.05.004

Díaz-Rodríguez AM, Parra Cota FI, Cira Chávez LA et al (2025) Microbial inoculants in sustainable agriculture: advancements, challenges, and future directions. Plants 14:191. https://doi.org/10.3390/plants14020191

Dow L, Gallart M, Ramarajan M et al (2023) Streptomyces and their specialised metabolites for phytopathogen control—comparative in vitro and in planta metabolic approaches. Front Plant Sci 14:1151912. https://doi.org/10.3389/fpls.2023.1151912

Duan B, Li L, Chen G et al (2021) 1-Aminocyclopropane-1-carboxylate deaminase-producing plant growth-promoting rhizobacteria improve drought stress tolerance in grapevine (Vitis vinifera L.). Front Plant Sci 12:706990. https://doi.org/10.3389/fpls.2021.706990

El-Saadony MT, Saad AM, Mohammed DM et al (2024) Drought-tolerant plant growth-promoting rhizobacteria alleviate drought stress and enhance soil health for sustainable agriculture: a comprehensive review. Plant Stress 14:100632. https://doi.org/10.1016/j.stress.2024.100632

Etesami H (2025) The dual nature of plant growth-promoting bacteria: benefits, risks, and pathways to sustainable deployment. Curr Res Microb Sci 9:100421. https://doi.org/10.1016/j.crmicr.2025.100421

Fatima A, Aftab U, Shaaban KA et al (2019) Spore forming Actinobacterial diversity of Cholistan Desert Pakistan: polyphasic taxonomy, antimicrobial potential and chemical profiling. BMC Microbiol 19:49. https://doi.org/10.1186/s12866-019-1414-x

Ghosh R, Barman S, Mukherjee R, Mandal NC (2016) Role of phosphate solubilizing Burkholderia spp. for successful colonization and growth promotion of Lycopodium cernuum L. (Lycopodiaceae) in lateritic belt of Birbhum district of West Bengal, India. Microbiol Res 183:80–91. https://doi.org/10.1016/j.micres.2015.11.011

Gopalakrishnan S, Vadlamudi S, Bandikinda P et al (2014) Evaluation of Streptomyces strains isolated from herbal vermicompost for their plant growth-promotion traits in rice. Microbiol Res 169:40–48. https://doi.org/10.1016/j.micres.2013.09.008

Guardado-Fierros BG, Tuesta-Popolizio DA, Lorenzo-Santiago MA et al (2024) Comparative study between Salkowski reagent and chromatographic method for auxins quantification from bacterial production. Front Plant Sci 15:1378079. https://doi.org/10.3389/fpls.2024.1378079

Harjai K, Sabharwal N (2017) Biofilm formation and quorum sensing in rhizosphere. In: Ahmad I, Husain FM (eds) Biofilms in plant and soil health, 1st edn. Wiley, pp 111–130
Hölker U, Lenz J (2005) Solid-state fermentation — are there any biotechnological advantages? Curr Opin Microbiol 8:301–306. https://doi.org/10.1016/j.mib.2005.04.006
Human ZR, Moon K, Bae M et al (2016) Antifungal Streptomyces spp associated with the infructescences of Protea spp in South Africa. Front Microbiol 7. https://doi.org/10.3389/fmicb.2016.01657
Janowska B, Andrzejak R (2022) Cytokinins and gibberellins stimulate the flowering and post-harvest longevity of flowers and leaves of Calla lilies (Zantedeschia Spreng.) with colourful inflorescence spathes. Agronomy 12:1859. https://doi.org/10.3390/agronomy12081859
Jiang P, Zhang Y, Hu R et al (2023) Advanced surface engineering of titanium materials for biomedical applications: from static modification to dynamic responsive regulation. Bioact Mater 27:15–57. https://doi.org/10.1016/j.bioactmat.2023.03.006
Johnson JS, Spakowicz DJ, Hong B-Y et al (2019) Evaluation of 16S rRNA gene sequencing for species and strain-level microbiome analysis. Nat Commun 10:5029. https://doi.org/10.1038/s41467-019-13036-1
Kabir E, Kim K-H, Kwon EE (2023) Biochar as a tool for the improvement of soil and environment. Front Environ Sci 11:1324533. https://doi.org/10.3389/fenvs.2023.1324533
Khalil AKA, Elgamouz A, Nazir S et al (2024) Preparation and characterization of clay based ceramic porous membranes and their use for the removal of lead ions from synthetic wastewater with an insight into the removal mechanism. Heliyon 10:e24939. https://doi.org/10.1016/j.heliyon.2024.e24939
Khan S, Srivastava S, Karnwal A, Malik T (2023) Streptomyces as a promising biological control agents for plant pathogens. Front Microbiol 14:1285543. https://doi.org/10.3389/fmicb.2023.1285543
Khirennas O, Mokrani S, Behira B et al (2023) Isolation, identification and screening of saharan actinomycete strain Streptomyces fimbriatus AC31 endowed with antimicrobial activity. Int J Mol Cell Med 12. https://doi.org/10.22088/IJMCM.BUMS.12.1.51
Kour D, Negi R, Khan SS et al (2024) Microbes mediated induced systemic response in plants: a review. Plant Stress 11:100334. https://doi.org/10.1016/j.stress.2023.100334
Kumar P, Rani S, Dahiya P et al (2022) Whole genome analysis for plant growth promotion profiling of Pantoea agglomerans CPHN2, a non-rhizobial nodule endophyte. Front Microbiol 13:998821. https://doi.org/10.3389/fmicb.2022.998821
Kumar V, Singh CS, Bakshi S et al (2023) Physicochemical and bioactive constituents, microbial counts, and color components of spray-dried Syzygium cumini L. pulp powder stored in different packaging materials under two controlled environmental conditions. Front Nutr 10:1258884. https://doi.org/10.3389/fnut.2023.1258884
Kumawat KC, Razdan N, Saharan K (2022) Rhizospheric microbiome: bio-based emerging strategies for sustainable agriculture development and future perspectives. Microbiol Res 254:126901. https://doi.org/10.1016/j.micres.2021.126901
Li H-P, Han Q-Q, Liu Q-M et al (2023) Roles of phosphate-solubilizing bacteria in mediating soil legacy phosphorus availability. Microbiol Res 272:127375. https://doi.org/10.1016/j.micres.2023.127375
Ma Y (2019) Seed coating with beneficial microorganisms for precision agriculture. Biotechnol Adv 37:107423. https://doi.org/10.1016/j.biotechadv.2019.107423
Mattedi A, Sabbi E, Farda B et al (2023) Solid-state fermentation: applications and future perspectives for biostimulant and biopesticides production. Microorganisms 11:1408. https://doi.org/10.3390/microorganisms11061408
Meenakshi S, Hiremath J, Meenakshi MH, Shivaveerakumar S (2024) Actinomycetes: isolation, cultivation and its active biomolecules. J Pure Appl Microbiol 18:118–143. https://doi.org/10.22207/JPAM.18.1.48
Miethke M, Marahiel MA (2007) Siderophore-based iron acquisition and pathogen control. Microbiol Mol Biol Rev 71:413–451. https://doi.org/10.1128/MMBR.00012-07

Mondal H, Thomas J (2022) Isolation and characterization of a novel actinomycete isolated from marine sediments and its antibacterial activity against fish pathogens. Antibiotics 11:1546. https://doi.org/10.3390/antibiotics11111546

Newitt J, Prudence S, Hutchings M, Worsley S (2019) Biocontrol of cereal crop diseases using streptomycetes. Pathogens 8:78. https://doi.org/10.3390/pathogens8020078

Nichio BTDL, Chaves RBR, Pedrosa FDO, Raittz RT (2025) Exploring diazotrophic diversity: unveiling Nif core distribution and evolutionary patterns in nitrogen-fixing organisms. BMC Genomics 26:81. https://doi.org/10.1186/s12864-024-10994-9

Niranjana VS, Ponnan S, Mukundan A et al (2025) Emerging trends in silane-modified nanomaterial–polymer nanocomposites for energy harvesting applications. Polymers 17:1416. https://doi.org/10.3390/polym17101416

Nonthakaew N, Panbangred W, Songnuan W, Intra B (2022) Plant growth-promoting properties of Streptomyces spp. isolates and their impact on mung bean plantlets' rhizosphere microbiome. Front Microbiol 13:967415. https://doi.org/10.3389/fmicb.2022.967415

Nozari RM, Ortolan F, Astarita LV, Santarém ER (2021) Streptomyces spp. enhance vegetative growth of maize plants under saline stress. Braz J Microbiol 52:1371–1383. https://doi.org/10.1007/s42770-021-00480-9

Nysanth NS, Sivapriya SL, Natarajan C, Anith KN (2022) Novel in vitro methods for simultaneous screening of two antagonistic bacteria against multiple fungal phytopathogens in a single agar plate. 3 Biotech 12:140. https://doi.org/10.1007/s13205-022-03205-3

Olanrewaju OS, Ayangbenro AS, Glick BR, Babalola OO (2019) Plant health: feedback effect of root exudates-rhizobiome interactions. Appl Microbiol Biotechnol 103:1155–1166. https://doi.org/10.1007/s00253-018-9556-6

Orozco-Mosqueda MDC, Santoyo G, Glick BR (2023) Recent advances in the bacterial phytohormone modulation of plant growth. Plants 12:606. https://doi.org/10.3390/plants12030606

Pang F, Li Q, Solanki MK et al (2024) Soil phosphorus transformation and plant uptake driven by phosphate-solubilizing microorganisms. Front Microbiol 15:1383813. https://doi.org/10.3389/fmicb.2024.1383813

Rajak P, Ganguly A, Adhikary S, Bhattacharya S (2023) Internet of things and smart sensors in agriculture: scopes and challenges. J Agric Food Res 14:100776. https://doi.org/10.1016/j.jafr.2023.100776

Ren F, Wu F, Gao L et al (2025) Proteomic and mechanistic insights into the efficiency of atmospheric and room-temperature plasma mutagenesis-driven bioconversion of corn stover by Trichoderma longibrachiatum. Fermentation 11:181. https://doi.org/10.3390/fermentation11040181

Risdian C, Mozef T, Wink J (2019) Biosynthesis of polyketides in Streptomyces. Microorganisms 7:124. https://doi.org/10.3390/microorganisms7050124

Saharan K, Sarma MVRK, Srivastava R et al (2010) Development of non-sterile inorganic carrier-based formulations of fluorescent pseudomonad R62 and R81 and evaluation of their efficacy on agricultural crops. Appl Soil Ecol 46:251–258. https://doi.org/10.1016/j.apsoil.2010.08.004

Santoyo G (2022) How plants recruit their microbiome? New insights into beneficial interactions. J Adv Res 40:45–58. https://doi.org/10.1016/j.jare.2021.11.020

Shah A, Nazari M, Antar M et al (2021) PGPR in agriculture: a sustainable approach to increasing climate change resilience. Front Sustain Food Syst 5:667546. https://doi.org/10.3389/fsufs.2021.667546

Sousa JADJ, Olivares FL (2016) Plant growth promotion by streptomycetes: ecophysiology, mechanisms and applications. Chem Biol Technol Agric 3:24. https://doi.org/10.1186/s40538-016-0073-5

Srivastava RK, Purohit S, Alam E, Islam MK (2024) Advancements in soil management: optimizing crop production through interdisciplinary approaches. J Agric Food Res 18:101528. https://doi.org/10.1016/j.jafr.2024.101528

Stavrakakis GM, Katsaprakakis DA, Braimakis K (2023) A computational fluid dynamics modelling approach for the numerical verification of the bioclimatic design of a public urban area in Greece. Sustainability 15:11642. https://doi.org/10.3390/su151511642

Suriyachadkun C, Chunhachart O, Srithaworn M et al (2022) Zinc-solubilizing *Streptomyces* spp. as bioinoculants for promoting the growth of soybean (*Glycine max* (L.) Merrill). J Microbiol Biotechnol 32:1435–1446. https://doi.org/10.4014/jmb.2206.06058

Tang J, Li Y, Zhang L et al (2023) Biosynthetic pathways and functions of indole-3-acetic acid in microorganisms. Microorganisms 11:2077. https://doi.org/10.3390/microorganisms11082077

Thepbandit W, Athinuwat D (2024) Rhizosphere microorganisms supply availability of soil nutrients and induce plant defense. Microorganisms 12:558. https://doi.org/10.3390/microorganisms12030558

Timofeeva A, Galyamova M, Sedykh S (2022a) Prospects for using phosphate-solubilizing microorganisms as natural fertilizers in agriculture. Plants 11:2119. https://doi.org/10.3390/plants11162119

Timofeeva AM, Galyamova MR, Sedykh SE (2022b) Bacterial siderophores: classification, biosynthesis, perspectives of use in agriculture. Plants 11:3065. https://doi.org/10.3390/plants11223065

Trivedi P, Pandey A, Palni LMS (2005) Carrier-based preparations of plant growth-promoting bacterial inoculants suitable for use in cooler regions. World J Microbiol Biotechnol 21:941–945. https://doi.org/10.1007/s11274-004-6820-y

Villavicencio-Vásquez M, Espinoza-Lozano F, Espinoza-Lozano L, Coronel-León J (2025) Biological control agents: mechanisms of action, selection, formulation and challenges in agriculture. Front Agron 7:1578915. https://doi.org/10.3389/fagro.2025.1578915

Wang Z, Solanki MK, Kumar A et al (2023) Promoting plant resilience against stress by engineering root microenvironment with Streptomyces inoculants. Microbiol Res 277:127509. https://doi.org/10.1016/j.micres.2023.127509

Woodruff HB (2014) Selman A. Waksman, Winner of the 1952 Nobel Prize for Physiology or Medicine. Appl Environ Microbiol 80:2–8. https://doi.org/10.1128/AEM.01143-13

Yang L, Shakeel Q, Xu X et al (2024) Optimized submerged batch fermentation for metabolic switching in Streptomyces yanglinensis 3–10 providing platform for reveromycin A and B biosynthesis, engineering, and production. Front Microbiol 15:1378834. https://doi.org/10.3389/fmicb.2024.1378834

Yu Y, Gui Y, Li Z et al (2022) Induced systemic resistance for improving plant immunity by beneficial microbes. Plants 11:386. https://doi.org/10.3390/plants11030386

Zhang P, Jin T, Kumar Sahu S et al (2019) The distribution of tryptophan-dependent Indole-3-acetic acid synthesis pathways in bacteria unraveled by large-scale genomic analysis. Molecules 24:1411. https://doi.org/10.3390/molecules24071411

Zhao X, Hou D, Xu J et al (2022) Antagonistic activity of fungal strains against fusarium crown rot. Plants 11:255. https://doi.org/10.3390/plants11030255

Commercial Biocontrol Agents and Biofertilizers from *Streptomyces*

12

Tanu Kumari, Palak Thakur, Ruchika Kumari, Vipasha Sharma, Randeep Singh, Rohit Sharma, and Ashun Chaudhary

Abstract

Pathogenic microorganism-induced plant diseases provide a significant challenge to agriculture, causing 30–40% of crop loss. It has been determined that using chemical pesticides has a negative impact on both the environment and human health. Consequently, it has become imperative to investigate natural alternatives to address these issues. The largest taxonomic group in microbial world that produces antibiotics is *Streptomyces*, a Gram-positive bacterium, which has been the subject of much investigation because of its propensity to produce a variety of antimicrobial substances, including organic compounds and metabolites. Researchers from worldwide investigated a number of techniques and strategies to extract novel bioactive chemicals from these microorganisms. *Streptomyces* can be used as biocontrol agents that contribute to the crop productivity and lower the use of agrochemicals against plant pathogens, promote plant growth, and reduce abiotic stresses that shed more light on profitable and environment-friendly method in promoting sustainable agriculture. The biofertilizers from *Streptomyces* effectively control plant pathogens through nitrogen fixation, solubilization of phosphate, and production of plant growth regulators, which lead to

Tanu Kumari and Palak Thakur contributed equally with all other contributors.

T. Kumari · P. Thakur · R. Kumari · V. Sharma · A. Chaudhary (✉)
Department of Plant Sciences, School of Life Sciences, Central University of Himachal Pradesh, Kangra, Himachal Pradesh, India
e-mail: ashun.chaudhary@hpcu.ac.in

R. Singh
PG Department of Zoology, Khalsa College, Amritsar, Punjab, India

R. Sharma
Department of Forest Products, College of Forestry, Dr. Yashwant Singh Parmar University of Horticulture and Forestry, Solan, Himachal Pradesh, India

T. Kaur et al. (eds.), *Streptomyces: Biological Candidates for Sustainable Agriculture*,
https://doi.org/10.1007/978-981-95-5803-2_12

improve the overall soil health. The *Streptomyces* produces major secondary metabolites that impact other microbes and plants, which play an important role in biological control. These prospects are reinforced by promising results in controlling phytopathogens, promoting sustainable agriculture, and integrating pest management strategies for agricultural use. The present chapter highlights key commercial biofertilizers and biocontrol agents derived from *Streptomyces,* ensuring enhanced efficacy, broader applicability, their mechanism of action such as inducing systematic resistance, and antibiotic and enzymatic degradation of pathogens. Additionally, it can explore how these bioproducts support organic farming, sustainable agriculture, and climate-resilient crop production.

Keywords

Biocontrol agent · Biofertilizer · *Streptomycetes* · Phytopathogens · Sustainable agriculture · IPM

12.1 Introduction

The worldwide population is expected to reach nine billion by 2050; therefore, agriculture production must increase. To meet this challenge, a wide range of strategies based on concepts from the twentieth-century green revolution are being implemented through research and technology programs (Sousa and Olivares 2016). To meet the expanding worldwide need for food, traditional agriculture has depended more on chemical pesticides and fertilizers, causing increased environmental damage (Pang et al. 2022). As a result, finding environment-friendly alternative is a significant priority in sustainable agriculture (Faddetta et al. 2023). Over 700 species exist in the *Streptomyces* genus, which belongs to the Actinobacteria phylum and is a group of neutrophilic, Gram-positive, facultative aerobic bacteria that inhabit various environments and habitats. They are soil saprophytes and can cause animal and plant diseases. These bacteria are non-motile, filamentous, and have a high guanine (G) and cytosine (C) content of about 70%. They are found in soil and account for around 10% of the microbial population. They develop during the vegetative phase by branching hyphal apices and creating a compact matrix. To treat various plant diseases, several strains of *Streptomyces* have been used. This genus has potential as a biocontrol and biofertilizer (Barbuto Ferraiuolo et al. 2021; Manikandan et al. 2023; Beroigui and Errachidi 2023). *Streptomyces* species have been identified from a variety of habitats because of their metabolic adaptability and cosmopolitan nature (Sousa and Olivares 2016). There have been reports of *Streptomyces* growing successfully at 50 °C (Mohammadipanah and Wink 2016). All observed soils in across the globe include *Streptomyces*; however, their abundance, function in biocenosis, and metabolic activity differ according to the ecological and geographic circumstances (Oskay 2009). *Streptomyces* sp. favors compost-rich soil.

Up to 10^{-6} *Streptomyces* colony-forming units may be found in one gram of fertile agricultural soil. *Streptomyces* accounts for up to 95% of all isolated actinomycetes when examining the distribution of several actinomycetes genera. Additionally, *Streptomyces* exist as endophytes in the proximity to plants (Hariprasad 2016). An environmentally friendly alternative for improving soil health and increasing agricultural yields is *Streptomyces* (Gopalakrishnan et al. 2015). *Streptomyces* strains with antagonistic traits can be employed as gene donors in generating resistant transgenic plants, as well as soil amendments as biofertilizers or biofungicides for disease control (Najmeh et al. 2014). In the green agriculture system, this boosts production, achieves agricultural sustainability, and supports ecosystem resilience and sustainability (Abdellatif et al. 2024). The varieties of *Streptomyces* have an impact on the fertility of soil as nutrient boosters because they are involved in nutrient management. They support plant growth through multiple mechanisms, including the plant growth regulators, phosphate and potassium solubilization, production of organic acids and siderophore, bioremediation and biodegradation, plant protection against biotic stress, PGP hormone production, and organic matter decomposition. (Mitra et al. 2022). Numerous studies have demonstrated that *Streptomyces* species provide various benefits, such as acting as biocontrol agents, efficiently colonizing the rhizosphere, and promoting plant growth, and with advancements in scientific tools and methods, plant growth-promoting bacteria (PGPB) were discovered, enabling their use as "biological fertilizers" for a safe and balanced approach to enhancing plant development (Al-Tammar and Khalifa 2023). PGPBs have been utilized extensively in the field, either alone or in combination with other microorganisms. Within integrated control systems, they have been shown to be successful in considerably increasing output, preventing disease outbreaks, and lowering or eliminating hazardous substances (Mohamad et al. 2022). *Streptomyces* species produce two-thirds of the 23,000 bioactive secondary metabolites generated by microorganisms (Pang et al. 2022). This genus also produces plenty of natural metabolites that have significant role as antibacterial, antiparasitic, antifungal, antioxidants, PGPR. This also generates various pharmaceuticals such as anticancer drugs, which include mitomycin, doxorubicin, dactinomycin, plicamycin, and daunorubicin, along with antibiotics, which include chloramphenicol, thienamycin, erythromycin, tetracycline, vancomycin, and streptomycin, immunosuppressive drugs, which include rapamycin, kanglemycin C, and tacrolimus, enzyme inhibitors, and various enzymes (Gebily et al. 2021; Ye et al. 2023). Nowadays, chemical fertilizers and botanicals are used to control pest and are utilized in farming with the goal of producing more food worldwide since they act as quick nourishment for plants, accelerating their growth. Constant chemical inoculation enriched with heavy metals degrades the fertility and quality of the soil and can lead to accumulation in plant tissues, which diminish both the edibility and nutritional content of fruit (Sneha et al. 2018). Scientists have tested blended fertilizers that combine biofertilizers with organic as well as chemical fertilizers (Soumare et al. 2020). The macrocyclic tetralactones are biosynthesized by *Streptomyces,* which are extensively used in agriculture and are useful neutralizing worms, mites, and insects. These include meilingmycin, milbemycin, avermectins, nanchangmycin, dianemycin, and piericidins (Pang et al. 2022). Particularly when considering their abundance in the biological management of soil-borne illnesses, *Streptomycetes* are essential for the sustainability of agriculture and serve as a gauge of the health of the soil (Oskay 2009). In order to manage some plant diseases, several products based on *Streptomyces sp* have been created and successfully tested. *Streptomyces griseoviridis* (MycostopR), for instance, has been used to handle *Fusarium oxysporum f.* sp. *lycopersici* and *Verticillium dahliae* in tomatoes as well as *Ceratocystis radicicola*, which makes date palms black. Additionally, *Streptomyces lydicus* WYEC108 has been used to make some commercialized products like Actinovate, Micro108, and Action Iron, while *Streptomyces saraceticus* KH400 has been used to make YAN TEN (Pacios-Michelena et al. 2021). Despite these progress and developments, there is still a lack of comprehensive data on how these *Streptomyces*-based products exhibit consistent results across different agroecosystems, on understanding their biocontrol and biofertilizer mechanism, and on their inclusion in sustainable farming system. Therefore, this study aims to discuss *Streptomycetes* as biocontrol agents and biofertilizers.

12.2 Biocontrol agents from *Streptomyces*

The use of living organisms to modify agricultural ecosystems for the purpose of controlling crop diseases or avoiding insect infestations is known as biological control (Boukhatem et al. 2022). The *Streptomyces* are known to generate secondary metabolites that have the potential to promote PGPB and biocontrol agents (Araujo 2022). Because of its renewable, affordable, and environmentally friendly capacity to ensure sustainable agriculture, the use of biocontrol agents is essential to modern agriculture (Manikandan et al. 2023). There are several ways that BCAs might work, including competing for resources and habitat, triggering host systematic resistance, or by producing hydrolytic enzymes, phytotoxins, or antimicrobial metabolites (Fig. 12.1). These ways can be transmitted by microbial consortia or by individual bacteria. In comparison to individual microorganisms, the combined modes of action that result from the interaction of many BCAs can function in concert to regulate disease more effectively (Dow et al. 2023). Various commercially available products from *Streptomyces* are shown in Table 12.1.

12.2.1 Antibiotic Production

Antibiotics are microbe natural products having low molecular weight that inhibit the development of other microorganisms (Dos Santos et al. 2012). Microorganisms employ the synthesis of antibiotics as one of their most potent weapons to eradicate phytopathogens and most researched biocontrol agents (Dasila et al. 2020). Numerous biocontrol strains are found to generate various antibiotics that have the ability to inhibit one or several diseases (Junaid et al. 2013). A diverse array of antibiotics have been derived from *Streptomyces* such as Vancomycin (glycopeptide) from

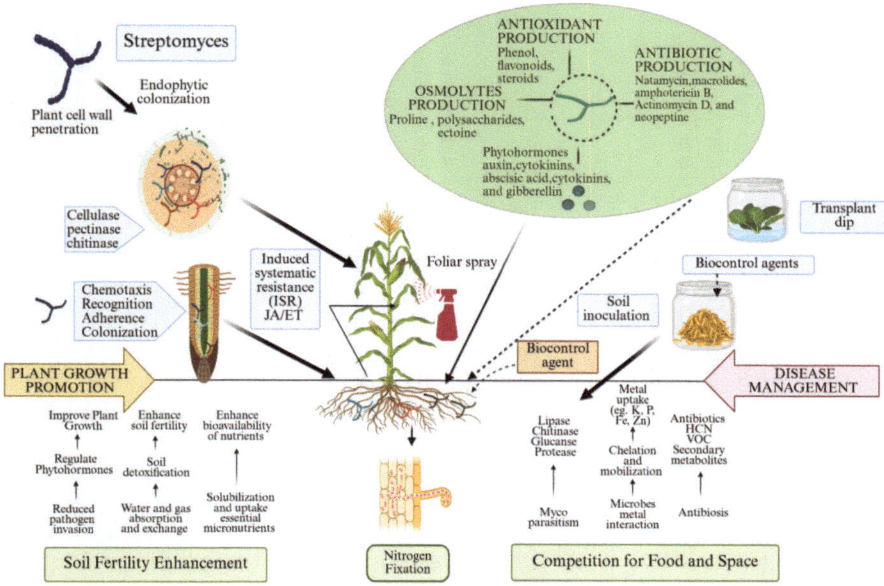

Fig. 12.1 The role of *Streptomycetes* in promoting sustainable agriculture by boosting plant development, soil fertility, and contributing to effective disease management. (Created in BioRender. Science, P. (2025) https://BioRender.com/56opt6g)

Streptomyces orientalis, Neomycin (aminoglycoside) from *Streptomyces fradiae*, clavulanic acid (β-lactam) from *Streptomyces clavuligerus,* tetracycline (tetracycline) from *Streptomyces aureofaciens*, Chloramphenicol (chloramphenicol) from *Streptomyces venezuelae*, nystatin (polyene) and clavulanic acid (β-lactam) from *Streptomyces clavuligerus* are examples (Law et al. 2018). Antibiotics target specific molecules, which are categorized based on their interaction with the cytoplasm, to inhibit bacterial growth by affecting vital cellular processes like DNA replication, RNA synthesis, cell wall creation, and protein production through physical contact, biochemical, molecular, and structural changes (Procópio et al. 2012). *Streptomyces* antibiotics mostly work by inhibiting ribosomal functional sites to target bacterial cellular components. Erythromycin, clindamycin, and chloramphenicol target the large ribosomal subunit (50S), whereas gentamicin, kanamycin, streptomycin, and tetracycline attach to the small ribosomal subunit (30S) to prevent protein synthesis. While carbapenems, cephalosporins, vancomycin, fosfomycin, bacitracin, and diaptomycin disrupt the integrity and synthesis of cell walls or membranes, ciprofloxacin and novobiocin hinder the translation of bacterial DNA (Quinn et al. 2020). Rice can be protected against bacterial infections by *Streptomyces* A20, which can generate streptomycin D, E, and F. Strong antagonistic action against tomato *Ralstonia solanacearum* is exhibited by actinomycin D, which was derived from *Streptomyces* sp. NEAU-HV9 (Pang et al. 2022). The rhizosphere of the medicinal plant *Curcuma longa* L. yielded a thermotolerant *Streptomyces* sp. TM32 strain that

Table 12.1 List of commercially available *Streptomyces* products as biocontrol agents across the globe

Product name	Strain	Biocontrol metabolite	Target pathogen/insect pest/target insect	Formulation available	Country	Mode of action	References
Agrimycin 17 and AS-50, Paushak, cuprimic 100 and 500, Dustret, cuprimicin 17, AAstrepto 17	*Streptomyces griseus*	Streptomycin	Bacterial diseases, such as canker and bacterial rots, Bacteria *Xanthomonas citri*, *X. oryzae*, and *Pseudomonas tabaci* of citrus, cotton, tobacco, olives, vegetables, potatoes, ornamentals, and stone fruit	Wettable powder	USA, India, China, Ukraine, Canada, and New Zealand	Reduce protein production due to interacting with 30S ribosomal subunit	Aggarwal et al. (2016); Vurukonda et al. (2018)
Kasumin, kasurab, kasugamycin, Kasu-rab-valida-sumi	*Streptomyces kasugaensis*	Kasugamycin	*Phtophthora sojae* causes soybean root rot, *Venturia* induces scab to apple, pear, *Cercospora* sp. leads to leaf spot in vegetables	Wettable powder, granule, dispersible powder, emulsifiable concentrate	Ukraine	It blocks amino acids to synthesize protein by preventing aminoacyl-tRNA from binding to ribosomes	Aggarwal et al. (2016); Vurukonda et al. (2018)
Astur	*Streptomyces avermitilis*	–	*Leptinotarsa decemlineata* attack on potato plants	–	Ukraine	–	Arasu et al. (2016)
Actofit	*Streptomyces avermitilis*	–	Complex of phytopathogens	–	Ukraine	–	Arasu et al. (2016)
Incide SP (Sri Biotec Laboratories India Ltd.)	*Streptomyces atrovirens*	–	Prevent insect infestations	Powder	India	–	Vurukonda et al. (2018); Reddy et al. (2016)

Arise, Proclaim, Affirm, Denim and shot-one	*Streptomyces avermitilis*	Emamectin benzotae (synthesized from abamectin)	Caterpillars (*lepidoptera*). It is used in maize (corn), soybeans, tea, cotton, peanuts, and vegetables	Wettable powder and emulsifiable concentrate	–	Emamectin works on the neurological system by causing an increase in chloride ion flow at the neuromuscular junction, which stops feeding and causes permanent paralysis	Aggarwal et al. (2016)
Matsuguard, Mesa, ultifora, Milbeknock and koromite	*Streptomyces hygroscopicus* subsp. *aureloacrimosus*	Milbemycin (a blend of milbemycin A3 and A4)	Leaf miner, citrus red mites, and kanzawa spider in eggplant, tea, and citrus	Wettable powder and emulsifiable concentrate	–	This improves GABA binding, which causes more chloride ions to enter the cell. This leads to hyperpolarization and the removal of signal transduction, which inhibits neurotransmission	Aggarwal et al. (2016); Vurukonda et al. (2018)
Agromec, Vibamec, Vertimec, Agri-Meck Avid, Clinch, Vamectin 1.8 EC, Abamex, Dynamec, Belpromec, Abacide, and Vapcomic	*Streptomyces avermitilis*	Abamectin / avermectins (mixture of avermectin B_{1a} and avermectin B_{1b})	Mites, fire ants, suckers, leaf miners, beetles and other insects in cotton, pome, vegetables (potatoes), ornamentals, citrus, and nut fruit	–	European union, USA, and World wide	It functions as the peripheral nervous system's γ-aminobutyric acid (GABA) receptor	Aggarwal et al. (2016); Vurukonda et al. (2018)
Mitecidin	*Streptomyces aureus*	Polynactin	European red mite (*Panoncychus ulmi*), Fruit orchards are prone to attack from mite (*Tetranychus cinnabarinus and T. urticae*	Wettable powder and emulsifiable concentrate	Japan	In the mitochondria, it causes basic cations like potassium to leak through the membrane's lipid layers	Aggarwal et al. (2016); Vurukonda et al. (2018)

(continued)

Table 12.1 (continued)

Product name	Strain	Biocontrol metabolite	Target pathogen/insect pest/target insect	Formulation available	Country	Mode of action	References
Avicta	*Streptomyces avermitilis*	Abamectin	*M. arenaria*, *M javenica*, *M incognita*, *Heterodera* sp. and *Pratylenchus* sp	–	USA, Brazil, south Africa, Paraguay, Argentina, and RSA	Antibiosis	Silva et al. (2022)
Tervigo	*Streptomyces avermitilis*	Avermectin	*Meloidogyne javenica*, *M. incognita*, and *M. arenaria*	–	Turkey, Morocco, Spain, South Korea, Italy	Antibiosis	Silva et al. (2022)
Cuprimic 100 and 500, Phytomycin, Mycoshield, Mycoject	*Streptomyces rimosus*	Oxytetracycline	Fire blight (*Erwinia amylovora*) and diseases caused by *Xanthomonas*, *Pseudomonas* sp., mycoplasmas	Water-soluble powder	–	It binds to the 30S and 50S bacterial ribosomal subunits, inhibits the binding of aminoacyl-tRNA and the termination factors RF1 and RF2 to the bacterial ribosome's A site, and inhibits the synthesis of proteins in bacteria	Aggarwal et al. (2016)
BLA-S	*Streptomyces griseochromogenes*	Blasticidin S	Rice blast (*Pyricularia oryzae*)	Dispersible powder, wettable powder, and emulsifiable concentrate	USA	In prokaryotes, it binds to the 50S ribosome and inhibits protein production, preventing peptidyl transfer and protein chain elongation	Aggarwal et al. (2016); Vurukonda et al. (2018)

Mycostop	*Streptomyces* K61	Mycostop	Root infections, *Alternaria*, *R. solani*, *Fusarium*, *Phytophthora*, and *Pythium* wilt causing damping-off	Wettable powder	Canada, USA, and EU	It colonizes plant roots and functions as a hyperparasite, destroying pathogenic fungi's cell walls. It produces metabolites that enable the pathogen to be killed	Aggarwal et al. (2016); Vurukonda et al. (2018); Narayanasamy (2013)
Delvolan	*Streptomyces chattanoogensis and Streptomyces natalensis*	Natamycin	*Fusarium oxysporum* causes basal rots on daffodils and ornamentals	Wettable powder	–	–	Aggarwal et al. (2016)
Mycin hustler, Valimun, Validacin, Valida, Sheathmar, Dantotsupadanvalida	*Streptomyces hygroscopicus*	Validamycin	*Rhizoctonia solani* and its variants cause diseases in cotton, ginger, sugar beet, rice, etc	Soluble concentrate, dispersible powder, powder for dry seed treatment	–	It inhibits trehalase in *R. solani* AG-1, but has no substantial impact on other glycohydrolytic enzymes. It induces abnormal branching of the pathogen tips and prevents further growth	Aggarwal et al. (2016)
Actinovate natural industries, Houston, USA	*Streptomyces lydicus* WYEC 108	Actinovate	Leaf diseases such as downy and powdery mildew, *Botrytis* and *Alternaria*, *Geotrichum*, *Postia*, and *Sclerotinia* Soil borne diseases, viz. *Phytophthora*,	Wettable powder, water dispersible granules	Canada, USA	The pathogen colonizes and grows around the plant structure, forming a synergistic connection with it by secreting essential and protective byproducts	Aggarwal et al. (2016); Vurukonda et al. (2018) Reddy et al. (2016); Narayanasamy (2013)

(continued)

Table 12.1 (continued)

Product name	Strain	Biocontrol metabolite	Target pathogen/insect pest/target insect	Formulation available	Country	Mode of action	References
			Verticillum, Fusarium, Pythium, and *Rhizoctonia*				
Mildiomycin	*Streptoverticillium rimofaciens*	Mildiomycin	Powdery mildew (*Uncinula necator, Erysiphe* sp., *Sphaerotheca* sp., and *Podosphaera* sp.) in ornamentals	Wettable powder	–	By inhibiting peptidyl transferase, it prevents fungus from producing proteins	Aggarwal et al. (2016)
Polyoxorim (Endrose, Stopit and Polyoxin Z) Polyoxin AL and Z, Polybelin	*Streptomyces cacaoi* var. *asoensis*	Polyoxorim	Fungal plant pathogens like *Sphaerotheca* sp., *Corynespora melonis, Botrytis cinerea, Cochilobolus miyabeanus, Sclerotinia sclerotinia, Alternaria alternata*, and others in apple, vines, vegetables, and ornamentals, as well as rice sheath blight, pear canker, apple, and *Helminthosporium*, inhibit cell wall biosynthesis and cause abnormal germ tube swelling	Wettable powder, paste, water-soluble granules, emulsifiable concentrate	–	The fungus becomes non-pathogenic because of its inhibition of cell wall biosynthesis and aberrant spore and hyphal tip germ tube enlargement	Aggarwal et al. (2016), Vurukonda et al. (2018)

Actin, Sri Biotech Laboratories India Ltd.	*Streptomyces atrovirens*	–	Antifungal	Liquid	India	–	Reddy et al. (2016); Vurukonda et al. (2018)
Safegrow KIBC Co Ltd.	*Streptomyces kasugaensis*	–	Sheath blight, large patch	–	South Korea	–	Vurukonda et al. (2018)
Mykocide KIBC Co. Ltd.; Yongin Gyeonggi-do	*Streptomyces colombiensis*	–	Grey mold, brown patch, powdery mildews	–	South Korea	–	Vurukonda et al. (2018); Ngamcharungchit et al. (2023)
Omycin	*Streptomyces kasugaensis*	–	Many fungal and bacterial plant pathogen control	–	India	–	Kaari et al. (2023)
Rhizovit	*Streptomyces rimosus*	–	*Phytophthora* sp., *R. solani*, *A. brassicola*, *Botrytis* sp., *Fusarium* sp., *Phomopsis* sp., and *Pythium* sp.	–	–	Antibiosis	Silva et al. (2022)

produces antibiotics (Quinn et al. 2020). In field tests, the natural strain of *Streptomyces melanosprofaciens* EF-76 showed biocontrol ability, but mutants lacking geldanamycin fail to suppress *Streptomyces* scabies, the pathogen responsible for potato scab (Dow et al. 2023). Kakadumycin is chemically similar to echinomycin, and produced by *Streptomyces* sp. NRRL330566, which was isolated from a fern-leaved Grevillea (*Grevillea pteridifolia*) tree (Pandey et al. 2020).

12.2.2 Induced Systematic Resistance (ISR)

ISR is a physiological condition that enhances plant defenses against biotic agents by responding to certain stimuli (Dasila et al. 2020). ISR occurs when plants activate their defense against invaders or when the host plant is primed to engage defense pathways in response to pathogen attacks (Viaene et al. 2016). It is a natural plant defense mechanism triggered by endophytic bacteria and other microorganisms. These bacteria release secondary metabolities such as VOCs, lipopeptide, and phytohormones, which trigger ethylene and jasmonic acid signaling pathways, enhancing plant immunity through systemic defense reactions like pathogenesis-related protein accumulation, callouse deposition, and cell wall lignification (Nimbulkar et al. 2025). ISR encourages a number of physiological changes, such as the build-up of defense-related molecules, localized cell death, and strengthening of the cell wall, which improve quickly and provide effective response to the future pathogenic attack (Lee et al. 2023). *Streptomyces* AcH 505 and *Arabidopsis* strains activate the SA/JA system to protect oak trees from powdery mildew and fungal diseases, respectively, as per Manikandan et al. (2023). A study found that using a single or a consortium of two *Streptomyces* strains (*Streptomyces shenzhenensis* TKSC3 and *Streptomyces* sp. SS8) in rice reduced ISR-mediated bacterial leaf streak (BLS) disease caused by *Xanthomonas oryzae* pv. *oryzicola* (Xoc) (Kumar et al. 2022a).

12.2.3 Competition for Nutrients

Biocontrol agents use competition to remove pathogens from the targeted zone (Ram et al. 2020). Microbes have to fight for available nutrients in nutrient-limited settings, such as soil and plant surfaces. To establish themselves in the environment, diseases and biocontrol agents must contend for resources and available space (Junaid et al. 2013). Beneficial bacteria employ antagonism via competing for available nutrients as one type of biocontrol strategy to survive pathogens and reduce illness (Wani and Gopalakrishnan 2019). By generating substances such as organic acids, hydrogen peroxide, bacteriocins, and siderophores that compete for nutrients and host adherence sites, *Streptomyces* strains demonstrate antimicrobial and antagonistic properties in opposition to pathogenic organisms in aquatic farming environments. This inhibits the growth of the pathogen (Khan et al. 2023).

12.2.4 Volatile Organic Compounds (VOCs)

Plants generate (VOCs), e.g., hydrocyanic acid (HCN), various acids, ketones, sulfides, alcohols, and aldehyde that diffuse in soil from micro- and macro-pores that protect them from phytopathogens, herbivores, and pests. Studies have shown that beneficial microbes produce HCN that reduces the harmful effect of pathogenic fungi (Maheshwari et al. 2020; Jacob and Sudini 2016). Hydrogen cyanide (HCN) is one of the several low-weight compounds that PGPB produces to combat viral infections (Paravar et al. 2023). *Streptomyces philanthi* volatile compound emitter inhibited the mycelial growth of rice fungal pathogens such as *Pyricularia grisea, Fusarium fujikuroi, B. oryzae,* and *Rhizoctonia solani* (Nimaichand et al. 2016).

12.2.5 Antimicrobial Production

Microbes such as bacteria, fungus, and actinomycetes are known to release antimicrobial compounds (AMCs), which are mostly classified as secondary metabolites (Tyagi et al. 2024). The use of antagonistic microorganisms to manage plant diseases, insects, and nematodes is a viable alternative to conventional pesticides, with considerable ecological and economic advantages (Ghosh and Mandal 2020). By generating secondary metabolites such as quinone, terpenoids, peptides, steroids, phenols, flavonoids and alkaloids, rhizospheric and endophytic bacteria enhance plant defense mechanism and may act as alternatives to fight drug resistance (Pandey et al. 2020). Actinobacterial strains can be effective biopesticides. *Streptomyces chrysomallus* P21 is effective against viral and fungal phytopathogens, while *Streptomyces avermitilis* CCM4697secretes avermectins, which are poisonous to acari. *Streptomyces hygroscopicus* subsp. TsKM V-4561 has insecticidal, bactericidal, and fungicidal activity. *Streptomyces ahygroscopicus* produces the antiviral substance, ε-poly-L-lysine (ε-PL) with a molecular weight of 3454–4352 Da, which is effective against TMV. When *Streptomyces olivaceus* metabolites were used separately, as spray and seed treatments, they were shown to be highly effective in reducing CMV infections in both open fields and in vitro conditions (Manjunatha et al. 2022). According to the study, *Streptomyces* species from agricultural soils that have been organically fertilized have the potential to be used as biocontrol and biofertilizer agents. Strong antimicrobial activity was demonstrated by six actinomycete isolates, including *Streptomyces olivochromogenes, Streptomyces avermitilis, Streptomyces rishiriensis, Streptomyces globisporus, Streptomyces sampsonii, and Streptomyces avidinii*, against phytopathogens that cause diseases in wheat and potatoes, such as *Fusarium poae* and *Clavibacter michiganensis* subsp. *sepedonicus*. Support is given to these strains in order to produce environmentally acceptable biopesticides and biofertilizers that will help manage plant diseases and promote sustainable farming methods (Charousova et al. 2016). Commercial applications for avermectins in crop protection include *Streptomyces avermitilis* fermentation products. They paralyze and kill insects by binding to their muscles and functioning as agonists for gamma aminobutyric acid (GABA)-gated chloride channels. They are effective against a diverse variety of insect species, including Lepidoptera, Hymenoptera, Isoptera, Diptera, Blattodea, and Coleoptera. Insecticides made from avermectins include amamectins and emamectins (Silva et al. 2022). Polyene antibiotics like nystatin and amphotericin B target ergosterol in fungal cells, disrupting their structural integrity and leading to cell death. Nystatin binds ergosterol, while amphotericin B creates transmembrane channels, enhancing membrane permeability and interfering with cellular processes, causing fungal cell death (Ali Mude et al. 2025). By feeding on harmful microorganisms, bacteria and fungus can hyperparasitize on pathogenic fungi. *Colletotrichum lindemuthianum* has been known to be parasitized by *S. griseus*, which results in internal parasitism of host hyphae and growth on the hyphae surface. As a result, the parasitized hyphae develop holes, blebs, and degraded cell walls (Subramanian et al. 2016). *Streptomyces termitum* ATC-2 produces aloesaponarin II, an antibacterial compound that effectively combats rice bacterial blight. *Streptomyces* sp. strain RM 365 inhibits *Pectobacterium carotovorum and P. atrosepticum,* causing potato soft rot, and *Xanthomonas campestris pv. glycines*, causing soybean bacterial pustule (Al-Quwaie 2024). A promising biocontrol agent for bacterial and fungal-induced plant diseases is *Streptomyces* sp. AN090126 (Le et al. 2022).

12.2.6 Biofilm Disruption

Biofilm formation occurs when microbial cells attach and grow on a surface, producing extracellular polymeric substances (EPS) around them (Ghosh and Mandal 2020). It was found that biofilms formed by *streptomyces* sp. CG252 is capable of eliminating $K_2Cr_2O_7$ from contaminated soil within just 3 days. However, the same strains in comparison to planktonic form only remove 80–90% of the material and take more than twice as long. These findings demonstrate *Streptomyces* sp. CG252 biofilms' potential for detoxification of heavy metal-contaminated environments (Buzón-Durán et al. 2020).

12.2.7 Hydrolytic Enzymes/Cell Wall-Degrading Enzymes

Soil actinobacteria are saprophytic that cause decomposition that leads to the release of cell wall-hydrolyzing enzymes, which include dehydrogenases, phosphatases, chitinases, nucleases, xylanases, amylases, proteinases, peptidases, peroxidases, cellulases, ligninases, β-glucanase, pectinases, hemicellulases, lipases, keratinase, and proteases that contribute significantly to biocontrol activity and shield plants from biotic stress and phytophathogen. Most significant is chitinases, and *Streptomyces* species have the capacity to control insect and fungal infestations (Dasila et al. 2020; Sathya et al. 2017). Purified chitinase derived from *Streptomyces lydicus* WYEC108 has the ability to degrade the cell walls of various plant-pathogenic fungi, such as Pythium, which can cause root rot in cereal crops (Lee et al. 2023).

12.2.8 Soil Fertility

Streptomycetes produce a volatile substance called geosmin, which gives soil its flavor or earthy smell after rain; it is released when bacteria die; the two-step mechanism takes place in which Geosmin synthase, a bifunctional enzyme of *Streptomyces coelicolor*, converts farnesyl diphosphate to geosmin. Soil fertility is frequently measured using geosmin, with more fertile soil being indicated by intensity. The human nose can detect up to five parts per trillion of geosmin. *Streptomycetes* also aid in the biogeocycling of nutrients such as iron, phosphorus, and nitrogen, and contribute to soil nutrients enhancement and boost the fertility of soil (Subramanian et al. 2016).

12.3 Biofertilizers from *Streptomyces*

The Indian biofertilizer market is projected to experience a significant compound annual growth rate (CAGR) expansion from FY2023 to FY2027, indicating a robust agricultural sector in the country (Kumar et al. 2022b). Biofertilizer bacteria enhance agricultural production by releasing growth regulators, fixing nitrogen, mobilizing phosphates, potassium, and zinc, and solubilizing soil minerals (Fig. 12.1). They are a great substitute for artificial fertilizers and maintain soil richness in micro- and macronutrient categories by breaking down potassium and phosphate, releasing chemicals for plant development, manufacturing antibiotics, and breaking down organic materials in the soil (Chakraborty and Akhtar 2021; El-Maraghy et al. 2021). Various commercially available products are in Table 12.2.

12.3.1 Phosphate Solubilization

Phosphate, a crucial macronutrient, aids in root development and plant maturity, but its insoluble form restricts the amount of soluble P in soil, making it essential for all plant growth (Goswami et al. 2020). Plants can only absorb H_2PO_4 and/or HPO_4^{-2} form, although it is present in bound form. P-solubilizing bacteria employ enzymes such as phosphor hydrolases, phosphonoacetate hydrolases, D-α-glycerophosphate, C-P lyase, acid phosphatases, and phytase to mineralize organic P and solubilize the bound form for usage by plants (Wani and Gopalakrishnan 2019). Phosphorus exists in soil in both organic and inorganic forms. Phosphate-solubilizing *Streptomyces* are vital for nutritional availability to the plants. Through the mechanism of phosphate solubilization, inorganic phosphorus compounds such as calcium phosphate (Ca-P), iron phosphate (Fe-P), and aluminum phosphate (Al-P) are made available to plants by microorganisms that secrete organic and inorganic acids, release siderophores, and produce H_2S. Furthermore, to degrade organic phosphorus molecules, such as orthophosphates and phosphates, these microbes release extracellular enzymes like phosphatases and phytases. The water-soluble form of phosphorus, mostly

Table 12.2 Commercially available *Streptomyces*-based biofertilizers

Company	*Streptomyces* strains	Formulation	Pathogenic control	References
Indogulf Bioag, India Revive (bio.)	*Streptomyces gelaticus*	Liquid	Sugar crop, orchards, aromatic crops fruits, and vegetables	Allouzi et al. (2022)
Mycostop® AgBio, Inc., USA	*Streptomyces strain K61*	Wettable powder	*Fusarium, Alternaria, Pythium, and Phytophthora*	Manikandan et al. (2023)
Bio-Nata Farmers Bio-Fertilizers and Organics	*Streptomyces lydicus*	Liquid	*Pythium, Rhizoctonia, Fusarium, Phytomatotricum, Aphanomyces, Botrytis, Postia, Sclerotinia, Verticillum, Monosporascus, Pyricularia, Monilinia, Colletotrichum, Alternaria solani, Goetrichum, Colletotrichum, Phytophthora, Cladosporium, and Armillaria*	
Green Marvel Greenlife BIO-tech Lab	*Streptomyces griseus*	Liquid	Pomegranate bacterial blight and citrus canker	
Microsat F® Cereal CCS Aosta Srl, Italy	*Streptomyces sp.* ST 60	–	Soybeans, tomatoes, beet, sunflowers, and cereals	Peter et al. (2020)
Omega® Varsha Bioscience and Technology India Pvt. Ltd., India	*Streptomyces sp.*	–	Vegetables and field crops	
Microsat F® Uno CCS Aosta Srl, Italy	*Streptomyces sp.* SB 14	Granule	Flowers, vegetables, and fruits	Vurukonda et al. (2018); Peter et al. (2020)
Microsat F® MO, CCS Aosta Srl	*Streptomyces* sp. *strains* SA51, SB14, and *SL 81*	Dry granule	Plant growth promoter	Vurukonda et al. (2018)

orthophosphoric acid (H_3PO_4), which is easily absorbed by plant roots, is produced by this mineralization process from these complicated forms (Chouyia et al. 2022, and Mahala et al. 2020). According to the study, *S.* mhcr0816 and *S. 1* mhce0811 released phosphate as a result of the effects of gluconic and malic acids (Olanrewaju and Babalola 2019).

12.3.2 Nitrogen Fixation

Nitrogen is a key limiting nutritional component for plant development in different environments. Using industrially generated nitrogen fertilizer is a popular method for increasing crop output (Gupta et al. 2012). Green fertilizer is made by applying *Streptomyces* as PGPR in the field either by itself or in combination with other nitrogen-fixing bacteria (Mahmood et al. 2024). It has been found that *Streptomyces* species are capable of fixing nitrogen. They have demonstrated this by growing on nitrogen-free culture, detecting acetylene reduction activity, and being confirmed by dilution measurement of the N^{15} isotope. Additionally, PCR amplification methods have been used to confirm the existence of nif genes linked to nitrogen fixation (Nimaichand et al. 2016). Under aerobic circumstances at 65 °C, *Streptomyces thermoautotrophicus* has been shown to grow chemolithoautotrophically with CO or H_2 and CO_2, using N_2 as their only nitrogen source (Taj and Rajkumar 2016). The pathway mechanism behind soil actinobacteria's nitrogen fixation process is yet unknown (Nimaichand et al. 2016). *Streptomyces* sp. can cause *Discaria trinervis* to produce more frankial nodulation (Swarnalakshmi et al. 2016). Different soybean varieties may exhibit increased nodulation and N_2 fixation as a result of the combined inoculation of bradyrhizobial strains (*Bradyrhizobium yuanmingense* MAS34) and endophytic actinomycetes (*Streptomyces grieseoflavus* P4) at a low density level (10^5 colony-forming units per g of seed). They suggest that this mixture should prove to be beneficial a biofertilizer for the nodulation, nutrients intake, nitrogen fixation, plant development, and crop production of soybeans (Solans et al. 2016; Swarnalakshmi et al. 2016; Htwe et al. 2018). The increased nodulation of chickpeas plant and soil nitrogen content while *Streptomyces* inoculants are applied, even in field settings, suggests their non-specific promotion as well as influence on rhizobia–legume symbiosis (Sathya et al. 2017). When *Streptomyces fumanus* was added to non-sterile soil, it competed with the native soil microbiota and may infiltrate the plant rhizosphere. The use of a *Streptomyces* gn-2 formulation has enhanced the rhizosphere microflora's composition and drawn saprophytic bacteria, such as oligotrophs and ammonificators. Increasing the growth and development of beneficial groups, including nitrogen-fixing bacteria, is facilitated by the presence of the biocontrol bacterium *Streptomyces fumanus* in the rhizosphere (Doolotkeldieva et al. 2015). *Rhizobium* sp. inoculation of strain *Streptomyces lydicus* WYEC108 increases root nodulation in peas by colonizing the surface cell layers, boosting the size of nodules, and the vitality of bacteroids by improving the absorption of iron and other critical nutrients. *Bradyrhizobium japonicum* was unable to develop nodules when *Streptomyces kanamyceticus* was present, but when the co-inoculation of *Streptomyces kanamyceticus* and antibiotic-resistant *B. japonicum* was done, the size and quantity of root nodules increased. This finding implies that antimicrobial compounds from *Streptomyces* originally prevented nodulation but could be overcome when resistance strains were present (Subramanian et al. 2016). Under field settings, *Streptomyces* inoculants have been shown to increase soil nitrogen content and chickpea nodulation, suggesting that they promote rhizobia–legume symbiosis in a non-specific manner. However, a thorough comprehension of these processes is required (Sathya et al. 2017). The study was conducted on *Streptomyces* from the South Dakota Badlands may actively fix nitrogen by the presence of the nifH gene and the incorporation of $^{15}N_2$, indicating that *Streptomyces* may play a new role in biological nitrogen fixation (Dahal et al. 2017).

12.3.3 Production of Phytohormones

Phytohormones are essential for plant growth and development. They respond to environmental stress by acting as molecular signals that could possibly be fatal or restrict plant development (Dasila et al. 2020). Biofertilizers generate phytohormones such as cytokinins, auxins, and gibberellins. These hormones promote seed germination, root elongation, and overall plant vigor, leading to better growth and productivity (Singh et al. 2024). In *Streptomyces atroolivaceus*, IAA causes cell differentiation, hyphal extension, and sporulation; in common beans, endophyte, while in *Streptomyces hygroscopicus*, TP_A045 is responsible for root elongation through the production of metabolites such as pteridicacids A and B (Sathya et al. 2017). *Centella asiatica*, a traditional ethno-medicinal plant, benefits from the formation of indole-acetic acid (IAA) by two *Streptomyces* sp. (71 g/mL and 197 g/mL) by encouraging cell division and elongation, which is made possible by actinobacteria (Rani et al. 2018). The five main intermediates used to synthesize IAA are tryptamine (TAM), tryptophan side-chain oxidase (TSO), indole-3-pyruvate (IPy), indole-3-acetamide (IAM), and indole-3-acetonitrile (IAN). Proteins in the L-amino acid oxidase gene cluster catalyze the oxidation of L-tryptophan by certain PGPR to IPy. Additionally, certain PGPR are capable of producing auxins that impact plant development, such as indole-3-butyric acid and indole-3-acetaldehyde (Pang et al. 2022). Elevated ethylene levels can harm crop growth by affecting defoliation and other cellular processes (Sarkar et al. 2017). Gibberellic acid (GA3) is a hormone found in plants that controls and stimulates growth. It can affect stems, leaves, and flowers by causing cell division and elongation. *Streptomyces neopeptinius* KNF2047T a kind of tomato seedling, produced significant quantities of GA3 and IAA, indicating that it has the ability to promote plant growth (Abdellatif et al. 2024). The actinomycete strain *Streptomyces filipinensis*, which produces both ACC deaminase and IAA, was shown to be more efficient than *Streptomyces atrovirens* in promoting plant development (Srivastava and Sharma 2014).

12.3.4 ACC Deaminase

ACC is produced by bacteria under stress or pathogen attack; deaminase transforms the ACC molecule into ammonia and ketobutyrate, inhibiting the synthesis of ethylene and fostering the development and lifespan of plants (Khan et al. 2023). When exposed to waterlogged circumstances, tomato plants infected with the endophytic *Streptomyces*, sp. GMKU 336 which produces 1-aminocyclopropane-1-

carboxylate deaminase (ACCD), demonstrated resistance (Johnson et al. 2021). *Streptomyces* sp. PGPA39 isolated from agricultural soil, produces ACC deaminase and aids in the growth of "MicroTom" tomato plants under salt stress (Shivlata and Satyanarayana 2017).

12.3.5 Siderophore Production

Siderophores are naturally occurring, low-molecular weight, Fe-chelating, high-affinity substances that are produced during iron famine by bacteria, cyanobacteria, fungus, and some plants (Tripathi et al. 2020). Soil ferrous iron (Fe^{2+}) is oxidized to produce ferric iron (Fe^{3+}), which is less accessible to microorganisms and plants. This decrease in Fe^{2+} availability leads to competition for iron absorption among plants, bacteria, and fungus. Bacterial strains produce siderophores, which bind to Fe^{3+} and form siderophore-Fe^{3+} complexes. These complexes are identified and moved across Gram-positive bacteria's cell membranes. Ferric iron attached to siderophores is converted to ferrous iron by membrane-bound or extracellular ferric-chelate reductases. Iron is released from siderophores into the cytoplasm through "gating" when cell membranes are connected. Siderophores can be recycled or decomposed after iron release (Khan et al. 2020). In order to protect plants against diseases, *Streptomyces* sp. create hydroxymate siderophores. *Streptomyces rochei* IDWR 19 produces a hydroxymate siderophore at a concentration of 34.17 mg/l (Johnson et al. 2021). The siderophore is produced and plant development is efficiently enhanced by *Streptomyces carpinensis, Streptomyces thermolilacinus, Streptomyces rochei carpinensis,* and *Streptomyces thermolilacinus*. Likewise, siderophore-producing *Streptomyces* was examined in saline soil, which enhanced plant development (Manikandan et al. 2023).

12.4 Formulation and Application Techniques

Biological control agent (BCA) research and commercialization require careful consideration of several crucial factors and formulations. Several key aspects that affect the formulation are stability, efficacy, shelf life, compatibility with agricultural equipment, production, and acquisition costs. Three primary components are commonly found in prepared BCA products. Firstly, active ingredients include secondary metabolites and synthetic microbial cells; secondly, adjuvants shield it from environmental challenges including UV light, high temperatures, and desiccation; and last, its carriers are inert materials that transport the active ingredient to the target location. BCA formulations are available in solid and liquid forms, with solid formulations offering higher stability and liquid formulations being easier and less costly; while liquid formulations include suspension concentration; emulsifiable concentrate; microemulsion; microencapsulated particles; and oil dispersion, dry formulation include wettable powders and water-dispersible granules. The different formulation types have varied viscosities and surface tensions; for instance,

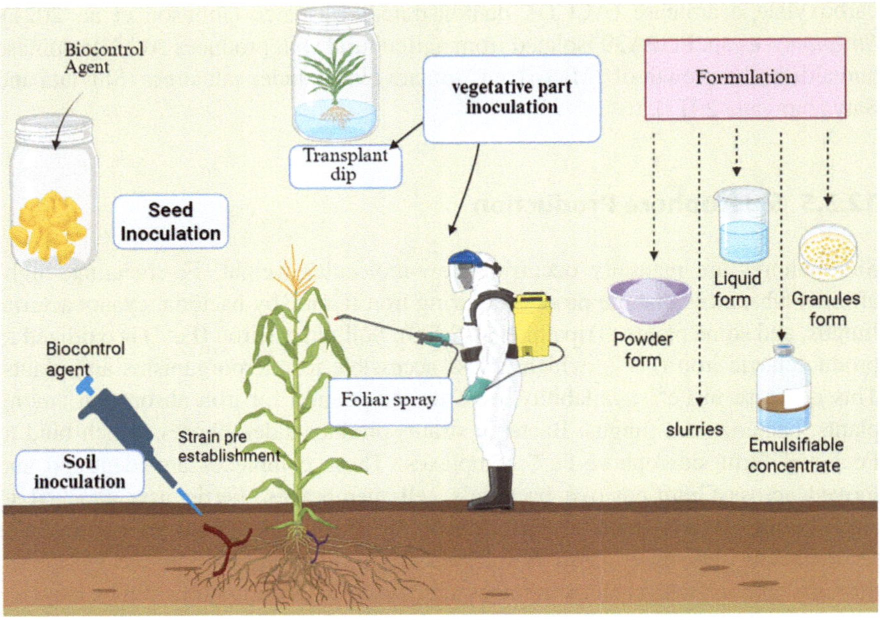

Fig. 12.2 The use of multiple formulation and application strategy in the field such as soil application, foliar spray, and seed inoculation. (Created in BioRender. Science, P. (2025). https://BioRender.com/sn99gke)

suspension concentrate has the highest viscosity while emulsifiable concentrate has the lowest surface tension during spraying (Preininger et al. 2018). The most effective biofertilizer is chosen based on its effective field performance and the kind of bioformulations used to meet the desired outcome (Jacob and Paranthaman 2023). A strain of effective biocontrol or biofertilizer must be produced in a way that makes it simple to apply to crops (Pirttilä et al. 2021). *Streptomyces* sp. strains, including Mycostop, considerably decreased Septoria leaf spot in field tests including potted plants and a poplar plantation (Fravel 2005).

Various plant protection techniques, including spraying on vegetative part like leaves, spraying, modification of soil, and seed treatment, have been shown (Fig. 12.2) to effectively control these pests and improve seed quality and production. The foliar and seed treatments are applied timely; plants develop strong and healthy seeds (Sarkar et al. 2017). Foliar spray can be applied to plants throughout the season, at key development phases, or under specific weather circumstances to encourage growth, increase resistance, or fight off diseases and pests (Preininger et al. 2018). Seed coating is a technique used by seed companies to improve microbial delivery seedling performance by processes such as seed pelleting, film coating, and seed encrusting. Seed treatment thickness and physical characteristics significantly impact seedling vigor and germination. Thin coatings may break or dissolve before planting, while thick, hard coatings may impede germination or cause abnormalities (Qiu et al. 2020). Different equipment like rotating pans, pelleting pans, and dry powder applicators are used for mixing liquids and dry powders or used alone. Many industries employ these machinery-based coatings, which are also referred to as dry coating, seed dressing, film coating, encrustments, and seed pelleting (Afzal et al. 2020). Among these, film coating is regarded as modern technique that applies a thin film of external material to seeds without altering their size, shape, or weight. It is similar to slurry coating, but with a less consistent and hard layer. Seed coating, a method of applying exogenous material to seeds, is increasingly popular for delivering beneficial microbes, enhancing crop adaptation and protection (Rocha et al. 2019). Soil inoculation involves introducing PBM into soil through soaking, soil integration, and microcapsules reduces chemical seed treatment by submerging seeds in a microbial solution. During germination, seeds produce exudates containing amino acids and carbohydrates, which microorganisms use as a source of nutrition in the soil (Paravar et al. 2023). The technique's commercialization could be expedited by creating formulations with longer shelf life, and wider action range in field acclimatization (Reddy et al. 2016). Grover et al. (2016) reported that seed germination of wheat was enhanced by soaking in *Streptomyces* sp., by adding *Streptomyces* sp. alone or along with *Azotobacter vinelandii* to the soil, which led to wheat growth under both regular and saline situations. Currently, various *Streptomyces* derived metabolites, such as streptomycin, kasugamycin, and polyoxin D, are widely applied as foliar spray to combat fungal and bacterial diseases. Products formulated from the *Streptomyces* strains such as Actinovate® and Mycostop® are effective in soil treatment (Rey and dumas 2017). Microbial biopesticides are a crucial and amazing tool in integrated pest management (IPM). There is now more interest in selecting beneficial microbes to create biological control agents (BCAs) (Tariq et al. 2020). Finally, the development and commercialization of biofertilizers and biocontrol agents aim to increase agricultural output, growth of plants, soil health, and phytopathogen biocontrol, while providing environmentally benign substitutes for synthetic agrochemicals such as chemical pesticides and fertilizers (Fig. 12.3).

12.5 Case Studies/Various Field Trials

Various case studies demonstrated that streptomycetes have the capacity to inhibit plant diseases and stimulate plant growth and make them commercially viable as biocontrol agents and biofertilizer. The study by (Nguyen et al. 2025) evaluated the growth-promoting and antagonistic properties of *Streptomyces* sp. VNUA74 strain, isolated from the banana rhizosphere in Vietnam. The strain showed strong antibacterial properties against various phytopathogens on bananas, including *Fusarium oxysporum* and many more. It also enhanced growth traits of banana plantlets in a pot experiment, suggesting that it could be used as a biocontrol and biofertilizer for sustainable agricultural crop production. The study conducted by (Sharma and Salwan 2018) shows that commercial applications of *Streptomyces* include the treatment of a variety of plant diseases. For more than 30 years, *Streptomyces* strain 5406 from China has been utilized to combat soil-borne infections and

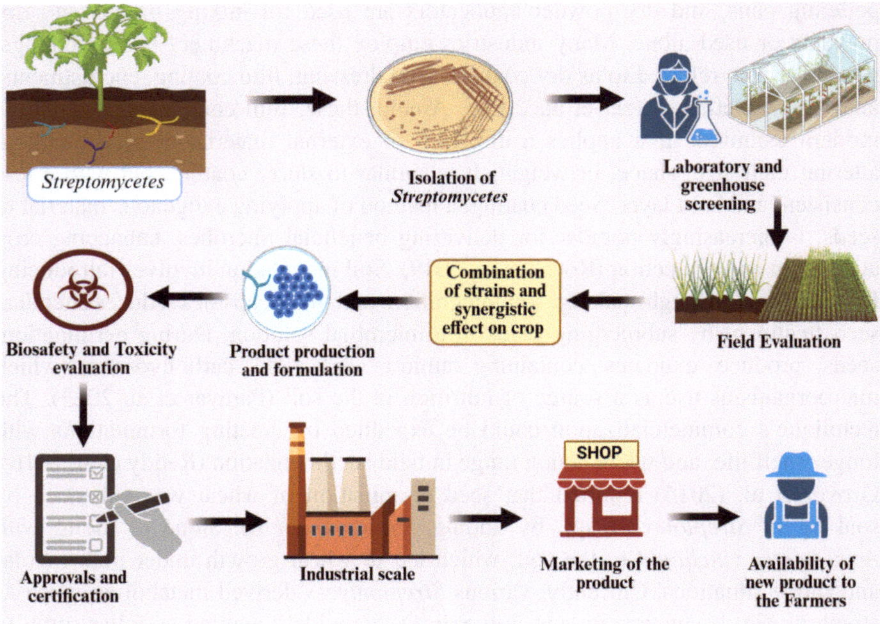

Fig. 12.3 A general flowchart outlining the action plan for commercialization process of *streptomycetes*-derived products. (Created in BioRender. Science, P. (2025). https://BioRender.com/p1it7q4)

cotton illnesses. Wei et al. 2024 reported that in the Chinese cabbage rhizosphere soil, *Streptomyces griseus* (A316) provides protection against rhizome illnesses by preventing the development of latent spores of rhizome pathogens. Thilagam and Hemalatha (2019) stated that *Streptomyces violaceoruber* may be used as a biofertilizer and biocontrol agent for chilli cultivation, as evidenced by a greenhouse experiment that showed that the application of *Streptomyces violaceoruber* fermentation broth decreased the incidence of chilli anthracnose and encouraged the growth of chilli seedlings, leading to increased germination, chlorophyll content, fresh weight, and plant height. Sharma and Salwan (2018) used combination of *Streptomyces rochei* and *Trichoderma harzianum* to treat pepper root rot brought on by *Phytophthora capsici,* while endophytic strains of Actinobacteria can suppress damping-off and crown and root rot of cucumbers more effectively than chemical fungicides similar to metalaxyl. Law et al. (2017) reported that several species of *Streptomyces* have been used in the greenhouse, growth chamber, and lab environment that are used to study the biological treatments for rice blast disease. Up to 88.3% the rice blast disease has been reduced by treating with the *Streptomyces* in the greenhouse experimental studies. Chaiharn et al. (2020) investigated that *Streptomyces palmae* PC12 exhibited strong antifungal activity against *Pyricularia* sp., the pathogen responsible for rice blast. In greenhouse conditions, its application improved rice plant growth by increasing plant height, root length, as well as root biomass in the RD6 variety. The strain significantly reduced disease severity and demonstrated potential as an effective biocontrol agent. Olanrewaju and Babalola (2019) reported that *Streptomyces kasugaensis* has the potential to promote plant development in *Pinus taeda* seedlings grown in a greenhouse and suppress the establishment of *Fusarium* and *Armillaria* pine rot.

12.6 Future Outcomes of *Streptomycetes*-Derived Products in Enhancing Sustainable Agriculture

As a rich biological resource, *Streptomyces* species have a substantial influence on agriculture, influencing crop yield, soil health, pest management, and environmental sustainability in several ways. With the increase in the awareness of sustainable agriculture, and use of *Streptomyces*-derived products in agricultural system, there is decrease in chemical inputs. The alternative to this is *Streptomyces*-based products, such as biofertilizers and biocontrol agents, which are anticipated to be used in agricultural operations. Various strains of the *Streptomyces* have their capacity to fix nitrogen, solubilize phosphates, and create phytohormones, improve plant growth and health, and increase yields, especially in the stress or low-fertility situations. *Streptomyces* are being used as natural substitutes for synthetic fertilizers and pesticides because of being cost-effective and environment-friendly. Their antibacterial activity leads to production of volatile organic compounds (VOCs) which helps to compete with the pathogens. *Streptomyces* promotes future agricultural sustainability by improving structure of soil and diversity by speeding up the nutrient cycle and organic matter decomposition. Climate-resilient agriculture and adapting agricultural systems to the effects of climate change depend on plants' capacity to decrease abiotic stresses like salinity and droughts by ACC deaminase activity along with other stress-responsive metabolites. Ready-to-use market formulations are expected to drive commercialization, facilitate the development of new products tailored to specific crops, regions, and farming practices within regulatory frameworks.

12.7 Conclusion

This chapter highlights that the plant diseases are the major problem for agriculture that cause crop losses all over the world. The emerging demand for sustainable farming has arisen that encourages the natural alternative and positive effect on the plant health. In the search of natural solutions, *Streptomyces* are a broad category of soil-dwelling, Gram-positive actinobacteria that have drawn attention for their advantageous agricultural applications due to their prevalence in soil and metabolic versatility. *Streptomycetes* sp. produce a wide range of antimicrobial compounds that show great biocontrol potential by reducing plant diseases, which lead to lower the dependency on chemical pesticides, and they promote safer and more environmentally friendly methods of managing pests. In addition to this, *Streptomycetes* also function as biofertilizers, improving plant nutrition by fixing nitrogen, solubilizing minerals, producing phytohormones, and lowering stress-induced ethylene levels by synthesizing ACC deaminase, auxins, and gibberellins. Furthermore, they play a crucial role in maintaining soil fertility. They encourage the nutrient cycling and production of geosmin, the earthy substance that gives good soil its earthy smell and has a positive impact on the structure and soil health. It is well documented that *Streptomycetes* produces a wide range of secondary metabolites with different bioactivities, such as plant growth stimulants, antibiotics, antifungals, and antiparasitic. In addition to this, they may be used as immunosuppressants and anticancer drugs. Agricultural products based on *Streptomycetes,* as evidenced from various field trials and case studies, help in increase crop yield; these products are applied through seed treatment, soil inoculation, and foliar sprays. Despite their promising potential, further study is necessary to enhance formulations of *Streptomycetes*, enhance application methods, and expand knowledge of how they interact with plants and soil microbiota. Such research will assist and guarantee consistent performance in the variety of agricultural systems and environmental conditions. To sum up, *Streptomycetes* hold immense potential to be a key component of sustainable agriculture. They provide a holistic approach in boosting agricultural output, reducing chemical inputs, and safeguarding environmental health. Integrating *Streptomycetes* into farming systems can promote long-term agricultural sustainability and ecological protection, as well as in enhancing global food security. Keeping in view its impact on agriculture, affecting crop production, soil health, pest control, and environmental sustainability across various dimensions, it could serve as a possible source of *Streptomycetes*-based products, including biofertilizers and biocontrol agents, which are anticipated to be used in agricultural operations. *Streptomycetes* promote future agricultural sustainability by improving structure of soil and diversity by speeding up the nutrient cycle and organic matter decomposition. Encouragement in the field application and as ready-to-use market formulations are expected to drive commercialization, facilitate the development of new products tailored to specific crops, regions, and farming practices within regulatory frameworks. The main aim of future research is to develop formulations that are both stable and efficient and having a long shelf life and user-friendly application methods. The integration of *Streptomycetes*-based solutions in the holistic agricultural systems, such as organic farming and IPM offers significant potential. With the promotion of greater yields through lowering the use of chemicals, improving soil biodiversity, and improving plant resilience. *Streptomycetes* may help boost crop yields while contributing to global food security and supporting both ecological protection and sustainable production.

Acknowledgments The authors are thankful for all of the assistance provided by the Central University of Himachal Pradesh throughout the writing of this chapter. This research did not receive any specific grants from any funding agencies.

References

Abdellatif AA, Gebily DA, Elmaghraby MM et al (2024) Transforming roles of actinobacteria in sustainable agriculture: from soil health and plant productivity perspective. In: Kaur S, Dwibedi V, Sahu PK (eds) Metabolomics, proteomics and gene editing approaches in biofertilizer industry. Springer, Singapore, pp 299–338

Afzal I, Javed T, Amirkhani M et al (2020) Modern seed technology: seed coating delivery systems for enhancing seed and crop performance. Agriculture (Basel) 10(11):526

Aggarwal N, Thind SK, Sharma S (2016) Role of secondary metabolites of actinomycetes in crop protection. In: Subramaniam G, Arumugam S, Rajendran V (eds) Plant growth promoting actinobacteria: a new avenue for enhancing the productivity and soil fertility of grain legumes. Springer, Singapore, pp 99–121

Ali Mude AS, Nageye YA, Bello KE (2025) Therapeutic application of *Streptomyces*: a choice alternative to modern therapies. Microb Infect Dis 6(2):916–928

Allouzi MM, Allouzi SM, Keng ZX et al (2022) Liquid biofertilizers as a sustainable solution for agriculture. Heliyon 8(12):e12609

Al-Quwaie DA (2024) The role of *Streptomyces* species in controlling plant diseases: a comprehensive review. Australas Plant Pathol 53(1):1–14

Al-Tammar FK, Khalifa AY (2023) An update about plant growth promoting *Streptomyces* species. J Appl Biol Biotechnol 10:1–10

Arasu MV, Esmail GA, Al-Dhabi NA et al (2016) Managing pests and diseases of grain legumes with secondary metabolites from actinomycetes. In: Subramaniam G, Arumugam S, Rajendran V (eds) Plant growth promoting actinobacteria: a new avenue for enhancing the productivity and soil fertility of grain legumes. Springer, Singapore, pp 83–98

Araujo R (2022) Advances in soil engineering: sustainable strategies for rhizosphere and bulk soil microbiome enrichment (Landmark Ed). Front Biosci 27(6):195

Barbuto Ferraiuolo S, Cammarota M, Schiraldi C et al (2021) *Streptomycetes* as platform for biotechnological production processes of drugs. Appl Microbiol Biotechnol 103:551–568

Beroigui O, Errachidi F (2023) *Streptomyces* at the heart of several sectors to support practical and sustainable applications: a review. Prog Microbes Mol Biol 6(1):a0000345

Boukhatem ZF, Merabet C, Tsaki H (2022) Plant growth promoting actinobacteria, the most promising candidates as bioinoculants? Front Agron 4:849911

Buzón-Durán L, Pérez-Lebeña E, Martín-Gil J et al (2020) Applications of *Streptomyces* sp. enhanced compost in sustainable agriculture. In: Meghvansi M, Varma A (eds) Biology of composts, Soil bology, vol 58. Springer, pp 257–291

Chaiharn M, Theantana T, Pathom-Aree W (2020) Evaluation of biocontrol activities of *Streptomyces* sp. against rice blast disease fungi. Pathogens 9(2):126

Chakraborty T, Akhtar N (2021) Biofertilizers: characteristic features and applications. In: Biofertilizers: study and impact. wiley, pp 429–489

Charousova I, Juraj ME, Alenárová E et al (2016) Effect of fertilization on biological activity of community of soil *streptomycetes*. J Cent Eur Agric 17(4):1134–1149

Chouyia FE, Ventorino V, Pepe O (2022) Diversity, mechanisms and beneficial features of phosphate-solubilizing *Streptomyces* in sustainable agriculture: a review. Front Plant Sci 13: 1035358

Dahal B, NandaKafle G, Perkins L et al (2017) Diversity of free-living nitrogen fixing *Streptomyces* in soils of the badlands of South Dakota. Microbiol Res 195:31–39

Dasila H, Joshi S, Sahgal M (2020) Dissecting structure and function of plant rhizomicrobiome: a genomic approach. In: Sharma SK, Singh UB, Sahu PK, Singh HV, Sharma PK (eds) Rhizosphere microbes, Microorganisms for sustainability, vol 23. Springer, Singapore, pp 73–103

Doolotkeldieva T, Bobusheva S, Konurbaeva M (2015) Effects of *Streptomyces* biofertilizer to soil fertility and rhizosphere's functional biodiversity of agricultural plants. Adv Microbiol 5(07): 555–571

Dos Santos PC, Fang Z, Mason SW et al (2012) Distribution of nitrogen fixation and nitrogenase-like sequences amongst microbial genomes. BMC Genom 13:162

Dow L, Gallart M, Ramarajan M et al (2023) *Streptomyces* and their specialised metabolites for phytopathogen control—comparative in vitro and in planta metabolic approaches. Front Plant Sci 14:1151912

El-Maraghy SS, Tohamy AT, Hussein KA (2021) Plant protection properties of the Plant Growth-Promoting Fungi (PGPF): Mechanisms and potentiality. Curr Res Environ Appl Mycol (J Fungal Biol) 11:391–415

Faddetta T, Polito G, Abbate L et al (2023) Bioactive metabolite survey of actinobacteria showing plant growth promoting traits to develop novel biofertilizers. Metabolites 13(3):374

Fravel DR (2005) Commercialization and implementation of biocontrol. Annu Rev Phytopathol 43(1):337–359

Gebily DA, Ghanem GA, Ragab MM et al (2021) Characterization and potential antifungal activities of three *Streptomyces* sp. as biocontrol agents against Sclerotinia sclerotiorum (Lib.) de Bary infecting green bean. Egypt J Biol Pest Control 31:1–5

Ghosh R, Mandal NC (2020) Use of plant growth–promoting Burkholderia species with rock phosphate–solubilizing potential toward crop improvement. In: Singh JS, Vimal SR (eds) Microbial services in restoration ecology. Elsevier, pp 139–156

Gopalakrishnan S, Srinivas V, Alekhya G et al (2015) Effect of plant growth-promoting *Streptomyces* sp on growth promotion and grain yield in chickpea (Cicer arietinum L). 3 Biotech 5: 799–806

Goswami M, Malakar C, Deka S (2020) Rhizosphere microbes for sustainable maintenance of plant health and soil fertility. In: Sharma SK, Singh UB, Sahu PK, Singh HV, Sharma PK (eds) Rhizosphere microbes, Microorganisms for sustainability, vol 23. Springer, Singapore, pp 35–72

Grover M, Bodhankar S, Maheswari M et al (2016) Actinomycetes as mitigators of climate change and abiotic stress. In: Subramaniam G, Arumugam S, Rajendran V (eds) Plant growth promoting actinobacteria: a new avenue for enhancing the productivity and soil fertility of grain legumes. Springer, Singapore, pp 203–212

Gupta G, Panwar J, Akhtar MS et al (2012) Endophytic nitrogen-fixing bacteria as biofertilizer. In: Lichtfouse E (ed) Sustainable agriculture reviews, Sustainable agriculture reviews, vol 11. Springer, Dordrecht, pp 183–221

Hariprasad KV (2016) Recent advancement in the development of biopesticides by Actinomycetes for the control of insect pests. In: Subramaniam G, Arumugam S, Rajendran V (eds) Plant growth promoting actinobacteria: a new avenue for enhancing the productivity and soil fertility of grain legumes. Springer, Singapore, pp 47–62

Htwe AZ, Moh SM, Moe K, Yamakawa T (2018) Effects of co-inoculation of *Bradyrhizobium japonicum SAY3-7* and *Streptomyces griseoflavus P4* on plant growth, nodulation, nitrogen fixation, nutrient uptake, and yield of soybean in a field condition. Soil Sci Plant Nutr 64(2): 222–229

Jacob S, Sudini HK (2016) Indirect plant growth promotion in grain legumes: role of actinobacteria. In: Subramaniam G, Arumugam S, Rajendran V (eds) Plant growth promoting actinobacteria: a new avenue for enhancing the productivity and soil fertility of grain legumes. Springer, Singapore, pp 17–32

Jacob SM, Paranthaman S (2023) Biofertilizers: an advent for eco-friendly and sustainable agriculture development. Vegetos 36(4):1141–1153

Johnson I, Kavitha R, Karthikeyan M (2021) Deciphering the biocontrol potential of *Streptomyces* sp. in plant disease management: a review. Int J Agric Environ Biotechnol 14:547–559

Junaid JM, Bhat TA, Bhat A et al (2013) Commercial biocontrol agents and their mechanism of action in the management of plant pathogens. Int J Mod Plant Anim Sci 1(12):39–57

Kaari M, Manikkam R, Annamalai KK et al (2023) Actinobacteria as a source of biofertilizer/biocontrol agents for bio-organic agriculture. J Appl Microbiol 134(2):lxac047

Khan A, Joshi M, Singh AV (2020) Rhizospheric microbial community: ecology, methods, and functions. In: Sharma SK, Singh UB, Sahu PK, Singh HV, Sharma PK (eds) Rhizosphere microbes, Microorganisms for sustainability, vol 23. Springer, Singapore, pp 127–148

Khan S, Srivastava S, Karnwal A, Malik T (2023) *Streptomyces* as a promising biological control agents for plant pathogens. Front Microbiol 14:1285543

Kumar S, Lad M, Mahera YA (2022b) Biofertilizer production in India: current status, classification and schemes. Int J Chem Sci 20(03):430

Kumar S, Sindhu SS, Kumar R (2022a) Biofertilizers: an ecofriendly technology for nutrient recycling and environmental sustainability. Curr Res Microb Sci 3:100094

Law JWF, Ser HL, Khan TM et al (2017) The potential of *Streptomyces* as biocontrol agents against the rice blast fungus, Magnaporthe oryzae (Pyricularia oryzae). Front Microbiol 8:3

Law JWF, Tan KX, Wong SH et al (2018) Taxonomic and characterization methods of streptomyces: a review. Prog Microbes Mol Biol 1(1):a0000009

Le KD, Yu NH, Park AR et al (2022) *Streptomyces* sp. AN090126 as a biocontrol agent against bacterial and fungal plant diseases. Microorganisms 10(4):791

Lee J, Kim S, Jung H et al (2023) Exploiting bacterial genera as biocontrol agents: mechanisms, interactions and applications in sustainable agriculture. J Plant Biol 66(6):485–498

Mahala DM, Maheshwari HS, Yadav RK et al (2020) Microbial transformation of nutrients in soil: an overview. In: Sharma SK, Singh UB, Sahu PK, Singh HV, Sharma PK (eds) Rhizosphere microbes, Microorganisms for sustainability, vol 23. Springer, Singapore, pp 175–211

Maheshwari HS, Agnihotri R, BhartiA et al (2020) Signaling in the rhizosphere for better plant and soil health. In: Sharma SK, Singh UB, Sahu PK, Singh HV, Sharma PK (eds) Rhizosphere microbes, Microorganisms for sustainability, vol 23. Springer, Singapore, pp 149–173

Mahmood SA, Al-Khateeb MT, Abedalsadda AJ et al (2024) Studying of new bio-fertilizer formula composed of *Streptomyces* and nitrogen fixing bacteria to sustain Vigna Radiate (L.). Iraqi J Ind Res 11(2):138–145

Manikandan A, Anandham R, Johnson I et al (2023) Bacillus and *Streptomyces* for nanagement of biotic stresses in plants for sustainable agriculture. In: Chhabra S, Prasad R, Maddela NR, Tuteja N (eds) Plant microbiome for plant productivity and sustainable agriculture, Microorganisms for sustainability, vol 37. Springer, Singapore, pp 263–288

Manjunatha L, Rajashekara H, Uppala LS et al (2022) Mechanisms of microbial plant protection and control of plant viruses. Plants 11(24):3449

Mitra D, Mondal R, Khoshru B et al (2022) Actinobacteria-enhanced plant growth, nutrient acquisition, and crop protection: advances in soil, plant, and microbial multifactorial interactions. Pedosphere 32(1):149–170

Mohamad OA, Liu YH, Huang Y et al (2022) The metabolic potential of endophytic actinobacteria associated with medicinal plant Thymus roseus as a plant-growth stimulator. Microorganisms 10(9):1802

Mohammadipanah F, Wink J (2016) Actinobacteria from arid and desert habitats: diversity and biological activity. Front Microbiol 6:1541

Najmeh S, Hosein SB, Sareh S et al (2014) Biological control of citrus green mould, Penicillium digitatum, by antifungal activities of *Streptomyces* isolates from agricultural soils. Afr J Microbiol Res 8:1501–1509

Narayanasamy P (2013) Development of formulations and commercialization of biological products. In: Biological management of diseases of crops, Progress in biological control, vol 16. Springer, Dordrecht, pp 129–187

Ngamcharungchit C, Chaimusik N, Panbangred W et al (2023) Bioactive metabolites from terrestrial and marine actinomycetes. Molecules 28(15):5915

Nguyen TT, Nguyen TT, Nguyen TH et al (2025) In silico and in vitro analyses reveal the potential use of *Streptomyces* parvulus VNUA74 as bioagent for sustainable banana production. Sci Rep 15(1):7049

Nimaichand S, Devi AM, Li WJ (2016) Direct plant growth-promoting ability of actinobacteria in grain legumes. In: Subramaniam G, Arumugam S, Rajendran V (eds) Plant growth promoting actinobacteria: a new avenue for enhancing the productivity and soil fertility of grain legumes. Springer, Singapore, pp 1–16

Nimbulkar P, Gupta G, Virkhare U et al (2025) Bacterial endophytes and their secondary metabolites: mechanisms of biosynthesis and applications in sustainable agriculture. J Umm Al-Qura Univ Appl Sci:1–13

Olanrewaju OS, Babalola OO (2019) *Streptomyces*: implications and interactions in plant growth promotion. Appl Microbiol Biotechnol 103:1179–1188

Oskay M (2009) Comparison of *Streptomyces* diversity between agricultural and non-agricultural soils by using various culture media. Sci Res Essays 4(10):997–1005

Pacios-Michelena S, Aguilar González CN, Alvarez-Perez OB et al (2021) Application of *Streptomyces* antimicrobial compounds for the control of phytopathogens. Front Sustain Food Syst 5: 696518

Pandey KD, Patel AK, Singh M et al (2020) Secondary metabolites from bacteria and viruses. In: Sinha R, Häder D-P (eds) Natural bioactive compounds. Acad Press, London, pp 19–40

Pang F, Solanki MK, Wang Z (2022) *Streptomyces* can be an excellent plant growth manager. World J Microbiol Biotechnol 38(11):193

Paravar A, Piri R, Balouchi H et al (2023) Microbial seed coating: an attractive tool for sustainable agriculture. Biotechnol Rep 37:e00781

Peter AJ, Amalraj EL, Talluri VR (2020) Commercial aspects of biofertilizers and biostimulants development utilizing rhizosphere microbes: global and Indian scenario. In: Sharma SK, Singh UB, Sahu PK, Singh HV, Sharma PK (eds) Rhizosphere microbes rhizosphere microbes: soil and plant functions, Microorganisms for sustainability, vol 23. Springer, Singapore, pp 655–682

Pirttilä AM, Tabas HMP, Baruah N et al (2021) Biofertilizers and biocontrol agents for agriculture: how to identify and develop new potent microbial strains and traits. Microorganisms 9(4):817

Preininger C, Sauer U, Bejarano A et al (2018) Concepts and applications of foliar spray for microbial inoculants. Appl Microbiol Biotechnol 102:7265–7282

Procópio RE, IRD S, Martins MK et al (2012) Antibiotics produced by *Streptomyces*. Braz J Infect Dis 16(12):466–471

Qiu Y, Amirkhani M, Mayton H et al (2020) Biostimulant seed coating treatments to improve cover crop germination and seedling growth. Agronomy 10(2):154

Quinn GA, Banat AM, Abdelhameed AM et al (2020) *Streptomyces* from traditional medicine: sources of new innovations in antibiotic discovery. J Med Microbiol 69(8):1040–1048

Ram RM, Rajput RS, Vaishnav A (2020) Management of Sclerotium rolfsii induced diseases in crops by Trichoderma species. In: Sharma SK, Singh UB, Sahu PK, Singh HV, Sharma PK (eds) Rhizosphere microbes: soil and plant functions, Microorganisms for sustainability, vol 23. Springer, Singapore, pp 593–617

Rani K, Dahiya A, Masih JC et al (2018) Actinobacterial biofertilizers: an alternative strategy for plant growth promotion. Int J Curr Microbiol Appl Sci 7(09):607–614

Reddy KR, Jyothi G, Sowjanya C et al (2016) Plant growth-promoting actinomycetes: mass production, delivery systems, and commercialization. In: Subramaniam G, Arumugam S, Rajendran V (eds) Plant growth promoting actinobacteria: a new avenue for enhancing the productivity and soil fertility of grain legumes. Springer, Singapore, pp 287–298

Rey T, Dumas B (2017) Plenty is no plague: Streptomyces symbiosis with crops. Trends Plant Sci 22(1):30–37

Rocha I, Ma Y, Souza-Alonso P et al (2019) Seed Coating: A Tool for Delivering Beneficial Microbes to Agricultural Crops. Front Plant Sci 10:1357

Sarkar A, Saha M, Meena VS (2017) Plant beneficial rhizospheric microbes (PBRMs): prospects for increasing productivity and sustaining the resilience of soil fertility. In: Meena V, Mishra P, Bisht J, Pattanayak A (eds) Agriculturally important microbes for sustainable agriculture. Springer, pp 3–29

Sathya A, Vijayabharathi R, Gopalakrishnan S (2017) Plant growth-promoting actinobacteria: a new strategy for enhancing sustainable production and protection of grain legumes. 3 Biotech 7: 102

Sharma V, Salwan R (2018) Biocontrol potential and applications of actinobacteria in agriculture. In: New and future developments in microbial biotechnology and bioengineering: actinobacteria: diversity and biotechnological applications. Elsevier, pp 93–108

Shivlata L, Satyanarayana T (2017) Actinobacteria in agricultural and environmental sustainability. In: Singh J, Seneviratne G (eds) Agro-environmental sustainability: volume 1: managing crop health. Springer, Cham, pp 173–218

Silva GD, Kitano IT, Ribeiro IADF et al (2022) The potential use of actinomycetes as microbial inoculants and biopesticides in agriculture. Front Soil Sci 2:833181

Singh S, Devi NB, Divya D et al (2024) Soil fertility management: role of organic amendments and bio-fertilizers: a review. Int J Res Agron 7(12):766–772

Sneha S, Anitha B, Sahair RA et al (2018) Biofertilizer for crop production and soil fertility. Acad J Agric Res 6(8):299–306

Solans M, Vobis G, Jozsa L et al (2016) Synergy of Actinomycete co-inoculation. In: Subramaniam G, Arumugam S, Rajendran V (eds) Plant growth promoting actinobacteria: a new avenue for enhancing the productivity and soil fertility of grain legumes. Springer, Singapore, pp 161–177

Soumare A, Boubekri K, Lyamlouli K et al (2020) From isolation of phosphate solubilizing microbes to their formulation and use as biofertilizers: status and needs. Front Bioeng Biotechnol 7:425

Sousa JAJ, Olivares FL (2016) Plant growth promotion by streptomycetes: ecophysiology, mechanisms and applications. Chem Biol Technol Agric 3:1–12

Srivastava MP, Sharma S (2014) Potential of PGPR bacteria in plant disease management. In: Sharma N (ed) Biological controls for preventing food deterioration, pp 87–116

Subramanian KS, Muniraj I, Uthandi S (2016) Role of actinomycete-mediated nanosystem in agriculture. In: Subramaniam G, Arumugam S, Rajendran V (eds) Plant growth promoting actinobacteria: a new avenue for enhancing the productivity and soil fertility of grain legumes. Springer, Singapore, pp 233–247

Swarnalakshmi K, Senthilkumar M, Ramakrishnan B (2016) Endophytic actinobacteria: nitrogen fixation, phytohormone production, and antibiosis. In: Subramaniam G, Arumugam S, Rajendran V (eds) Plant growth promoting actinobacteria: a new avenue for enhancing the productivity and soil fertility of grain legumes. Springer, Singapore, pp 123–145

Taj ZZ, Rajkumar M (2016) Perspectives of plant growth-promoting actinomycetes in heavy metal phytoremediation. In: Subramaniam G, Arumugam S, Rajendran V (eds) Plant growth promoting actinobacteria: a new avenue for enhancing the productivity and soil fertility of grain legumes. Springer, Singapore, pp 213–231

Tariq M, Khan A, Asif M et al (2020) Biological control: a sustainable and practical approach for plant disease management. Acta Agric Scand B Soil Plant Sci 70(6):507–524

Thilagam R, Hemalatha N (2019) Plant growth promotion and chilli anthracnose disease suppression ability of rhizosphere soil actinobacteria. J Appl Microbiol 126(6):1835–1849

Tripathi K, Kumar N, Singh M et al (2020) Fungal siderophore: biosynthesis, transport, regulation, and potential applications. In: Sharma SK, Singh UB, Sahu PK, Singh HV, Sharma PK (eds) Rhizosphere microbes: soil and plant functions, Microorganisms for sustainability, vol 23. Springer, Singapore, pp 387–408

Tyagi A, Lama Tamang T, Kashtoh H et al (2024) A review on biocontrol agents as sustainable approach for crop disease nanagement: applications, production, and future perspectives. Horticulturae 10(8):805

Viaene T, Langendries S, Beirinckx et al (2016) *Streptomyces* as a plant's best friend? FEMS Microbiol Ecol *92*(8):fiw119

Vurukonda SSKP, Giovanardi D, Stefani E (2018) Plant growth promoting and biocontrol activity of *Streptomyces* sp. as endophytes. Int J Mol Sci 19(4):952

Wani SP, Gopalakrishnan S (2019) Plant growth-promoting microbes for sustainable agriculture. In: Sayyed R, Reddy M, Antonius S (eds) Plant growth promoting rhizobacteria (PGPR): prospects for sustainable agriculture. Springer, Singapore, pp 19–45

Wei X, Xie B, Wan C et al (2024) Enhancing soil health and plant growth through microbial fertilizers: mechanisms, benefits, and sustainable agricultural practices. Agronomy 14(3):609

Ye JJ, Zou RJ, Zhou DD et al (2023) Insights into the phylogenetic diversity, biological activities, and biosynthetic potential of mangrove rhizosphere actinobacteria from Hainan Island. Front Microbiol 14:1157601

Applications of Streptomycetes-Derived Nanoparticles

13

Monika Yadav, Ashish Gupta, Sapna Devi, and Jaya Verma

Abstract

Streptomyces, a genus of filamentous bacteria, have emerged as a viable and important source of biosynthesized nanoparticles (NPs). *Streptomyces*-derived nanoparticles are attracting widespread attention due to their various bioactive characteristics and nanoscale interactions. These nanoparticles, which typically consist of metals such as gold, silver, copper oxide, and zinc oxide, have outstanding anticancer, antibacterial, antioxidant, and wound-healing characteristics, making them extremely valuable in biomedical applications. In agriculture, they operate as biopesticides and plant growth promoters, boosting sustainable agricultural techniques. They also play a crucial role in environmental applications, including bioremediation and wastewater treatment. In the industrial sector, they serve as nanocatalysts and antibacterial agents for food preservation. The multifunctionality of these nanoparticles demonstrates their promise for novel solutions in healthcare, agriculture, and environmental sustainability.

M. Yadav
Department of Applied Sciences and Humanities, Faculty of Engineering and Technology, Rama University, Kanpur, Uttar Pradesh, India

A. Gupta
Department of Botany, Banaras Hindu University, Varanasi, Uttar Pradesh, India

S. Devi
Department of Microbiology, School of Bioengineering and Biosciences, Lovely Professional University, Phagwara, Punjab, India

J. Verma (✉)
Department of Microbiology, Guru Nanak Dev University, Amritsar, Punjab, India

T. Kaur et al. (eds.), *Streptomyces: Biological Candidates for Sustainable Agriculture*,
https://doi.org/10.1007/978-981-95-5803-2_13

Keywords

Streptomyces · Nanoparticles · Antimicrobials · Antioxidants · Plant growth promoters

13.1 Introduction

Nanotechnology continues to develop as a significant topic due to its wide range of uses, making it a focus of intensive scientific study (Golinska et al. 2014). Nanotechnology has emerged as a key scientific achievement of the twenty-first century, including the synthesis, manipulation, and use of materials with dimensions less than 100 nanometers. This interdisciplinary field has numerous applications in various industries, including environmental protection, agriculture, biotechnology, food processing, pharmaceuticals, biomedicine, wastewater treatment, and ecological monitoring (Rassaei et al. 2011; Zahra et al. 2020). Recent discoveries show a considerable growth in the synthesis of metal-based nanoparticles (NPs) with regulated morphologies and different physicochemical characteristics (Sharma et al. 2019). Nanoparticles generally vary in size from 10 to 100 nanometers (10^9 m), providing unique and desired properties. The term "nano" comes from the Greek word for "tiny" or "extremely small" (Tikariha et al. 2012). Metallic NPs made of zinc (Zn), gold (Au), copper (Cu), magnesium (Mg), silver (Ag), and titanium (Ti) are considered crucial building blocks in nanotechnology due to their adaptable forms and sizes (Basavaraj et al. 2012; Saravana Kumar et al. 2015; Salem et al. 2020). Figure 13.1 shows that metallic nanoparticles are used in biomedical, environmental, agricultural, and industrial sectors.

Metallic NPs are used in a wide variety of applications, including analytical, biological probes, reprography, optoelectronics, display technologies (Sastry et al. 2003), water purification, and catalytic processes (Hutchison 2008; Sharma et al. 2019), photocatalysis (Fu and Fu 2015), biosensor development (Manimaran and Kannabiran 2017; Zheng et al. 2019; Ying et al. 2020), and biomedical diagnostics, particularly for detecting cancer cells (Salem et al. 2020; Menon et al. 2017). Additionally, they are utilized in antimicrobial formulations (Ameen et al. 2020; Tanwar et al. 2020; Tehri et al. 2020), targeted drug delivery systems (Li et al. 2011), gene therapy for hereditary disorders (Golinska et al. 2014), and environmental monitoring to detect ecotoxic metals and hazardous reagents (Menon et al. 2017; Ahmad et al. 2003). The biological synthesis of metal nanoparticles (NPs) has gained considerable attention due to their environmentally friendly nature and potential for sustainable nanomaterial production. Several biological entities, including plant extracts, complete plants, and microorganisms such as algae, yeast, bacteria, and actinobacteria, have been studied for this purpose (Zheng et al. 2019; Tehri et al. 2020). Bacteria have proved to be especially excellent bio-factories for nanoparticle synthesis, owing to their distinct metal resistance mechanisms that help in the bioreduction of metal ions into stable metallic nanoparticles (Tehri et al. 2020). Actinobacteria in this system are aerobic, Gram-positive filamentous

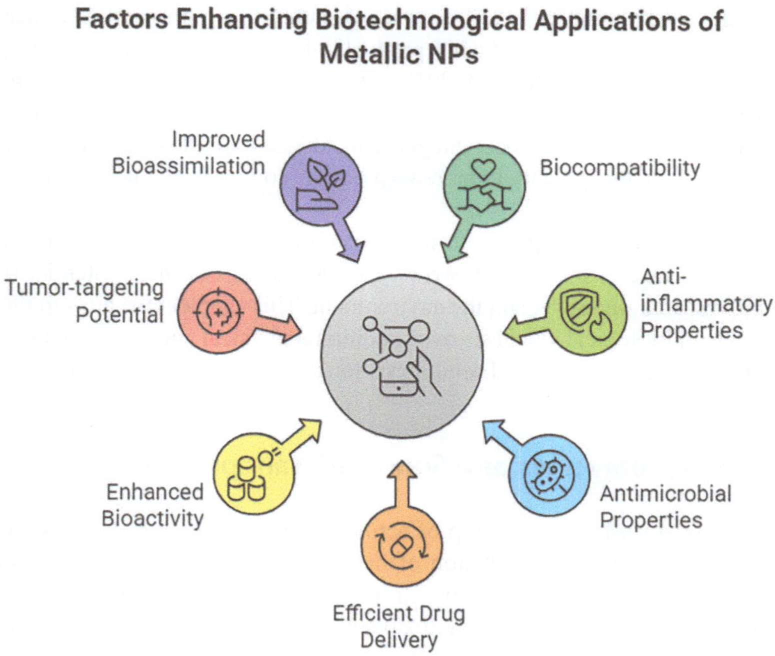

Fig. 13.1 Various applications associated with metallic nanoparticles

bacteria that are mostly categorized as Actinomycetales. These bacteria are well known because they produce a wide variety of secondary metabolites with important biological activities, such as anticancer, antibiotic, antioxidant, antiparasitic, and anti-biofouling capabilities (Zotchev 2012; Manivasagan et al. 2013).

Recent developments have enabled the production of nanomaterials from various metals, including titanium (Ti), silver (Ag), copper (Cu), gold (Au), and zinc (Zn). These nanoparticles are often formed through physical, chemical, or biological methods. Physical synthesis techniques frequently produce low yields, while chemical methods typically require large amounts of precursor chemicals and solvents, which can generate harmful intermediates. Biological synthesis, however, is viewed as a more sustainable and environmentally friendly option because it uses cleaner processes that do not generate toxic byproducts. Additionally, biosynthetic methods are known to produce nanoparticles with high chemical purity, excellent monodispersity, and precisely controlled morphological features (Ying et al. 2020; Ameen et al. 2020). The functional applications of metal-based nanoparticles in nanotechnology, particularly in antimicrobial research, depend significantly on the specific metal used, with varying activity among different microbial strains (Tanwar et al. 2020; Tehri et al. 2020). Notably, silver (Ag) nanoparticles have been extensively studied and traditionally chosen in biological synthesis due to their strong antibacterial properties, especially in treating infectious diseases.

Currently, nanoparticles have been manufactured using physical, chemical, and biological processes, each producing products of varying sizes, shapes, and physicochemical qualities (Akhtar et al. 2013). While physical and chemical approaches may produce high-quality nanoparticles, there is a growing emphasis on biologically based synthesis processes. These biogenic techniques are cleaner, safer, and more sustainable, with the potential for cost-effective commercial production (Vahabi et al. 2011; Golinska et al. 2014).

The purpose of this chapter is to investigate the biological production of metal-based nanoparticles using *Streptomyces* species and assess their potential uses in scientific research, industry, and the environment. This chapter also aims to illustrate the benefits of biological synthesis over traditional physical and chemical processes, highlighting its environmental friendliness, long-term viability, and cost efficiency.

13.2 Streptomycetes as a Source of Nanoparticles

Streptomyces, a well-studied group of actinobacteria, has emerged as a significant biological system for the production of metal nanoparticles (NPs) (Gahlawat and Choudhury 2019). These filamentous, Gram-positive, soil-dwelling bacteria are well known for their capacity to create a diverse range of antibiotics and secondary compounds. It has also been used effectively in the green synthesis of different metal nanoparticles, such as silver (Ag), gold (Au), copper (Cu), zinc (Zn), and manganese (Mn), as indicated in Table 13.1 (Waghmare et al. 2014; Omran and Omran 2020). Copper nanoparticles (Cu NPs) have gained popularity due to their many uses in medicinal, environmental, and industrial domains (Fouda and Shaheen 2017; Issaabadi et al. 2017; Ismail et al. 2019). *Streptomyces rochei* MHM13, *Streptomyces coelicolor,* and *Streptomyces* sp. LK3, *Streptomyces griseorubens* AU2, and *Streptomyces viridodiastaticus* SSHH-1 are among the *Streptomyces* species that have demonstrated strong abilities in Ag NPs biosynthesis (Abd-Elnaby et al. 2016; Karthik et al. 2014; Manikprabhu and Lingappa 2013). These species have demonstrated tremendous promise in ecologically sustainable nanotechnology, highlighting the potential of *Streptomyces* as efficient microbial platforms for the environmentally friendly manufacturing of metal nanoparticles.

13.3 Application of *Streptomyces*-Derived Nanoparticles

These biosynthesized NPs exhibit distinctive characteristics compared to those produced via conventional chemical and physical methods, primarily due to the absence of external stabilizing or capping agents during the synthesis process. As a result, they offer a broader spectrum of potential applications across various disciplines (Zarina and Nanda 2014).

Nanoparticles (NPs) have demonstrated remarkable potential across a wide array of applications, including magnetic devices, photocatalysis, electrocatalysis, microelectronics, anticorrosive coatings, biomedicine, and reprography. In particular, the

Table 13.1 Different types of metallic NPs synthesized by *Streptomyces*

Streptomyces spp.	Type of nanoparticle	References
Streptomyces sp. VITBT7	Silver	Subashini and Kannabiran (2013)
Streptomyces sp. NK52	Gold	Prakash et al. (2013)
Streptomyces sp. MBRC-91	Silver	Manivasagan et al. (2015)
Streptomyces sp.	Silver	Abdeen et al. (2014)
Streptomyces parvulus SSNP11	Silver	Prakasham et al. (2014)
Streptomyces sp. HBUM171191	Zinc/Manganese	Waghmare et al. (2011)
Streptomyces rochei	Silver	Selvakumar et al. (2012)
Streptomycetes viridogens HM10	Gold	Balagurunathan et al. (2011)
Streptomyces hygroscopicus	Gold	Waghmare et al. (2014)
Streptomyces sp. ERI-3	Silver	Zonooz and Salouti (2011)
Streptomyces aureofaciens MTCC356	Silver	Sundaramoorthi et al. (2011)
Streptomyces spp. I, II; *Rhodococcus* spp.	Silver	Sukanya et al. (2013)
Streptomyces sp. LK3	Silver	Karthik et al. (2014)
Streptomyces albidoflavus CNP10	Silver	Golinska et al. (2014)
Streptomyces microflavus strain FSHJ31	Selenium	Forootanfara et al. (2014)
Streptomyces parvulus DPUA1549	Silver	de Amorim Silva et al. (2017)
Streptomyces owasiensis DPUA1748	Silver	de Amorim Silva et al. (2017)
Streptomyces hygroscopicus	Gold	Balagurunathan et al. (2011)
Streptomyces narbonensis SSHH-1E	Silver	El-Naggar et al. (2016)
Streptomyces viridogens	Gold	de Amorim Silva et al. (2017)
Streptomyces capillispialisca-1	Copper	Saad et al. (2018)
Streptomyces plicatus	Zinc oxide	Kalaba et al. (2021)
Streptomyces MHM38	Copper oxide	Bukhari et al. (2021)
Streptomyces catenulae	Silver	Khalil et al. (2022)
Streptomyces griseus and S. albus	Silver	Sambangi and Gopalakrishnan (2023)
Streptomyces enissocaesilis BS1	Silver	Shaaban et al. (2024)
Streptomyces albogriseolus	Gold	El-Naggar et al. (2024)

biotechnological applications of metallic nanoparticles (NPs) have expanded significantly due to their biocompatibility, anti-inflammatory, and antimicrobial properties, efficient drug delivery capabilities, enhanced bioactivity and bioavailability, tumor-targeting potential, and improved bioassimilation (Li et al. 2011; Salem et al. 2020).

Among the biological systems employed, *Streptomyces*-derived nanoparticles have shown considerable promise. Their unique bioactive properties suggest potential applications in various sectors, including food safety, environmental remediation, medical therapeutics, and agricultural practices (Fig. 13.2).

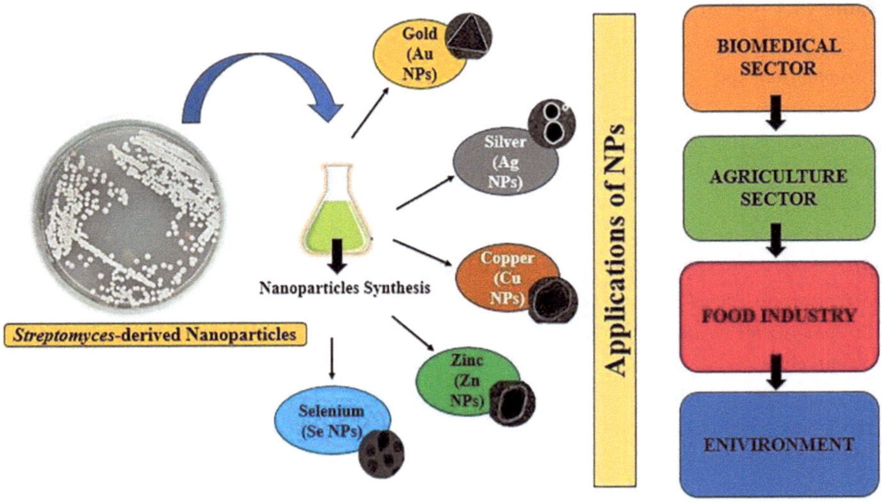

Fig. 13.2 Application of *Streptomyces*-derived nanoparticles

13.3.1 Streptomycetes-Derived NPs in Biomedical Applications

Streptomyces hold a crucial position in the pharmaceutical and medical industries primarily due to their remarkable ability to synthesize a diverse range of clinically important antibiotics, including vancomycin, gentamicin, amphotericin, chloramphenicol, and neomycin (Alani et al. 2012). This antibiotic-producing potential makes them indispensable microbial resources for large-scale industrial applications. Beyond their well-established role in antibiotic production, actinobacteria have recently emerged as highly effective biological platforms for the green synthesis of metallic nanoparticles (NPs). Many of these biosynthesized NPs exhibit broad-spectrum antimicrobial properties, as highlighted in Table 13.2.

Silver Nanoparticles Silver nanoparticles (AgNPs) produced by *Streptomyces* species have demonstrated strong antibacterial activity, including efficacy against multidrug-resistant bacterial strains (Shirley et al. 2010). *Streptomyces* sp. 09PBT005 generated AgNPs with remarkable cytotoxic and antibacterial properties (Saravana Kumar et al. 2015). Furthermore, extracellular AgNPs produced by *Streptomyces viridochromogenes* have shown substantial antibacterial activity against both Gram-negative and Gram-positive bacteria, as well as pathogenic fungi like *Candida albicans* (El-Naggar et al. 2016). *Streptomyces* sp. SS2's secondary metabolites have been successfully exploited in the production of silver nanoparticles (AgNPs), providing an ecologically beneficial and sustainable method. These biologically synthesized AgNPs demonstrated notable antibacterial activity against several pathogenic strains, including *Staphylococcus aureus*, *Bacillus subtilis*, *Staphylococcus epidermidis*, *Escherichia coli*, and *Vibrio cholera* (Mohanta and Behera 2014). Manikprabhu and Lingappa (2013) and Li et al. (2011) reported

Table 13.2 Biomedical applications of metallic NPs synthesized from *Streptomyces*

Streptomyces sp.	NPs	Size (nm)	Antimicrobial activity	References
*Streptomyces parvulus*SSNP11	Ag	2–12	Antibacterial activity	Sowani et al. (2016)
Streptomyces sp. VITPK1	Ag, Au	20–45	Antifungal activity	Sanjenbam et al. (2014)
Streptomyces sp. VITBT7	Ag	20–70	Antifungal and antibacterial activity	Subashini and Kannabiran (2013)
Streptomyces coelicolor	Ag	20–50	Anti-MRSA activity	Manikprabhu and Lingappa (2013)
Streptomyces sp. VITSTK7	Ag	20–70	Anti-fungal activity	Thenmozhi et al. (2013)
Streptomyces sp. JAR1	Ag	68	Antibacterial and antifungal activity	Chauhan et al. (2013)
Streptomyces sp.	Zn	–	Antibacterial nanopackaging	Rajamanickam et al. (2012)

that extracellularly synthesized AgNPs from *Streptomyces* sp. 09PBT005 and *Streptomyces aureofaciens* displayed significant antibacterial activity. Other strains, including *Streptomyces coelicolor* klmp33, *S. rochei*, *S. hygroscopicus*, and *Streptomyces* sp. ERI-3, were also demonstrated to biosynthesize nanoparticles with excellent antibacterial characteristics, highlighting their expanding relevance in the biomedical and healthcare industries. *Streptomyces* sp. VITSJK10 biosynthesizes AgNPs with anti-ESBL (extended-spectrum β-lactamase) activity, proving effective against MDR infections. Similarly, *S. coelicolor* klmp33 has been shown to create a pigment that aids in silver nitrate reduction, resulting in the generation of AgNPs with antibacterial action against ESBL-producing *E. coli* strains (Manikprabhu and Lingappa 2013). Moreover, Usha et al. (2010) revealed that silver nanoparticles produced by *Streptomyces* species may be used as antimicrobial coatings on textiles, protecting against *E. coli, S. aureus*, and *Aspergillus niger*.

Additionally, silver nanoparticles from Streptomycetes also showed anticancerous activity. AgNPs produced by *Streptomyces rochei* HMM13 demonstrated antibacterial and anticancer activities. Cytotoxicity studies found that cancer cell lines such as MCF-7, HCT-116, HepG2, A549, and PC-3 are more susceptible to these nanoparticles, whereas HELA, CACO, and HEP-2 cell lines are more resistant (Waghmare et al. 2014).

Anticancerous Activity Their antimicrobial mechanisms typically involve interruption of the bacterial cell membrane, leading to compromised structural integrity and inhibition of essential cellular functions in pathogenic microorganisms (Subbaiya et al. 2017). Moreover, the increased antibacterial effectiveness is primarily due to the nanoparticles' high surface area-to-volume ratio, which allows for efficient contact with bacterial membranes, eventually leading to cell lysis. *Streptomyces* *atrovirens* has been shown to create extracellular silver nanoparticles (AgNPs) that have powerful anticancer properties. These AgNPs exhibit dose-dependent cytotoxicity against the MCF-7 breast cancer cell line (Subbaiya et al. 2017).

Synergistic Effect with Antibiotics The increasing threat of antimicrobial resistance poses a significant challenge to the healthcare sector. In this context, Ag NPs biosynthesized by *Streptomyces* sp. JAR1 has shown promising antimicrobial potential (Chauhan et al. 2013). Several studies have further evaluated the synergistic effects of biosynthesized AgNPs in combination with conventional antibiotics, revealing an enhanced antimicrobial response against clinically relevant pathogens. Prakasham et al. (2014) investigated the combined efficacy of AgNPs and antibiotics against both Gram-negative (*Klebsiella pneumoniae*, *Pseudomonas putida*) and Gram-positive (*Bacillus subtilis*, *Salmonella typhi*) bacterial strains. Synergistic experiments verified that traditional antibiotics, such as ampicillin, streptomycin, cefalexin, and penicillin, demonstrated considerably increased effectiveness when conjugated with these biosynthesized silver nanoparticles (Prakasham et al. 2014). These data indicated that when combined with conventional antibiotics, these nanoparticles enhanced antimicrobial effectiveness and helped suppress the emergence of antibiotic resistance in pathogens.

Selenium Nanoparticles On the other hand, biosynthesized selenium nanoparticles (SeNPs) from *Streptomyces microflavus* strain FSHJ31 have also been reported to exhibit significant anticancer activity (Forootanfara et al. 2014). Ahmad et al. (2015) studied the impacts of biologically produced selenium nanorods (SeNrs) synthesized from *Streptomyces bikiniensis* strain EssamA-1 and showed their cytotoxic abilities against human cancer cell lines HepG2 and MCF-7, resulting in significant cell death. In a related study, Ramya et al. (2015) documented that extracellular SeNPs synthesized by *Streptomyces minutiscleroticus* M10A62 exerted notable antiproliferative effects against HepG2 and HeLa cell lines. These findings underscore the therapeutic promise of *Streptomyces*-mediated SeNPs in cancer treatment.

Copper and Zn Nanoparticles *Streptomyces* species have also been shown to reduce copper sulfate, resulting in extracellular production of copper nanoparticles (Cu NPs). These Cu NPs, when incorporated into fabrics, successfully prevent microbiological contamination. Similarly, *Streptomyces* reduced copper sulfate and zinc nitrate, resulting in the creation of metal oxide nanoparticles having antifungal and antibacterial characteristics. When utilized as fabric coatings, biosynthesized nanoparticles operate as barriers, reducing the spread of pathogens. The green production of zinc oxide nanoparticles (ZnO NPs) utilizing *Streptomyces* species offers an environmentally friendly technique with potential uses in cancer treatment. *Streptomyces* serves as a natural reducing agent, resulting in the formation of ZnO NPs with considerable anticancer activity against A549 lung cancer cells. Additionally, these nanoparticles have broad-spectrum antibacterial action against a variety of microbial species.

Gold Nanoparticles In addition to silver and zinc nanoparticles, actinobacteria, such as *Streptomyces* sp. NK52, have been shown to produce gold nanoparticles (AuNPs), which have significant biological activities such as inhibiting lipid peroxidation and scavenging free radicals (Prakash et al. 2013; Manivasagan et al. 2015). El-Naggar et al. (2024) reported antitumor activity of gold nanoparticles obtained from *Streptomyces albogriseolus.*

13.3.2 Streptomycetes-Derived NPs in the Food Industry

Nanoparticles (NPs) have an important role in the food sector, notably in two areas: improving food packaging and processing (Espitia et al. 2012; Shukla 2012). Silver nanoparticles (AgNPs) have a significant potential to break bacterial biofilms due to their nanoscale penetrating capabilities. As a result, AgNPs are being investigated for their potential function in food packaging decontamination and sanitation operations (Huang et al. 2015). Furthermore, zinc oxide nanoparticles (ZnO NPs) have been identified as potentially useful substances for the production of food containers and nutritious coatings, with potential benefits in use as food packaging (Rajamanickam et al. 2012; Prasad et al. 2014). The use of nanoparticles in packaging methods has been found to minimize microbiological contamination and improve the shelf life of food goods by maintaining their freshness (Salem and Fouda 2021). *Streptomyces* species produce antibacterial covering materials by extracellular production of zinc nanoparticles and metal oxides. These coatings can increase shelf life, preserve quality, and ensure food safety, making them suitable for use in food packaging and processing. Studies have revealed that silver nanoparticles may be integrated into both biodegradable and non-biodegradable polymers to create antibacterial food packaging. Such packaging not only increases the shelf life of perishable items but also improves their safety. Additionally, rigorous testing of these nanoparticle-infused products is required to determine their economic viability.

13.3.3 Streptomycetes-Derived NPs in the Agriculture Sector

Streptomyces species are well known for their capacity to create a diverse range of bioactive chemicals. These metabolites have important roles in plant growth promotion (PGP), mineral and nutrient absorption, and resistance to numerous plant diseases and pests (Aggarwal et al. 2016). Several investigations have shown that *Streptomyces* has PGP potential and defense-inducing capacities in numerous agriculturally important crops, particularly rice, sorghum, pigeonpea, and chickpea (Wani and Gopalakrishnan 2019). *Streptomyces* have a wide range of applications in agriculture as plant growth promoters that stimulate the formation of indole acetic acid (IAA, a plant hormone) and biocontrol agents (antifungals and biopesticides).

Nanotechnology has the potential to significantly enhance agricultural product quality, reduce environmental and health concerns associated with pesticides, and increase production and food security. Nanoscale materials have unique features that make them ideal for making creative solutions to assist agriculture and related sectors. Nanotechnology can improve soil quality and agricultural productivity by stimulating plant growth using nanoparticle-based fertilizers. Furthermore, using nanoparticle-based transporters and compounds for pesticides and fertilizers can reduce their overall consumption while maintaining agricultural output (Tikariha et al. 2021).

Metal and metal oxide nanoparticles (NPs) have become known for their strong antifungal properties, thus providing a viable strategy to treating fungal phytopathogens that cause major agricultural losses. Recent research has shown that biologically generated copper and copper oxide NPs, particularly those manufactured with *Streptomyces* spp., have significant antifungal activity. These nanoparticles have been shown to inhibit the growth of pathogenic fungi such as *Pythium ultimum*, *Alternaria alternata*, *Fusarium oxysporum*, and *Aspergillus niger*, highlighting their potential use in sustainable plant disease management strategies (Saad et al. 2018; Hassan et al. 2019). Metal-based nanoparticles and oxides have recently been applied directly to soils to improve nutrient uptake, reduce the requirement for chemical fertilizers and pesticides and, ultimately, increase agricultural yields (Kumar et al. 2019). For example, *Streptomyces* sp. LK3's extracellular biosynthesis of silver nanoparticles (AgNPs) has demonstrated noteworthy efficiency in farming, especially due to its acaricidal properties against *Haemaphysalis bispinosa*, with LC50 values of 16.45 mg/L and 16.10 mg/L, respectively (Karthik et al. 2014). Zinc oxide (ZnO) nanoparticles were found to be highly effective (96.72%) against the nematode *Meloidogyne incognita* (Elqady et al. 2025).

13.3.4 Streptomycetes-Derived NPs in Environmental Applications

The biosynthesis of nanoparticles using Streptomycetes offers ecological benefits over traditional chemical methods, supporting sustainability goals. This green synthesis reduces ecological impact through energy-efficient processes, generating less toxic waste, and improving nanomaterial biocompatibility. By leveraging the metabolic potential of these actinobacteria, researchers have developed eco-friendly methods that address challenges in conventional nanomanufacturing while aiding innovative environmental remediation applications.

13.3.4.1 Reduced Chemical Toxicity and Hazardous Byproducts

Traditional nanoparticle synthesis largely relies on toxic reducing agents such as sodium borohydride and stabilizing compounds like polyvinylpyrrolidone, which produce hazardous waste requiring specialized disposal methods (Navada et al. 2025). In contrast, the synthesis mediated by Streptomycetes uses biological reductants, including NADH-dependent enzymes and sulfur-containing proteins, found in bacterial cell-free supernatants or biomass (Baygar and Ugur 2017; Nainangu et al. 2024). For instance, *Streptomyces griseorubens* AU2 synthesizes silver nanoparticles (AgNPs) extracellularly through metabolites secreted during its growth phase, thereby eliminating the need for synthetic capping agents (Baygar and Ugur 2017). The absence of solvent residues in biosynthetic procedures further augments environmental safety. Chemical synthesis customarily necessitates organic solvents, such as toluene or hexane, while Streptomycetes function in aqueous media at neutral pH (Baygar and Ugur 2017; Nainangu et al. 2024). A comparative investigation reveals that biogenic AgNPs synthesized from *Streptomyces chiangmaiensis* SSUT88A exhibit lower LC50 values (130 μg/mL in brine shrimp assays) relative to their chemically synthesized counterparts, indicating reduced ecotoxicological impact (Nainangu et al. 2024).

13.3.4.2 Energy Efficiency and Low-Carbon Production

The synthesis mediated by Streptomycetes operates under ambient conditions (25–30 °C, atmospheric pressure), contrasting sharply with the high-energy requirements of physical methods such as laser ablation or chemical vapor deposition (Ying et al. 2022; Baygar and Ugur 2017). For example, the biosynthesis of AgNPs by *S. griseorubens* AU2 occurs at 28 °C with mild agitation (130 rpm), using 85% less energy compared to thermal decomposition techniques that necessitate 300 °C (Baygar and Ugur 2017). This process leverages the bacterial enzymatic machinery—particularly nitrate reductases and cytochrome P450 systems—to catalyze the reduction of metal ions without requiring external energy inputs (Baygar and Ugur 2017; Nainangu et al. 2024).

13.3.4.3 Sustainable Resource Utilization

Streptomycetes flourish on low-cost agricultural and industrial byproducts, thereby facilitating circular economy models. Research has employed molasses, lignocellulosic waste, and even heavy metal-contaminated water as growth substrates for strains capable of nanoparticle production (El-Naggar et al. 2017; Nainangu et al. 2024). For instance, *Streptomyces rochei* SSCM102, derived from mangrove ecosystems, exhibits optimal growth in saline conditions (3–5% NaCl), permitting nanoparticle production utilizing brackish water resources without competing for freshwater (Nainangu et al. 2024).

13.3.4.4 Biodegradability and Ecosystem Compatibility

Streptomycetes-derived nanoparticles demonstrate superior environmental compatibility attributed to the presence of proteinaceous capping layers that facilitate microbial degradation. In contrast to synthetic polymers employed in chemical synthesis, these biological stabilizers undergo decomposition into non-toxic peptides and sugars under natural environmental conditions (El-Naggar et al. 2017; Baygar and Ugur 2017). Empirical field trials illustrate the complete degradation of biogenic AgNPs within 60 days in soil, as opposed to the protracted duration of over 240 days observed for chemically stabilized nanoparticles (El-Naggar et al. 2017). The biological provenance of these nanomaterials concurrently diminishes the risks associated with bioaccumulation. In aquatic ecosystems, TiO_2 nanoparticles synthesized by *Streptomyces* exhibit a 55% reduction in bioaccumulation within *Daphnia magna* relative to commercial counterparts, a phenomenon attributed to the protein corona, which mitigates surface reactivity (Elqady et al. 2025).

13.4 Applications in Environmental Remediation

13.4.1 Water Treatment Technologies

Silver nanoparticles (AgNPs) are widely known for their powerful antimicrobial properties, effective against both aerobic and anaerobic bacteria. Their small, consistent size makes them highly stable and resistant to oxidation. They are particularly effective at killing bacteria such as *E. coli* and *Staphylococcus aureus.* A promising method for water treatment involves combining AgNPs with activated carbon (AC), an excellent adsorbent (El-Naggar et al. 2017). This combination improves both disinfection and contaminant removal. Research has also shown that nanoparticles derived from Streptomycetes are effective in water purification. For example, a composite of AgNPs and cellulosic fibers can remove up to 99.9% of fecal coliforms through a combination of adsorption and antimicrobial action (El-Naggar et al. 2017). Furthermore, nanoparticles derived from other sources also show great promise. Iron oxide nanoparticles from the marine bacterium *Streptomyces* sp. SMGL39 were found to degrade 78% of tetracycline residues in wastewater (Nainangu et al. 2024).

13.4.2 Soil Decontamination

Naturally produced nanoparticles improve the effectiveness of using plants to clean up pollutants, and at the same time, they help to lessen the toxic effects of heavy metals on those plants. A study from 2017 by El-Naggar et al. showed that silver nanoparticles (AgNPs) can lock arsenic in place within polluted soil. This method reduced the amount of arsenic that leaked out of the soil by 65% compared to using traditional chemical stabilizers.

Recently in a study by Elqady et al. (2025), titanium dioxide (TiO_2) nanoparticles, made from the bacterium *Streptomyces rubrolavendulae,* increased the efficiency of sunflower plant (*Helianthus annuus*) to absorb heavy metals by 40%. This happens because the nanoparticles make the soil around the plant's roots more acidic.

13.4.3 Air Pollution Control

Nanoparticles created using bacteria from the genus *Streptomyces* could be used to clean up the air. Research by Baygar and Ugur in 2017 showed that CeO_2 (cerium oxide) nanoparticles, which were made with *S. griseorubens* AU2, were able to convert 89% of harmful nitrogen oxide gases into other, less harmful compounds at room temperature. A recent study by Nainangu et al. in 2024 found that silver nanoparticles (AgNPs) coated with proteins could break down 74% of fine particulate matter (PM2.5) when activated by visible light, a process known as photocatalysis.

13.4.4 Carbon Sequestration Potential

Streptomycetes help achieve carbon neutrality in two main ways:

I. Biosynthetic carbon capture: They capture carbon during a process called biosynthetic carbon capture. As the bacteria grow to produce nanoparticles, their metabolic processes absorb 2.3–3.1 grams of carbon dioxide for every gram of biomass produced (Baygar and Ygur 2017).
II. Nanoparticle-enhanced sequestration: It is made possible by iron oxide nanoparticles, which are created with the aid of certain microbes. These nanoparticles increase the amount of organic carbon in agricultural soil by 22% by boosting microbial activity (Elqady et al. 2025). Field studies conducted have confirmed that combining *Streptomycetes*-based nanoparticle synthesis with crop farming can create a nanomanufacturing process with a net-negative carbon footprint.

13.5 Industrial Benefits of Using Streptomycetes-Derived Nanoparticles

13.5.1 Eco-Friendly and Sustainable Production

Biosynthesis of nanoparticles using *Streptomyces* offers a green, eco-friendly, and non-toxic method compared to traditional chemical and physical techniques. The conventional methods often involve harsh chemicals, high energy use, and toxic waste, but the biological approach leverages natural microbial processes to reduce environmental impact and improve safety for workforces (Kerdtoob et al. 2024; Kalaba et al. 2021; Loqman et al. 2022; Pallavi et al. 2022).

13.5.2 Cost-Effectiveness and Scalability

Streptomycetes are well-suited for the large-scale, economical production of nanoparticles because they can grow on inexpensive media and in mild conditions. The biosynthesis process is typically a simple, one-step procedure that can be easily scaled up for industrial use (Kerdtoob et al. 2024; Kalaba et al. 2021; Pallavi et al. 2022).

13.5.3 Diverse Bioactivity and Multifunctionality

Nanoparticles made from *Streptomyces* have many useful biological properties, such as fighting viruses, bacteria, and fungi, as well as acting as antioxidants and inhibiting tumors (Gupta et al. 2024; Gupta et al. 2025). Because of their diverse functions, these nanoparticles can be used in many fields like medicine, farming, food preservation, and cleaning up the environment (Anjum et al. 2024; Kalaba et al. 2021).

13.5.4 Enhanced Biocompatibility and Safety

Compared to nanoparticles made with chemicals, biologically synthesized nanoparticles are often more compatible with living systems and are less toxic. This makes them particularly useful for applications in medicine, food, and agriculture, where safety and meeting regulatory requirements are top priorities (Kerdtoob et al. 2024; Kalaba et al. 2021).

13.5.5 Controlled Synthesis and Unique Properties

Streptomycetes are capable of precisely controlling the reduction of metal ions, which allows for the creation of nanoparticles with specific sizes, shapes, and surface properties. These tailored properties are essential for ensuring the nanoparticles to work effectively in specific industries, including catalysis, drug delivery, and biosensing (Kerdtoob et al. 2024; Kalaba et al. 2021).

13.5.6 Mitigation of Antibiotic Resistance

Nanoparticles derived from Streptomycetes are effective against antibiotic resistance. They introduce new antimicrobial agents that work in novel ways. This helps address a major global health issue and aids in creating new types of therapies (Anjum et al. 2024). Because of these advantages, Streptomycetes are considered a valuable industrial resource for producing nanoparticles in a way that is sustainable, safe, and adaptable. (Anjum et al. 2024; Kerdtoob et al. 2024; Kalaba et al. 2021; Pallavi et al. 2022).

13.6 Challenges and Future Directions

The method of creating nanoparticles using Streptomycetes is good for the environment, but to produce them on a large scale, several problems need to be solved:

I. Genetic instability: After being grown in the lab for a long time, the strains produce fewer nanoparticles. Storing them using cryopreservation can help keep the strains productive (Nainangu et al. 2024).
II. Metabolic engineering: The use of CRISPR technology can boost the expression of nitrate reductase, which in turn enhances the production of silver nanoparticles by 3.7 times more efficient (Baygar and Ugur 2017).

III. Circular system design: Combining nanoparticle production with biogas generation from the leftover bacterial material makes the whole process more cost-effective (El-Naggar et al. 2017).

13.7 Conclusion

Streptomyces offers a sustainable platform for the green synthesis of metallic nanoparticles (NPs) such as gold, silver, copper oxide, and zinc oxide. These biologically synthesized NPs exhibit remarkable bioactivities, including antimicrobial, anticancer, antioxidant, and wound-healing, making them valuable in biomedical applications. In agriculture sector, these nanoparticles can be used as natural pesticides and help plants to grow better, which supports environmentally friendly agriculture by cutting down on synthetic chemicals. In environmental clean-up, the nanoparticles help in cleaning up pollution and treating dirty water, particularly when combined with materials like activated carbon. Industrially, they are used as tiny catalysts to speed up chemical reactions and as antimicrobial agents to keep food fresh. Together, all these applications show that these nanoparticles have the potential to help solve major worldwide problems associated with health, farming, environment, and manufacturing. Furthermore, nanoparticles made from streptomycetes are becoming key tools for helping industries shift to sustainable practices, especially when they can be used for things like earning carbon credits and meeting green chemistry standards. If the nanotechnology industry replaces just 30% of its chemically made nanoparticles with these biologically produced ones, it could cut yearly carbon emissions by 12.7 million metric tons. This is the same as taking 2.7 million cars off the road. Making nanoparticles this way doesn't just reduce environmental damage; it also helps restore ecosystems. This marks a major change in how we use nanotechnology. Further detailed studies are needed which will lead to new, sustainable solutions in the field of nanotechnology.

References

Abdeen S, Geo S, Praseetha PK, Dhanya RP (2014) Biosynthesis of silver nanoparticles from Actinomycetes for therapeutic applications. Int J Nano Dimens 5:155–162

Abd-Elnaby HM, Abo-Elala GM, Abdel-Raouf UM, Hamed MM (2016) Antibacterial and anticancer activity of extracellular synthesized silver nanoparticles from marine Streptomyces rochei MHM13. Egypt J Aquat Res 42(3):301–312

Aggarwal N, Thind SK, Sharma S (2016) Role of secondary metabolites of actinomycetes in crop protection. In: Plant growth promoting actinobacteria: a new avenue for enhancing the productivity and soil fertility of grain legumes, pp 99–121

Ahmad A, Senapati S, Khan MI, Kumar R, Sastry M (2003) Extracellular biosynthesis of monodisperse gold nanoparticles by a novel extremophilic actinomycete, *Thermomonospora* sp. Langmuir 19(8):3550–3553

Ahmad MS, Yasser MM, Sholkamy EN, Ali AM, Mehanni MM (2015) Anticancer activity of biostabilized selenium nanorods synthesized by *Streptomyces bikiniensis* strain Ess_amA-1. Int J Nanomedicine 10:3389–3401

Akhtar MS, Panwar J, Yun YS (2013) Biogenic synthesis of metallic nanoparticles by plant extracts. ACS Sustain Chem Eng 1(6):591–602

Alani F, Moo-Young M, Anderson W (2012) Biosynthesis of silver nanoparticles by a new strain of Streptomyces sp. compared with Aspergillus fumigatus. World J Microbiol Biotechnol 28: 1081–1086

Ameen F, AlYahya S, Govarthanan M, ALjahdali N, Al-Enazi N, Alsamhary K, Alharbi SA (2020) Soil bacteria *Cupriavidus* sp. mediates the extracellular synthesis of antibacterial silver nanoparticles. J Mol Struct 1202:127233

Anjum MS, Khaliq S, Ashraf N, Anwar MA, Akhtar K (2024) Bioactive Streptomycetes: a powerful tool to synthesize diverse nanoparticles with multifarious properties. J Basic Microbiol 64(9):e2400129

Balagurunathan R, Radhakrishnan M, Rajendran RB, Velmurugan D (2011) Biosynthesis of gold nanoparticles by actinomycete *Streptomyces viridogens* strain HM10. Indian J Biochem Biophys 48:331–335

Basavaraj U, Praveenkumar N, Sabiha TS, Rupali S, Samprita B (2012) Synthesis and characterization of silver nanoparticles. Int J Pharm Bio Sci 2(3):10–14

Baygar T, Ugur A (2017) Biosynthesis of silver nanoparticles by *Streptomyces griseorubens* isolated from soil and their antioxidant activity. IET Nanobiotechnol 11(3):286–291

Baygar T, Ugur A (2017) In vitro evaluation of antimicrobial and antibiofilm potentials of silver nanoparticles biosynthesised by Streptomyces griseorubens. IET Nanobiotechnology 11(6):677–681

Bukhari SI, Hamed MM, Al-Agamy MH, Gazwi HS, Radwan HH, Youssif AM (2021) Biosynthesis of copper oxide nanoparticles using *Streptomyces* MHM38 and its biological applications. J Nanomater (1):6693302

Chauhan R, Kumar A, Abraham J (2013) A biological approach to the synthesis of silver nanoparticles with *Streptomyces* sp. JAR1 and its antimicrobial activity. Sci Pharm 81(2): 607–618

de Amorim Silva T, Quelemes PV, de Araújo AR, de Moraes ACM, Azevedo RB, da Silva DA, Teixeira MFS (2017) Extracellular biogenic synthesis of silver nanoparticles by Actinomycetes from amazonic biome and its antimicrobial efficiency. Afr J Biotechnol 16(43):2072–2082

El-Naggar NEA, El-Sawah AA, Elmansy MF et al (2024) Process optimization for gold nanoparticles biosynthesis by *Streptomyces albogriseolus* using artificial neural network, characterization and antitumor activities. Sci Rep 14(1):4581

El-Naggar MY, Ramadan W, El-Hamamsy RA (2017) The application of mediated biosynthesized green silver nanoparticles by *Streptomyces griseorubens* in water treatment. J Pure Appl Microbiol 11(2):749–757

El-Naggar NEA, Mohamedin A, Hamza SS, Sherief AD (2016) Extracellular biofabrication, characterization, and antimicrobial efficacy of silver nanoparticles loaded on cotton fabrics using newly isolated Streptomyces sp. SSHH-1E. J Nanomater 2016(1):3257359

Elqady EM, El-Said E, Tharwat AA, El-Khashab LAA, Mostafa IM, Hamed FZ, El-Enain IMA (2025) Biogenic synthesis of titanium nanoparticles by *Streptomyces rubrolavendulae* for sustainable management of *Icerya aegyptiaca* (Douglas). Sci Rep 15(1):1380

Espitia PJP, Soares NDFF, Coimbra JSDR, de Andrade NJ, Cruz RS, Medeiros EAA (2012) Zinc oxide nanoparticles: synthesis, antimicrobial activity and food packaging applications. Food Bioprocess Technol 5(5):1447–1464

Forootanfara H, Zare B, Fasihi-Bam H, Amirpour-Rostami S, Ameri A, Shakibaie M, Torabi Nami M (2014) Biosynthesis and characterization of selenium nanoparticles produced by terrestrial actinomycete *Streptomyces microflavus* strain FSHJ31. Res Rev 3(2):47–53

Fouda A, Shaheen TI (2017) Silver nanoparticles: biosynthesis, characterization and application on cotton fabrics. Microbiol Res J Int 20(1):1–14

Fu L, Fu Z (2015) *Plectranthusamboinicus* leaf extract–assisted biosynthesis of ZnO nanoparticles and their photocatalytic activity. Ceram Int 41(2):2492–2496

Gahlawat G, Choudhury AR (2019) A review on the biosynthesis of metal and metal salt nanoparticles by microbes. RSC Adv 9(23):12944–12967

Golinska P, Wypij M, Ingle AP, Gupta I, Dahm H, Rai M (2014) Biogenic synthesis of metal nanoparticles from actinomycetes: biomedical applications and cytotoxicity. Appl Microbiol Biotechnol 98(19):8083–8097

Gupta A, Pandey BC, Gautam RK, Tiwari I, Kumari N (2025) Antibacterial activity of green-synthesized silver nanoparticles. In: Antimicrobial materials and coatings. Woodhead Publishing, pp 545–568

Gupta A, Pandey BC, Verma J, Tiwari I, Sahu AN, Manhas RK, Kumari N (2024) Biosynthesis of silver nanoparticles from flower extract of *Dillenia indica* and its efficacy as antibacterial and antioxidant. Microb Pathog 193:106779

Hassan SED, Fouda A, Radwan AA, Salem SS, Barghoth MG, Awad MA, El-Gamal MS (2019) Endophytic actinomycetes *Streptomyces*spp mediated biosynthesis of copper oxide nanoparticles as a promising tool for biotechnological applications. JBIC J Biol Inorg Chem 24(3):377–393

Huang J, Lin L, Sun D, Chen H, Yang D, Li Q (2015) Bio-inspired synthesis of metal nanomaterials and applications. Chem Soc Rev 44(17):6330–6374

Hutchison JE (2008) Greener nanoscience: a proactive approach to advancing applications and reducing implications of nanotechnology. ACS Nano 2(3):395–402

Ismail M, Gul S, Khan MI, Khan MA, Asiri AM, Khan SB (2019) Green synthesis of zerovalent copper nanoparticles for efficient reduction of toxic azo dyes Congo red and methyl orange. Green Process Synth 8(1):135–143

Issaabadi Z, Nasrollahzadeh M, Sajadi SM (2017) Green synthesis of the copper nanoparticles supported on bentonite and investigation of its catalytic activity. J Clean Prod 142:3584–3591

Kalaba MH, Moghannem SA, El-Hawary AS et al (2021) Green synthesized ZnO nanoparticles mediated by *Streptomyces plicatus*: characterizations, antimicrobial and nematicidal activities and cytogenetic effects. Plants 10(9):1760

Karthik L, Kumar G, Kirthi AV, Rahuman AA, Bhaskara Rao KV (2014) Streptomyces sp. LK3-mediated synthesis of silver nanoparticles and its biomedical application. Bioprocess Biosyst Eng 37:261–267

Kerdtoob S, Chanthasena P, Rosyidah AL, Limphirat W, Penkhrue W, Ganta P, Nantapong N (2024) *Streptomyces monashensis* MSK03-mediated synthesis of gold nanoparticles: characterization and antibacterial activity. RSC Adv 14(7):4778–4787

Khalil MA, El-Shanshoury AE, Alghamdi MA et al (2022) *Streptomyces catenulae* as a novel marine actinobacterium mediated silver nanoparticles: characterization, biological activities, and proposed mechanism of antibacterial action. Front Microbiol 13:833154

Kumar A, Singh AK, Choudhary KK (eds) (2019) Role of plant growth promoting microorganisms in sustainable agriculture and nanotechnology. Woodhead Publishing

Li X, Xu H, Chen ZS, Chen G (2011) Biosynthesis of nanoparticles by microorganisms and their applications. J Nanomater 2011(1):270974

Loqman A, Outammassine A, El Garraoui O, El Bali B, Ouhdouch Y, Outzourhit A, Loqman S (2022) *Streptomyces thinghirensis* sp. nov. as a promising path for green synthesis of silver nanoparticles: high eradication of multidrug-resistant bacteria and catalytic activity. J Environ Chem Eng 10(6):108889

Manikprabhu D, Lingappa K (2013) Antibacterial activity of silver nanoparticles against methicillin-resistant *Staphylococcus aureus* synthesized using model *Streptomyces* sp. pigment by photo-irradiation method. J Pharm Res 6(2):255–260

Manimaran M, Kannabiran K (2017) Actinomycetes-mediated biogenic synthesis of metal and metal oxide nanoparticles: progress and challenges. Lett Appl Microbiol 64(6):401–408

Manivasagan P, Kang KH, Kim DG, Kim SK (2015) Production of polysaccharide-based bioflocculant for the synthesis of silver nanoparticles by *Streptomyces* sp. Int J Biol Macromol 77:159–167

Manivasagan P, Venkatesan J, Senthilkumar K, Sivakumar K, Kim SK (2013) Biosynthesis, antimicrobial and cytotoxic effect of silver nanoparticles using a novel *Nocardiopsis* sp. MBRC-1. Biomed Res Int 2013:287638

Menon S, Rajeshkumar S, Kumar V (2017) A review on biogenic synthesis of gold nanoparticles, characterization, and its applications. Resour Effic Technol 3(4):516–527

Mohanta YK, Behera SK (2014) Biosynthesis, characterization and antimicrobial activity of silver nanoparticles by Streptomyces sp. SS2. Bioprocess Biosyst Eng 37:2263–2269

Nainangu P, Mothilal SN, Subramanian K, Thanigaimalai M, Kandasamy R, Srinivasan GP, Antonyraj APM (2024) Characterization and antibacterial evaluation of eco-friendly silver nanoparticles synthesized by halophilic *Streptomyces rochei* SSCM102 isolated from mangrove sediment. Mol Biol Rep 51(1):730

Navada KM, Shetty AR, Rai R, Kumar S, SC, G., & Aroor, G. (2025) Sustainable synthesis, amplified efficacy: doped ZnO nanostructures driving photocatalytic excellence for wastewater remediation—a comprehensive review. Chem Ecol 41(2):215–251

Omran BA, Omran BA (2020) Prokaryotic microbial synthesis of nanomaterials (the world of unseen). In: Nanobiotechnology: a multidisciplinary field of science. Springer, pp 37–79

Pallavi SS, Rudayni HA, Bepari A, Niazi SK, Nayaka S (2022) Green synthesis of silver nanoparticles using *Streptomyces hirsutus* strain SNPGA-8 and their characterization, antimicrobial activity, and anticancer activity against human lung carcinoma cell line A549. Saudi J Biol Sci 29(1):228–238

Prakash D, Mahale V, Bankar A, Nawani N, Zinjarde S, Kapadnis B (2013) Biosynthesis of colloidal gold nanoparticles by *Streptomyces* sp. NK52 and its anti-lipid peroxidation activity. Colloids Surf B: Biointerfaces 110:52–57

Prakasham RS, Kumar BS, Kumar YS, Kumar KP (2014) Production and characterization of protein encapsulated silver nanoparticles by marine isolate *Streptomyces parvulus* SSNP11. Indian J Microbiol 54(3):329–336

Prasad R, Kumar V, Prasad KS (2014) Nanotechnology in sustainable agriculture: present concerns and future aspects. Afr J Biotechnol 13(6):705–713

Rajamanickam U, Mylsamy P, Viswanathan S, Muthusamy P (2012) Biosynthesis of zinc nanoparticles using actinomycetes for antibacterial food packaging. In: International conference on nutrition and food sciences, Singapore, p 39

Ramya S, Shanmugasundaram T, Balagurunathan R (2015) Biomedical potential of actinobacterially synthesized selenium nanoparticles with special reference to anti-biofilm, anti-oxidant, wound healing, cytotoxic and anti-viral activities. J Trace Elem Med Biol 32: 30–39

Rassaei L, Marken F, Sillanpää M, Amiri M, Cirtiu CM, Sillanpää M (2011) Nanoparticles in electrochemical sensors for environmental monitoring. TrAC Trends Anal Chem 30(11): 1704–1715

Saad EL, Salem SS, Fouda A, Awad MA, El-Gamal MS, Abdo AM (2018) New approach for antimicrobial activity and bio-control of various pathogens by biosynthesized copper nanoparticles using endophytic actinomycetes. J Radiat Res Appl Sci 11(3):262–270

Salem SS, El-Belely EF, Niedbała G, Alnoman MM, Hassan SED, Eid AM, Shaheen TI, Elkelish A, Fouda A (2020) Bactericidal and in-vitro cytotoxic efficacy of silver nanoparticles (Ag-NPs) fabricated by endophytic actinomycetes and their use as coating for the textile fabrics. Nanomaterials 10(10):2082.

Salem SS, Fouda A (2021) Green synthesis of metallic nanoparticles and their prospective biotechnological applications: an overview. Biol Trace Elem Res 199(1):344–370

Sambangi P, Gopalakrishnan S (2023) *Streptomyces*-mediated synthesis of silver nanoparticles for enhanced growth, yield, and grain nutrients in chickpea. Biocatal Agric Biotechnol 47:102567

Sanjenbam P, Gopal JV, Kannabiran K (2014) Anticandidal activity of silver nanoparticles synthesized using *Streptomyces* sp. VITPK1. J Mycol Med 24(3):211–219

Saravana Kumar P, Balachandran C, Duraipandiyan V, Ramasamy D, Ignacimuthu S, Al-Dhabi NA (2015) Extracellular biosynthesis of silver nanoparticle using *Streptomyces* sp. 09 PBT 005 and its antibacterial and cytotoxic properties. Appl Nanosci 5(2):169–180

Sastry M, Ahmad A, Khan MI, Kumar R (2003) Biosynthesis of metal nanoparticles using fungi and actinomycetes. Curr Sci 85(2):162–170

Selvakumar P, Viveka S, Prakash S, Jasminebeaula S, Uloganathan R (2012) Antimicrobial activity of extracellularly synthesized silver nanoparticles from marine derived *Streptomyces rochei*. Int J Pharm Biol Sci 3(3):188–197

Shaaban MT, Mohamed BS, Zayed M, El-Sabbagh SM (2024) Antibacterial, antibiofilm, and anticancer activity of silver-nanoparticles synthesized from the cell-filtrate of *Streptomyces enissocaesilis*. BMC Biotechnol 24(1):8

Sharma D, Kanchi S, Bisetty K (2019) Biogenic synthesis of nanoparticles: a review. Arab J Chem 12(8):3576–3600

Shirley AD, Dayanand A, Sreedhar B, Dastager SG (2010) Antimicrobial activity of silver nanoparticles synthesized from novel Streptomyces species. Dig J Nanomater Biostruct 5(2): 447–451

Shukla K (2012) Nanotechnology and emerging trends in dairy foods: the inside story to food additives and ingredients. Int J Nanosci Nanotechnol 1(1):41–58

Sowani H, Mohite P, Munot H, Shouche Y, Bapat T, Kumar AR et al (2016) Green synthesis of gold and silver nanoparticles by an actinomycete Gordonia amicalis HS-11: mechanistic aspects and biological application. Process Biochem 51(3):374–383

Subashini J, Kannabiran K (2013) Antimicrobial activity of *Streptomyces* sp. VITBT7 and its synthesized silver nanoparticles against medically important fungal and bacterial pathogens. Pharm Lett 5(3):192–200

Subbaiya R, Saravanan M, Priya AR, Shankar KR, Selvam M, Ovais M, Barabadi H (2017) Biomimetic synthesis of silver nanoparticles from Streptomyces atrovirens and their potential anticancer activity against human breast cancer cells. IET Nanobiotechnol 11(8):965–972

Sukanya MK, Saju KA, Praseetha PK, Sakthivel G (2013) Therapeutic potential of biologically reduced silver nanoparticles from actinomycete cultures. J Nanosci 2013:940719

Sundaramoorthi C, Devarasu S, Vengadesh PK (2011) Antimicrobial and wound healing activity of silver nanoparticles synthesized from *Streptomyces aureofaciens*. Int J Pharm Res Dev 12:69–75

Tanwar J, Sharma M, Parmar A, Tehri N, Verma N, Gahlaut A, Hooda V (2020) Antibacterial potential of silver nanoparticles against multidrug-resistant bacterial isolates from blood cultures. Inorg Nanomet Chem 50(11):1150–1156

Tehri N, Kaur R, Maity M, Chauhan A, Hooda V, Vashishth A, Kumar G (2020) Biosynthesis, characterization, bactericidal and sporicidal activity of silver nanoparticles using the leaves extract of *Litchi chinensis*. Prep Biochem Biotechnol 50(9):865–873

Thenmozhi M, Kannabiran K, Kumar R, Khanna VG (2013) Antifungal activity of *Streptomyces* sp. VITSTK7 and its synthesized Ag2O/Ag nanoparticles against medically important *Aspergillus* pathogens. J Mycol Med 23(2):97–103

Tikariha S, Singh S, Banerjee S, Vidyarthi AS (2012) Biosynthesis of gold nanoparticles, scope and application: a review. Int J Pharm Sci Res 3(6):1603–1610

Tikariha H, Pavagadhi S, Mayalagu S, Poh MCH, Swarup S (2021) Hybrid genome assembly for predicting functional potential of a novel Streptomyces strain as plant biomass valorisation agent. Indian Journal of Microbiology 61(3):283–290

Usha R, Prabu E, Palaniswamy M, Venil CK, Rajendran R (2010) Synthesis of metal oxide nano particles by Streptomyces sp. for development of antimicrobial textiles. Global J Biotechnol Biochem 5(3):153–160

Vahabi K, Mansoori GA, Karimi S (2011) Biosynthesis of silver nanoparticles by fungus *Trichoderma reesei* (a route for large-scale production of AgNPs). Insci J 1(1):65–79

Waghmare SS, Deshmukh AM, Sadowski Z (2014) Biosynthesis, optimization, purification and characterization of gold nanoparticles. Afr J Microbiol Res 8(2):138–146

Waghmare SS, Deshmukh AM, Kulkarni SW, Oswaldo LA (2011) Biosynthesis and characterization of manganese and zinc nanoparticles. Univers J Environ Res Technol 1(1)

Wani SP, Gopalakrishnan S (2019) Plant growth-promoting microbes for sustainable agriculture. In: Plant growth promoting rhizobacteria (PGPR): prospects for sustainable agriculture, pp 19–45

Ying J, Zheng Y, Zhang H, Fu L (2020) Room temperature biosynthesis of gold nanoparticles with *Lycorisaurea* leaf extract for the electrochemical determination of aspirin. Rev Mex Ing Quim 19(2):585–592

Ying S, Guan Z, Ofoegbu PC, Clubb P, Rico C, He F, Hong J (2022) Green synthesis of nanoparticles: current developments and limitations. Environ Technol Innovation 26:102336

Zahra Z, Habib Z, Chung S, Badshah MA (2020) Exposure route of TiO2 NPs from industrial applications to wastewater treatment and their impacts on the agro-environment. Nano 10(8): 1469

Zarina A, Nanda A (2014) Green approach for synthesis of silver nanoparticles from marine Streptomyces-MS 26 and their antibiotic efficacy. J Pharm Sci Res 6(10):321

Zheng Y, Huang Y, Shi H, Fu L (2019) Green biosynthesis of ZnO nanoparticles by *Plectranthusamboinicus* leaf extract and their application for electrochemical determination of norfloxacin. Inorg Nanomet Chem 49(9):277–282

Zonooz NF, Salouti M (2011) Extracellular biosynthesis of silver nanoparticles using cell filtrate of *Streptomyces* sp. ERI-3. Sci Iran 18(6):1631–1635

Zotchev SB (2012) Marine actinomycetes as an emerging resource for the drug development pipelines. J Biotechnol 158(4):168–175

Extreme Environment *Streptomyces*: Potential Sources for Novel Metabolites 14

Nsovo Nyeleti Mayimele and Manikandan Gurusamy

Abstract

Streptomyces are a group of Gram-positive, filamentous bacteria that have the ability to produce a great diversity of structurally unique secondary metabolites with a potential for biological activity. There is evidence of newly discovered *Streptomyces* species with a broad range of biosynthetic capabilities from frigid climates and geothermal springs while working in harsh conditions. These *Streptomyces* strains that thrive in harsh environments are categorised as extremophiles because they have evolved specialised metabolic pathways and the capacity to withstand high salinities, high temperatures, or high-pressure situations. The advent of next-generation sequencing technologies, modifications to metabolite screening protocols, and other precise techniques have made it possible to identify a number of cryptic biosynthetic gene clusters, most of which are unknown under typical laboratory conditions. In this chapter, we demonstrate how strains of *Streptomyces* from extreme environments yield novel bioactive leads. We also identified the challenges of studying these organisms and listed solutions that are environmentally sensitive and innovative approaches to biosampling and cultivation strategies, which will indicate the metabolic potential of microorganisms in extreme environments. The ultimate goal is the discovery of

N. N. Mayimele (✉)
Department of Pharmaceutical Sciences, Tshwane University of Technology, Arcadia Campus, Pretoria, South Africa

Pharmaceutical and Biotech Advancement in Africa (PBA2), Department of Pharmaceutical Sciences, Tshwane University of Technology, Arcadia Campus, Pretoria, South Africa
e-mail: MayimeleNN@tut.ac.za

M. Gurusamy
Pharmaceutical and Biotech Advancement in Africa (PBA2), Department of Pharmaceutical Sciences, Tshwane University of Technology, Arcadia Campus, Pretoria, South Africa

T. Kaur et al. (eds.), *Streptomyces: Biological Candidates for Sustainable Agriculture*,
https://doi.org/10.1007/978-981-95-5803-2_14

new next-generation therapeutics and sustainable food production methods. Genomics, metabolomics, and synthetic biology are reaching a turning point in their efforts to activate novel metabolic pathways from microorganism biospecimens.

Keywords

Streptomyces · Secondary metabolites · Biological activity · Microorganisms · Extreme environments

14.1 Introduction

Streptomyces represent a phylogenetically distinct group of Gram-positive, filamentous bacteria within the Actinobacteria phylum, comprising over 500 validated species of the genus *Streptomyces* (Quinn et al. 2020). These soil-dwelling microorganisms exhibit a complex life cycle that is characterised by vegetative mycelial growth followed by the formation of aerial hyphae and spore chains, a developmental program regulated by signalling cascades involving gamma-butyrolactones and other secondary messengers. Their morphological complexity, which rivals that of filamentous fungi, enables sophisticated environmental interactions and resource utilisation strategies. As dominant members of terrestrial microbial communities, streptomycetes contribute to biogeochemical cycles through the secretion of diverse extracellular enzymes capable of degrading recalcitrant organic compounds, including chitin, cellulose, and lignin (De Simeis and Serra 2021).

The ecological impact of streptomycetes extends beyond nutrient cycling to include intricate biotic interactions mediated by their extraordinary metabolic capabilities. These bacteria are renowned for producing volatile organic compounds that impart the characteristic "earthy" aroma to soil, primarily through biosynthesis of geosmin and 2-methylisoborneol (Alam et al. 2022). More significantly, their secondary metabolism represents one of nature's richest sources of bioactive molecules, yielding approximately two-thirds of all clinically used antibiotics, along with numerous antifungal, antiparasitic, and anticancer agents. Recent genomic analyses have revealed that the biosynthetic potential of these organisms far exceeds what was previously recognised through traditional cultivation methods, with each strain harbouring numerous cryptic gene clusters encoding putative specialised metabolites (Shu and Huang 2022). This untapped chemical diversity has driven intensified exploration of streptomycetes from unconventional habitats, particularly extreme environments where evolutionary pressures may have selected for novel biosynthetic pathways (Coelho et al. 2022). This chapter aims to focus on the isolation of *Streptomyces* from extreme environments, genomic insights, ecological and biotechnological significance of extreme *Streptomyces,* and also it discusses the novel metabolites produced by *Streptomyces* with diverse biological activities and challenges in studying extreme environment *Streptomyces.*

14.2 Extreme Environments

Streptomyces has historically been an important source of significant bioactive secondary metabolites, such as antibiotics, immunosuppressive drugs, anticancer drugs, and other bioactive compounds, isolated from environmental sources (Mann 2001; Berdy 2005; Cragg et al. 2005). Sadly, in the last few decades, the constant rediscovery of the same or similar compounds that were previously described or known from terrestrial Streptomyces has occurred. This is why sourcing potential microorganisms from previously unexplored or underutilised natural environments appears to be a winning condition in searching new bioactive molecules (Pathom-Aree et al. 2006; Romanenko et al. 2013; Skropeta and Wei 2014). In relation to this, there has been much recent focus on more extreme environment habitats such as deep-sea, desert, cryo, and volcanic environments for the potential isolation of new Streptomyces species.

The study of extremophilic microorganisms has altered our understanding of life's boundaries on earth, revealing adaptations to conditions previously considered inhospitable. Extremophiles thrive in environments that are characterised by intense physicochemical stresses, including deep-sea hydrothermal vents with pressures exceeding 100 MPa, hyperarid deserts with water activity below 0.6, hypersaline lakes approaching NaCl saturation, and polar ice caps with permanent freezing temperatures (Sivalingam et al. 2019). These extreme habitats impose strong selective pressures that drive the evolution of specialised cellular mechanisms, including protein stabilisation strategies, membrane lipid modifications, and unique osmoregulatory pathways. Among bacteria, streptomycetes demonstrate particular resilience in marginal environments, with documented populations in deep marine sediments, acidic mine drainage sites, and high-altitude geothermal springs (Odoh et al. 2021).

Marine-derived streptomycetes have attracted particular interest due to their demonstrated capacity to produce structurally novel secondary metabolites with potent bioactivities. These strains employ sophisticated adaptation strategies to cope with osmotic stress, including the accumulation of compatible solutes like ectoine and hydroxyectoine, as well as modifications to cell wall architecture and membrane fluidity (Komaki 2023). The increasing recognition of extensive redundancy in known secondary metabolite pathways, revealed through large-scale genome mining projects, has indicated the need to explore underexplored ecological niches for truly novel chemistry (Ossai et al. 2022). Contemporary bioprospecting approaches now combine advanced cultivation techniques with culture-independent metagenomic analysis to access the substantial portion of microbial diversity that remains uncultivated. Targeted isolation efforts focusing on extreme environments have yielded streptomycete strains with unusual biosynthetic capabilities, including the production of halogenated compounds in hypersaline lakes and pressure-resistant antibiotics in deep-sea sediments (Zhang et al. 2018).

14.3 Isolation and Cultivation Techniques

Streptomyces represent one of the most prolific sources of bioactive natural products, yet their isolation and cultivation remain challenging due to their growth requirements and low culturability under standard laboratory conditions (Donald et al. 2022; Kapinusova et al. 2023). Traditional cultivation approaches often fail to recover diverse *Streptomyces* populations, necessitating innovative strategies to access their full biosynthetic potential. Advances have focused on modifying growth media by incorporating environmental substrates—such as soil extracts amended with 1% dextran and 10% vitamins—which have successfully supported the growth of previously unculturable strains (Donald et al. 2022; Shepherdson et al. 2023). The addition of mineral supplements like crushed coral and iron oxides has proven particularly effective, likely by mimicking the oligotrophic conditions of natural habitats.

Emerging techniques combine cultivation with molecular detection methods to improve recovery rates. Enzyme-substrate arrays, monoclonal antibodies targeting *Streptomyces* polysaccharides, and stable isotope probing (32SB hybridisation) provide complementary approaches to genome-based detection (Adam et al. 2018). These methods help bridge the gap between microbial presence observed through metagenomics and actual culturability. For extreme environments, in situ cultivation techniques using diffusion chambers or soil substrate membrane systems have successfully recovered novel taxa that resist standard isolation methods (Lewis et al. 2021). Such approaches are revolutionising our ability to study previously 'unculturable' *Streptomyces* populations and their metabolic potential.

14.3.1 Sampling Methods in Extreme Environments

This has led to the continuous screening of several *Streptomyces* strains in the hopes of eventually isolating new molecules for therapeutic use (Olanrewaju and Babalola 2019), but their ability to produce these molecules depends greatly on cultural conditions. Cultural conditions include culture media, inoculum size and type, incubation time, temperature, and pH, all of which impact the biosynthesis of secondary metabolites (El-Naggar et al. 2016). In light of this, it is important to develop optimal conditions for consistent and maximal production of bioactive metabolites (Reddy et al. 2011).

Antarctic cold deserts have emerged as promising sources of novel *Streptomyces* strains due to their extreme conditions and low anthropogenic disturbance (Quinn and Dyson 2024). A recent study of Herring Island soils in East Antarctica recovered 30 *Streptomyces* isolates using soil substrate culturing techniques, with 86% showing antimicrobial activity against pathogenic bacteria and fungi (Benaud et al. 2021). The isolates demonstrated potent activity against Gram-positive pathogens (*Bacillus subtilis*, *Staphylococcus aureus*) and phytopathogens (*Ralstonia solanacearum*), as well as antifungal effects against *Aspergillus niger* and *Dendrodochium coniospora*—important agents of postharvest spoilage (Xie and Pathom-Aree 2021).

Table 14.1 Categories of cultivation strategies for *Streptomyces*

Category	Description
Simple methods	Including dilution plating on minimal media, which remains effective for many soil-dwelling strains. These low-cost approaches typically yield common, fast-growing species
Combinatorial approaches	Pairing selective isolation conditions with targeted growth media. For cave samples, peat extract baits have successfully recovered slow-growing actinobacteria, though full characterisation remains incomplete (Boubekri et al. 2022)
Environmental simulation	Recreating native physicochemical conditions in vitro. A notable success involved subculturing temperate forest soils in lignin–hydrolysate media for 1–2 months, yielding 12 novel mycobacterial strains among diverse actinobacteria (Adam et al. 2018)
High-throughput systems	Microfluidic cultivation devices and chip-based technologies that enable single-cell isolation and growth monitoring

These antarctic strains represent potentially novel taxa, as indicated by phylogenetic analyses of limited reference sequences (Benaud et al. 2022). With recovery rates of 1.63×10^4 CFU/g soil, this environment appears particularly rich in culturable actinobacteria compared to other extreme habitats. The success of soil substrate culturing—where environmental samples are used directly as growth substrates—highlights the importance of maintaining native microenvironments during isolation attempts (Masand et al. 2018). This approach preserves critical microbe–substrate interactions that are often disrupted in conventional plating methods.

14.3.2 Cultivation Strategies for *Streptomyces*

Cultivation strategies for Streptomyces can be categorised by complexity and resource requirements (Xie and Pathom-Aree 2021). The most effective strategies often combine multiple approaches (Table 14.1). For example, pretreatment of samples with mild heat or antibiotics reduces fast-growing competitors, while supplementation with signalling molecules like γ-butyrolactones can induce sporulation in dormant strains (Kapinusova et al. 2023). As cultivation efforts expand to more extreme environments—from deep caves to Polar Regions—these innovative techniques are revealing *Streptomyces* diversity that was previously inaccessible to natural product discovery pipelines.

14.4 Isolation of *Streptomyces* from Extreme Environments (Case Studies)

The growing crisis of antibiotic resistance has revitalised interest in exploring extreme environments for novel *Streptomyces* strains with untapped biosynthetic potential (Sivalingam et al. 2019). Recent advances in genome sequencing have revealed that less than 10% of *Streptomyces* secondary metabolite potential has been characterised, with the majority of genetic diversity residing in unsequenced strains from underexplored habitats (Ordóñez-Enireb et al. 2022).

14.4.1 *Streptomyces* from Antarctic Regions

The Antarctic continent, containing 90% of Earth's ice, hosts unique microbial communities adapted to permanent cold, high salinity, and UV radiation (Nopnakorn and Nopnakorn 2018). Despite these harsh conditions, novel species like *Streptomyces fildesensis* demonstrated bioactivity at low temperatures. This psychrophilic strain produces broad-spectrum antimicrobial compounds active against multidrug-resistant pathogens, a characteristic linked to its membrane lipid composition and cold-active enzymes (Ordóñez-Enireb et al. 2022). Antarctic *Streptomyces* exhibit several adaptive features including the following:

- Enhanced membrane fluidity through unsaturated fatty acids
- Cold-shock proteins maintaining protein folding
- Antifreeze compounds preventing ice crystal formation
- UV-protective pigments like melanin and carotenoids

Biotechnological screening of these strains has yielded cell-free extracts with potent activity against *Staphylococcus aureus* (90% inhibition) and *Candida albicans* (75% inhibition), demonstrating their pharmaceutical potential (Rafiee et al. 2024).

14.4.2 *Streptomyces* from Hot Springs

Thermophilic *Streptomyces* from hot springs (50–80 °C) have evolved sophisticated mechanisms to maintain cellular integrity under extreme heat (Rafiee et al. 2024). Key adaptations include the following:

- Heat-stable enzymes with reinforced tertiary structures
- Saturated membrane lipids reducing fluidity
- DNA repair systems counteracting thermal denaturation
- Compatible solutes protecting against oxidative stress

Notably, *Streptomyces* from Yellowstone's geothermal features produce thermostable polymerases used in PCR, while Indonesian hot spring isolates yield novel polyketides active against pancreatic cancer cells (IC50 < 5 μM). The geochemical diversity of hot springs creates microhabitats supporting distinct *Streptomyces* populations—acidic springs (pH 2–3) harbour acidophiles producing halogenated compounds, while alkaline springs (pH 9–11) host alkaliphiles synthesising novel β-lactams (Sivalingam et al. 2019).

14.4.3 *Streptomyces* from Deep-Sea Environments

The deep ocean (>1000 m depth) represents Earth's largest biome, with hydrostatic pressures reaching 110 MPa in trenches like the Mariana (Jackson et al. 2018). Piezophilic *Streptomyces* from these environments exhibit:

- Pressure-regulated gene expression
- Increased unsaturated fatty acids in membranes
- Barotolerant secondary metabolism
- Novel siderophores for iron acquisition

A breakthrough came with *Streptomyces abyssalis* from the Mariana Trench, which produces abyssomicins—potent antibiotics that inhibit folate biosynthesis in multi-drug resistant *Staphylococcus aureus* (MRSA) (MIC 0.25 μg/mL). Deep-sea sediments yield 10–100× more microbial diversity than surface waters, with recent metagenomic studies suggesting that >90% of deep-sea *Streptomyces BGCs* remain uncharacterised (Jackson et al. 2018). Current exploration focuses on hadal zones (>6000 m), where novel genera like *Marinactinospora* produce antitumor macrolides stable at 100 MPa.

14.5 Genomic Insights

Recent advances in genome-sequencing technologies have revolutionised our understanding of *Streptomyces* diversity and biosynthetic potential. Analysis of 748 high-quality genomes from public databases revealed substantial genomic variation, with sizes ranging from 4.50 to 12.75 Mb and encoding 3790–12,258 protein-coding genes (Shi et al. 2021; Liao et al. 2022). These genomes, primarily from *Streptomyces* and related genera (*Komagataeibacter, Nonomuraea*), contain numerous uncharacterised biosynthetic gene clusters (BGCs) for secondary metabolites. Notably, comparative genomics of marine versus terrestrial strains show distinct adaptive strategies—marine isolates exhibit genome reduction, lower GC content, and specialised BGCs for ribosomally synthesised peptides (RiPPs), along with unique salinity-response genes absent in soil-dwelling relatives (Tian et al. 2016). Such adaptations highlight how extreme environments drive metabolic innovation in streptomycetes.

While genomic mining has accelerated natural product discovery, challenges remain in analysing degraded samples like formalin-fixed tissues. Recent methods enable metabolite recovery from these specimens by cleaving crosslinks without molecular decomposition, preserving peptides and small molecules for analysis (Vijayaram et al. 2022; Velotta et al. 2022). This approach is particularly valuable for studying historical samples, though conformational changes may still hinder some detection efforts.

Table 14.2 Genomics of extreme *Streptomyces*

Streptomyces	Description
Halophiles	Enriched in ion transporters and osmolyte synthesis genes
Thermophiles	Feature heat-stable enzymes and DNA repair systems
Psychrophiles	Contain cold-shock proteins and antifreeze metabolites

14.5.1 Genome-Sequencing Technologies

The proliferation of *Streptomyces* genome sequences has transformed natural product research, enabling biosynthesis prediction, bioprospecting, and metabolic engineering (Atanasov et al. 2021). Large-scale sequencing initiatives have revealed thousands of cryptic BGCs, leading to novel scaffolds through combinatorial biosynthesis. However, gaps persist in understanding operon regulation and pathway functionality, especially in understudied species like Nocardia (Satam et al. 2023). Limited bioinformatic tools further complicate rational genome mining, though phylogeny-guided strategies have successfully identified bioactive compounds in rare taxa (Cibichakravarthy and Arul Jose 2021).

For appropriate genomic comparisons, coverage-based richness estimators are critical. Analyses of partial 16S rRNA and psbA genes suggest that sampling coverage below 10% yields unreliable diversity estimates (Zhang et al. 2021). Improved methods—such as multi-locus sequencing of environmental samples—are needed to accurately profile *Streptomyces* communities in extreme habitats.

14.5.2 Comparative Genomics of Extreme *Streptomyces*

Extremophilic *Streptomyces* exhibit unique genomic adaptations that underpin their metabolic versatility. A phylogenomic analysis of 112 genomes, including 50 from marine and thermophilic species, revealed clade-specific BGCs for terpenoids, oligosaccharides, and antibiotics like pleuromutilin (Tian et al. 2016). Marine species (*Streptomyces maris*) and thermophiles (*Streptomyces thermotolerans*) harbour novel BGCs absent in terrestrial strains, correlating with their ability to produce structurally unique metabolites (Table 14.2).

These genomic insights, coupled with heterologous expression of extreme-environment BGCs, are expanding the pipeline for drug discovery (Bauman et al. 2021). Future studies should prioritise sequencing neglected extremophiles to uncover further chemical diversity.

14.6 Ecological and Biotechnological Significance of Extreme *Streptomyces*

The genus *Streptomyces* represents one of the most extensively studied microbial taxa for antibiotic discovery, particularly from terrestrial sources such as soil and marine sediments (Sivalingam et al. 2019). In recent years, marine environments, including hydrothermal vents and hypersaline ecosystems, have emerged as promising sources of novel *Streptomyces* species with unique biosynthetic potential (Jagannathan et al. 2021; Girão et al. 2022). These extreme habitats, characterised by high salinity, elevated temperatures, and alkaline pH, exert selective pressures that drive the evolution of specialised metabolic pathways. While most *Streptomyces* species are soil-dwelling bacteria common in temperate and tropical regions, recent taxonomic studies highlight the distinct phylogenetic and biosynthetic diversity of extremophilic *Streptomyces* strains (Donald et al. 2022; Quinn and Dyson 2024).

Extremophilic *Streptomyces* isolates exhibit notable ecological and genomic differences compared to conventional strains. Although they occupy harsh environments, most are not true extremophiles but rather extremotolerant, with over 60% originating from brackish or freshwater systems (Mishra 2023; Ahdab and Lienhard 2021). Genomic analyses reveal their propensity to produce unique secondary metabolites, including γ-butyrolactones, 4-pyridones, and cyanobactins, which are rarely found in terrestrial strains. Research has particularly focused on isolates from saline lakes, deep-sea vents, and Polar Regions, uncovering novel bioactive compounds with pharmaceutical potential.

14.6.1 Role in Soil Ecosystems

Actinobacteria represent one of Earth's oldest microbial lineages, with an evolutionary history spanning over 400 million years (Meenakshi et al. 2024; Hayat et al. 2021). Among these, *Streptomyces* species dominate soil ecosystems, contributing approximately 2% of global biomass (Seshadri et al. 2022; Cheng et al. 2022). Their filamentous growth and spore-forming capabilities enable them to degrade complex organic matter, including lignin and cellulose, playing a critical role in nutrient cycling (Tan et al. 2022; Grgas et al. 2023).

With over 650 described species, *Streptomyces* remains the most prolific source of bioactive natural products, including antibiotics, antifungals, anticancer agents, and immunosuppressants (Alam et al. 2022). However, more than half of these species remain uncharacterised, representing a vast reservoir of untapped biosynthetic potential (Herbrík et al. 2020). Phylogenetic studies suggest that current taxonomic classifications may require revision, as genomic analyses reveal deeper evolutionary divergences among extremophilic strains.

14.6.2 Interactions with Other Microorganisms

Actinobacteria, particularly *Streptomyces*, *Micromonospora*, and *Nocardia*, are renowned for their ability to produce bioactive metabolites with diverse pharmacological properties (Selim et al. 2021). Despite their importance, the biochemical diversity of these microorganisms remains understudied compared to other bacteria and fungi. Extreme environments, such as Polar Regions, deserts, and deep-sea ecosystems, harbour unique actinobacterial communities that produce novel compounds with ecological and therapeutic relevance (Shu and Huang 2022).

Recent explorations of sulphur-rich hot springs, hypersaline lakes, and caves have yielded *Streptomyces* strains with unprecedented biosynthetic capabilities (Hui et al. 2021; Donald et al. 2022). These habitats, characterised by stable yet extreme conditions, host microbial communities with high rates of bioactive metabolite production. Notably, 78–100% of cultured actinobacteria from these niches exhibit antagonistic properties, producing compounds with novel structures and ecological functions. Continued research in these underexplored environments promises to uncover new drug leads and expand our understanding of microbial chemical ecology.

14.7 Novel Metabolites of Extreme *Streptomyces* with Diverse Biological Activities

Streptomyces are renowned for their exceptional capacity to produce structurally diverse secondary metabolites with pharmacological potential. Historically understudied, streptomycetes from extreme environments, including alkaline salt lakes, polar soils, and subterranean aquifers, are now recognised as rich sources of novel biosynthetic pathways (Sukmarini 2022). In recent years, the search for potential *Streptomyces* species has largely focused on extreme environment habitats including deep-sea, desert, cryo, and volcanic environments. Until recently, the majority of compounds from the genus *Streptomyces* have been isolated using culture-dependent methods instead of the metagenomic approach. We have shown that there is, indeed, good rationale for studying species-specific behaviour using culture-dependent approaches and, indeed, to consider a number of strategically correct approaches including 'one strain many compounds' (Goodfellow et al. 2018), to understand the unique capabilities of the species and isolate potential novel compounds. Thus, it is hoped that cultivation-based approaches will generate unprecedented knowledge for the development of new antibiotics, genome studies, and combinatorial biosynthesis. As noted in this section, *Streptomyces* is a promising genus in search of new potential therapeutic compounds from underexplored extreme environments, thereby developing next-generation drugs to combat the globally increasing problems of drug resistance, and an abundance of new drugs with no or few side effects.

Recent biosynthetic analyses of *Streptomyces* strains from extreme environments, such as the Thar Desert, have uncovered novel cyclodepsipeptides, polytheonamides, dibenzo-p-dioxins, bacteriocins, and β-lactams, many of which exhibit promising health benefits (Sukmarini 2022). For instance, *Streptomyces* sp. SMH9 produces non-ribosomal peptides with antibacterial properties, while *Streptomyces kasuaensis* yields sinalbin, a compound with demonstrated anticancer activity (Masand et al. 2018). These metabolites also exhibit antioxidant, diuretic, analgesic, and anti-inflammatory properties, underscoring their therapeutic versatility. Over the past decade, more than 100 novel bioactive compounds have been characterised from streptomycetes, expanding their pharmacological repertoire.

Notable examples include cystostatin from *Streptomyces* sp. IV, which induces tumour cell apoptosis through mitochondrial DNA disruption, and antiviral metabolites from *Streptomyces viridosporus* that inhibit pseudorabies virus replication (Quinn et al. 2020). Additionally, streptomycetes produce specialised anti-insect toxins, such as ribosome-inactivating proteins from *Streptomyces albus* that target rice weevils, chitinases lethal to *Bombyx mori* larvae, and thiopeptin from *Streptomyces azureus* with antifilarial properties. A recently discovered alkaliphilic *Streptomyces* strain exhibits unique metal-accumulation capabilities while producing industrially valuable enzymes like amylases and proteases (De Simeis and Serra 2021).

14.7.1 Types of Metabolites Produced

Streptomyces generate an extraordinary array of secondary metabolites, with approximately 22,500 documented to date. Actinobacteria account for 10,100 of these compounds, over 7000 of which originate from *Streptomyces* species (De Simeis and Serra 2021; Donald et al. 2022). Clinically, these bacteria produce more than 70% of natural-origin antibiotics, including glycopeptides (e.g. vancomycin), polyketides (e.g. erythromycin), aminoglycosides (e.g. streptomycin), and β-lactams (e.g. penicillin derivatives) (Selim et al. 2021). Beyond antimicrobials, their metabolites span cytotoxic agents (e.g. doxorubicin), immunosuppressants (e.g. rapamycin), and plant growth promoters (e.g. auxins) (Fig. 14.1). Cyclic dipeptides, a structurally unique class of metabolites, demonstrate stability and bioactivity. These compounds, composed of two amino acid residues, exhibit antimicrobial, antioxidant, and anticancer properties while modulating bacterial virulence (Selim et al. 2021). Their resistance to proteolytic degradation and ability to interact with diverse molecular targets make them ideal candidates for drug development. Other notable metabolites include terpenoids, alkaloids, and macrolides, which interfere with essential cellular processes such as cell wall biosynthesis and protein translation in pathogens (Masand et al. 2018).

14.7.2 Biological Activities of Metabolites

Extreme-environment streptomycetes exhibit broad-spectrum bioactivity against Gram-positive and Gram-negative bacteria, fungi, and cancer cells. Disk diffusion assays of crude extracts from strains like *Streptomyces alboflavus* and *Streptomyces alboviridis* reveal potent inhibition against 17 microbial strains, with ethyl acetate and chloroform fractions showing particularly strong efficacy (Masand et al. 2018). Bioassay-guided fractionation and advanced spectroscopic techniques (e.g. NMR, TOF-MS, HPLC) have identified key metabolites, including geldanamycin—an

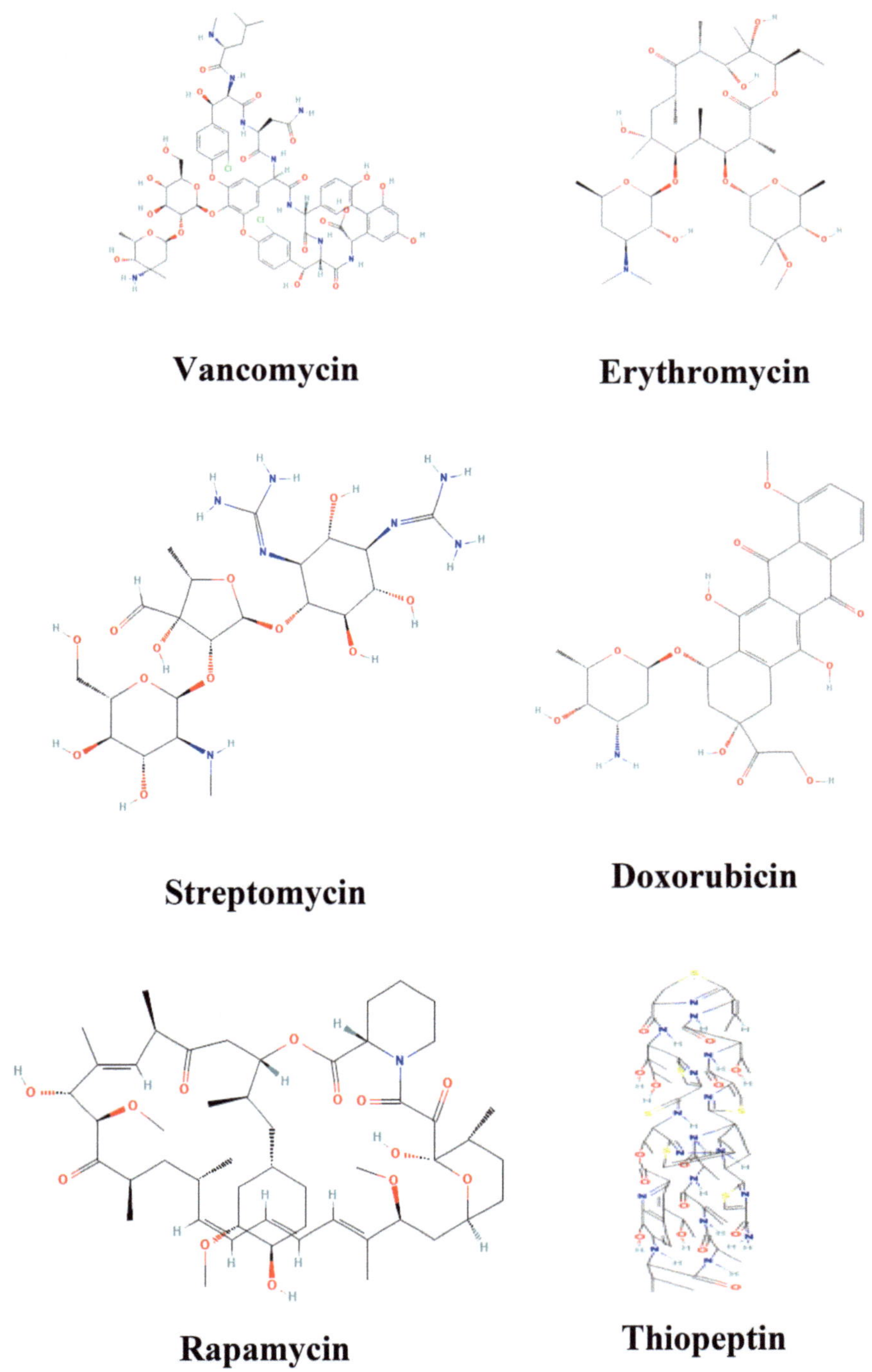

Fig. 14.1 Some of the antibiotics obtained from *Streptomyces* spp.

Hsp90 inhibitor—and actinomycin D, which intercalates DNA at GpC sites to block transcription (Sivalingam et al. 2019).

Metabolites targeting tubulin (e.g. epothilones) disrupt microtubule assembly, suppressing cancer cell proliferation. Notably, *S. alboflavus* and *S. alboviridis* produce 12 novel compounds with unique scaffolds and mechanisms, offering solutions to drug resistance (Donald et al. 2022). Their phylogenetically distinct lineages (95–96% 16S rRNA similarity to known species) underscore the untapped potential of extremophilic streptomycetes in drug discovery (Table 14.3).

14.8 Biotechnological Applications of *Streptomyces*

Streptomyces represent one of the most biotechnologically valuable bacterial genera, producing over two-thirds of all clinically used antibiotics and numerous other bioactive compounds (Cibichakravarthy and Arul Jose 2021). These actinobacteria possess an extraordinary biosynthetic capacity encoded by extensive secondary metabolite gene clusters, many of which remain cryptic under standard laboratory conditions. Modern genome mining approaches have revealed that *Streptomyces* species maintain an ancient polygenic repertoire of biosynthetic genes, though evolutionary genome reduction has led to the loss of certain ancestral pathways (Masand et al. 2018). Their ecological adaptability enables the colonisation of diverse extreme environments—from Polar Regions to deep-sea hydrothermal vents—making them rich sources of novel compounds with applications across medicine, agriculture, and industry (Fig. 14.2).

14.8.1 Pharmaceutical Applications

The pharmaceutical potential of streptomycetes is unparalleled, with their metabolites spanning virtually all therapeutic categories (Sivalingam et al. 2019). These bacteria produce the following:

- Antibiotics (e.g. aminoglycosides, tetracyclines, and β-lactams)
- Anticancer agents (e.g. doxorubicin and bleomycin)
- Immunosuppressants (e.g. rapamycin)
- Antidiabetics (e.g. acarbose)
- Neuroprotective compounds

Notably, analysis of 2210 polyketide synthase (PKS) genes revealed that *Streptomyces* species account for nearly 60% of known adenylation domain-containing PKSs, highlighting their biosynthetic dominance (Rappaport and Oliverio 2023). Recent discoveries from extreme environments are particularly promising—acidic habitats (pH <2.5) have yielded novel antifungal metabolites to address the critical shortage of effective antifungals. However, bioprospecting efforts face logistical

Table 14.3 *Streptomyces* compounds with their biological activities (agricultural, pharmaceutical, and industry)

S. No	Name of the species	Compounds	Biological activities	Applications	References
1	*Streptomyces* sp. NTK 937	Caboxamycin	Antibacterial, antifungal, and antitumor	Pharmaceutical	Sivalingam et al. (2019)
2	*Streptomyces* sp. CNR-698	Ammosamides	Cytotoxic	Pharmaceutical	Sivalingam et al. (2019)
3	*Streptomyces niveus* SCSIO 3406	Marfuraquinocins	Cytotoxic and antibacterial	Pharmaceutical	Sivalingam et al. (2019)
4	*Streptomyces scopuliridis* SCSIO ZJ46	Desotamides B–D	Antibacterial	Pharmaceutical	Sivalingam et al. (2019)
5	*Streptomyces fradiae* MM456M-mF7	Fradiamines A and B	Antibacterial	Pharmaceutical	Sivalingam et al. (2019)
6	*Streptomyces atratus* SCSIO ZH16	Atratumycin	Antituberculosis	Pharmaceutical	Sivalingam et al. (2019)
7	*Streptomyces* sp. C38	Atacamycins	Enzyme inhibitor and antiproliferative	Industrial/ Pharmaceutical	Nachtigall et al. (2011)
8	*Streptomyces* sp. strain DB634	Abenquines	Enzyme inhibitor for phosphodiesterase type 4b	Industrial/ Pharmaceutical	Schulz et al. (2011)
9	*Streptomyces clavuligerus*	Cephamycin	Antimicrobial activity	Pharmaceutical	Akintunde (2014)
10	*Streptomyces venezuelae*	Chloramphenicol	Antimicrobial activity	Pharmaceutical	
11	*Streptomyces aureofaciens*	Tetracycline	Antimicrobial activity	Pharmaceutical	
12	*Streptomyces kanamyceticus*	Kanamycin	Antimicrobial activity	Pharmaceutical	
13	*Streptomyces spectablis*	Spectinomycin	Antimicrobial activity	Pharmaceutical	
14	*Streptomyces cinnamonensis*	Monensin	Antimicrobial activity	Pharmaceutical	
15	*Streptomyces lavendulae*	Mitomycin C	Antimicrobial activity	Pharmaceutical	
16	*Streptomyces griseus*	Streptomycin	Antimicrobial activity	Pharmaceutical	Ikeda et al. (1999), Nicault et al. (2021)
17	*Streptomyces avermitilis*	Polypeptide avermectin	Antimicrobial activity	Pharmaceutical	

18	*Streptomyces cattleya*	Fluorometabolites	Antimicrobial activity	Pharmaceutical	
19	*Streptomyces* sp. AM-2504	Virantmycin B	Active against Dengue virus	Pharmaceutical	Hirota-Takahata et al. (2017)
20	*Streptomyces yanglinensis*	Antifungal substances	Inhibit the formation of *Aspergillus flavus* mycelium	Agricultural	Shakeel et al. (2018), Lyu et al. (2020)
21	*Streptomyces* sp.	Kasugamycin	Foliar fungicides and bactericides	Agricultural	Rey and Dumas (2017)
22	*Streptomyces* sp.	Streptomycin			
23	*Streptomyces* sp.	Polyoxin D			

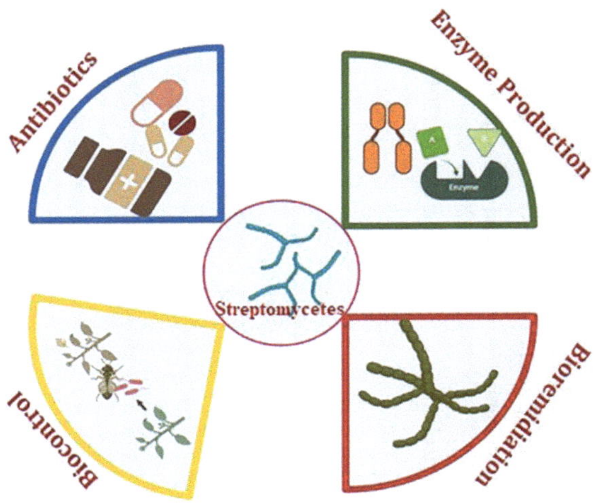

Fig. 14.2 Biotechnological applications of Streptomycetes. (Created in https://BioRender.com)

Table 14.4 Streptomycetes biocontrol properties and mechanisms

Biocontrol property	Mechanisms
Antibiosis	Production of antifungal/antibacterial metabolites
Enzymatic activity	Chitinases, proteases, and pectinases degrade pathogen cell walls
Competition	Efficient nutrient acquisition and niche colonisation
Induced systemic resistance	Priming plant defence mechanisms

challenges, as extreme environment isolates often require specialised preservation and transport conditions to maintain viability.

14.8.2 Agricultural Applications

In agriculture, streptomycetes serve as potent biocontrol agents and plant growth promoters through multiple mechanisms (Le et al. 2022; Torres-Rodriguez et al. 2022) (Table 14.4).

Field applications of *Streptomyces* spp. also demonstrate efficacy in reducing disease incidences:

- Strain SC43 reduced cucumber wilt incidence by 78% through soil application (Omar et al. 2022).
- *Streptomyces* sp. O15 controlled Fusarium wilt via chlamydospore inhibition.
- Tomato bacterial wilt (*Ralstonia solanacearum*) suppression exceeded 65%.

Plant Growth Promotion:

- Nutrient solubilisation (P, K, and Zn)
- Nitrogen fixation

- Phytohormone production
- Organic acid secretion

Field trials show also show consistent benefits:

- 25–40% yield increases in mango and citrus
- 30% reduction in postharvest diseases
- Improved lentil seed yield by 22%

These multifunctional properties make Streptomyces ideal components for sustainable agriculture systems, reducing reliance on chemical inputs while improving soil health and crop productivity (El-Saadony et al. 2022). Ongoing research focuses on developing formulation technologies to enhance field stability and efficacy of *Streptomyces*-based bioinoculants.

14.9 Challenges in Studying Extreme Environment of *Streptomyces*

The study of extremophilic *Streptomyces* presents unique scientific challenges that require integrated approaches combining traditional microbiology with modern molecular techniques. While genome sequencing has revolutionised our understanding of biosynthetic potential, culture-based methods remain essential for discovering truly novel compounds (Sivalingam et al. 2019). A critical limitation emerges from metabolic redundancy—the average *Streptomyces* genome contains 20–30 biosynthetic gene clusters (BGCs), yet typically only 10–20% are expressed under laboratory conditions. This discrepancy between genomic potential and observable metabolites creates a 'silent majority' of undiscovered compounds that may require extreme environmental triggers for activation. Current predictive algorithms analysing BGC phylogeny and network properties show promise, but still fail to account for approximately 40% of secondary metabolite diversity in extremophiles (Jackson et al. 2018). This field urgently needs high-throughput computational approaches that can simulate native extreme conditions while allowing metabolite screening at scale.

14.9.1 Technical Limitations

Traditional isolation media, such as Starch Casein nitrate Agar (SCNA), recover less than 1% of environmental actinobacterial diversity, creating what is known as the 'great plate count anomaly' (Ordóñez-Enireb et al. 2022). However, modern solutions or techniques also face their own constraints.

Culture Challenges:

- Most extremophiles require specialised media mimicking native conditions (e.g. high salt, pressure, or temperature).
- Slow growth rates (doubling times >24 h for many psychrophiles).
- Need co-cultivation with symbiotic organisms in many cases.

Molecular Technical Hurdles:

- 16S rRNA primer bias underestimates true diversity by 15–30%.
- Metagenomic libraries often miss GC-rich actinobacterial DNA.
- Heterologous expression systems fail to activate 60–70% of extreme-environment BGCs.

Recent advances in microfluidics-based cultivation and diffusion chambers have improved recovery rates to 10–15% for some extreme habitats (Rafiee et al. 2024). However, comprehensive catalogues of benthic actinobacterial communities remain incomplete, particularly for:

- High-pressure deep-sea sediments
- Hypersaline lakes (>20% salinity)
- Permafrost environments
- Acidic (pH <3) geothermal systems

14.10 Conclusion and Future Prospects

The exploration of biological resources for commercial applications raises ethical concerns that must be carefully addressed. Bioprospecting activities, when conducted without proper protocols, risk becoming biopiracy—the unauthorised commercialisation of biological resources traditionally used by indigenous communities (Masand et al. 2018). Ethical bioprospecting requires adherence to four key principles established by the Convention on Biological Diversity (CBD):

- Prior informed consent from source communities must provide explicit permission before collection.
- Commercial profits must be fairly distributed to originating countries/communities.
- Collection methods must not threaten species survival.
- Resources cannot be used in prohibited applications or locations.

These principles present substantial implementation challenges, particularly in extreme environments where jurisdictional boundaries may be unclear (Benaud et al. 2022). The pharmaceutical industry faces additional ethical dilemmas when prospecting in vulnerable ecosystems like tropical rainforests, Polar Regions, and hyper-arid deserts—environments that contain up to 80% undocumented microbial diversity with high biogeographic endemism.

The exploration of *Streptomyces* species from extreme environments has revolutionised the understanding of microbial diversity and biosynthetic potential. These extremotolerant actinobacteria were once considered limited to terrestrial habitats but have now been identified in a variety of extreme ecological niches, including Polar Regions, deep-sea sediments, geothermal springs, hypersaline lakes, and arid deserts, where they exhibit unique adaptations and produce structurally novel secondary metabolites. The environmental pressures in these habitats have acted as evolutionary catalysts, promoting metabolic innovation and leading to the biosynthesis of bioactive compounds with unparalleled pharmaceutical and industrial potential.

Technological advances in genomics and metabolite screening have uncovered a wealth of cryptic biosynthetic gene clusters, many of which remain silent under conventional laboratory conditions. These findings highlight the need to mimic native conditions through innovative isolation and cultivation strategies in order to unlock the full metabolic potential of extremophilic *Streptomyces*. Despite the progress in genome mining and bioinformatics, substantial challenges remain, particularly in the activation and expression of dormant pathways, and the ethical challenges that are surrounding bioprospecting in vulnerable ecosystems. For the future, integrative approaches that combine high-throughput sequencing, metabolomics and synthetic biology, alongside environmentally sensitive and ethically grounded bioprospecting practices, are essential for harnessing the unexplored diversity of extreme-environment *Streptomyces*. Such efforts will contribute to the discovery of next-generation therapeutics and sustainable agricultural solutions while creating opportunities for the understanding of microbial life at extreme environments.

References

Adam D, Maciejewska M, Naômé A, Martinet L, Coppieters W, Karim L, Baurain D, Rigali S (2018) Isolation, characterisation, and antibacterial activity of hard-to-culture actinobacteria from cave moonmilk deposits. Antibiotics 7(2):28

Ahdab YD, Lienhard JH (2021) Desalination of brackish groundwater to improve water quality and water supply. In: Global groundwater. Elsevier, pp 559–575

Akintunde OG (2014) Production of an antibiotic-like activity by Streptomyces sp. COUK1 under different growth conditions (Master's thesis, East Tennessee State University)

Alam K, Mazumder A, Sikdar S, Zhao YM, Hao J, Song C, Wang Y, Sarkar R, Islam S, Zhang Y, Li A (2022) *Streptomyces*: the biofactory of secondary metabolites. Front Microbiol 13:968053

Atanasov AG, Zotchev SB, Dirsch VM, Supuran CT (2021) Natural products in drug discovery: advances and opportunities. Nat Rev Drug Discov 20(3):200–216

Bauman KD, Butler KS, Moore BS, Chekan JR (2021) Genome mining methods to discover bioactive natural products. Nat Prod Rep 38(11):2100–2129

Benaud N, Edwards RJ, Amos TG, D'Agostino PM, Gutiérrez-Chávez C, Montgomery K, Nicetic I, Ferrari BC (2021) Antarctic desert soil bacteria exhibit high novel natural product potential, evaluated through long-read genome sequencing and comparative genomics. Environ Microbiol 23(7):3646–3664

Benaud N, Chelliah DS, Wong SY, Ferrari BC (2022) Soil substrate culturing approaches recover diverse members of Actinomycetota from desert soils of Herring Island, East Antarctica. Extremophiles 26(2):24

Berdy J (2005) Bioactive microbial metabolites. J Antibiot 58(1):1–26

Boubekri K, Soumare A, Mardad I, Lyamlouli K, Ouhdouch Y, Hafidi M, Kouisni L (2022) Multifunctional role of Actinobacteria in agricultural production sustainability: a review. Microbiol Res 261:127059

Cheng R, Li X, Jiang L, Gong L, Geslin C, Shao Z (2022) Virus diversity and interactions with hosts in deep-sea hydrothermal vents. Microbiome 10(1):235

Cibichakravarthy B, Jose PA (2021) Biosynthetic potential of *Streptomyces* rationalises genome-based bioprospecting. Antibiotics 10(7):873

Coelho LP, Alves R, Del Río ÁR, Myers PN, Cantalapiedra CP, Giner-Lamia J, Schmidt TS, Mende DR, Orakov A, Letunic I, Hildebrand F (2022) Towards the biogeography of prokaryotic genes. Nature 601(7892):252–256

Cragg GM, Kingston DG, Newman DJ (2005) Anticancer agents from natural products. CRC press

De Simeis D, Serra S (2021) Actinomycetes: a never-ending source of bioactive compounds—an overview on antibiotics production. Antibiotics 10(5):483

Donald L, Pipite A, Subramani R, Owen J, Keyzers RA, Taufa T (2022) *Streptomyces*: still the biggest producer of new natural secondary metabolites, a current perspective. Microbiol Res 13(3):418–465

El-Naggar NE, Mohamedin AH, Sherief AA, Hussien SM (2016) Optimization of fermentation conditions for production of bioactive metabolites effective against *Staphylococcus epidermidis* by a newly isolated *Nocardiopsis chromatogenes* strain SH89 using the response surface methodology. Global J Pure Appl Sci 10(1):823–839

El-Saadony MT, Saad AM, Soliman SM, Salem HM, Ahmed AI, Mahmood M, El-Tahan AM, Ebrahim AA, Abd El-Mageed TA, Negm SH, Selim S (2022) Plant growth-promoting microorganisms as biocontrol agents of plant diseases: mechanisms, challenges and future perspectives. Front Plant Sci 13:923880

Girão M, Ribeiro I, Carvalho MDF (2022) Actinobacteria from marine environments: a unique source of natural products. In: Natural products from Actinomycetes: diversity, ecology and drug discovery. Springer Singapore, Singapore, pp 1–45

Goodfellow M, Nouioui I, Sanderson R, Xie F, Bull AT (2018) Rare taxa and dark microbial matter: novel bioactive actinobacteria abound in Atacama Desert soils. Antonie Van Leeuwenhoek 111(8):1315–1332

Grgas D, Rukavina M, Bešlo D, Štefanac T, Crnek V, Šikić T, Habuda-Stanić M, Landeka Dragičević T (2023) The bacterial degradation of lignin—a review. Water 15(7):1272

Hayat S, Ashraf A, Aslam B, Asif R, Muzammil S, Zahoor MA, Afzal M, Saqalein M, Siddique MH, Muzammil A (2021) Actinobacteria: potential candidate as plant growth. In: Plant stress physiology, p 387

Herbrík A, Corretto E, Chroňáková A, Langhansová H, Petrásková P, Hrdý J, Čihák M, Krištůfek V, Bobek J, Petříček M, Petříčková K (2020) A human lung-associated *Streptomyces* sp. TR1341 produces various secondary metabolites responsible for virulence, cytotoxicity and modulation of immune response. Front Microbiol 10:3028

Hirota-Takahata Y, Kurosawa E, Ishimoto Y, Iwadate Y, Kizuka M, Chiba J et al (2017) Vestaines, novel vasoactive compounds, isolated from *Streptomyces* sp. SANK 63697. J Antibiot 70(2): 179–186

Hui MLY, Tan LTH, Letchumanan V, He YW, Fang CM, Chan KG, Law JWF, Lee LH (2021) The extremophilic actinobacteria: from microbes to medicine. Antibiotics 10(6):682

Ikeda H, Nonomiya T, Usami M, Ohta T, Ōmura S (1999) Organization of the biosynthetic gene cluster for the polyketide anthelmintic macrolide avermectin in *Streptomyces* avermitilis. Proc Natl Acad Sci 96(17):9509–9514

Jackson SA, Crossman L, Almeida EL, Margassery LM, Kennedy J, Dobson AD (2018) Diverse and abundant secondary metabolism biosynthetic gene clusters in the genomes of marine sponge derived *Streptomyces* spp. isolates. Mar Drugs 16(2):67

Jagannathan SV, Manemann EM, Rowe SE, Callender MC, Soto W (2021) Marine actinomycetes, new sources of biotechnological products. Mar Drugs 19(7):365

Kapinusova G, Lopez Marin MA, Uhlik O (2023) Reaching unreachables: obstacles and successes of microbial cultivation and their reasons. Front Microbiol 14:1089630

Komaki H (2023) Recent progress of reclassification of the genus *Streptomyces*. Microorganisms 11(4):831

Le KD, Yu NH, Park AR, Park DJ, Kim CJ, Kim JC (2022) *Streptomyces* sp. AN090126 as a biocontrol agent against bacterial and fungal plant diseases. Microorganisms 10(4):791

Lewis WH, Tahon G, Geesink P, Sousa DZ, Ettema TJ (2021) Innovations to culturing the uncultured microbial majority. Nat Rev Microbiol 19(4):225–240

Liao B, Hu H, Xiao S, Zhou G, Sun W, Chu Y, Meng X, Wei J, Zhang H, Xu J, Chen S (2022) Global pharmacopoeia genome database is an integrated and mineable genomic database for traditional medicines derived from eight international pharmacopoeias. Sci China Life Sci 65:1–9

Lyu A, Yang L, Wu M, Zhang J, Li G (2020) High efficacy of the volatile organic compounds of *Streptomyces* yanglinensis 3-10 in suppression of Aspergillus contamination on peanut kernels. Front Microbiol 11:142

Mann J (2001) Natural products as immunosuppressive agents. Nat Prod Rep 18(4):417–430

Masand M, Sivakala KK, Menghani E, Thinesh T, Anandham R, Sharma G, Sivakumar N, Jebakumar SR, Jose PA (2018) Biosynthetic potential of bioactive Streptomycetes isolated from arid region of the Thar desert, Rajasthan (India). Front Microbiol 9:687

Meenakshi S, Hiremath J, Meenakshi MH, Shivaveerakumar S (2024) Actinomycetes: isolation, cultivation and its active biomolecules. J Pure Appl Microbiol 18(1):118

Mishra RK (2023) Fresh water availability and its global challenge. Br J Multidiscipl Adv Stud 4(3):1–78

Nachtigall J, Kulik A, Helaly S, Bull AT, Goodfellow M, Asenjo JA et al (2011) Atacamycins A–C, 22-membered antitumor macrolactones produced by *Streptomyces* sp. C38. J Antibiot 64(12): 775–780

Nicault M, Zaiter A, Dumarcay S, Chaimbault P, Gelhaye E, Leblond P, Bontemps C (2021) Elicitation of antimicrobial active compounds by *Streptomyces*-fungus co-cultures. Microorganisms 9(1):178

Nopnakorn P, Nopnakorn P (2018) Polar Actinomycetes and their secondary metabolites. PoLAR 8(10)

Odoh CK, Obi CJ, Francis A, Unah UV, Egbe K, Akpi UK, Lerum N, Wanderi K (2021) Extremophilic fungi and their role in control of pathogenic microbes. In: Recent trends in mycological research: volume 1: agricultural and medical perspective, pp 219–249

Olanrewaju OS, Babalola OO (2019) *Streptomyces*: implications and interactions in plant growth promotion. Appl Microbiol Biotechnol 103(3):1179–1188

Omar AF, Abdelmageed AH, Al-Turki A, Abdelhameid NM, Sayyed RZ, Rehan M (2022) Exploring the plant growth-promotion of four *Streptomyces* strains from rhizosphere soil to enhance cucumber growth and yield. Plants 11(23):3316

Ordóñez-Enireb E, Cucalón RV, Cárdenas D, Ordóñez N, Coello S, Elizalde P, Cárdenas WB (2022) Antarctic fungi with antibiotic potential isolated from Fort William Point, Antarctica. Sci Rep 12(1):21477

Ossai J, Khatabi B, Nybo SE, Kharel MK (2022) Renewed interests in the discovery of bioactive actinomycete metabolites driven by emerging technologies. J Appl Microbiol 132(1):59–77

Pathom-Aree W, Stach JE, Ward AC, Horikoshi K, Bull AT, Goodfellow M (2006) Diversity of actinomycetes isolated from challenger deep sediment (10,898 m) from the Mariana Trench. Extremophiles 10(3):181–189

Quinn GA, Dyson PJ (2024) Going to extremes: progress in exploring new environments for novel antibiotics. npj Antimicrob Resist 2(1):8

Quinn GA, Banat AM, Abdelhameed AM, Banat IM (2020) *Streptomyces* from traditional medicine: sources of new innovations in antibiotic discovery. J Med Microbiol 69(8):1040–1048

Rafiee Z, Jalili Tabaii M, Moradi M, Harirchi S (2024) Unveiling antibacterial potential and physiological characteristics of thermophilic bacteria isolated from a hot spring in Iran. Microorganisms 12(4):834

Rappaport HB, Oliverio AM (2023) Extreme environments offer an unprecedented opportunity to understand microbial eukaryotic ecology, evolution, and genome biology. Nat Commun 14(1): 4959

Reddy NG, Ramakrishna DPN, Rajagopal SV (2011) Optimization of culture conditions of *Streptomyces rochei* (MTCC 10109) for the production of antimicrobial metabolites. Egypt J Biol 13:21–29

Rey T, Dumas B (2017) Plenty is no plague: *Streptomyces* symbiosis with crops. Trends Plant Sci 22(1):30–37

Romanenko LA, Tanaka N, Kalinovskaya NI, Mikhailov VV (2013) Antimicrobial potential of deep surface sediment associated bacteria from the Sea of Japan. World J Microbiol Biotechnol 29(7):1169–1177

Satam H, Joshi K, Mangrolia U, Waghoo S, Zaidi G, Rawool S, Thakare RP, Banday S, Mishra AK, Das G, Malonia SK (2023) Next-generation sequencing technology: current trends and advancements. Biology 12(7):997

Schulz D, Beese P, Ohlendorf B, Erhard A, Zinecker H, Dorador C, Imhoff JF (2011) Abenquines A–D: Aminoquinone derivatives produced by *Streptomyces* sp. strain DB634. J Antibiot 64(12):763–768

Selim MSM, Abdelhamid SA, Mohamed SS (2021) Secondary metabolites and biodiversity of actinomycetes. J Genet Eng Biotechnol 19(1):72

Seshadri R, Roux S, Huber KJ, Wu D, Yu S, Udwary D, Call L, Nayfach S, Hahnke RL, Pukall R, White JR (2022) Expanding the genomic encyclopedia of Actinobacteria with 824 isolate reference genomes. Cell Genom 2(12)

Shakeel Q, Lyu A, Zhang J, Wu M, Li G, Hsiang T, Yang L (2018) Biocontrol of aspergillus flavus on peanut kernels using *Streptomyces* yanglinensis 3-10. Front Microbiol 9:1049

Shepherdson EM, Baglio CR, Elliot MA (2023) *Streptomyces* behavior and competition in the natural environment. Curr Opin Microbiol 71:102257

Shi W, Sun Q, Fan G, Hideaki S, Moriya O, Itoh T, Zhou Y, Cai M, Kim SG, Lee JS, Sedlacek I (2021) gcType: a high-quality type strain genome database for microbial phylogenetic and functional research. Nucleic Acids Res 49(D1):D694–D705

Shu WS, Huang LN (2022) Microbial diversity in extreme environments. Nat Rev Microbiol 20(4): 219–235

Sivalingam P, Hong K, Pote J, Prabakar K (2019) Extreme environment *Streptomyces*: potential sources for new antibacterial and anticancer drug leads? Int J Microbiol 2019(1):5283948

Skropeta D, Wei L (2014) Recent advances in deep-sea natural products. Nat Prod Rep 31(8): 999–1025

Sukmarini L (2022) Marine bacterial ribosomal peptides: recent genomics-and synthetic biology-based discoveries and biosynthetic studies. Mar Drugs 20(9):544

Tan F, Cheng J, Zhang Y, Jiang X, Liu Y (2022) Genomics analysis and degradation characteristics of lignin by *Streptomyces* thermocarboxydus strain DF3-3. Biotechnol Biofuels Bioprod 15(1): 78

Tian X, Zhang Z, Yang T, Chen M, Li J, Chen F et al (2016) Comparative genomics analysis of *Streptomyces* species reveals their adaptation to the marine environment and their diversity at the genomic level. Front Microbiol 7:998

Torres-Rodriguez JA, Reyes-Pérez JJ, Quiñones-Aguilar EE, Hernandez-Montiel LG (2022) Actinomycete potential as biocontrol agent of phytopathogenic fungi: mechanisms, source, and applications. Plants 11(23):3201

Velotta JP, McCormick SD, Whitehead A, Durso CS, Schultz ET (2022) Repeated genetic targets of natural selection underlying adaptation of fishes to changing salinity. Integr Comp Biol 62(2): 357–375

Vijayaram S, Sun YZ, Zuorro A, Ghafarifarsani H, Van Doan H, Hoseinifar SH (2022) Bioactive immunostimulants as health-promoting feed additives in aquaculture: a review. Fish Shellfish Immunol 130:294–308

Xie F, Pathom-Aree W (2021) Actinobacteria from desert: diversity and biotechnological applications. Front Microbiol 12:765531

Zhang X, Li SJ, Li JJ, Liang ZZ, Zhao CQ (2018) Novel natural products from extremophilic fungi. Mar Drugs 16(6):194

Zhang L, Chen F, Zeng Z, Xu M, Sun F, Yang L, Bi X, Lin Y, Gao Y, Hao H, Yi W (2021) Advances in metagenomics and its application in environmental microorganisms. Front Microbiol 12:766364

Recent Advances in *Streptomyces* Biotechnology for Agricultural Applications 15

Neethu Asokan, B. P. Rakshitha, Diksha Jindal, Rajeshwari Ullagaddi, Balasubramanian Velramar, and Sushma Chauhan

Abstract

Among actinobacteria, *Streptomyces* have emerged as potential candidates for green agriculture due to their various metabolites, and metabolic capabilities. They are known prolific producers of great variety of metabolites such as antibiotics, phytohormones, enzymes along with other plant growth-promoting entities. They possess a tremendous credit of stimulating plant growth and suppressing plant diseases and also by enhancing soil health. According to the recent breakthrough studies, *Streptomyces* have also been used in the development of biofertilizers, biopesticides, biocontrol agents, biostimulants, and other metabolites, which can help in enhancing soil fertility. Their activities include plant growth-promoting hormones (such as indole acetic acid (IAA), gibberellins, and cytokinins), nitrogen fixation, phosphate solubilization, and siderophore production. They are also very effective against controlling a wide range of plant diseases due to their ability to produce antimicrobial compounds that not only drive away fungi, bacteria, viruses but also enable plants to develop their own defense mechanisms. In addition to this, *Streptomyces* enrich the soil through the digestion of organic material, replenishment of nutrients, and the

N. Asokan (✉) · B. P. Rakshitha · R. Ullagaddi
Department of Life Sciences, Sri Satya Sai University for Human Excellence, Kalaburagi, Karnataka, India

D. Jindal
Adesh Institute of Pharmacy and Biomedical Sciences, Adesh University, Bathinda, Punjab, India

B. Velramar
Amity Institute of Biotechnology, Amity University Chhattisgarh, Raipur, Chhattisgarh, India

S. Chauhan
Department of Chemical and Biochemical Engineering, Dongguk University, Seoul Campus, Seoul, Republic of Korea

T. Kaur et al. (eds.), *Streptomyces: Biological Candidates for Sustainable Agriculture*,
https://doi.org/10.1007/978-981-95-5803-2_15

release of enzymes that can breakdown nutrients to simpler form. Advanced genetic engineering and Omics technology have enhanced agriculture by enabling strain development that can tolerate extreme conditions like stress, produce beneficial metabolites and biocontrol agents to address issues such as land degradation, change in climate, and the increasing demand for food security. The chapter will dwell on the current developments in employing *Streptomyces* in farming practices. It discusses their role in enhancing plant growth, disease control, as well as soil vitality. Finally, the chapter also covers the possible advantages and the challenges of using such kind of beneficial organisms to achieve more sustainable agriculture practices.

Keywords

Streptomyces · Sustainable agriculture · Strain improvement · Plant growth-promoting *Streptomyces* · Biofertilizers · Biopesticides

15.1 Introduction

Actinobacteria are among the most extensive and varied groups of bacteria identified to date. They are Gram-positive filamentous bacteria that fall under the Phylum Actinobacteria and order Actinomycetales. These are widespread organisms and found in various environments, including terrestrial soils, aquatic habitats (marine and fresh water ecosystems) (Chater et al. 2010). The genome of this organism has a high G + C content. The morphological feature, along with the ability of certain individuals to produce spores for asexual reproduction, has led to the classification of actinobacteria as microorganisms that bridge the gap between bacteria and fungi (Bhatti et al. 2017). However, the genetic and cellular characteristics of actinobacteria align them more closely with bacteria, with traits such as chromosome contained within a prokaryotic nucleoid and a cell wall made of peptidoglycan. The exploration of actinobacteria as a potential source of bioproducts began in the 1920s with the initial documentation of antibiosis by a species within this group. Later, in the 1940s, streptomycin—the first effective therapeutic compound for tuberculosis—and other compounds with antibiotic qualities generated by actinobacteria were discovered. Currently, actinobacterial are estimated to account for 64% of all naturally occurring antibiotics. In actuality, over the past decade, there has been a significant increase in actinobacteria-related publications, particularly those that concentrate on species that have potential applications in supporting plant growth and biocontrol. A variety of biological insecticides, such as those made by *Streptomyces*, have proven to be a successful substitute for synthetic pesticides in the fight against dangerous insects (Bhatti et al. 2017). But in order to employ biopesticides effectively and solve associated issues like pest resistance, a thorough understanding of both the mechanisms of action of the bioactive products and the biology of the generating microorganisms is required.

Streptomyces anchor huge genomes with biosynthetic gene clusters (BGCs) proficient for the construction of several metabolites with great scope in agriculture. Genome mining approaches integrated with bioinformatics tools have augmented the unearthing of novel bioactive compounds like siderophores, pheromones, growth hormones, antimicrobials, and biocontrol agents that are beneficial in agriculture. The technology has enabled to identify the strain-specific pathways and harness for sustainable farming practices.

CRISPR-Cas9 and other genetic engineering tools for strain improvement and activation of silent pathways have been advanced for the targeted manipulation of genes and regulatory elements in *Streptomyces* for strain improvement and activation of silent BGCs. This improvement has resulted in the development of effective antibiotics, phytohormones, and nutrient solubilizers, which further enhance crop health and adaptability of the developed strains in diverse soil and climate.

Synthetic biology provides strong strategies to reprogram *Streptomyces* for optimized production of agriculturally relevant metabolites. Designing of synthetic promoters, ribosome-binding sites, and regulatory pathways has promoted higher yield and stability. Dynamic pathway regulation and optimization have allowed tuning of metabolite production to meet agricultural demands. These have also enabled feasibility of producing biofertilizers and biopesticides at industrial scales, thereby supporting eco-friendly farming systems.

Omics technologies like transcriptomics, proteomics, and metabolomics provide holistic insights into the physiology and biosynthetic capacity of *Streptomyces*. The Omics technology has helped researchers to understand regulation and improve metabolite yields. Transcriptomic analyses enable the detection of gene expression involved in hormone production, nutrient cycling, and disease suppression. Similarly, proteomics has caused the understanding of metabolite pathways while metabolomics provide detailed profiles of the bioactive compounds. Omics strategies are used in agriculture to enhance the reliability and productivity of *Streptomyces*-based inoculants making them effective in promoting plant growth and immunity. Formulation advances improving stability, shelf-life, and application efficacy of Streptomyces-based biofertilizers and biopesticides.

15.2 *Streptomyces:* An Overview

The *Streptomyces* genus, belonging to the phylum actinobacteria, accommodates a phylogenetically diverse group of Gram-positive, filamentous microorganisms characterized by complex morphological differentiation and an exceptionally highly sophisticated secondary metabolism. These microorganisms are pervasive in non-aquatic and aquatic environments, with precise ecological dominance in soil ecosystems, where they contribute significantly to the decomposition of organic substrates and biogeochemical cycling (Chater et al. 2010). Their life cycle includes distinct developmental stages, together with vegetative substrate mycelia and reproductive aerial hyphae, ultimately producing resilient spores that ensure survival under desiccating conditions, and facilitate significant dispersal. Genomically,

Streptomyces species are outstanding by their huge, linear chromosomes displaying elevated G + C content (normally 65–78%), which encode a large array of specialized biosynthetic gene clusters. These genetic determinants underpin their high-quality capability for secondary metabolite production and adaptive responses to environmental challenges. Such substances are typically not a part of primary metabolism, and they are most often a byproduct of metabolism that has no function in major stages of the cell processess. With more than 900 species described, *Streptomyces* is not only the largest genus of actinobacteria but also one of the most genetically and metabolically versatile microbial taxa identified till date (van der Meij et al. 2017). Historically, they have been celebrated for their role in antibiotic production, accounting for over two-thirds of naturally derived antibiotics used in clinical medicine (van der Meij et al. 2017). Today, however, *Streptomyces* are now increasingly recognized for their broader ecological contributions, particularly in sustainable agriculture. An important soil bacteria belonging to the actinobacteria family are *Streptomyces*. Actinobacteria being the source of nearly two-thirds of naturally occurring antibiotics has *Streptomyces* genus accounting for roughly 75% of these isolates (Silva et al. 2022). Furthermore, *Streptomyces* are known to produce around 7600 bioactive chemicals (Berdy 2005). Hence, *Streptomyces* is currently considered the primary producer of antibiotics produced in pharmaceuticals.

15.3 Bioactive Compounds Derived

Streptomyces are known for their unprecedented biosynthesis potential. They produce a wide range of bioactive secondary metabolites that are useful for ecological, pharmaceutical, and agricultural functions (Barka et al. 2016). These include clinically important antibiotics (e.g. streptomycin, tetracycline), antifungal agents, antivascular and antitumor compounds, immune suppressants, and many other pharmacologically important compounds (Barka et al. 2016). The presence of large biosynthetic gene clusters (BGCs), such as polyketide synthases (PKS), non-ribosomal peptide synthases (NRPs) and hybrid systems, enables the production of structurally complex molecules with high bioactivity, and facilitates their remarkable metabolic diversity. *Streptomyces* contributes significantly in agriculture sector mainly to the production of plant hormones like the indole-3-acetic acid, gibberellins, and cytokinins, which can regulate plant growth and development (Jog et al. 2014). They also produce siderophores to improve plant iron absorption, and enzymes such as cellulases, glucanases, and chitinases that damage cell walls of phytopathogenic fungi, thereby contributing to plant protection (Jog et al. 2014). Latest findings also highlight the role of volatile organic compounds(VOCs) derived from *Streptomyces* in mediating exchanges between different biological kingdoms, affecting rhizosphere root structure, inducing systemic resistance, and shaping microbial community dynamics (Vurukonda et al. 2018). Figure 15.1 depicts the various mechanisms (direct and indirect) exhibited by agriculturally important *Streptomyces* spp.

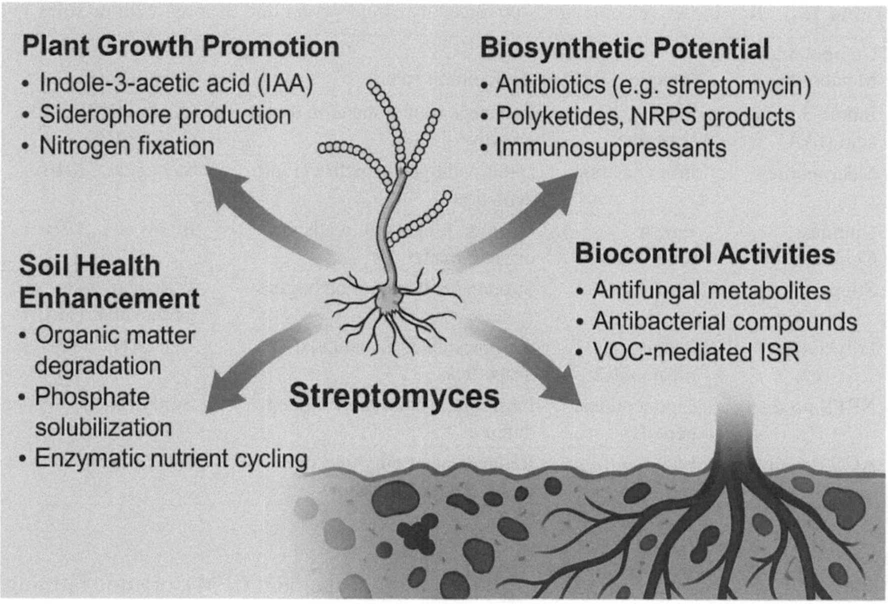

Fig. 15.1 Diverse mechanisms (direct and indirect) exhibited by *Streptomyces* spp. in enhancing plant growth

15.4 *Streptomyces* in Agriculture

Application of *Streptomyces* in agriculture sector has moved beyond their antimicrobial capabilities. *Streptomyces* have been reported to help in nutrient mobilization via nitrogen fixation, phosphate solubilization, and siderophore-mediated micronutrient uptake and, thus, have emerged as potent plant growth-promoting rhizobacteria (PGPR) (Gopalakrishnan et al. 2013) (Table 15.1). Such mechanisms contribute to improved crop yields and sustainability under abiotic stresses such as drought, high temperature, salinity, heavy metals, and poor soil fertility. Wide production of antimicrobial and antagonistic agents like lytic enzymes, signaling molecules, the inhibition of phytopathogens like fungi (*Fusarium*, *Rhizoctonia*, *Alternaria*, *Colletotrichum*, and *Magnaporthe*), bacteria (*Xanthomonas*, *Erwinia*), and also some plant viruses have been made possible. *Streptomyces* are also reported to induce systemic resistance (ISR) in the host plants, which activates the innate protection systems of the plants and limits the occurrence of diseases (Barka et al. 2016; Vurukonda et al. 2018). Moreover, their ability to build sustainable interactions within the rhizosphere microbiome again contributes to the long-term sustainability of agricultural ecosystems.

With more preference to eco-friendly and resourceful agricultural practices, *Streptomyces* have gained popularity for developing biocontrol agents, nutrient metabolites, biofertilizers, biopesticides, and soil health improvement products.

Table 15.1 Key bioactive compounds produced by *Streptomyces* and their agricultural roles

Compound/ Metabolite	Function	Agricultural role	Reference
Indole-3-acetic acid (IAA)	Plant hormone	Enhances root elongation and growth	Glick (2012), Barka et al. (2016)
Siderophores	Iron chelation	Limit pathogen growth via iron depletion	Saha et al. (2016)
Chitinase, Cellulase	Enzyme production	Degrade fungal cell walls and organic matter	Bharti et al. (2015)
Streptomycin	Antibiotic	Suppresses bacterial pathogens	Waksman and Lechevalier (1953)
Polyketides	Secondary metabolites	Antimicrobial, nematicidal properties	Nett et al. (2009)
NRPS products	Lipopeptides, peptides	Inhibit fungi and induce plant defense	Bérdy (2012)
ACC deaminase	Enzyme	Reduces plant ethylene stress under abiotic conditions	Glick (2014)

These support the principles of integrated pest management (IPM), organic farming, and climate-resilient agriculture, offering an effective solution to the challenges of unsustainable food production, which are instigated by overdependence on chemicals, desertification of the soil, and the increasing need to achieve sustainable food production systems.

15.5 Plant Growth-Promoting *Streptomyces*

Streptomyces have received prominence in agricultural field due to their multifaceted mechanisms that facilitate plant growth (Gopalakrishnan et al. 2020). These soil-residing filamentous microorganisms colonize the rhizosphere and root surfaces, establishing mutualistic interactions with variety of crop species. Their plant growth-promoting (PGP) activities can be broadly classified into direct mechanisms, which affect plant physiology, and indirect mechanisms, which suppress pathogenic microbes and alleviate stresses (Gopalakrishnan et al. 2020).

Phytohormone production, considerably indole-3-acetic acid (IAA), by *Streptomyces* is one of the most significant contributions in plant growth promotion. IAA promotes root morphology by enhancing root elongation, root emergence, and root hair improvement, ultimately improving water and nutrient acquisition (Glick 2012; Barka et al. 2016). Few species also synthesize other hormones that include gibberellins and cytokinins, which undoubtedly positively affect seed germination and shoot elongation (Jog et al. 2014).

Streptomyces also show another important trait of phosphate solubilization by secreting organic acids (e.g., gluconic, citric, and oxalic acids) that convert insoluble phosphates into forms that are readily available to plants (Bharti et al. 2015). Phosphorus bioavailability is inherently low in tropical and sub-tropical soils.

Although *Streptomyces* are not classical nitrogen-fixing microorganisms, some strains exhibit associative nitrogen-fixing competencies, especially in co-culture systems with diazotrophs. These indirect interactions enhance nitrogen assimilation and plant biomass (Saha et al. 2016; Vurukonda et al. 2018). *Streptomyces* are also efficient producers of siderophores, low-molecular-weight iron-chelating compounds, which beat phytopathogens for iron in the rhizosphere. This limits pathogen proliferation while facilitating iron uptake by the plant (Saha et al. 2016). Another significant feature is the production of 1-aminocyclopropane-1-carboxylic acid (ACC) deaminase, which degrades the ACC, a precursor of ethylene (Glick 2014). By reducing ethylene levels under abiotic stresses (e.g., drought, salinity, and heavy metals), these microorganisms mitigate plant growth inhibition and help maintain physiological balance (Glick 2014; Vurukonda et al. 2018). Jointly, these mechanisms not only enhance plant nutrient uptake and energy efficiency but also contribute to soil health and stabilization of microbial communities, reinforcing *Streptomyces* as crucial agents for sustainable agricultural development.

15.5.1 Production

The production scale of *Streptomyces* for agricultural or farm applications relies on their ability to synthesize and secrete bioactive compounds in a consistent and economically viable manner. Due to their filamentous architecture, amenability to genetic manipulation, and fitness to fermentation procedures, *Streptomyces* species are held as the best microbial factories to produce plant-beneficial metabolites like phytohormones, antibiotics, enzymes, siderophore, and volatile organic compounds (VOCs) (Barka et al. 2016; Fazal et al. 2020).

There are two strategies that have been adopted in large-scale production, such as solid-state fermentation (SSF) and submerged fermentation (SmF). SSF employs low-moisture substrates such as wheat bran or rice husk, which favors high yield of metabolites, and mimics soil-like conditions, which are favorable for sporulation and extracellular compounds production (Gomes et al. 2018). On the contrary, SmF provide greater control of parameters such as aeration, pH, and nutrient availability, and is also valuable to optimize these parameters to enhance metabolite production and streamlining subsequent downstream processing.

The composition of *Streptomyces*-based bioinoculants usually involves the addition of carriers such as talcum powder, peat, vermiculite, or alginate balls in order to extend the stability of microbial population, facilitate ease of human use, and promote rhizospheric settlement or colonization (Reddy et al. 2014). Microencapsulation procedures and nanocarrier-based formulations are emerging approaches to protect spores and retain soundness or viability under various environmental conditions (Amaresan et al. 2021). Synthetic biology and metabolic engineering have further enhanced their production efficiency. Techniques such as heterologous expression of biosynthetic gene cluster (BGCs), pathway refactoring, and CRISPR-Cas9-based genome editing have significantly increased the biotic potential of selected strains of *Streptomyces* (Zhu et al. 2022). These strategies in conjunction with AI-assisted bioprocess modeling provide a rich path toward future generation of bioproduction systems. Collectively, these innovative technologies and bioprocessing systems have redesigned soil *Streptomyces* to reliable microbial biomanufacturers for developing biofertilizers, biostimulants, and biocontrol agents in sustainable agriculture.

15.5.2 Applications

Streptomyces species hold great significance for their wide applications in the agriculture sector (Table 15.2). The primary application of *Streptomyces* in agricultural field is their use as biofertilizers. Field research on cereals, legumes, and vegetables has demonstrated significant progresses in biomass yield, especially when Streptomyces-based inoculants are used, particularly in nutrient-poor soils. Their ability to solubilize phosphates, enhance nitrogen uptake, and release plant hormones allows crops to utilize soil nutrients more efficiently (Gopalakrishnan et al. 2013). Such biofertilizers are often used in the form of seed dressings, root dips, or soil amendments, depending on crop and field implications. Furthermore, *Streptomyces* also serve as successful biocontrol agents against a broad spectrum of plant pests and pathogens. They are mostly employed to fight against diseases that are caused by *Fusarium, Rhizoctonia, Pythium, Alternaria, and Phytophthora* spp. (Abbasi et al. 2021). Their biocontrol mechanisms include antibiosis, competition, and stimulation of plant defense mechanisms (Sadeghi et al. 2017). They are equipped with the ability to produce antibiotics, siderophores, and lytic enzymes (chitinases, cellulases, and glucanases), which allows them to restrain the development of pathogens both in vitro and in situ (Viaene et al. 2016). *Streptomyces*-based commercial formulations such as Mycostop (which is derived of *Streptomyces griseoviridis*) and Actinovate (derived from *Streptomyces lydicus*) are commercially available and are widely used as part of integrated pest management (IPM)

Table 15.2 *Streptomyces* species and their agricultural applications

Application	Mechanism	Strains	References
Biofertilizer	IAA Production, nitrogen fixation, and phosphate solubilization	*Streptomyces coelicolor*, *Streptomyces griseus*	Gopalakrishnan et al. (2013)
Biological control agent	Enzyme secretion, antibiotic production, and pathogen inhibition	*S. lydicus*, *S. griseoviridis*	Sadeghi et al. (2017), Viaene et al. (2016)
Plant probiotics	Stress tolerance induction and ACC deaminase activity	*Streptomyces avermitilis*, *Streptomyces rochei*	Glick (2014), Vurukonda et al. (2018)
Commercial formulations	Foliar spray, soil application, and seed treatment	*S. lydicus* (Actinovate), *S. griseoviridis* (Mycostop)	Mitra et al. (2020)

programs. These products are applied to many different systems such as greenhouse crops, field vegetables, and turf grass, and offer benefits over agrochemicals due to their low toxicity, eco-friendliness, and compatibility with organic farming (Mitra et al. 2020).

Moreover, *Streptomyces* species are also considered as biostimulants to enhance tolerance/resistance to abiotic stresses such as drought, salinity, and heat. They stabilize plant metabolism by secreting ACC deaminase, and by inducting systemic tolerance response, thereby resulting in the mitigation of yield losses under stressful conditions (Glick 2014; Vurukonda et al. 2018). This application has special significance in the context of global climatic change, where building crop resilience is essential for ensuring food security.

The application of *Streptomyces*-based technologies in seed priming, soil fertility enhancement, and disease suppression program in modern agricultural system indicates a drift toward the development of nature-based, biologically informed farming practices. With a broad spectrum of applications, *Streptomyces* are expected to play an increasingly important role in future agriculture, further supported by developments in genomics, precision farming, and field-scale validations.

15.6 *Streptomyces* as Biocontrol Agents

Streptomyces plays a significant role in the plant growth promotion by protecting the crops from harmful pests and pathogens (Silva et al. 2022). *Streptomyces* produce a large variety of bioactive metabolites, including antibiotics, enzyme inhibitors, pesticides, and herbicides, which offer promise for applications in agriculture as plant protection and plant growth-promoting products (Vurukonda et al. 2021). The secondary metabolites produced by *Streptomyces* such as blasticidin-s, Kasugamycin, Streptomycin, Oxytetracycline, Validamycin, Polyoxins, natamycin, actinovate, mycostop, abamectin/avermectins, emamectin benzoate, polynactins, and milbemycin have shown significant potential in controlling insect pests and plant pathogens due to their high specificity, biodegradability, and less environmental toxicity (Boykova et al. 2023). Therefore, microbial-derived products serve as an effective substitute for minimizing the usage of agrochemicals, combining high efficiency with a sustainable approach that benefits both the planet and human health.

Applications of commercial formulations containing *S. griseoviridis* (Mycostop®) and *S. lydicus* (Actinovate®) have already proven their success in the agricultural systems (Goudjal et al. 2013). Furthermore, with advances in genomics and metabolomics, it has been found that *Streptomyces* harbor abundant hidden gene clusters suggesting more discoveries (Zhang et al. 2019).

15.6.1 Antifungal Activity

Crops often suffer significant losses due to fungal pathogens, but *Streptomyces* cope with these threats using various strategies. While chemical fungicides are highly effective, they can be harmful to people and other living creatures, and may also promote the development of fungicide-resistant strains. *Streptomyces,* with their ability to produce a plethora of diverse secondary metabolites, serve as powerful biocontrol agents against plant pathogenic fungi (Yuan and Crawford 1995). These soil-dwelling microbes thrive under competition in the rhizosphere, deploying an antifungal arsenal that includes polyene macrolides such as natamycin and candicidin; aromatic polyketides; non-ribosomal peptides; and volatile organic compounds (Bubici et al. 2019). *S. lydicus* and *S. violaceusniger* produce different polyenes (e.g., natamycin, candicidin), toyocamycin, isoafricanol, and spirofungin that cause damage to the fungal membrane (Yuan and Crawford 1995; Ma et al. 2020; Gao et al. 2023).

During metabolic reactions, standard pathways create compounds called VOCs, such as fatty acids and alcohols, which then act as antifungal gases (Elleuch et al. 2021). Chitinases, glucanases, and proteases are carefully produced by the *Streptomyces* and released outside to help degrade the cells of other organisms (Elleuch et al. 2021). Enzymatic degradation works in combination with antifungals and can cause plants to develop resistance across their whole system. Strains such as *S. lydicus* and *S. rochei* have shown effective control of *Fusarium oxysporum*, *Botrytis cinerea*, and *Sclerotinia sclerotiorum* under laboratory and greenhouse conditions (Le et al. 2022).

Strains in the *Streptomyces* family, such as *Streptomyces* sp. CB-75, reduce the amount of *Fusarium oxysporum* by about 85% in the lab and 73% in plant conditions (Chen et al. 2018; Bubici et al. 2019). *Sclerotinia sclerotiorum* was not able to grow in the presence of VOCs released by *S. albulus* in dual-culture assays (Elleuch et al. 2021). *Streptomyces* such as *S. rochei* and *S. sampsonii* produce chitinases, glucanases, proteases, and cellulases that help destroy the walls of fungi (Elleuch et al. 2021). Green house studies also show that *Streptomyces* spp. have the ability to reduce the risks of *Fusarium*, *Sclerotinia*, and *Botrytis* diseases by up to 80% (Chen et al. 2018; Stoica et al. 2023). In both greenhouse and field settings, a group of five *Streptomyces* strains exhibited antagonistic potential against *Fusarium oxysporum*. *Streptomyces* species, such as *S. coelicolar*, *S. girseus*, *S. albus*, *S. antibiotics*, and *S. champavatii*, shown to have antagonistic action against *Rhizoctonia solani* on tomatoes in comparison to the uninoculated control plants (Prastiti et al. 2023).

Streptomyces are able to manufacture antifungal compounds with NRPS, PKS, or hybrid NRPS–PKS systems found inside BGCs (Seipke and Hutchings 2019; Alam et al. 2022a and b). Some antifungal macrolides, like polyenes (such as candicidin and natamycin), begin with PKS modules adding several acyl-CoA precursors, and then undergo glycosylation and oxidation, resulting in chemicals that disrupt bacterial membranes (Seipke and Hutchings 2019). Toyocamycin and its kind, called nucleoside analogs, get their core from combining a modified ribose with a pyrrolopyrimidine via NRPS-free mechanisms, and nucleoside synthases are responsible for the assembly. Gene expression is regulated by transcription factors that respond to what is going on in the environment, and two-component systems notice changes, such as stress or new nutrients, and then begin the process of turning genes on (Rodríguez et al. 2023).

Because of their multiple modes of antifungal activity, *Streptomyces* are very effective in controlling fungi while reducing the likelihood of resistance development. Further, advancements in genome mining and metabolomics have made it possible to discover new BGCs for making antifungal metabolites and developing new biopesticides. These attributes show the importance of *Streptomyces* as green or eco-friendly substitutes or alternatives to chemicals fungicides to support sustainable agriculture.

15.6.2 Antibacterial Activity

Streptomyces have also been recognized for their use against plant bacterial pathogens. Bacterial phytopathogens such as *Ralstonia solanacearum*, *Pseudomonas syringae*, and *Xanthomonas campestris* are posing great threat to food availability worldwide. These bacteria can harm different crops and usually resist chemical treatments that kill bacteria. While *Streptomyces* were well known for making antibiotics of medical use, they are now recognized as effective biocontrol agents in agriculture. They are able to manufacture different kinds of antibacterials, for example, aminoglycosides, tetracyclines, macrolides, as well as peptide antibiotics such as nigericin, manumycin, and kasugamycin (Le et al. 2022; Stoica et al. 2023). They influence bacterial activity in three ways: by blocking the synthesis of proteins, disrupting the membrane, and disturbing how the cell wall is made. In addition, many of these compounds work against several pests and are not harmful to plants. *S. griseoviridis* and *S. lydicus* have been assessed and found to control bacterial wilt, soft rot, and leaf spot diseases just like chemical-based methods (Le et al. 2022). Apart from attacking other organisms, *Streptomyces* use indirect ways, including making siderophores that capture iron and reduce its availability to pathogens, and by producing lytic enzymes for bacteria and nucleic acid. Certain strains influence the microorganisms living around plants, raising populations of desirable bacteria and fungi that beat out pathogenic organisms.

More novel antibacterial compounds that aid agriculture are being found due to progress in high-throughput screening, metabolomics, and synthetic biology. Besides, *Streptomyces* are able to colonize plant roots and sometimes act as endophytes, which increase their chances of being used as bioinoculants. Since the focus on antibiotics and chemicals is growing, *Streptomyces* provide a sustainable and accurate way to fight bacterial diseases in crops.

(a) *Antibiotic Production*

Streptomyces are able to make compounds named nigericin, buramycin, manumycin, griseoviridin, and kasugamycin, which prevent the infection of bacteria

Ralstonia solanacearum and *Pseudomonas* spp. (Le et al. 2022; Stoica et al. 2023). Paclobutrazol shows results in protecting plants from diseases that spread in the soil (Le et al. 2022). Proteases break down bacterial cells and siderophores hold iron that stops the growth of these pathogens.

The formation of antibacterial antibiotics by *Streptomyces* species depends mostly on the action of NRPS and PKS pathways. In these systems, simple metabolic products are converted via enzymes into antibiotics with different parts being added in a highly ordered manner (Alam et al. 2022a, b). Novobiocin, a kind of aminocoumarin, gives a clear example of this mechanism. These are formed by intermediates present in the shikimate pathway, namely prephenate, tyrosine, and glucose-1-phosphate. Due to the actions of about 23 genes in a BGC, also called a biosynthetic gene cluster, the building blocks comprising the active antibiotic are put together through oxidation, amide bond formation, and cyclization enzyme reactions (Bode et al. 2002). Macrolides and tetracyclines are samples of polyketide antibiotics formed through the process carried out by Type I PKS complexes. These enzymes carry out different functions, and are made up of several catalytic domains grouped in modules. The addition of each acyl unit to the chain formed by the modules builds the macrocyclic lactone known as tylactone (the precursor of tylosin) that is transformed further by adding sugar, methyl groups, and hydroxyl groups (Weissman and Leadlay 2005).

On the other hand, NRPS systems manufacture peptide-based antibiotics daptomycin and antimycin by gradually adding amino acids using adenylation, thiolation, and condensation. Nonproteinogenic amino acids like kynurenine and the addition of lipid tails can be found in NRPSs, just like in daptomycin according to Baltz (2009). Antibiotic production in *Streptomyces* is regulated by small signaling molecules, like γ-butyrolactones. With the onset of the stationary phase, these molecules send signals that prompt antibiotic production by regulating transcriptional regulators such as AdpA (Horinouchi 2002). Also, two-component systems sense different stimuli and translate them into effects on the expression of BGCs (van der Heul et al. 2018). Furthermore, links between genes and signals from the environment allow multiple pathways to be expressed together. Nutrient deprivation and high levels of stress are recognized to move bacteria into secondary metabolism, resulting in increased production of antibiotics (Zhou et al. 2016). Nonactin, a biologically active substance generated by many *Streptomyces* species, is a macrotetrolide antibiotic made up of four nonactic acid subunits (Boykova et al. 2023). Nonactic acid and its derivatives—monactin, dynactin, trinactin, and tetranactin—have been shown to be effective against bacteria, fungi, insects, and mites. It was also demonstrated that *Streptomyces globisporus*-generated nonactin had insecticidal effects against the Colorado potato beetle (*Leptinotarsa decemlineata*). The inhibition of growth of several pathogenic bacteria, notably *Agrobacterium tumefaciens*, by *Streptomyces cavourensis* TN638, which produces a combination of nonactin and other derivatives of nonactin acid, was outlined. Several species of *Streptomyces* synthesized tetranactin, which inhibited the growth of a variety of bacteria and fungi and had acaricidal effects against the two-spotted spider mite (*Tetranychus urticae*).

15.6.3 Antiviral Activity

Even though research on them is not very extensive, *Streptomyces* display great potential against viruses. Plant viruses are a problem that receives little attention, though they are very destructive to crops. Common strategies, including working with insects and virus-resistant crops, do not work well, and farmers mostly do not use chemical treatments against viruses. In this situation, using *Streptomyces* for antiviral purposes is a reliable and promising intervention. While *Streptomyces* are better recognized for their roles in fighting fungi and bacteria, they have just been found to suppress viruses such as Tobacco mosaic virus (TMV), Potato virus Y (PVY), and Zucchini yellow mosaic virus (ZYMV) (Ali et al. 2023; Wang et al. 2024a, b). Some antiviral mechanisms create bioactive compounds such as anisomycin, ε-poly-L-lysine, cytosine peptidemycin, and GP-1 that stop viral replication and help prevent its movement (Stoica et al. 2023).

Molecular techniques and omics approaches applied on concealed antiviral genes clusters can cause the identification of possible brand-new antiviral chemicals. Also, the validation of microbial application in the field, improvement of their formulation, and their incorporation in programs to control the virus and safeguard crops in the long term is worth consideration.

(a) *Antiviral Metabolites*

Wang et al. (2024a, b) reported inhibition of TMV and Potato Virus Y (PVY), in in vitro and green house conditions, by anisomycin produced *by S. albulus* SN40 that inhibit viral helicase enzyme and also induce development of resistance in host plant (Wang et al. 2024a, b). According to Manjunatha et al., (2022) and Zhang et al., (2013), cytosine peptidemycin and 3-poly-L-lysine produced together by *S. hygroscopicus* not only exhibit ant-viral properties but can also induce the mechanism of systemic-acquired resistance in plants to become activated. Stoica et al. (2023) identified that one of the glycoproteins (GP-1 glycoprotein) of the LoneStar DL strain of *Streptomyces* sp. ZX01 almost fully inhibited TMV.

Some antiviral peptides called cytosine peptidemycin and ε-poly-L-lysine are formed when nonribosomal peptide synthetases assemble canonical and non-canonical amino acids into regular or cycle peptides. These molecules might change after assembly, such as being polymerized or getting an acetyl group on one side, which improves their ability to fight viruses (Stoica et al. 2023). When certain strains are exposed to culture conditions with plant cells, they make antiviral proteins called glycoproteins, among which GP-1 prevents viruses from getting in or replicating inside the host (Wang et al. 2024a, b). The synthesis of these compounds is supervised by inner-cell messenger systems, and γ-butyrolactones help to activate the dormant or unrevealed antiviral gene clusters by lifting the ban placed on regulators (Horinouchi, 2002). BGCs for antiviral purposes are often turned on by events like skipping vegetative growth and starting to grow in the air as well as by resources or signals from the avirulent host. Having regulatory genes, efflux systems, and transporters within BGCs allows antiviral compounds to be created as well as released and located within plant tissue to aid defense and symbiosis (Wang et al. 2024a, b).

(b) *Boosting a Plant's Defense Mechanisms*

Apart from acting against viruses, *Streptomyces* can also boost the plants' natural resistance and antioxidant systems. Treating plants with metabolites or filtered extracts from *Streptomyces* bacteria has shown a decrease in disease rates, increase in the level of chlorophyll, and activation of more phenols and peroxidases (Ali et al. 2023). Treating zucchini plants with *S. griseus*, *S. rochei*, and *S. sampsonii* in greenhouses resulted in >95% reduction of ZYMV, increase in antioxidant enzymes, and thereby improved plant health (Ali et al. 2023). Potato plants treated with filtrates obtained from *S. ambofaciens* and similar species also showed lowered incidence of PVY (Stoica et al. 2019).

Plant's immune response by salicylic acid and jasmonic acid increases the plant's ability to guard against systemic infections. With continued research, researchers locate more compelling facts concerning the possible use of *Streptomyces* to cure viral diseases.

15.7 Technological Advances

Improvements in genomic, metabolomic, and synthetic biology technologies are causing *Streptomyces* to regain attention as biocontrol agents. Previously, a limited number of metabolic routes were used because most of the biosynthetic gene clusters (BGCs) remained inactive or not easily recognizable. However, using modern approaches such as antiSMASH and EvoMining has made it possible to find numerous hidden BGCs that produce novel types of antimicrobial molecules (Cruz-Morales 2022). These types of compounds are polyketides, non-ribosomal peptides, hybrids containing NRPS and PK, and metabolites made of arsenic. With the help of CRISPR/Cas9 and other genetic engineering methods, it is now possible to precisely edit or overproduce the pathway genes that activate silent BGCs and make more of the desired metabolites (according to Cobb et al. 2015). With such platforms, researchers are able to work with tough strains by turning to *S. coelicolor* or *E. coli* model organisms. Because of combined use of transcriptomics and metabolite, it has become easier for researchers to find relationships between genes and metabolic products.

Moreover, formulation science has played a major role in the commercial success of Streptomycete-based biopesticides. With the help of microencapsulation, emulsifiable concentrates, and granule-based delivery, the problems of shelf-life, UV light sensitivity, and inconsistent field use are overcome. *Streptomyces griseoviridis* (Mycostop®) and *Streptomyces lydicus* (Actinovate®) serve as proof that lab findings can be applied to useful agricultural products. Future approaches include using AI-guided tools to find new metabolites, building sensors that detect pathogens as they are controlled, and applying streptomycete mixtures to different agricultural areas in a targeted way. Over time, these improvements have made *Streptomyces* valuable for new forms of sustainable farming.

(a) *Genomics and Bioinformatics*

Industrial approaches to genome mining are sometimes able to bring out cryptic BGCs. EvoMining helped scientists find different ways of synthesizing compounds such as those involved in producing peptide bonds and arsenic-containing molecules (Cruz-Morales 2022).

(b) *Genetic Engineering*

CRISPR/Cas systems have made it possible to activate unknown biosynthetic pathways and increase the amount of useful secondary metabolites produced by the microorganisms (Cobb et al. 2015). Excessive expression of genes has resulted in higher amounts of toyocamycin and spirofungin (Ma et al. 2020).

(c) *Developing the Treatment*

While *S. lydicus* WYEC108 and *S. griseoviridis* K61 are efficient in control, they have problems regarding stability and regulations (Yuan and Crawford 1995; Le et al. 2022). Collectively, these technological innovations have made practical translation of *Streptomyces* and their product research to wider field-level applications thereby bridging Actinobacterial biotechnology and sustainable agriculture.

15.8 Applications in Agriculture

Use as Biofertilizers *Streptomyces* is mostly introduced into the soil as a biofertilizer by treating seeds and drenching the soil. Using 10^8 CFU of suspension per milliliter and a 45–60-minute seed inoculation has aided in better germination, better establishment of roots, and an improvement in yield of chickpea and sorghum, according to the scientific studies (Gopalakrishnan et al. 2015; Alekhya and Gopalakrishnan 2016). Combining soil drenching at sowing and later stages with root dip has further helped the microbes colonize the rhizosphere, so that the legume crops continue to grow well and produce increased yields in field tests (Gopalakrishnan et al. 2022). *Streptomyces* inoculants have brought positive results when used in rice and wheat during greenhouse trials and pot experiments. In particular, *Streptomyces corchorusii* UCR3-16 enhanced plant height by 20%, biomass by 25%, grain yield by 15%, available phosphorus levels by 30%, and the activity of dehydrogenase enzymes in the soil by 40%. At the same time, it was able to suppress the sheath blight pathogens affecting rice (Zhang et al. 2016). Perlite-based formulations also increased the resistance of wheat to drought and improved its survival (Delgado-García et al. 2022). After two years of applying a consortium of 14 strains in chickpea, nodulation, the amount of stover and grain, rhizosphere microbial nitrogen, and soil organic carbon showed significant increases (Gopalakrishnan et al. 2015, 2022). Zhang et al. (2016) reported that rice plants treated with *Streptomyces* UCR3-16 showed a decreased disease incidence of 40% in sheath blight and 12% increased yield (Zhang et al. 2016). Similarly, up to seven rotational strains in inoculants improved tillering and grain yield in sorghum during different seasons (Alekhya and Gopalakrishnan 2016). From these data, it is confirmed that *Streptomyces*-based biofertilizers in agriculture are useful alternatives to chemical fertilizers.

Streptomyces biofertilizers work by various mechanisms including solubilization of more phosphate for root uptake, and IAA production, ACC deaminase production to reduce stress ethylene levels, production of siderophores for enhanced iron uptake, and formation of ammonia, leading to improved root development (Zhang et al. 2016).

Many delivery methods, such as seed coatings, soil drenches, granules, and biofilm carriers, make these products flexible, versatile, and suitable for both small and large applications. Experiments are underway using foliar sprays with culture filtrate to benefit from systemic resistance. *Streptomyces*-based biofertilizers can adjust well into holistic farming practices, function synergistically with organic amendments, co-inoculation with *Rhizobium*, and remain effective even when chemical inputs are reduced.

Formulations Biofertilizers developed from *Streptomyces* spp. should ensure the viability of the microbes for their long shelf-life. Research has shown that peat is the main carrier, reaching $\geq 10^6$ CFU/g viability for at least 12 months because it is well hydrated and naturally protects organisms (Gopalakrishnan and Pandian 2016). Alternatively, talc and talc-starch blends are budget-friendly options, yet cells stored in them survive for a shorter length of time than those kept in petroleum jelly (Gopalakrishnan and Pandian 2016). Recent studies of perlite-based solid-carrier biofilms show that microbes become more resistant to drying and sunlight, and they also remain viable in more numbers in the rhizosphere and boost the plant's growth (Delgado-García et al. 2022). Multi-strain inoculants containing cow dung and peat have shown noticeable results, keeping about 6×10^5 CFU/g of bacteria for legumes (Sutthisa et al. 2023).

Generally, *Streptomyces*-based biofertilizers are made with high concentrations of spores (over 10 Billion per milliliter) in nutrient media, then carriers are sterilized, spores are mixed, the product is dried, and stored at 4 °C–25 °C (Gopalakrishnan et al. 2015; Gopalakrishnan and Pandian 2016). Storage requires checking the counts of CFU, moisture levels, and pH to ensure good quality. Using skim milk, trehalose, and glycerol as additives and relying on new microencapsulation and biofilm-forming methods are ways to increase a microbe's ability to handle stresses, including those from the field (Delgado-García et al. 2022). Co-inoculating multiple microbial strains makes biofertilizers' performance better, since the different bacteria support each other in moving nutrients to plants, preventing diseases, and living near plant roots. The majority of consortia formulations reach at least ten million viable bacteria in every gram. It is still a key goal in research to keep strain compatibility high and activities within mixed systems under control (Gopalakrishnan et al. 2022).

Use as Biopesticides or Biocontrol Agents In addition to biofertilizers, biopesticides are also gaining importance in sustainable agriculture. Biopesticides utilizing microbial biological control agents (MBCAs) exhibit remarkable outcomes against agricultural plant pathogens, particularly due to the specific targeting provided by their natural molecular mechanisms. These mechanisms can reduce pest populations and restore ecological balance in the environment (Prastiti et al. 2023). Among the diverse microorganisms found in agricultural settings, actinobacteria stand out due to their unique bioactive properties (Prastiti et al. 2023). Their prolific production of antimicrobial compounds and other metabolites provides a competitive advantage, enhances soil fertility, offers plants defense against infections exerted by harmful pathogens, and supports their survival under environment stresses (Olanrewaju and Babalola 2019). For example, endophytic *Streptomyces* from citrus were found to have antifungal activity against *Guignardia sublineolum*, and *Pythium* sp. Endophytic *Streptomyces* were also found to control *Sclerotium rolfsii* and *Phytophthora* in chickpea. Regardless of the chickpea genotypes employed, it was discovered that certain endophytic *Streptomyces*, both as a single inoculation and in a consortium, lessened the brutality of *Botrytis cinerea*, which causes Botrytis gray mold disease. The inhibition of plant diseases is significantly aided by antioxidant enzymes. It has been found that *Streptomyces* increase antioxidant enzymes in plants infected with pathogen. For example, it was discovered that the Chickpea leaves injected with endophytic *Streptomyces* exhibited increased levels of antioxidant enzymes, including polyphenol oxidase, catalase, superoxide dismutase, glutathione reductase, phenylalanine ammonia-lyase, ascorbate peroxidase, and guaiacol peroxidase, compared to the uninoculated control (Abbasi et al. 2021).

Additionally, many reports also documented the role of secondary metabolites of *Streptomyces* spp. in the biological control of insect pests including *S. litura* (Arasu et al. 2013), *S. liltoralis* (Bream et al. 2001), *Ostrinia furnacalis* (Bayot-Custodio et al. 2014), *H. armigera* (Osman et al. 2007; Arasu et al. 2013), *Aphis gossypii* (Zhang et al. 2010) *Mythimna separate* (Wang et al. 2011), and *Tribolium castaneum* (Anwar et al. 2016). Several secondary metabolites, with insecticidal activity, such as avermectins, prasinons, milbemycin, doramectin, tetranectin, dianemycin, nanchangmycin, macrotetrolides faerifungin, and spinosad have been obtained from genus *Streptomyces* species and established as potential protective agents against a variety of pest insects (Montesinos 2003; Omura 2008).

15.9 *Streptomyces* in Soil Health

One of the most significant differences between soils and a pile of inert rock particles is that soil is a home to a diverse population of microorganisms that get their energy by oxidizing organic remains of plants and animals. Eventually, the plants growing in the soil benefit from the products of microbial action because the microorganisms continuously oxidize the remains of dead plants, releasing nitrogenous and mineral compounds in plants available form (Bhatti et al. 2017). The major divisions of soil microorganisms include bacteria, actinobacteria, fungi, and algae. The characteristic filamentous actinobacteria, primarily members of the *Streptomyces* and *Micromonospora* groups, grow as extensively branched hyphae that break up into spores, either by the tip of the hyphae producing one or two spores or other mechanisms. They use carbon and nitrogen sources like cellulose, hemicellulose, proteins, and lignin, and are often abundant under pastures and dominate the surface layers of grass lands and thatch (Silva et al. 2022). Because actinobacteria are slow grower than bacteria and fungi, they initially attack fresh organic residues, subsequently degrading more resistant material. They form a variety of dark black to brown pigments that add to the dark color of soil humus and are responsible for the breakdown of humid substances. By producing hydrolytic enzymes, actinobacteria play a critical role in recycling organic materials in the environment. Actinobacteria are specialized decomposers of cellulose and lignin, which are present in wood, paper, and insect exoskeletons, as well as chitin. Plants can once again access nutrients as a result of the breakdown of these components. Actinobacteria that are thermophilic (adapted to high temperatures) and thermotolerant are primarily in charge of decomposition throughout the composting process. The breakdown of organic waste at high temperatures during the composting process is mostly caused by thermophilic (high temperature-adapted) and thermotolerant actinobacteria. The organic material self-heats during the first stage of composting due to the intense rise in microbial activity. High composting temperatures aid in the destruction of dangerous microorganisms, such as coliforms, viruses, and weed seeds. Actinobacteria primarily lead aerobic lives, therefore, they require well-aerated compost for their optimal decomposing activity.

Thus, due to diverse metabolic activities in soil or rhizosphere, *Streptomyces* spp. enhance PGP traits, such as root volume, root growth (length), root hair development, nodulation, plant hormone concentrations, shoot growth, tiller number, panicle number, grain/fruit yield, and stover yield in cereals, legumes, and horticultural crops. These effects are mediated by multitude of mechanisms (biologic nitrogen fixation, mineral solubilization (phosphorus, potassium, and zinc), iron chelation, and the release of plant growth hormones) exhibited by *Streptomyces* spp. Under field conditions, *Streptomyces*' spp. PGP potential has been reported in tomato, wheat, rice, beans, peas, pigeon peas, and chickpeas. PGP features depend on *Streptomyces* ability of colonizing host plant roots at the proper location and time (Vurukonda et al. 2021). The presence of a significant population, root-colonizing ability, PGP capability, and rhizosphere competence are all necessary traits for a successful host–*Streptomyces* relationship.

15.10 Conclusion

The study of microbial biodiversity for the creation of biotechnological solutions is a global trend aligned with this approach. Actinobacteria are mostly found in soil, where they make up over 30% of all microorganisms. Their bacterial counts range from 10^6 to 10^9 per gram. Actinobacterial enzymes may hydrolyze a variety of substrates, including cellulose, lignin, and chitin, which are present in wood wastes and insect exoskeletons. Plants and other soil creatures can use the nutrients that are released during this process. There are countless chemical structures produced by *Streptomyces* species. While pharmacologically active chemicals, enzyme inhibitors, insecticides, and herbicides were also described, the mainstream of these compounds are antibiotics (antimicrobial, anticancer, or antiviral medicines). The bioactive metabolites of actinobacteria may find use in agricultural fields as phytoprotection products because of their many significant characteristics. More research is focused on exploring potent species of *Streptomyces* and exploiting them for the benefit of agriculture and environment.

References

Abbasi S, Spor A, Sadeghi A, Safaie N (2021) *Streptomyces* strains modulate dynamics of soil bacterial communities and their efficacy in disease suppression caused by Phytophthora capsici. Sci Rep 11(1):9317

Alam K, Mazumder A, Sikdar S, Zhao Y-M, Hao J, Song C, Wang Y, Sarkar R, Islam S, Zhang Y, Li A (2022a) Streptomyces: the biofactory of secondary metabolites. Front Microbiol 13: 968053

Alam MT, Merlo ME, Hodgson DA, Wellington EM, Takano E, Breitling R (2022b) Genome-scale modeling and transcriptome analysis of *Streptomyces coelicolor* for antibacterial compound biosynthesis. Microb Cell Factories 21(1):19

Alekhya G, Gopalakrishnan S (2016) Plant growth-promotion by *Streptomyces* spp. in sorghum (*Sorghum bicolor* L.). Afr J Biotechnol 15(33):182–194. [Google Scholar]

Ali R et al (2023) Antiviral activities of three *Streptomyces* spp. against Zucchini yellow mosaic virus infecting squash. Egypt J Biol Pest Control 32:50

Amaresan N, Kumar K, Annapurna K, Kumar D, Sankaranarayanan A (2021) Biocontrol mechanisms of plant growth promoting microorganisms. In: Plant growth promoting microorganisms for sustainable agricultural development. Springer, Singapore

Anwar S, Ali B, & Sajid I (2016) Screening of rhizospheric actinomycetes for various in-vitro and in-vivo plant growth promoting (PGP) traits and for agroactive compounds. Frontiers in microbiology 7:1334

Arasu MV, Al-Dhabi NA, Saritha V, Duraipandiyan V, Muthukumar C, Kim SJ (2013) Antifeedant, larvicidal and growth inhibitory bioactivities of novel polyketide metabolite isolated from *Streptomyces* sp. AP-123 against *Helicoverpa armigera* and *Spodoptera litura*. BMC Microbiol 13:105

Baltz RH (2009) Daptomycin: mechanisms of action and resistance, and biosynthetic engineering. Curr Opin Chem Biol 13(2):144–151

Barka EA, Vatsa P, Sanchez L et al (2016) Taxonomy, physiology, and natural products of Actinobacteria. Microbiol Mol Biol Rev 80:1–43

Bayot-Custodio AN, Alcantara EP, Zulaybar TO (2014) Draft genome sequence of insecticidal *Streptomyces* sp. strain PCS3-D2, isolated from mangrove soil in Philippines. Genome Announc 2(3):e00448–e00414. https://doi.org/10.1128/genomeA.00448-14

Berdy J (2005) Bioactive microbial metabolites. The Journal of antibiotics 58(1):1–26

Bérdy J (2012) Thoughts and facts about antibiotics: where we are now and where we are heading. J Antibiot 65:385–395. https://doi.org/10.1038/ja.2012.27

Bharti N, Pandey SS, Barnawal D, Patel VK, Kalra A (2015) Plant growth promoting rhizobacteria *Dietzia natronolimnaea* modulates the expression of stress responsive genes providing protection of wheat from salinity stress. Sci Rep 5:17550. https://doi.org/10.1038/srep17550

Bhatti AA, Haq S, Bhat RA (2017) Actinobacteria benefaction role in soil and plant health. Microb Pathog 111:458–467

Bode HB, Bethe B, Höfs R, Zeeck A (2002) Big effects from small changes: possible ways to explore nature's chemical diversity. Chembiochem 3(7):619–627

Boykova I, Yuzikhin O, Novikova I, Ulianich P, Eliseev I, Shaposhnikov A et al (2023) Strain *Streptomyces* sp. P-56 produces nonactin and possesses insecticidal, acaricidal, antimicrobial and plant growth-promoting traits. Microorganisms 11(3):764

Bream AS, Ghazal SA, El–Aziz ZKA, Ibrahim SY (2001) Insecticidal activity of selected actinomycetes strains against the Egyptian cotton leaf worm *Spodoptera littoralis* (Lepidoptera: Noctuidae). Meded Fac Landbouwkd Toegep Biol Wetensch Univ Gent 66(2a):503–544

Bubici G et al (2019) Antifungal effects and biocontrol potential of lipopeptide-producing *Streptomyces* against banana Fusarium wilt. Front Microbiol 14:1177393

Chater KF, Biro S, Lee KJ et al (2010) The complex extracellular biology of *Streptomyces*. FEMS Microbiol Rev 34:171–198

Chen X et al (2018) Biocontrol of banana Fusarium wilt by *Streptomyces* sp. CB-75. Front Plant Sci 9:705

Cobb RE, Wang Y, Zhao H (2015) High-efficiency multiplex genome editing of *Streptomyces* species using an engineered CRISPR/Cas system. ACS Synth Biol 4(6):723–728

Cruz-Morales P (2022) Functional and evolutionary genomics of the *Streptomyces* metabolism [doctoral thesis, University of Wisconsin]. bioRxiv

Delgado-García KG, Sánchez-López A, Maier R-M, Martínez-Vicente Á (2022) *Streptomyces* spp. biofilmed solid inoculant improves microbial survival and plant-growth efficiency of *Triticum aestivum*. Appl Sci 12(22):11425

Elleuch L et al (2021) Characterization and potential antifungal activities of three *Streptomyces* spp. against *Sclerotinia sclerotiorum*. Egypt J Biol Pest Control 31:37

Fazal A, Thankachan D, Harris E, & Seipke RF (2020) A chromatogram-simplified Streptomyces albus host for heterologous production of natural products. Antonie Van Leeuwenhoek 113(4): 511–520

Gao L, Kumaravel K, Xiong Q, Liang Y, Ju Z, Jiang Y, & Zhang J (2023) Actinomycins produced by endophyte Streptomyces sp. GLL-9 from navel orange plant exhibit high antimicrobial effect against Xanthomonas citri susp. citri and Penicillium italicum. Pest Management Science 79(11):4679–4693

Glick BR (2012) Plant growth-promoting bacteria: mechanisms and applications. Scientifica 2012: 963401. https://doi.org/10.6064/2012/963401

Glick BR (2014) Bacteria with ACC deaminase can promote plant growth and help to feed the world. Microbiological research 169(1):30–39

Gomes RC, de Souza PM, da Silva R (2018) Solid-state fermentation: a promising technique for bioprocessing. Biotechnol Lett 40(9–10):1299–1308

Gopalakrishnan S, Pandian SK (2016) Bio-formulations for plant growth-promoting *Streptomyces* sp. SASAP J 9(3):45–54

Gopalakrishnan S, Humayun P, Kiran BK, Kannan IG, Vidya MS, Deepthi K, Rupela O (2013) Evaluation of actinomycete isolates for biological control of *Fusarium wilt* and plant growth-promotion in chickpea. Int J Microbiol 2013:538–540

Gopalakrishnan S, Velayutham M, Meenakshi V, Velazhahan R (2015) Effect of plant growth-promoting *Streptomyces* sp. on chickpea growth and nutrient availability. *3*. Biotech 5:1219–1232

Gopalakrishnan S, Srinivas V, Prasanna SL (2020) Streptomyces. In: Beneficial microbes in agro-ecology. Academic Press, pp 55–71

Gopalakrishnan S et al (2022) *Streptomyces* consortia-mediated plant growth-promotion in chick-pea. Front Microbiol. [Google Scholar]

Goudjal Y, Toumatia O, Sabaou N, Barakate M, Mathieu F, & Zitouni A (2013) Endophytic actinomycetes from spontaneous plants of Algerian Sahara: indole-3-acetic acid production and tomato plants growth promoting activity. World Journal of Microbiology and Biotechnology 29(10):1821–1829

Horinouchi S (2002) A microbial hormone, A-factor, as a master switch for morphological differentiation and secondary metabolism in *Streptomyces griseus*. Front Biosci 7:d2045–d2057

Jog R, Pandya M, Nareshkumar G et al (2014) Mechanism of phosphate solubilization and antifungal activity of *Streptomyces* spp. isolated from wheat roots and rhizosphere. Plant Soil 376:397–410

Le KD, Yu NH, Park AR, Park D-J, Kim C-J, Kim J-C (2022) *Streptomyces* sp. AN090126 as a biocontrol agent against bacterial and fungal plant diseases. Microorganisms 10(4):Article 791

Ma Z et al (2020) Titer improvement of toyocamycin in *Streptomyces diastatochromogenes*. Front Microbiol 11:555

Manjunatha L, Rajashekara H, Uppala LS, Ambika DS, Patil B, Shankarappa KS, & Mishra AK (2022) Mechanisms of microbial plant protection and control of plant viruses. Plants 11(24): 3449

Mitra D, Roy SS, Banerjee S (2020) Actinobacteria for sustainable agriculture: recent advances and future prospects. In: New and future developments in microbial biotechnology and bioengineer-ing. Elsevier

Montesinos E (2003) Development, registration and commercialization of microbial pesticides for plant protection. Int Microbiol 6:245–252

Nett M, Ikeda H, Moore BS (2009) Genomic basis for natural product biosynthetic diversity in the actinobacteria. Nat Prod Rep 26(11):1362–1384. https://doi.org/10.1039/b817069j

Olanrewaju OS, Babalola OO (2019) *Streptomyces*: implications and interactions in plant growth promotion. Appl Microbiol Biotechnol 103:1179–1188

Omura S (2008) Ivermectin: 25 years and still going strong. Int J Antimicrob Agents 31:91–98

Osman G, Mostafa S, Mohamed SH (2007) Antagonistic and insecticidal activities of some *Streptomyces* isolates. Pak J Biotechnol 4(1–2):65–71

Prastiti RD, Indrawan AD, Suryaminarsih P, Mujoko T, Widjajani BW (2023) Survivability and benefit evaluation of *Streptomyces* sp. and Trichoderma sp. as active ingredients of biopesticides and soil fertility enhancer in shallot fields at Wates Village Tulungagung. In: IOP conference series: earth and environmental science (Vol. 1131, No. 1). IOP Publishing, p 012011

Reddy MS, Kumar S, Babita K, Raghuwanshi R (2014) Selection of carriers for development of efficient bioformulations. Indian J Microbiol 54(4):419–426

Rodríguez-Alonso G, Toledo-Marcos J, Serrano-Aguirre L, Rumayor C, Pasero B, Flores A, & Arroyo M (2023) A novel lipase from Streptomyces exfoliatus DSMZ 41693 for biotechnologi-cal applications. International Journal of Molecular Sciences 24(23):17071

Sadeghi A, Karimi E, Dahaji PA, Javid MG, Dalvand Y, Askari H (2017) The effect of plant growth-promoting *Streptomyces* sp. on the growth and yield of cucumber under field conditions. Soil Biol Biochem 105:1–7

Saha M, Sarkar S, Sarkar B, Sharma BK, Bhattacharjee S, Tribedi P (2016) Microbial siderophores and their potential applications: a review. Environ Sci Pollut Res Int 23(5):3984–3999. https://doi.org/10.1007/s11356-015-4294-0

Seipke RF, Hutchings MI (2019) The regulation and biosynthesis of antimycins. Beilstein J Org Chem

Silva GDC, Kitano IT, Ribeiro IADF, Lacava PT (2022) The potential use of actinobacteria as microbial inoculants and biopesticides in agriculture. Front Soil Sci 2:833181

Stoica I, Niu F, Ni Z (2023) Mechanisms of microbial plant protection and control of plant viruses. Microbial Rev 12(3):145–162

Sutthisa W, Thukham-mee W, Trachoo N, Vangnai AS (2023) Development of a biofertilizer from *Streptomyces* strains using cow dung–peat based formulation. Trop Nat Hist 23(1):47–58

van der Heul HU, Bilyk BL, McDowall KJ, Seipke RF, van Wezel GP (2018) Regulation of antibiotic production in actinobacteria: new perspectives from the post-genomic era. Nat Prod Rep 35(6):575–604

Van der Meij A, Worsley SF, Hutchings MI, van Wezel GP (2017) Chemical ecology of *Streptomyces*. Nat Prod Rep 34(7):845–868. https://doi.org/10.1039/C7NP00010C

Viaene T, Langendries S, Beirinckx S, Maes M, Goormachtig S (2016) *Streptomyces* as a plant's best friend? FEMS Microbiol Ecol 92(8):fiw119

Vurukonda SSKP, Giovanardi D, Stefani E (2018) Plant growth promoting and biocontrol activity of *Streptomyces* spp. as endophytes. Int J Mol Sci 19:952

Vurukonda SSKP, Giovanadri D, & Stefani E (2021) Growth Promotion and Biocontrol Activity of Endophytic Streptomyces spp. Prime Archives in Molecular Sciences 2nd Edition 1:1–55

Waksman SA, Lechevalier HA (1953) Guide to the classification and identification of the actinobacteria and their antibiotics. Williams & Wilkins, Baltimore

Wang XJ, Zhang J, Wang JD, Huang SX, Chen YH, Liu CX, Xiang WS (2011) Four new doramectin congeners with acaricidal and insecticidal activity from *Streptomyces avermitilis* NEAU1069. Chem Biodivers 8:2117–2125

Wang L et al (2024a) Antiviral mechanisms of anisomycin from *Streptomyces albulus* SN40 on Potato Virus Y. J Agric Food Chem 72(7):3506–3519

Wang W, Zhang L, Stoica CE (2024b) *Streptomyces*-derived antiviral secondary metabolites and their biosynthetic gene clusters: recent insights and applications. Front Microbiol 15:1112349

Weissman KJ, Leadlay PF (2005) Combinatorial biosynthesis of reduced polyketides. Nat Rev Microbiol 3(12):925–936

Yuan WM, Crawford DL (1995) Characterization of *Streptomyces lydicus* WYEC108 as a biocontrol agent against fungus root rot. Appl Environ Microbiol 61(11):3949–3954

Zhang X, Shi Y, Wang X, Zhang W, Lou K (2010) Isolation, identification and insecticidal activity of endophyte from *Achnatherum inebrians*. Acta Microbiol Sin 50(4):530–536

Zhang N, Song Z, Xie Y, Cui P, Jiang H, Yang T, & Liu X (2013) Identification and characterization of antifungal active substances of Streptomyces hygroscopicus BS-112. World Journal of Microbiology and Biotechnology 29(8):1443–1452

Zhang X, Wang Q, Chen X, Liu R, Gao L (2016) Biocontrol and PGP activities of *Streptomyces corchorusii* UCR3-16 in rice; powder formulation and field trials. Biol Control 99:9–17

Zhang X, Hindra, Elliot MA (2019) Unlocking the trove of metabolic treasures: activating silent biosynthetic gene clusters in bacteria and fungi. Curr Opin Microbiol 51:9–15

Zhou H, Li Y, Tang Y (2016) Engineering natural product biosynthetic pathways in *Streptomyces* for chemical diversity and pharmaceutical discovery. Nat Prod Rep 33(6):783–810

Zhu H, Sandiford SK, van Wezel GP (2022) Triggers and cues that activate antibiotic production in actinobacteria. J Ind Microbiol Biotechnol 49(4):kuac022

Role of Genetic Engineering in *Streptomycetes*

16

Deepika Sharma, Ramandeep Saini, and Nainci Dhiman

Abstract

Streptomyces species, a genus of actinobacteria, are well known for their complex life cycle and excellent potential to produce natural products, which include many clinically and agriculturally important antibiotics. Despite their enormous biotechnological potential, the genetic engineering of *Streptomyces* remains challenging due to limited genetic tools and the presence of silent biosynthetic gene clusters (BGCs) that require activation for product recovery. Research in genetic engineering and synthetic biology has reached an advanced level, supporting gene mining, strain improvement, strain optimization, and targeted activation of secondary metabolic pathways. Strategies include the development of clustered regularly interspaced short palindromic repeats (CRISPR)-mediated genome editing, the design of synthetic genetic components, and the improvement of strain productivity through regulatory genes and enhanced metabolic activity. Moreover, overcoming growth-related limitations in submerged fermentation has further enhanced the industrial applications. These innovations have significantly enhanced the potential of *Streptomyces*-derived bioactive compounds though strain-dependent variations and construct-specific hurdles persist. The present chapter provides insights into sequencing, bioinformatics, and molecular toolkits that are essential for unlocking the complete biosynthetic capacity of *Streptomyces*, thereby enabling more efficient antibiotic production and novel drug discovery.

D. Sharma (✉) · R. Saini · N. Dhiman
Department of Biotechnology, Chandigarh Group of Colleges, Landran, Mohali, India
e-mail: deepika.rise@cgc.edu.in

T. Kaur et al. (eds.), *Streptomyces: Biological Candidates for Sustainable Agriculture*,
https://doi.org/10.1007/978-981-95-5803-2_16

Keywords

Streptomyces Engineering · Genome Editing · Biosynthetic gene clusters · CRISPR Technology

16.1 Introduction

The genus *Streptomyces* is an extremely significant and well-studied group of filamentous actinobacteria, mainly found in terrestrial soil and decaying vegetation (Meenakshi et al. 2024). *Streptomyces* species are Gram-positive, aerobic, and characterized by a unique mycelial network of spore-bearing aerial hyphae. *Streptomyces* species have a unique potential to generate an extraordinary range of biologically active secondary metabolites due to their unique metabolic diversity and large linear chromosomes with sizes of 6.5–12 Mbp (Harir et al. 2018). *Streptomyces* produce more than two-thirds of currently used naturally occurring antibiotics in medicine, and some of the most prominent ones are streptomycin, tetracycline, chloramphenicol, erythromycin, and rifamycin (Alam et al. 2022). Streptomycetes have biotechnological significance not only because of the vast amounts of antibiotics they produce but also because of the anticancer, immunosuppressants, antihelmintics, and antifungals they can produce (Jaroszewicz et al. 2021). Their natural products are important medicines, agricultural products, and industrial products. The existence of multiple biosynthetic gene clusters (BGCs) in their genomes underlies this unprecedented biosynthetic potential, although most of these gene clusters are either not expressed or poorly expressed under standard laboratory growth conditions. Decoding of these silent BGCs remains one of the most interesting challenges and opportunities in natural product discovery (Rutledge and Challis 2015). The current chapter focuses on essential molecular toolkits for unlocking the complete biosynthetic potential of Streptomycetes, thereby enabling more efficient antibiotic production and novel drug discovery.

16.1.1 Ecological and Biological Uniqueness

Streptomyces play a major role in the breakdown of complex organic substances in soil, such as lignin and chitin, thereby contributing to nutrient cycling and the maintenance of soil health (Javed et al. 2021). They cause the earthy odor of soil as a result of eosin production, a volatile organic compound. Biologically, their life cycle is complex and includes mycelium growth on the substrate, differentiation of aerial mycelium, and sporulation, a process similar to that of multicellular differentiation in higher organisms. This complexity contributes to their ability to regulate the timing and coordination of secondary metabolite production, which typically coincides with the onset of nutrient limitation and sporulation (Schlimpert and Elliot 2023).

16.1.2 Secondary Metabolites and BGCs

One of the characteristic features of *Streptomyces* is their ability to synthesize a wide variety of secondary metabolites through BGCs. These clusters code for the enzymes involved in the biosynthesis of polyketides, non-ribosomal peptides, terpenes, aminoglycosides, and other more intricate compounds (Dilshad et al. 2021). On average, there are 20–40 BGCs in one *Streptomyces* genome, though only a limited subset is expressed under standard conditions. These cryptic gene clusters hold immense untapped potential for the discovery of new drugs. Improved sequencing technologies and bioinformatics analysis have greatly facilitated the detection and annotation of these clusters. Genome mining strategies and combinatorial use of computational tools with experimental verification have enabled the prediction of chemical scaffolds from uncharacterized BGCs, and have renewed the interest in *Streptomyces*-based drug discovery (Dinglasan et al. 2025).

16.1.3 Industrial and Pharmaceutical Relevance

Streptomyces strains continue to be a cornerstone of industrial biotechnology as the major microbial source of natural antibiotics and other therapeutic drugs. Their large-scale production through submerged fermentation is, however, affected by pellet formation, poor or limited oxygen transfer, and variable morphological features that hinder productivity (Ferraiuolo et al. 2021). Strain engineering and fermentation optimization have been adopted to overcome these challenges, leading to significant increase in yields of metabolites.

In response to the increasing global danger of antimicrobial resistance (AMR), there is a critical need to uncover new classes of antibiotics (Salam et al. 2023). *Streptomyces*, with their rich genomic reservoir of unexplored BGCs, offer a convincing direction toward the future generation of antimicrobials. Apart from antibiotics, these versatile microbes are also being studied extensively for other non-antibiotic applications, such as the production of enzymes, bioremediation, promotion of plant growth, and their role as microbial cell factories involved in synthesis of complex natural compounds such as flavonoids and alkaloids (Kumar et al. 2021).

16.1.4 Integration with Bioinformatics and Omics Technologies

The integration of next-generation sequencing (NGS), transcriptomics, proteomics, and metabolomics has further revolutionized our understanding about *Streptomyces* biology (Saini and Mishra 2024). These omics tools allow for an integrated perception of metabolic pathways, regulatory networks, and environmental responses, giving essential information for rational strain engineering. The identification and functional prediction of BGCs are made possible by bioinformatics tools such as antiSMASH, PRISM, and BAGEL, while machine learning methods are being established to predict metabolite structures and regulatory behaviors (Vignolle 2022).

16.1.5 Biotechnological Challenges and Advancements

Despite being prolific producers of bioactive compounds, there are a number of factors that currently limit the biotechnological potential of streptomycetes. These include challenges in genetic manipulation due to large GC-rich genomes, intricate regulatory systems, low titers of some metabolites, and difficulties involved in the up-scaling of fermentation (Krysenko 2024). Conventional transformation and cloning methods are inefficient, hampering strain improvement processes and the induction of cryptic BGCs (Alam et al. 2021).

Emerging technologies in synthetic biology and genome engineering hold great promise to overcome these challenges. CRISPR–Cas technologies now facilitate accurate genome editing, such as targeted deletions, insertions, and activation of silent BGCs (Wang et al. 2023). Modular genetic elements like synthetic promoters and optimized ribosome-binding sites have been developed to allow fine-tuned expression of genes. Chassis engineering and pathway refactoring simplify strains to increase productivity and lower metabolic burden (Liu et al. 2022). Furthermore, the heterologous expression of BGCs in genetically amenable hosts, combined with classical mutagenesis and adaptive evolution, is paving the way for efficient production of novel metabolites and unlocking new frontiers in antibiotic and therapeutic discovery (Hemmerling and Piel 2022).

16.2 Challenges in Genetic Engineering of *Streptomyces*

Even with the huge potential of *Streptomyces* as prolific producers of bioactive compounds, their genetic manipulation is highly challenging. These challenges arise both from their peculiar biological features and from technological constraints that hinder efficient manipulation of their genomes. This section discusses in detail the major hurdles in genetic manipulation of *Streptomyces*, with special focus on silent BGCs, transformation efficiency, and the paucity of efficient molecular tools, as mentioned in Table 16.1.

16.2.1 Silent or Cryptic BGCs

One of the most formidable challenges in *Streptomyces* genetic engineering is the existence of a high number of silent or cryptic BGCs that encode enzymes for biosynthesis of different secondary metabolites with great pharmaceutical value (Zhang et al. 2023). In most cases, these BGCs are transcriptionally inactive under normal laboratory conditions because of intricate and tightly controlled regulatory processes, including stringent promoter activity, epigenetic modifications, and the

Table 16.1 Challenges in genetic engineering of *Streptomyces* and strategic approaches

S. No.	Challenge area	Description	Implications	Potential solutions / recent advances
1.	Silent or cryptic BGCs	Many biosynthetic gene clusters remain transcriptionally inactive under lab conditions due to complex regulation	Limits discovery of novel bioactive compounds	CRISPR activation (CRISPRa), synthetic promoter insertion, use of transcriptional activators, genome mining, and high-throughput expression screening
2.	Genetic complexity and high GC content	High GC content (>70%) leads to secondary structure formation, complicating PCR, sequencing, and cloning. Linear chromosomes with TIRs add further complexity	Reduces transformation and recombination efficiency	Use of GC-rich optimized polymerases, synthetic gene design, CRISPR–Cas systems adapted for high GC genomes, and linear chromosome editing kits (e.g., Red/ET recombinant engineering)
3.	Inefficient transformation	Transformation is strain-dependent; protoplast methods are fragile, conjugation is labor-intensive, and electroporation has limited efficiency	Time-consuming strain screening and low editing success	Use of conjugation-compatible vectors, electroporation-optimized protocols, phage-mediated delivery systems, and development of universal chassis hosts for DNA uptake
4.	Limited genetic tools	Lack of versatile vectors, robust reporters, and inducible promoters; conventional systems like GFP or lacZ often fail	Inhibits reliable gene expression and pathway control	Synthetic biology toolkits with standardized parts, *Streptomyces*-specific promoters, dual-reporter systems (e.g., luxCDABE-lacZ), and CRISPR interference systems (CRISPRi)
5.	Regulatory network complexity	Intricate regulation by global and pathway-specific regulators links metabolism with development and stress responses	Hard to isolate and optimize a single biosynthetic pathway without side effects	Modular gene circuits, orthogonal regulatory elements, machine learning-guided pathway optimization, and use of riboswitches or feedback-resistant regulators

(continued)

Table 16.1 (continued)

S. No.	Challenge area	Description	Implications	Potential solutions / recent advances
6.	Scale-up limitations	Morphological variability, pellet formation, and unstable gene expression during industrial fermentation reduce yield and consistency	Discrepancy between lab-scale and industrial-scale results	Adaptive laboratory evolution (ALE), chassis engineering for stable morphology, continuous culture systems, and real-time metabolic feedback monitoring
7.	Codon usage bias	Codon preferences differ significantly from commonly used systems like *E. coli*, leading to poor expression of heterologous genes	Low protein yield and misfolding of foreign genes	Codon optimization tools for *Streptomyces*, gene synthesis with tailored codons, and use of native *Streptomyces* ribosome-binding site libraries
8.	Plasmid instability	Plasmids may be lost over time, especially under non-selective industrial conditions	Results in inconsistent expression and loss of engineered traits	Chromosomal integration of expression cassettes, toxin–antitoxin-based plasmid stabilization systems, and use of low-copy, stable vectors

requirement for particular environmental signals (Zhgun 2023). Techniques to activate these silent clusters—such as heterologous expression in more manageable hosts, promoter designing to substitute indigenous regulatory sequences with stronger or inducible ones, and manipulation of global regulator genes—have shown promising results but are still hampered by the complexity of gene regulation in *Streptomyces* (Ji et al. 2024). The absence of effective high-throughput screening tools to assess metabolite production from these activated clusters further impedes advancements, rendering identification and application of new bioactive compounds a labor-intensive and time-consuming process.

16.2.2 Genetic Complexity and High GC Content

The genomic organization of *Streptomyces* species presents significant challenges to genetic manipulation because of their enormous genome size (usually 6.5–12 Mbp) and extremely high GC content (well above 70%), which negatively affect DNA stability and the effectiveness of standard molecular manipulation techniques (Murray 2024). The high GC content hinders PCR amplification by favoring the generation of stable secondary structures, decreases transformation efficiencies, and complicates DNA sequencing, synthesis, and cloning (Akiyama et al. 2025). Compounding to this complexity is the linear arrangement of *Streptomyces* chromosomes in contrast to the circular chromosomes found in most bacteria. These linear chromosomes contain terminal inverted repeats (TIRs) along with covalently attached terminal proteins, which interfere with homologous recombination and complicate cloning or genome-editing approaches (Deng et al. 2023). These genomic features necessitate specialized tools and methodologies, making genetic manipulation of *Streptomyces* challenging for enhanced metabolite production or pathway engineering.

16.2.3 Inefficient Transformation Techniques

Streptomyces transformation, an essential process for introducing exogenous DNA, is notoriously inefficient and extremely strain-specific, representing one of the biggest hurdles in genetic engineering attempts (Robinson et al. 2021). Prototoplast transformation and intergeneric conjugation with *Escherichia coli* are routinely employed methods, but both have serious limitations. Protoplast-based methods are highly sensitive, have low transformation efficiencies, and need stringent osmotic conditions, whereas conjugation, though more stable, is labor-intensive and constrained by plasmid size and compatibility (Hopwood and Chater 2020). Emerging technologies such as electroporation and phage-mediated transformation are promising but still have problems with reproducibility and strain specificity (Kristich et al. 2007). Although the introduction of the DNA can be done successfully, its incorporation into the genome is more often random or unstable and may lead to loss of plasmid and unstable expression of the desired characters. This imprecision reduces the success rate of the desired genome modifications, and also necessitates laborious screening processes to identify viable transformants, which in turn slows down the overall pace of genetic improvements in *Streptomyces*.

16.2.4 Lack of Robust and Versatile Genetic Tools

The available genetic toolkit to work with *Streptomyces* is still limited and not as versatile as it is in well-characterized model organisms such as *E. coli* or *Saccharomyces cerevisiae*. *Streptomyces* plasmid vectors are limited and usually deficient in desirable properties such as efficient replication origins, broad host range vectors, and high-strength promoters (Put et al. 2024). In *Streptomyces*, reporter systems commonly used to analyze gene expression, e.g., GFP or lacZ, frequently do not work due to codon usage bias, protein folding mistakes, or regulatory incompatibilities (Finger-Bou 2023). It is not uncommon that selectable markers are only viable with a limited selection of antibiotics, which may be inappropriate due to regulation concerns or cross-resistance issues. This small and inconsistent genetic toolbox holds back efforts to achieve precise and predictable gene expression. There is a pressing need to develop tightly regulated inducible promoters, synthetic transcriptional regulatory circuits, and standardized modular genetic components that can reliably be used in the distinct cellular environment of *Streptomyces* (Seshadri et al. 2025).

16.2.5 Regulatory Network Complexity

An extremely complex and multilayered regulatory system governs the secondary metabolism in *Streptomyces*. This regulatory system is composed of both general and pathway-specific regulators that are sensitive to a wide range of environmental cues such as nutrient levels, stress, and intercellular signaling molecules (Falguera 2024). This intricate regulation tightly links secondary metabolite production with the organism's developmental stages, particularly sporulation, making it difficult to manipulate metabolite biosynthesis independently (Lajtai-Szabó et al. 2022). Consequently, rational strain improvement is made difficult by the fact that changing a single regulatory component can have far-reaching and generally unforeseen effects on many different biosynthetic pathways. For example, the induction of a specific targeted secondary metabolic pathway may suppress another pathway or trigger a substantial metabolic burden, with a consequent loss in overall productivity (Noble 2024). This regulatory complexity represents a significant hurdle to achieving targeted and effective boost of secondary metabolite production in *Streptomyces*.

16.2.6 Industrial Scale-Up Limitations

While promising results have been obtained with genetically manipulated *Streptomyces* strains at the laboratory level, up-scaling these achievements to industrial fermentors presents a myriad of additional challenges. Morphological problems like pellet formation, abnormal oxygen transfer, and cell clumping can seriously limit metabolite yields and undermine the reproducibility of the fermentation process (Thakur et al. 2023). Genetic changes made during strain engineering may also be unstable in the long-term fermentation runs, and inducible expression systems that work well under the controlled conditions in laboratory often fail to perform consistently under the more heterogeneous conditions of large-scale industrial operations (Whitford et al. 2021). To overcome these drawbacks, the newer strain optimization strategies such as adaptive laboratory evolution (ALE), chassis strain engineering, and high-throughput mutagenesis are under active research. However, the efficiencies of these approaches remain variable, and attaining reproducible, high-yield production at an industrial level continues to be a principal obstacle to the biotechnological exploitation of *Streptomyces* (Zuchowski 2024).

Although *Streptomyces* holds exceptional promise for the production of novel antibiotics and other secondary metabolites, multiple genetic and technological barriers continue to limit their full exploitation. Overcoming these challenges will require a combination of synthetic biology innovations, improved transformation protocols, vigorous genetic toolkits, and a deeper understanding of their complex regulatory mechanisms. The integration of omics technologies, machine learning, and genome mining will be crucial in addressing these obstacles and unlocking the untapped potential of *Streptomyces*.

16.3 Advances in Genetic and Synthetic Biology Approaches

The introduction of accurate genome-editing methods, synthetic biology, and rational design approaches has greatly changed the genetic engineering landscape of *Streptomyces* and other microorganisms to achieve the goal of overcoming the intrinsic limitations (David et al. 2021). A combination of these novel methods has ushered in a new era of *Streptomyces* strain optimization for enhanced metabolite production, activation of cryptic BGCs, and improved industrial use (Covington et al. 2021). The greatest innovations that have transformed streptomycetes biotechnology are the CRISPR-based genome editing, the synthetic gene circuits, pathway refactoring, and heterologous expression systems.

16.3.1 CRISPR-Based Genome Editing in *Streptomyces*

CRISPR–Cas technology, initially discovered as a bacterial antiviral system, has emerged as a revolutionary genome-editing tool and is currently a powerful tool for targeted genome editing/modification in *Streptomyces* strains (Karthik et al. 2022). CRISPR-based systems including CRISPR/Cas9, CRISPR/Cas12, and CRISPR/Cas13 allow researchers to delete, insert, or edit genes with high precision and specificity. In *Streptomyces*, this technology has been applied widely to increase metabolite production by targeting regulatory genes or metabolic pathways, allowing activation of cryptic BGCs, and facilitating gebe knockouts or activation of genes to probe biosynthetic pathways (Kim et al. 2025). CRISPR-based systems allow genome-wide editing, which can be used to systematically search the genome to find potential targets that can increase production or offer new traits to the organism. The accuracy, convenience, and effectiveness of CRISPR-based genome editing have entrenched its position as a cornerstone of contemporary *Streptomyces* genetic engineering.

The process of genome editing in *Streptomyces* strains using CRISPR–Cas involves a number of steps as shown in Fig. 16.1. The first step in strain engineering is target gene selection in which genes or regions in the *Streptomyces* genome are chosen to edit, most commonly to increase the output of metabolites or activate silent BGCs (Tao et al. 2018). In the second step, the CRISPR–Cas system is then programmed by producing or designing guide RNAs to direct the Cas enzyme to the gene of choice or selected site. These CRISPR–Cas tools are then transferred into *Streptomyces* cells using techniques like electroporation or conjugation. Genome editing takes place in the cells and causes double-stranded breaks on the location of the target gene, which are then repaired by knocking out or altering the gene.

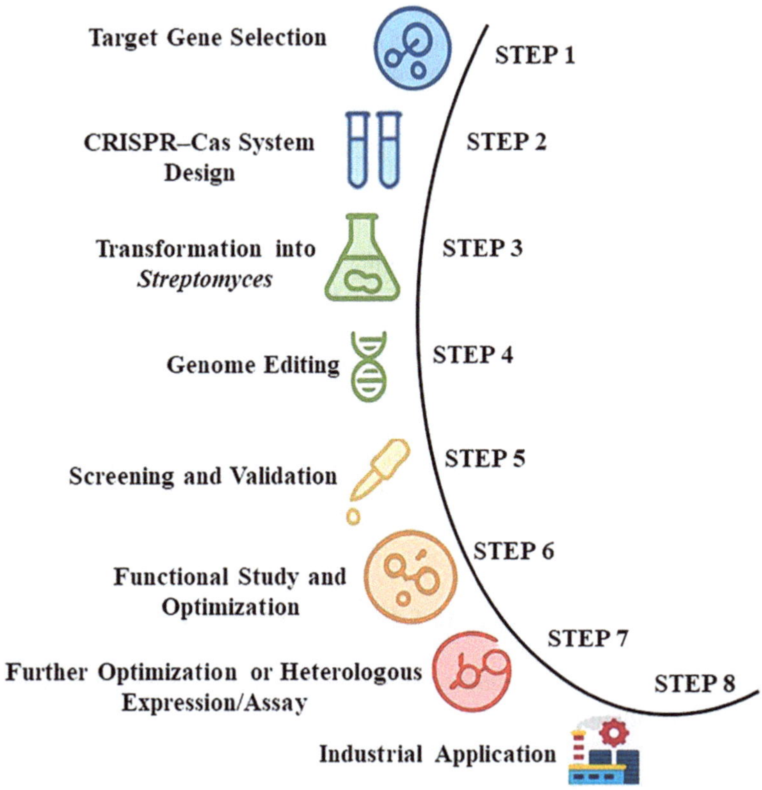

Fig. 16.1 CRISPR–Cas-based genome-editing workflow for *Streptomyces* strain improvement

Following genome editing, screening and verification are conducted using PCR or sequencing to make sure that the desired genetic modification has been made. A functional investigation and optimization step follows the subsequent successful verification, which identifies how the genetic alterations affect the performance of the strain and optimizes the conditions to maximize metabolite yield. Other optimization or heterologous expression of foreign biosynthetic pathways might be used when necessary to add new compounds or activate otherwise silenced pathways (Li et al. 2018). The enhanced strain is then scaled up for industrial applications to make large quantities of bioactive metabolites. After successful genome editing, the modified *Streptomyces* strains can effectively synthesize useful metabolites in an industrial environment. This process starts with the selection of target genes, and then designing and transformation of CRISPR–Cas system into *Streptomyces* cells. Genome editing, screening, and validation are then followed by functional optimization, additional strain improvements, and finally, scale-up of the optimized strain has positioned CRISPR technology at the forefront of industrial and pharmaceutical Streptomyces strain development.

16.3.2 Synthetic Biology Tools and Modular Genetic Parts

Synthetic biology is an interdisciplinary research area at the intersection of biology, engineering, and computer science, and aims to design and synthesize new biological systems. In the case of *Streptomyces* biotechnology, synthetic biology has been used to design highly functional modular genetic constructs that are precise to engineer and can be easily integrated into microbial genomes (Aubry et al. 2019). The approach enables directed manipulation of metabolism and regulatory pathways, and increases the ability to optimize *Streptomyces* strains for increased production of bioactive products, such as antibiotics and anticancer drugs.

As presented in Table 16.2, synthetic gene circuits, modular genetic parts, and gene synthesis technologies are the most significant synthetic biology tools to revolutionize *Streptomyces* biotechnology. The synthetic gene circuits are created to control the expression of genes in response to specific environmental or cellular stimuli in such a way that the biosynthetic pathways may be dynamically regulated in *Streptomyces* (Katz et al. 2018). This enables the scientists to manipulate the expression of several genes simultaneously, which is essential to optimize the synthesis of secondary metabolites. Modular genetic elements—standardized and interchangeable DNA building blocks as promoters, terminators and ribosome-binding sites (RBS)—can be used in combination to assemble a larger constructs (Blazquez et al. 2023). The predictability and efficiency of biosynthetic pathway engineering in *Streptomyces* are achieved using this modularity, providing greater control over metabolic flux. The technology also allows the creation of new synthetic pathways in the laboratory that can be transferred to produce new metabolites, naturally within the organism.

16.3.3 Pathway Refactoring and Rational Design

Pathway refactoring involves reengineering of BGCs to enhance functionality, regulatory control, and expression in a chosen microbial host. Pathway refactoring in *Streptomyces* biotechnology refers to the elimination of native regulatory components and their replacement with well-characterized synthetic analogs, such as strong constitutive or inducible promoters, standardized ribosome-binding sites, and modular terminators. This allows fine-tuning of gene expression and synchronous biosynthetic enzyme production, which eventually results in higher metabolite yields (Hsu et al. 2023). Rational design refers to purposeful modification of genetic and regulatory components based on in-depth knowledge of metabolic pathways, enzyme kinetics, and transcriptional control. The application of genome-based metabolic modes and bioinformatics approaches enables scientists to recognize important bottlenecks, competing metabolic branches, or inefficient regulatory nodes of native BGCs (Li et al. 2024). Such knowledge enables the anticipation and application of certain genetic interventions to enhance metabolic output.

Key strategies in pathway refactoring and rational design include systematic enhancement of the expression and efficiency of biosynthetic pathways in

Table 16.2 Synthetic biology tools and modular genetic parts used in *Streptomyces* biotechnology

S. No.	Synthetic biology tool	Description	Application in *Streptomyces* biotechnology
1.	Synthetic gene circuits	Engineered systems that control gene expression in response to specific inputs	Used to dynamically regulate biosynthesis pathways, enabling fine-tuned control of metabolite production
2.	Modular genetic parts	Standardized, interchangeable DNA components like promoters, RBS, and terminators	Facilitates the construction of custom genetic constructs for targeted metabolic engineering
3.	Gene synthesis	The process of designing and synthesizing entire biosynthetic pathways	Enables the creation of novel pathways for the production of non-native or modified metabolites
4.	CRISPR-based genome editing	Precise genome editing using CRISPR–Cas systems	Allows for accurate modification of *Streptomyces* genomes, activating or silencing genes and optimizing pathways
5.	Promoters	DNA sequences that initiate gene transcription	Used to control the expression levels of genes in *Streptomyces*, optimizing metabolic outputs
6.	Ribosome-binding sites (RBS)	Sequences that control translation initiation	Allows for fine-tuned regulation of gene expression at the translational level in engineered *Streptomyces* strains
7.	Terminator sequences	Sequences that signal the end of transcription	Ensures proper termination of transcription, preventing read-through and enhancing expression control
8.	Plasmids and vectors	DNA molecules used to introduce foreign genes into *Streptomyces* cells	Used for the stable integration and expression of synthetic or modified biosynthetic pathways in *Streptomyces*
9.	Biosynthetic gene clusters (BGCs)	Groups of genes involved in the synthesis of a specific metabolite	The activation or modification of cryptic BGCs in *Streptomyces* to produce novel or enhanced metabolites
10.	Synthetic biology toolkits	Collections of predefined biological parts for constructing genetic circuits	Enables rapid assembly of genetic constructs for pathway optimization and strain engineering

Streptomyces. It entails achieving adequate levels of gene expression via promoter engineering and ensuring effective translation of proteins by codon optimization, particularly in heterologous hosts (Xie et al. 2024). Rapid combinatorial assembly of synthetic pathways with high-throughput DNA assembly tools, such as Golden Gate and Gibson Assembly, enables construction of multienzyme biosynthetic pathways. Additionally, spatial co-localization of enzymes via scaffold proteins enables enhanced substrate channeling and flux within these multienzyme biosynthetic

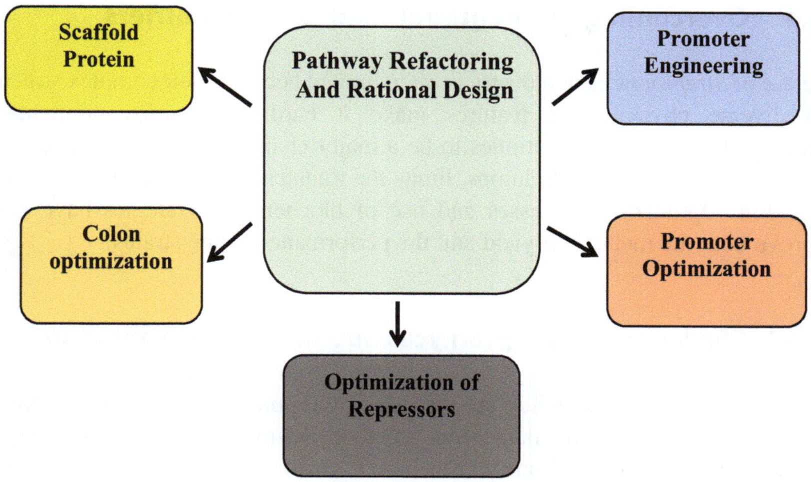

Fig. 16.2 Strategies in pathway refactoring and rational design

pathways, which in combination maximize metabolite production, as shown in Fig. 16.2.

16.3.4 Heterologous Expression and Chassis Development

Heterologous expression involves transferring and expressing biosynthetic clusters from one organism (e.g., rare or uncultivable *Streptomyces* strains) into genetically tractable and high-yielding hosts (Sword et al. 2024). This approach is particularly valuable for exploring cryptic or silent BGCs that remain unexpressed under conventional laboratory conditions in their native hosts. *Streptomyces coelicolor*, *Streptomyces lividans*, and *Streptomyces albus* have been extensively employed as heterologous expression hosts due to their well-characterized genetics, rapid growth, and efficient secretion systems. The ideal host, or "chassis," should offer the following characteristics, including high transformation efficiency, minimal endogenous secondary metabolite background to reduce interference, compatibility with a broad range of genetic parts, robust expression systems, and scalable fermentation capacity.

Advances in chassis development have led to the engineering of "clean" *Streptomyces* strains with deleted native BGCs to create metabolically streamlined hosts that are more receptive to synthetic pathways. Synthetic biology has enabled the design of synthetic mini-BGCs or hybrid pathways that combine genes from different organisms for producing novel compounds (Qiu et al. 2024).

16.4 Overcoming Fermentation and Growth Barriers

The use of *Streptomyces* in industry is often limited because their complex structures and diverse physiological features make it hard to optimize fermentation. Submerged fermentation continues to be a major challenge because it leads to the formation of thick mycelial clumps, limits the transfer of oxygen, and causes foam formation. Optimizing processes and use of biochemical strategies have helped improve both the metabolite yield and the performance of the strains.

16.4.1 Challenges in Streptomyces Submerged Fermentation

Bioactive secondary metabolites are produced in response to intricate environmental cues, and are abundant in filamentous microorganisms. One of the best ways to stimulate the production of valuable metabolic products is to mimic natural stimuli through laboratory co-cultivation. Although co-cultures' biosynthetic results are well documented, the bioprocessing components of these approaches are often disregarded (Boruta 2021).

Co-cultivation methodological challenges: Basic bioprocess-related steps can strongly influence co-culture results, even though current knowledge of microbial interactions is typically still insufficient to offer firm foundations for logical and prediction-based planning of co-cultivation runs. The experimental design should take into account the following crucial elements of co-cultivation methodology:

Co-culture initiation approach: Initiating a co-culture requires careful planning, which is also essential when designing co-cultivation runs. Before the process begins, a lot of "when and how?" questions must be addressed. The most popular method for controlling the co-culture results and preventing the producer from being suppressed by its microbial partner is to modify the relative amounts of the participating microorganisms. This is typically accomplished practically by altering the inoculation volumes (Arora et al. 2020).

Experimental configuration: Monoculture controls must be run if the purpose of the experiment is to demonstrate how well the co-cultivation strategy induces or increases the synthesis of secondary metabolites. However, there are some problems that are not covered when designing the controls for secondary metabolite production in submerged co-cultures, primarily due to broth carryover. In addition to having specific amounts of nutrients that were not utilized during preculture, the liquid that goes with the partner strain can also be enhanced with the molecules that the strain secretes. The study may be viewed as an assessment of a successful black box that is in charge of increasing the levels of metabolites or producing new molecules if the co-culture is examined as a system that enables the achievement of specific production-related goals not possible through traditional cultivation. Instead of proving that the observed effects are solely caused by the presence of an associated microorganism, the experiment in this instance aims to show the effectiveness of co-cultivation strategy (Diender et al. 2021).

Conditions for the co-cultivation medium and process: Reports on culture medium development or optimization were not included with co-cultivation studies that demonstrated success in terms of secondary metabolites production in *Streptomyces*. The medium selection was apparently made during the preliminary stages and was not thought to be one of the most difficult parts of the research project (previously formulated media compositions were commonly used). In the case of co-cultures, the knowledge gained from conducting standard monocultures could therefore be applied directly (Maglangit et al. 2020). In the event that at least one of the co-cultured partners came from the sea, the media were altered to accommodate their needs and contained a few extra ingredients. During the co-cultivation, the pH was allowed to fluctuate freely without the use of any pH control techniques. Importantly, as was already mentioned, no prior work addressed the process of rigorously optimizing or modifying the medium composition to affect the co-culture's outcome and eventually develop a productive co-cultivation system with *Streptomyces* as the producer (Schaberle et al. 2014).

16.4.2 Methods of Optimization for Higher Yields

In fermentation, the process optimization and control are crucial for making sure that a process can continue to generate a profit while keeping the necessary level of quality. However, living cell activity is difficult to understand and very nonlinear, and its control and monitoring are rather complicated (Chai et al. 2022). The exploration of increased production of both new and existing bioactive metabolites from different microbes can be accomplished through the optimization of media components and physicochemical parameters. This is typically accomplished by a variety of traditional techniques, such as one-factor-at-a-time (OFAT). Another empirical method that is widely used for fermentation process modeling, optimization, and analysis is response surface methodology (RSM) (Latha et al. 2017).

One factor or variable is varied at a time with other variables held constant in OFAT experiments, a classical medium optimization method. Concentrations of the constituents of the medium chosen were then varied in the range desired. The OFAT is the most common among the researchers in producing the medium composition and is being used in the initial phases of many studies because of its ease and convenience. In initial stages of medium development in the synthesis of new metabolites or existing compounds using new sources, this technique is still applied today (Singh et al. 2017). RSM refers to a statistical technique of model building that considers the effect of numerous independent variables to establish the best value of variables to achieve the desired outcomes. The concepts upon which the experiment is based were developed by the statistician Ronald Aylmer Fisher during 1920s. RSM has also been used successfully to model and optimize variables in biochemical and biotechnological fermentation processes. RSM has become a trendy approach to the optimization and analysis of different microbial products recently due to the availability of multiple computer programs such as Design Expert, Satistica, and Minitab that have made the implementation of RSM easier (Breig and Luti 2021; Kanwar et al. 2024).

16.4.3 The Application of Anti-Foaming Agents and Surfactants in Fermentation

Foaming is a crucial unit operation in a number of food processing industries. Foam is also known as liquid gas dispersion in which the gas acts as the dispersion medium and the liquid as the disperse phase. Foam is a dispersion made up of gas bubbles divided by thin liquid layers. The food industry must effectively control the complex phenomenon of foam generation and stabilization (Deotale et al. 2023). Chemical preservatives known as defoamers or antifoaming agents lessen the production of foam and eliminate foam that has formed in industrial process liquids. Chemical agents can be used in two ways to control foam: They are either added to the solution after foam production to control foaming (defoamers) or applied prior to foam production to control foaming (antifoaming agents). They are also widely employed in biotechnology, including anaerobic digesters, bioprocess industries, cell culture, microbial fermentation, and the production of biogas. Nevertheless, antifoaming agents can alter surface characteristics, such as lipid contents, which alter permeability and impact the growth of both prokaryotic and eukaryotic organisms (Khan et al. 2022). Biosurfactants are amphipathic molecules that have emulsifying qualities and lower surface and interfacial tensions. They are produced by microorganisms, plants, and animals. Depending on the microorganisms producing them, their nutritional sources, and their physical characteristics, fermentation processes generate microbial surfactants. These biomolecules have numerous advantages as compared to synthetic surfactants, such as low toxicity, high biodegradability, and the ability to be made from renewable raw materials (Barbosa et al. 2023).

16.5 Unlocking Biosynthetic Potential Through Omics and Toolkits

The intricate and repeated steps in *Streptomyces* biosynthesis require advanced approaches to explore their hidden abilities. Many scientists are now using multi-omics technologies, sequencing, and synthetic biology toolkits to discover, activate, and engineer silent BGCs.

16.5.1 Omics Technologies: Revolutionizing Drug Discovery

Proteomics and genomics are two potent technologies that are transforming the process of finding new drugs. Whole-genome studies of an organism include DNA sequencing and patterns of gene expression, and it is called as genomics. The study of every protein in an organism, including its structure, interactions, and functions, is known as proteomics. Proteomics and genomics can both be used to find new therapeutic targets. A molecule that plays a role in the disease process is a drug target. Scientists can discover possible drug targets that can be altered to treat diseases by understanding the molecular underpinnings of those conditions (Sundrasekar and Sahgal 2024). Transcriptomics studies is referred to expression and function of all RNA molecules in cells, and the full potential of spatial transcriptomics data analysis may be unlocked through interdisciplinary approaches. Epigenomics studies the epigenetic markers involved in the regulation of gene expression, such as DNA methylation and histone modifications. Metabolomics studies includes the types, concentrations, and changes of all metabolites within cells, improving our ability to translate intercellular communication. Investigators can identify various tumor subtypes and create individualized treatment plans for each subtype by using multi-omics analysis (Wang et al. 2024).

16.5.2 Technologies for Genome Exploration Through Sequencing

Next-generation sequencing (NGS) technology has transformed genomics by increasing understanding of DNA and its structure and function. As a result of advances in technology, there is now extensive research on genetic information, and scientists can study it like never before. This is made possible due to the high throughput, cost effectiveness, and rapid evolution of technologies. NGS has enabled many new fields, basic biology, clinical diagnostics genomics, transcriptomics, epigenomics, metagenomics, and other omics studies. The latest technology such as Illumina, Pacific Biosciences, and Oxford Nanopore NGS platforms have advanced genomics tremendously by making the parallel sequencing of DNA fragments (millions to billions) possible. This offers new possibilities for studying dissimilarity within genes, expression of genes, changes not related to DNA sequences, and even the diversity of microorganisms. NGS technologies allow finding variants responsible for disease development, novel drug targets, and biologically complex phenomena like tumor heterogeneity and the processes underlying development (Satam et al. 2023).

16.5.3 Bioinformatics: Interpreting Quiet Gene Clusters

Sequences must be categorized or grouped, but the majority of vectors are numerical. Clustering sequences directly is still the best method, even though there are some tools that can convert DNA, RNA, or protein sequences to numeric vectors (Zou et al. 2020). Bioinformatics techniques have mostly been trained on finding "backbone" or "signature" enzymes, such as nonribosomal peptide synthetase, polyketide synthase, and hybrid, because they have mostly concentrated on identifying gene clusters in bacteria and fungi. One way to think of these enzymes is as the "founders" of a particular cluster of metabolic genes. The remaining genes in any given cluster might have pathway-specific regulators, "tailoring" or "decorating" enzymes (like oxidoreductases and methyltransferases), and possibly a gene that allows tolerance to the potentially toxic pathway end product. The emphasis has shifted to the identification of additional signature enzymes that catalyze the synthesis of compounds like terpenes, butyrolactones, siderophores, and melanins as techniques advance and computational prediction pipelines gain strength (Chavali and Rhee 2018).

16.5.4 Techniques for Silent Gene Cluster Activation

Natural products (NPs) are a significant source of drug molecules, such as immunosuppressive, antifungal, anticancer, antibiotic, and antiparasitic substances. The discovery of NPs is largely due to *Streptomyces*, and the genes that produce NPs are generally encoded in a continuous region of the genome known as BGCs (Liu et al. 2021). Co-culture techniques, ribosome engineering, constitutive or inducible promoter insertion, and BGC expression in a heterologous host were some of the earliest techniques. The following are the most recent reported strategies that have been developed, as well as the improvements made upon the preceding methods. Among them are reporter-directed random or transposon mutagenesis and high-throughput elicitor screening (HiTES) to classify small-molecule inducers, and CRISPR–Cas9 to place an active promoter on a dormant BGC to activate it (Mao et al. 2018). Cas9 binds to a specific CRISPR sequence in a bacterial genome with the result of a double-stranded break with the subsequent homologous recombination. That is how the CRISPR–Cas9 system works. This has been applied to NP discovery to silence repressor genes in the native organism and replace it with constitutive promoters in front of SBGCs (silent BCGs) (Tomm et al. 2019). Elicitors are small molecules that occur in libraries and that are employed to induce the expression of non-expressed BGCs. A HiTES elicitor screening technique recognizes specific small chemical compounds in the library that are needed to activate the cryptic BGCs. The HiTES method consists of two main steps: screening of the elicitors to stimulate the BGCs and a genetic readout reporter assay to detect new cryptic metabolites. Under this method, a reporter gene, e.g., LacZ or eGFP, is integrated into the BGCs and the strain is grown in a 96-well plate. In the event that the elicitor induces the growth of the strain following incubation, the color or fluorescence observed in the culture plate will reveal the extent to which high-throughput screening of small compound elicitors is vital in increasing the activity of the silent BGCs (Meena et al. 2024).

16.5.5 *Streptomyces* Engineering Using CRISPR and Synthetic Biology

A CRISPR–Cas system is comprised of Cas protein (endonuclease) and guide RNA with a spacer sequence that directs the endonuclease to its associated protospacer sequence via a scaffold. The Cas endonuclease breaks double-stranded DNA using HNH (complementary strand) and RuvC-like (non-complementary strand), which occurs only if the correct protospacer adjacent motif (PAM) is present, thus gate-keeping the target for the CRISPR–Cas complex to the protospacer. Genome

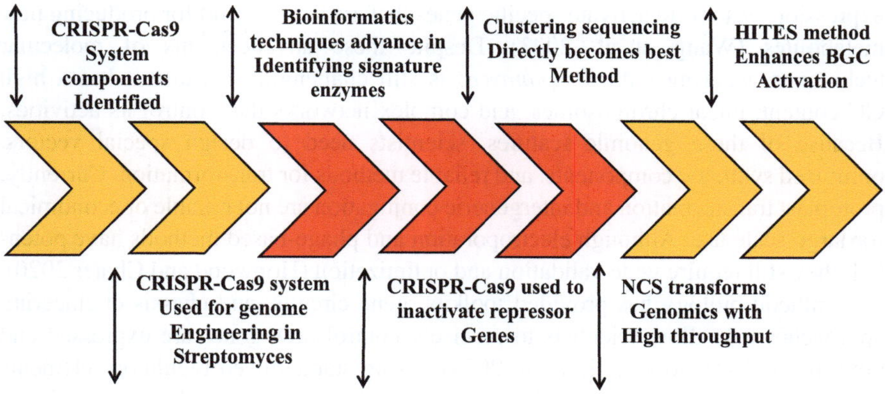

Fig. 16.3 Advancement in biosynthetic potential exploration

engineering with CRISPR–Cas is based on the fact that double-stranded break is repaired either by non-homologous end-joining (NHEJ) or homology-directed repair (HDR). In the NHEJ pathway, a few base pairs are added or removed indiscriminately at the DNA break site (Doudna and Charpentier 2014). On the contrary, sharp genome editing is achieved through DNA repair by means of the HDR pathway, which necessitates donor DNA incorporation into the cleavage site. Most *Streptomyces* species lack the NHEJ genes of bacteria, which means that the editing mediated by HDR is more suitable for genetic engineering using CRISPR–Cas systems (Tong et al. 2015). Traditionally, accurate and scarless genome editing in *Streptomyces* was accomplished through double-crossover events, a process that involved extensive and tedious screening procedures. Considering the devastating double-stranded break that the CRISPR–Cas system induces in bacteria until repaired, the organisms that survive after the transformation of the CRISPR–Cas plasmid are more likely to be edited as intended. As a result, there is a considerable reduction in the effort and time needed for the screening processes (Lee et al. 2024). An overview of recent advancement in exploration of biosynthetic potential using CRISPR–Cas system and synthetic biology is presented in Fig. 16.3.

16.6 Discussion

The development of CRISPR-based systems, synthetic biology and integrated omics technologies has brought major changes in *Streptomyces* genetic engineering (Kim et al. 2025). Although this genus is a key for finding new drugs, especially antibiotics, much of its biosynthetic ability is still hidden because of ongoing challenges in biology and technology. A major opportunity in *Streptomyces* biotechnology is to activate silent or hidden BGCs, as they represent a huge and unexplored portion of the genome. Traditionally, the difficulties to study these BGCs in the laboratory slowed down the search for new drugs. However, nowadays, scientists can use tools such as CRISPR–Cas9, synthetic promoters, and heterologous expression systems to activate specific genes that are very helpful for producing new metabolites (Wang et al. 2023). Despite these advancements of molecular techniques, working with *Streptomyces* is still challenging because it has a high GC content, linear chromosomes, and complex networks that control its activities. Because of these genomic features, scientists need to design special vectors, optimized synthetic components, and reliable methods for transformation. Currently, protoplast transformation and intergeneric conjugation are not reliable or economical for large-scale use. Although electroporation and phage-based methods have potential, they still require wide validation and optimization (Hopwood and Chater 2020).

Synthetic biology has provided toolkits, gene circuits, and chassis engineering approaches that allow scientists to precisely control how genes are expressed and how metabolism works (Qiu et al. 2024). Using standardized regulatory elements and spatial organizing enzymes in a certain way has demonstrated great promise in improving the production of secondary metabolites. Furthermore, advanced DNA assembly platforms such as Gibson or Golden Gate cloning have increased the speed of building biosynthetic pathways, allowing for more reliable strain actions. The use of omics methods in *Streptomyces* allows us to see the entire picture of its metabolic and regulatory systems. Together, these datasets and machine learning models are helping to predict designs for biosynthetic pathways and to enhance strains' performance. In particular, NGS tools and pipelines like antiSMASH and PRISM have made the process of identifying and characterizing BGCs both quicker and more accurate (Lee et al. 2024; Sharma et al. 2025).

Fermentation scaling is still a significant obstacle in bioprocessing. Because *Streptomyces* grows in long chains and tends to form pellets, submerged fermentation can be difficult (Ferraiuolo et al. 2021). Approaches such as combination of cultures, statistical methods, and use of anti-foaming agents and biosurfactants have led to better fermentation outcomes and improved robustness. Nonetheless, applying lab-scale improvements to industrial settings is still difficult due to challenges with genetic stability, reproducibility, and cost. *Streptomyces* engineering is rapidly evolving, and emerging tools are unlocking genetic potential. Emerging tools application will depend on overcoming strain-specific limitations, improving transformation efficiency, scaling fermentation processes, and ensuring regulatory predictability. A multidisciplinary approach that combines molecular biology, systems engineering, computational biology, and industrial biotechnology will be essential to convert *Streptomyces* into next-generation microbial factories for sustainable drug discovery (Kim et al. 2025).

16.7 Conclusion

The chapter comprehensively addresses the multifaceted role of genetic and synthetic biology approaches in enhancing the biosynthetic capacity of *Streptomyces*. Being highly productive producers of antibiotics and other bioactive molecules, *Streptomyces* strains harbor many silent BGCs that are not transcriptionally expressed under normal conditions. Current tools including CRISPR–Cas systems and synthetic biology platforms have transformed the capacity to unleash and engineer these BGCs. Some of the prominent advances include CRISPR-mediated genome editing, synthetic gene circuits, pathway refactoring, and heterologous expression systems that enhance strain performance and metabolite yield. Despite these advancements, a number of limitations still exist like inefficient transformation, high GC content, regulatory complexity, and scale-up challenges. Strategic advancements like adaptive evolution, modular toolkits, and high-throughput elicitor screening are facilitating to overcome these limitations. Further, co-cultivation and statistical optimization techniques (e.g., OFAT, RSM) are improving fermentation efficiency. Omics tools and machine learning are also opening the way for predictive biosynthesis and rational design of new metabolites. Together, these developments have the enormous potential to convert *Streptomyces* from traditional antibiotic manufacturers into next-generation biofactories that can meet the immediate worldwide demand for new therapeutics amid increasing antimicrobial resistance.

References

Akiyama M, Tashiro M, Huang Y, Uehara M, Kanzaki T, Itaya M, et al (2025) Generative AI-driven artificial DNA design for enhancing inter-species gene activation and enzymatic degradation of PET. bioRxiv, 2025-05

Alam K, Hao J, Zhang Y, Li A (2021) Synthetic biology-inspired strategies and tools for engineering of microbial natural product biosynthetic pathways. Biotechnol Adv 49:107759

Alam K, Mazumder A, Sikdar S, Zhao YM, Hao J, Song C et al (2022) Streptomyces: the biofactory of secondary metabolites. Front Microbiol 13:968053

Arora D, Gupta P, Jaglan S, Roullier C, Grovel O, Bertrand S (2020) Expanding the chemical diversity through microorganisms co-culture: current status and outlook. Biotechnol Adv 40: 107521

Aubry C, Pernodet JL, Lautru S (2019) Modular and integrative vectors for synthetic biology applications in Streptomyces spp. Appl Environ Microbiol 85(16):e00485–e00419

Barbosa FG, Castro-Alonso MJ, Rocha TM, Sánchez-Muñoz S, de Arruda GL, Viana MCA et al (2023) An overview of developments and challenges in the production of biosurfactant by fermentation processes. In: Biosurfactants and Sustainability: From Biorefineries Production to Versatile Applications, John Wiley and Sons: Hoboken, NJ pp 117–142

Blázquez B, León DS, Torres-Bacete J, Gómez-Luengo Á, Kniewel R, Martínez I et al (2023) Golden standard: a complete standard, portable, and interoperative MoClo tool for model and non-model proteobacteria. Nucleic Acids Res 51(19):e98–e98

Boruta T (2021) A bioprocess perspective on the production of secondary metabolites by *Streptomyces* in submerged co-cultures. World J Microbiol Biotechnol 37(10):171

Breig SJM, Luti KJK (2021) Response surface methodology: a review on its applications and challenges in microbial cultures. Mater Today Proc 42:2277–2284

Chai WY, Teo KTK, Tan MK, Tham HJ (2022) Fermentation process control and optimization. Chem Eng Technol 45(10):1731–1747

Chavali AK, Rhee SY (2018) Bioinformatics tools for the identification of gene clusters that biosynthesize specialized metabolites. Brief Bioinform 19(5):1022–1034

Covington BC, Xu F, Seyedsayamdost MR (2021) A natural product chemist's guide to unlocking silent biosynthetic gene clusters. Annu Rev Biochem 90(1):763–788

David F, Davis AM, Gossing M, Hayes MA, Romero E, Scott LH, Wigglesworth MJ (2021) A perspective on synthetic biology in drug discovery and development—current impact and future opportunities. SLAS Discov 26(5):581–603

Deng L, Zhao Z, Liu L, Zhong Z, Xie W, Zhou F et al (2023) Dissection of 3D chromosome organization in Streptomyces coelicolor A3 (2) leads to biosynthetic gene cluster overexpression. Proc Natl Acad Sci U S A 120(11):e2222045120

Deotale SM, Dutta S, Moses JA, Anandharamakrishnan C (2023) Foaming and defoaming–concepts and their significance in food and allied industries: a review. Discov Chem Eng 3(1):9

Diender M, Olm IP, Sousa DZ (2021) Synthetic co-cultures: novel avenues for bio-based processes. Curr Opin Biotechnol 67:72–79

Dilshad R, Jamil N, Batool R (2021) Biosynthetic gene clusters in bacteria: a review: bacterial secondary metabolites producing genes. Proc Pak Acad Sci B Life Environ Sci 58(3):29–42

Dinglasan JLN, Otani H, Doering DT, Udwary D, Mouncey NJ (2025) Microbial secondary metabolites: advancements to accelerate discovery towards application. Nat Rev Microbiol:1–17

Doudna JA, Charpentier E (2014) The new frontier of genome engineering with CRISPR-Cas9. Science 346(6213):1258096

Falguera JVT (2024) Stress responses affecting the sporulation and specialized metabolism programs of Streptomycetes (Doctoral dissertation, University of Toronto (Canada))

Ferraiuolo SB, Cammarota M, Schiraldi C, Restaino OF (2021) *Streptomycetes* as platform for biotechnological production processes of drugs. Appl Microbiol Biotechnol 105:551–568

Finger-Bou M (2023) Not so giant steps on tuning protein production and diminished genomes (Doctoral dissertation, Wageningen University and Research)

Harir M, Bendif H, Bellahcene M, Fortas Z, Pogni R (2018) Streptomyces secondary metabolites. In: Basic biology and applications of actinobacteria. IntechOpen

Hemmerling F, Piel J (2022) Strategies to access biosynthetic novelty in bacterial genomes for drug discovery. Nat Rev Drug Discov 21(5):359–378

Hopwood DA, Chater KF (2020) Streptomycetes. In: Genetics and breeding of industrial microorganisms. CRC Press, pp 7–42

Hsu SY, Lee J, Sychla A, Smanski MJ (2023) Rational search of genetic design space for a heterologous terpene metabolic pathway in Streptomyces. Metab Eng 77:1–11

Jaroszewicz W, Bielaska P, Lubomska D, Kosznik-Kwanicka K, Golec P et al (2021) Antibacterial, antifungal and anticancer activities of compounds produced by newly isolated Streptomyces strains from the Szczelina Chocho? Owska Cave (Tatra Mountains, Poland). Antibiotics 10(10):1212

Javed Z, Tripathi GD, Mishra M, Dashora K (2021) Actinomycetes–the microbial machinery for the organic-cycling, plant growth, and sustainable soil health. Biocatal Agric Biotechnol 31:101893

Ji CH, Je HW, Kim H, Kang HS (2024) Promoter engineering of natural product biosynthetic gene clusters in actinomycetes: concepts and applications. Nat Prod Rep 41:672

Kanwar K, Sharma D, Singh H, Pal M, Bandhu R, Azmi W (2024) In vitro effects of alginate lyase SG4$^+$ produced by *Paenibacillus lautus* alone and combined with antibiotics on biofilm formation by mucoid *Pseudomonas aeruginosa*. Braz J Microbiol 55(2):1189–1203

Karthik L, Arivarasan VK, Vignesh MR, Anitha P (2022) CRISPR ERA: current applications and future perspectives on Actinobacteria. In: Actinobacteria: microbiology to synthetic biology, Springer Nature: Singapore. pp 181–202

Katz L, Chen YY, Gonzalez R, Peterson TC, Zhao H, Baltz RH (2018) Synthetic biology advances and applications in the biotechnology industry: a perspective. J Ind Microbiol Biotechnol 45(7):449–461

Khan SS, Ullah I, Wang Q, Xu H, Tiso T, Blank LM, Liu L (2022) Thermal and ultrasound foam control in biotechnology: a mini-review. Discov Chem Eng 2(1):7

Kim DG, Gu B, Cha Y, Ha J, Lee Y, Kim G et al (2025) Engineered CRISPR-Cas9 for Streptomyces sp. genome editing to improve specialized metabolite production. Nat Commun 16(1):874

Kristich CJ, Salomon CE, Dunny GM (2007) Genetic exchange in gram-positive bacteria. In: Methods for general and molecular microbiology, ASM Press (Wiley): Washington, D.C. pp 756–799

Krysenko S (2024) Synthetic biology approaches for engineering of secondary metabolism in *Streptomyces*: manipulation of crosstalk between primary and secondary metabolism. MDPI Preprints. Basel (Switzerland)

Kumar M, Sarma DK, Shubham S, Kumawat M, Verma V, Nina PB et al (2021) Futuristic non-antibiotic therapies to combat antibiotic resistance: a review. Front Microbiol 12:609459

Lajtai-Szabó P, Hülber-Beyer É, Nemestóthy N, Bélafi-Bakó K (2022) The role of physical support in secondary metabolite production by Streptomyces species. Biochem Eng J 185:108495

Latha S, Sivaranjani G, Dhanasekaran D (2017) Response surface methodology: a non-conventional statistical tool to maximize the throughput of *Streptomyces* species biomass and their bioactive metabolites. Crit Rev Microbiol 43(5):567–582

Lee Y, Hwang S, Kim W, Kim JH, Palsson BO, Cho BK (2024) CRISPR-aided genome engineering for secondary metabolite biosynthesis in *Streptomyces*. J Ind Microbiol Biotechnol 51: kuae009

Li L, Wei K, Zheng G, Liu X, Chen S, Jiang W, Lu Y (2018) CRISPR-Cpf1-assisted multiplex genome editing and transcriptional repression in Streptomyces. Appl Environ Microbiol 84(18): e00827–18

Li H, Gao S, Shi S, Zhao X, Ye H, Luo Y (2024) Rational construction of genome-minimized Streptomyces host for the expression of secondary metabolite gene clusters. Synth Syst Biotechnol 9(3):600–608

Liu Z, Zhao Y, Huang C, Luo Y (2021) Recent advances in silent gene cluster activation in *Streptomyces*. Front Bioeng Biotechnol 9:632230

Liu J, Wang X, Dai G, Zhang Y, Bian X (2022) Microbial chassis engineering drives heterologous production of complex secondary metabolites. Biotechnol Adv 59:107966

Maglangit F, Fang Q, Kyeremeh K, Sternberg JM, Ebel R, Deng H (2020) A co-culturing approach enables discovery and biosynthesis of a bioactive indole alkaloid metabolite. Molecules 25(2): 256

Mao D, Okada BK, Wu Y, Xu F, Seyedsayamdost MR (2018) Recent advances in activating silent biosynthetic gene clusters in bacteria. Curr Opin Microbiol 45:156–163

Meena SN, Wajs-Bonikowska A, Girawale S, Imran M, Poduval P, Kodam KM (2024) High-throughput mining of novel compounds from known microbes: a boost to natural product screening. Molecules 29(13):3237

Meenakshi S, Hiremath J, Meenakshi MH, Shivaveerakumar S (2024) Actinomycetes: isolation, cultivation and its active biomolecules. J Pure Appl Microbiol 18(1):118

Murray K (2024) Developing a new efficient CRISPR-Cas9 base homology directed repair oriented gene editing plasmid for Streptomyces (Doctoral dissertation, Oxford Brookes University)

Noble K (2024) Characterising the two-component system-regulated biosynthesis of Formicamycin in Streptomyces formicae (Doctoral dissertation, University of East Anglia)

Put H, Gerstmans H, Vande Capelle H, Fauvart M, Michiels J, Masschelein J (2024) Bacillus subtilis as a host for natural product discovery and engineering of biosynthetic gene clusters. Nat Prod Rep 41(7):1113–1151

Qiu S, Yang B, Li Z, Li S, Yan H, Xin Z et al (2024) Building a highly efficient Streptomyces super-chassis for secondary metabolite production by reprogramming naturally-evolved multifaceted shifts. Metab Eng 81:210–226

Robinson SL, Piel J, Sunagawa S (2021) A roadmap for metagenomic enzyme discovery. Nat Prod Rep 38(11):1994–2023

Rutledge PJ, Challis GL (2015) Discovery of microbial natural products by activation of silent biosynthetic gene clusters. Nat Rev Microbiol 13(8):509–523

Saini P, Mishra P (2024) Revolution in microbial bioprospecting via the development of omics-based technologies. In: Bioprospecting of microbial resources for agriculture, environment and bio-chemical industry. Springer Nature Switzerland, Cham, pp 171–191

Salam MA, Al-Amin MY, Salam MT, Pawar JS, Akhter N, Rabaan AA, Alqumber MA (2023) Antimicrobial resistance: a growing serious threat for global public health. Healthcare 11(13): 1946. Multidisciplinary Digital Publishing Institute

Satam H, Joshi K, Mangrolia U, Waghoo S, Zaidi G, Rawool S et al (2023) Next-generation sequencing technology: current trends and advancements. Biology 12(7):997

Schäberle TF, Orland A, König GM (2014) Enhanced production of undecylprodigiosin in *Streptomyces* coelicolor by co-cultivation with the corallopyronin A-producing myxobacterium, *Corallococcus coralloides*. Biotechnol Lett 36:641–648

Schlimpert S, Elliot MA (2023) The best of both worlds—*Streptomyces coelicolor* and *Streptomyces venezuelae* as model species for studying antibiotic production and bacterial multicellular development. J Bacteriol 205(7):e00153–e00123

Seshadri K, Abad AN, Nagasawa KK, Yost KM, Johnson CW, Dror MJ, Tang Y (2025) Synthetic biology in natural product biosynthesis. Chem Rev 125(7):3814–3931

Sharma D, Kanwar K, Manhas RK (2025) Genomic analysis of *Streptomyces amritsarensis* MTCC 11845 and characterization of novel lanthipeptide 2A. Indian J Exp Biol *63*(3):229–238

Singh V, Haque S, Niwas R, Srivastava A, Pasupuleti M, Tripathi C (2017) Strategies for fermentation medium optimization: an in-depth review. Front Microbiol 7:2087

Sundarasekar J, Sahgal G (2024) Role of genomics and proteomics in drug discovery. In: Concepts in pharmaceutical biotechnology and drug development. Springer Nature Singapore, Singapore, pp 207–246

Sword TT, Abbas GS, Bailey CB (2024) Cell-free protein synthesis for nonribosomal peptide synthetic biology. Front Nat Prod 3:1353362

Tao W, Yang A, Deng Z, Sun Y (2018) CRISPR/Cas9-based editing of *Streptomyces* for discovery, characterization, and production of natural products. Front Microbiol 9:1660

Thakur M, Kumar P, Rajput D, Yadav V, Dhaka N, Shukla R, Dubey KK (2023) Genome-guided approaches and evaluation of the strategies to influence bioprocessing assisted morphological engineering of *Streptomyces* cell factories. Bioresour Technol 376:128836

Tomm HA, Ucciferri L, Ross AC (2019) Advances in microbial culturing conditions to activate silent biosynthetic gene clusters for novel metabolite production. J Ind Microbiol Biotechnol 46(9–10):1381–1400

Tong Y, Charusanti P, Zhang L, Weber T, Lee SY (2015) CRISPR-Cas9 based engineering of actinomycetal genomes. ACS Synth Biol 4(9):1020–1029

Vignolle GA (2022) Modeling novel bioinformatics approaches to investigate bioactive substance production based on genomics and transcriptomics (Doctoral dissertation, Technische Universität Wien)

Wang D, Jin S, Lu Q, Chen Y (2023) Advances and challenges in CRISPR/Cas-based fungal genome engineering for secondary metabolite production: a review. J Fungi 9(3):362

Wang Z, Zhao Y, Zhang L (2024) Emerging trends and hot topics in the application of multi-omics in drug discovery: a bibliometric and visualized study. Curr Pharm Anal 21:20

Whitford CM, Cruz-Morales P, Keasling JD, Weber T (2021) The design-build-test-learn cycle for metabolic engineering of streptomycetes. Essays Biochem 65(2):261–275

Xie H, Su YT, Bu QT, Li YP, Zhao QW, Du YL, Li YQ (2024) Stepwise increase of fidaxomicin in an engineered heterologous host *Streptomyces albus* through multi-level metabolic engineering. Synth Syst Biotechnol 9(4):766–774

Zhang L, Wang C, Chen K, Zhong W, Xu Y, Molnár I (2023) Engineering the biosynthesis of fungal nonribosomal peptides. Nat Prod Rep 40(1):62–88

Zhgun AA (2023) Fungal BGCs for production of secondary metabolites: main types, central roles in strain improvement, and regulation according to the piano principle. Int J Mol Sci 24(13): 11184

Zou Q, Lin G, Jiang X, Liu X, Zeng X (2020) Sequence clustering in bioinformatics: an empirical study. Brief Bioinform 21(1):1–10

Zuchowski R (2024) Communities of niche-optimized strains (CoNoS)-a novel concept for improving biotechnological production. Forschungszentrum Jülich GmbH, Zentralbibliothek

MIX
Papier aus verantwortungsvollen Quellen
Paper from responsible sources
FSC® C105338

If you have any concerns about our products,
you can contact us on
ProductSafety@springernature.com

In case Publisher is established outside the EU,
the EU authorized representative is:
Springer Nature Customer Service Center GmbH
Europaplatz 3, 69115 Heidelberg, Germany

Printed by Libri Plureos GmbH
in Hamburg, Germany